T0223947

Lecture Notes in Computer Science 9965

Commenced Publication in 1973
Founding and Former Series Editors:
Gerhard Goos, Juris Hartmanis, and Jan van Leeuwen

Augusto Sampaio · Farn Wang (Eds.)

Theoretical Aspects of Computing – ICTAC 2016

13th International Colloquium
Taipei, Taiwan, ROC, October 24–31, 2016
Proceedings

Springer

Editors
Augusto Sampaio
Centro de Informática
Universidade Federal de Pernambuco
Recife, Pernambuco
Brazil

Farn Wang
Department of Electrical Engineering
National Taiwan University
Taipei
Taiwan

ISSN 0302-9743   ISSN 1611-3349 (electronic)
Lecture Notes in Computer Science
ISBN 978-3-319-46749-8   ISBN 978-3-319-46750-4 (eBook)
DOI 10.1007/978-3-319-46750-4

Library of Congress Control Number: 2016952524

LNCS Sublibrary: SL1 – Theoretical Computer Science and General Issues

This Springer imprint is published by Springer Nature
The registered company is Springer International Publishing AG
The registered company address is: Gewerbestrasse 11, 6330 Cham, Switzerland

Preface

This volume contains the papers presented at ICTAC 2016, the 13th International Colloquium on Theoretical Aspects of Computing, held during October 24–31, 2016, in Taipei, Taiwan, ROC.

The International Colloquium on Theoretical Aspects of Computing (ICTAC) is a series of annual events founded in 2004 by the United Nations International Institute for Software Technology. Its purpose is to bring together practitioners and researchers from academia, industry, and government to present research results and exchange experiences and ideas. Beyond these scholarly goals, another main purpose is to promote cooperation in research and education between participants and their institutions from developing and industrial regions.

The city of Taipei, where this edition of ICTAC took place, is the capital and the largest city of Taiwan, ROC (Republic of China). The National Taiwan University (NTU), host of the colloquium, is the top university in Taiwan with 17,000 undergraduate students and 15,000 graduate students. The Department of Electrical Engineering of NTU, where most ICTAC 2016 sessions have been held, is the leading research group in Taiwan and has technical contributions to Taiwan's world-renowned giants, including TSMC, MediaTek, Quanta Computers, etc. ICTAC 2016 was sponsored by Microsoft Research, Springer, EasyChair, the Ministry of Science and Technology (ROC), the Ministry of Education (ROC), the Ministry of Economical Affairs (ROC), the Taipei Municipal Government, and the National Taiwan University.

In this edition of ICTAC, we had four invited speakers: Hsu-Chun Yen, from the National Taiwan University; Leonardo de Moura, from Microsoft Research, USA; Heike Wehrheim, form Universität Paderborn, Germany; and Wen-Lian Hsu, from Academia Sinica, Taiwan. They delivered keynote speeches as well as tutorials.

ICTAC 2016 received 60 submissions from 26 different countries. Each submission was reviewed by at least three members of the Program Committee, along with help from external reviewers. Out of these 60 submissions, 23 regular papers were accepted. The committee also accepted one short paper and one tool paper. Apart from the paper presentations and invited talks, ICTAC 2016 continued the tradition of previous ICTAC conferences in holding a five-course school on important topics in theoretical aspects of computing.

We thank all the authors for submitting their papers to the conference, and the Program Committee members and external reviewers for their excellent work in the review, discussion, and selection process. We are indebted to all the members of the Organizing Committee, including Dr. Churn-Jung Liau (Academia Sinica), Prof. Jonathan Lee (NTU), Dr. Yu-Fang Chen (Academia Sinica), and Prof. Fang Yu (National Cheng-Chi University), for their hard work in all phases of the conference, as well as to Filipe Arruda and Gustavo Carvalho, who helped enormously with managing EasyChair and several other operational aspects of the reviewing and proceedings creation process. We also acknowledge our gratitude to the Steering Committee for their constant support.

We are also indebted to EasyChair that greatly simplified the assignment and reviewing of the submissions as well as the production of the material for the proceedings. Finally, we thank Springer for their cooperation in publishing the proceedings, and for the sponsorship of the two best paper awards.

August 2016   Augusto Sampaio
                                                         Farn Wang

Organization

Program Committee

Bernhard K. Aichernig	TU Graz, Austria
Farhad Arbab	CWI and Leiden University, The Netherlands
Mauricio Ayala-Rincon	Universidade de Brasilia, Brazil
Mario Benevides	Universidade Federal do Rio de Janeiro, Brazil
Ana Cavalcanti	University of York, UK
Yu-Fang Chen	Academia Sinica, Taiwan
Gabriel Ciobanu	Romanian Academy, Institute of Computer Science, Iasi, Romania
Hung Dang Van	UET, Vietnam National University, Vietnam
Ana De Melo	University of Sao Paulo, Brazil
Rocco De Nicola	IMT - Institute for Advanced Studies Lucca, Italy
Razvan Diaconescu	IMAR, Romania
Jin Song Dong	National University of Singapore, Singapore
José Luiz Fiadeiro	Royal Holloway, University of London, UK
John Fitzgerald	Newcastle University, UK
Marcelo Frias	Buenos Aires Institute of Technology, Argentina
Martin Fränzle	Carl von Ossietzky Universität Oldenburg, Germany
Lindsay Groves	Victoria University of Wellington, NZ
Kim Guldstrand Larsen	Computer Science, Aalborg University, Denmark
Zhenjiang Hu	NII, Japan
Jie-Hong Roland Jiang	National Taiwan University, Taiwan
Cliff Jones	Newcastle University, UK
Luis Lamb	Federal University of Rio Grande do Sul, Brazil
Martin Leucker	University of Lübeck, Germany
Zhiming Liu	Birmingham City University, UK
Dominique Mery	Université de Lorraine, LORIA, France
Alexandre Mota	Universidade Federal de Pernambuco, Brazil
Mohammadreza Mousavi	Halmstad University, Sweden
Tobias Nipkow	TU München, DE
Jose Oliveira	Universidade do Minho, Portugal
Catuscia Palamidessi	Inria, France
Paritosh Pandya	TIFR, India
António Ravara	Universidade Nova de Lisboa, Portugal
Camilo Rueda	Universidad Javeriana, Colombia
Jacques Sakarovitch	CNRS/Telecom ParisTech, France
Augusto Sampaio	Federal University of Pernambuco, Brazil
Sven Schewe	University of Liverpool, UK

Martin Schäf	SRI International, USA
Emil Sekerinski	McMaster University, Canada
Hiroyuki Seki	Nagoya University, Japan
Tetsuo Shibuya	The University of Tokyo, Japan
Andrzej Tarlecki	Warsaw University, Poland
Kazunori Ueda	Waseda University, Japan
Frank Valencia	LIX, Ecole Polytechnique, France
Farn Wang	National Taiwan University, Taiwan
Jim Woodcock	University of York, UK
Hsu-Chun Yen	National Taiwan University, Taiwan
Shoji Yuen	Nagoya University, Japan
Naijun Zhan	Institute of Software, Chinese Academy of Sciences, China
Lijun Zhang	Institute of Software, Chinese Academy of Sciences, China
Huibiao Zhu	East China Normal University, China

Additional Reviewers

Accattoli, Beniamino
Aman, Bogdan
Bistarelli, Stefano
Bollig, Benedikt
Bouyer-Decitre, Patricia
Bucchiarone, Antonio
Ciancia, Vincenzo
Dang, Duc-Hanh
De Moura, Flavio L.C.
Decker, Normann
Foster, Simon
Goulão, Miguel
Hafemann Fragal,
    Vanderson
Hanazumi, Simone
Helouet, Loic
Hsu, Tzu-Chien
Höfner, Peter
Hölzl, Johannes
Imai, Keigo
Jensen, Peter Gjøl
Lengal, Ondrej
Li, Guangyuan

Li, Qin
Liu, Bo
Liu, Wanwei
Lluch Lafuente, Alberto
Mikučionis, Marius
Montrieux, Lionel
Moreira, Nelma
Nantes-Sobrinho, Daniele
Naumowicz, Adam
Nguyen, Thi Huyen Chau
Nishida, Naoki
Nyman, Ulrik
Ody, Heinrich
Olsen, Petur
Popescu, Andrei
Rocha, Camilo
Rot, Jurriaan
Sanchez, Huascar
Sato, Shuichi
Scheffel, Torben
Schumi, Richard
Serre, Olivier
Singh, Neeraj

Song, Fu
Strathmann, Thomas
Sun, Meng
Taromirad, Masoumeh
Ter Beek, Maurice H.
Thomsen,
    Michael Kirkedal
Thorn, Johannes
Ting, Gan
Tiplea, Ferucio
Truong, Hoang
Tu, Kuan-Hua
Tutu, Ionut
Ventura, Daniel
Wang, Hung-En
Wang, Shuling
Wiedijk, Freek
Wolter, Uwe
Worrell, James
Xie, Wanling
Yokoyama, Tetsuo
Zhao, Hengjun

Invited Papers

Verification of Concurrent Programs
on Weak Memory Models

Oleg Travkin and Heike Wehrheim

Institut für Informatik, Universität Paderborn, 33098, Paderborn, Germany
{oleg82,wehrheim}@uni-paderborn.de

Abstract. Modern multi-core processors equipped with weak memory models seemingly reorder instructions (with respect to program order) due to built-in optimizations. For concurrent programs, weak memory models thereby produce interleaved executions which are impossible on sequentially consistent (SC) memory. Verification of concurrent programs consequently needs to take the memory model of the executing processor into account. This, however, makes most standard software verification tools inapplicable.

In this paper, we propose a technique (and present its accompanying tool WEAK2SC) for *reducing* the verification problem for weak memory models to the verification on SC. The reduction proceeds by generating – out of a given program and weak memory model (here, TSO or PSO) – a new program containing all reorderings, thus already exhibiting the additional interleavings on SC. Our technique is *compositional* in the sense that program generation can be carried out on single processes without ever needing to inspect the state space of the concurrent program. We formally prove compositionality as well as soundness of our technique.

WEAK2SC takes standard C programs as input and produces program descriptions which can be fed into automatic model checking tools (like SPIN) as well as into interactive provers (like KIV). Thereby, we allow for a wide range of verification options. We demonstrate the effectiveness of our technique by evaluating WEAK2SC on a number of example programs, ranging from concurrent data structures to software transactional memory algorithms.

Petri Nets and Semilinear Sets
(Extended Abstract)

Hsu-Chun Yen

Department of Electrical Engineering, National Taiwan University,
Taipei, 106, Taiwan, Republic of China
yen@cc.ee.ntu.edu.tw

Abstract. Semilinear sets play a key role in many areas of computer science, in particular, in theoretical computer science, as they are characterizable by Presburger Arithmetic (a decidable theory). The reachability set of a Petri net is not semilinear in general. There are, however, a wide variety of subclasses of Petri nets enjoying semilinear reachability sets, and such results as well as analytical techniques developed around them contribute to important milestones historically in the analysis of Petri nets. In this talk, we first give a brief survey on results related to Petri nets with semilinear reachability sets. We then focus on a technique capable of unifying many existing semilinear Petri nets in a coherent way. The unified strategy also leads to various new semilinearity results for Petri nets. Finally, we shall also briefly touch upon the notion of *almost semilinear sets* which witnesses some recent advances towards the general Petri net reachability problem.

The Lean Theorem Prover

Leonardo de Moura

Microsoft Research
leonardo@microsoft.com

Abstract. Lean is a new open source theorem prover being developed at Microsoft Research and Carnegie Mellon University, with a small trusted kernel based on dependent type theory. It aims to bridge the gap between interactive and automated theorem proving, by situating automated tools and methods in a framework that supports user interaction and the construction of fully specified axiomatic proofs. The goal is to support both mathematical reasoning and reasoning about complex systems, and to verify claims in both domains. Lean is an ongoing and long-term effort, and much of the potential for automation will be realized only gradually over time, but it already provides many useful components, integrated development environments, and a rich API which can be used to embed it into other systems.

In this talk, we provide a short introduction to the Lean theorem prover, describe how mathematical structures are encoded in the system, quotient types, the type class mechanism, and the main ideas behind the novel meta-programming framework available in Lean. More information about Lean can be found at http://leanprover.github.io. The interactive book "Theorem Proving in Lean"[1] is the standard reference for Lean. The book is available in PDF and HTML formats.

---

[1] http://leanprover.github.io/tutorial.

Contents

Calculi

Specifications

Composition and Transformation

Automata

Temporal Logics

Tool and Short Papers

Invited Papers

Verification of Concurrent Programs on Weak Memory Models

Oleg Travkin[✉] and Heike Wehrheim[✉]

Institut für Informatik, Universität Paderborn, 33098 Paderborn, Germany
{oleg82,wehrheim}@uni-paderborn.de

Abstract. Modern multi-core processors equipped with weak memory models seemingly reorder instructions (with respect to program order) due to built-in optimizations. For concurrent programs, weak memory models thereby produce interleaved executions which are impossible on sequentially consistent (SC) memory. Verification of concurrent programs consequently needs to take the memory model of the executing processor into account. This, however, makes most standard software verification tools inapplicable.

In this paper, we propose a technique (and present its accompanying tool WEAK2SC) for *reducing* the verification problem for weak memory models to the verification on SC. The reduction proceeds by generating – out of a given program and weak memory model (here, TSO or PSO) – a new program containing all reorderings, thus already exhibiting the additional interleavings on SC. Our technique is *compositional* in the sense that program generation can be carried out on single processes without ever needing to inspect the state space of the concurrent program. We formally prove compositionality as well as soundness of our technique.

WEAK2SC takes standard C programs as input and produces program descriptions which can be fed into automatic model checking tools (like SPIN) as well as into interactive provers (like KIV). Thereby, we allow for a wide range of verification options. We demonstrate the effectiveness of our technique by evaluating WEAK2SC on a number of example programs, ranging from concurrent data structures to software transactional memory algorithms.

1 Introduction

With the advent of multi-core processors, we have recently seen new types of bugs in concurrent programs coming up[1]. These bugs are due to the weak memory semantics of multi-core processors, which in their architectures are streamlined towards high performance. In executions of concurrent programs, weak memory causes program statements to seemingly be executed in an order different from the

---

[1] See T. Lane. Yes, waitlatch is vulnerable to weak-memory-ordering bugs, http://www.postgresql.org/message-id/24241.1312739269@sss.pgh.pa.us, 2011, or D. Dice. A race in LockSupport park() arising from weak memory models, https://blogs.oracle.com/dave/entry/a_race_in_locksupport_park, 2009.

© Springer International Publishing AG 2016
A. Sampaio and F. Wang (Eds.): ICTAC 2016, LNCS 9965, pp. 3–24, 2016.
DOI: 10.1007/978-3-319-46750-4_1

program order. Two such weak memory models are TSO (total store order, the memory model of the Intel x86 [24,38]) and PSO (partial store order, the memory model of the SPARC processor [16]). On the contrary, concurrent executions adhering to program order are said to be *sequentially consistent* (SC) [29], a notion long ago introduced by Leslie Lamport. As Lamport has already realized in 1979, on multi-core processors concurrent programs might exhibit behaviour which is not explainable by their program order. However, it is only today that multi-core processor come at large use, and consequently correctness of concurrent programs needs to be checked with respect to weak memory models.

Over the years, a lot of verification tools for concurrent programs assuming sequential consistency have been developed (e.g., Spin [23], the tools of the CSeq family [25] or CBMC [26]). For weak memory models, these cannot be readily used. Thus, instead a number of specialised analysis tools have recently come up, e.g. [3,11,19,33,40,41,44]. The technique which we propose in this paper takes the opposite approach: instead of developing new tools for weak memory models, we generate a new program which already contains all the statement reorderings which a weak memory model might construct. Thereby, we reduce verification on weak memory models to verification on sequential consistency, and hence enable re-use of existing model checkers for concurrent programs. Our technique is parametric in the memory model, and applicable to all memory models with an *operational* definition (as opposed to an axiomatic definition like given in [32,38]). Here, we provide a definition for three memory models: TSO, PSO and SC.

The generation of the SC-program proceeds via the construction of a so-called *store-buffer graph* which abstractly represents the contents of the store-buffer (a queue associated to a core). This construction does, however, only need to be done for single processes (within a concurrent program): the techniques is *compositional*, and hence we can avoid having to deal with the state space of the parallel system during SC-program generation.

We have implemented this technique within our new tool WEAK2SC. It takes C programs as input and first of all produces an intermediate representation using the LLVM compiler framework [30]. On the intermediate representation, we have the right granularity of operations allowing us to precisely see read and write operations (on which reorderings take place). Out of this, we separately construct a store-buffer graph for every process. The store buffer graphs in a sense act as control flow graphs of the SC programs. Currently, we automatically generate SC programs in two forms: as Promela code (and thus input to the model checker Spin [23]) and as logical formulae (input to the interactive prover KIV [22]). In principle, however, code generation is possible for every verification tool allowing for nondeterminstic programs. So far, our tool covers the TSO and PSO memory models. In this and in the full automation of our transformations, we extend the approach of [43]. We have carried out a number of experiments on mutual exclusion algorithms, concurrent data structures and a software transactional memory algorithm. The experiments show that we can indeed detect errors in programs which just appear in weak memory models, and we can verify correctness of corrected (i.e., fenced) versions.

2   Memory Models – TSO, PSO and SC

Weak memory models describe the semantics of concurrent programs when executed on multi-core processors. Multi-core processors are highly optimized for performance, but these optimizations are – if one takes a correctness point of view – only sound for sequential programs. As long as sequential programs are run on multi-core processors, their semantics, i.e., the executions they exhibit, stay the same. When concurrent programs are run on multi-core processors, their executions show interleaved behaviour which is not explainable by the program order of their sequential parts. Weak memory models seem to *reorder* program instructions.

A large amount of research so far has gone into finding out what reorderings which memory model will produce [6]. For the two weak memory models we consider here (TSO and PSO), the reorderings can best be explained operationally by looking at the way access to shared memory takes place. In both TSO and PSO, cores have local *store buffers* which cache values written by processes running in these cores. Caching allows for faster access: if a variable needs to be read which is cached, access to shared memory can be avoided and instead the local copy in the store buffer is taken. Unlike real caches, no coherence protocols are run on store buffers to prevent them from getting inconsistent. Occasionally, the contents of the store buffer is *flushed* into main memory. Flushing can be enforced by adding memory barriers to the program code, so called *fence* operations. TSO and PSO deviate in the way they organize their store buffers: TSO employs a single FIFO queue and PSO a FIFO queue per shared variable. In both memory models, flushing nondeterministically flushes one entry of the queue(s) to shared memory. For PSO, this means that variable values are potentially flushed into main memory in an order different from the order of writing them to the store buffer.

For a more detailed explanation, consider the three examples programs in Fig. 1. They give three so-called *litmus tests* (taken from [24]) which exemplify a specific form of reordering. For each litmus test, we give the initial values of shared variables (x and y), the code of two processes and possible final values of registers. All these outcomes are not feasible on SC.

- The program on the left has a *write-read reordering*: the writing to x and y is first of all cached in the store buffers, and thus the two reads can still read the initial values of x and y from shared memory (possible for PSO and TSO).
- The program in the middle has an *early-read*: the writes to x and y are cached in the store buffers, and the reads (of the same process) will read these variables, whereas the reads from other processes will see the values in shared memory (valid for PSO and TSO).
- The program on the right has a *write-write reordering*: in PSO (and not in TSO), the write to y can be flushed to memory before the write to x, and thus from the point of view of process 2 these two writes are reordered.

For our reduction to SC, we will directly use this operational explanation of reorderings via the contents of store buffers. More precisely, we will construct a

a) *Initially* : $x = 0 \wedge y = 0$

Process 1	Process 2
$1 : write(x, 1);$	$1 : write(y, 1);$
$2 : read(y, r1);$	$2 : read(x, r2);$
$3 :$	$3 :$

$r1 = 0 \wedge r2 = 0$ possible

c) *Initially* : $x = 0 \wedge y = 0$

Process 1	Process 2
$1 : write(x, 1);$	$1 : read(y, r1);$
$2 : write(y, 1);$	$2 : read(x, r2);$
$3 :$	$3 :$

$r1 = 1 \wedge r2 = 0$ possible

b) *Initially* : $x = 0 \wedge y = 0$

Process 1	Process 2
$1 : write(x, 1);$	$1 : write(y, 1);$
$2 : read(x, r1);$	$2 : read(y, r3);$
$3 : read(y, r2);$	$3 : read(x, r4);$
$4 :$	$4 :$

$r1 = r3 = 1 \wedge r2 = r4 = 0$ possible

Fig. 1. Litmus tests for (a) write-read reordering (left), (b) for early-reads (middle) and (c) for write-write reordering (right)

symbolic contents of the store buffer for programs, and out of this generate the new SC program. We next start with formally defining the semantics of programs on SC, PSO, and TSO, as to be able to compare these later (which we need for proving soundness of our reduction technique).

3   Memory Model Semantics

For programs, we assume a set *Reg* of registers local to processes and a set of variables *Var*, shared by processes. For simplicity, both take as values just integers[2]. A set of labels $\mathcal{L}$ is used to denote program locations. The following definition gives the grammar of programs.

Definition 1. *A* sequential program *(or process) P is generated by the following grammar:*

$$P ::= \ell : read(x, r) \mid \ell : write(x, r) \mid \ell : write(x, n) \mid$$
$$\ell : r := expr \mid \ell : fence \mid \ell : skip \mid P_1; P_2 \mid$$
$$\ell : if \ (bexpr) \ then \ P_1 \ else \ P_2 \ fi \mid$$
$$\ell : while \ (bexpr) \ do \ P_1 \ od \mid \ell : goto \ \ell'$$

where $x \in Var$, $n \in \mathbb{Z}$, $r \in Reg$, $\ell, \ell' \in \mathcal{L}$ and bexpr is a boolean and expr an arithmetic expression over Var, Reg and $\mathbb{Z}$.

A concurrent program *S is defined as $[P_1 || \ldots || P_n]$ where all $P_i, 1 \le i \le n$, are sequential programs.*

---

[2] This restriction is of course lifted in our implementation.

$$type_{TSO} \;\widehat{=}\; (Var \times \mathbb{Z})^*$$
$$init_{TSO} \;\widehat{=}\; sb = \langle\rangle$$
$$write_{TSO}(x,n) \;\widehat{=}\; sb' = sb \,^\frown\, \langle(x,n)\rangle \qquad \{skip\}$$
$$write_{TSO}(x,r) \;\widehat{=}\; sb' = sb \,^\frown\, \langle(x,reg(r))\rangle \qquad \{skip\}$$
$$read_{TSO}(x,r) \;\widehat{=}\; (x \notin sb \wedge readM_{TSO}(x,r)) \vee \quad \{rd(x,r)\}$$
$$(x \in sb \wedge readL_{TSO}(x,r)) \qquad \{(r := n)\}$$
$$(\text{where } n = lst_{TSO}(x,sb))$$
$$readM_{TSO}(x,r) \;\widehat{=}\; reg' = reg[r \mapsto mem(x)]$$
$$readL_{TSO}(x,r) \;\widehat{=}\; reg' = reg[r \mapsto lst_{TSO}(x,sb)]$$
$$fence_{TSO} \;\widehat{=}\; sb = \langle\rangle \qquad \{skip\}$$
$$flush_{TSO} \;\widehat{=}\; \exists(x,n) : sb = \langle(x,n)\rangle \,^\frown\, sb' \qquad \{wr(x,n)\}$$
$$\wedge\, mem' = mem[x \mapsto n]$$

where $lst_{TSO}(x,sb) = n$ iff $\exists\, sb_{pre}, sb_{suf} : sb = sb_{pre} \,^\frown\, \langle(x,n)\rangle \,^\frown\, sb_{suf} \wedge x \notin sb_{suf}$.

Fig. 2. Memory model TSO (Color figure online)

Out of the program text, we can derive a function $succ : \mathcal{L} \to \mathcal{L}$ denoting the *successor* of a label ℓ in the program. Similarly, we use functions $succ_T$ and $succ_F$ for the successors in if and while statements (on condition being true, false, respectively). We assume the first statement in a sequential program to have label ℓ_0.

Processes have a local state represented by a function $reg : Reg \to \mathbb{Z}$ (registers) together with the value of a program counter, and concurrent programs in addition have a shared global state represented by a function $mem : Var \to \mathbb{Z}$ (shared variables). We use the notation $mem[x \mapsto n]$ to stand for the function mem' which agrees with mem up to x which is mapped to n (and similar for other functions). A memory model is fixed by stating how the writing to and reading from global memory takes place. Memory models use *store buffers* to

$$type_{PSO} \;\widehat{=}\; (Var \to \mathbb{Z}^*)$$
$$init_{PSO} \;\widehat{=}\; \forall v \in Var : sb(v) = \langle\rangle$$
$$write_{PSO}(x,n) \;\widehat{=}\; sb'(x) = sb(x) \,^\frown\, \langle n \rangle \qquad \{skip\}$$
$$write_{PSO}(x,r) \;\widehat{=}\; sb'(x) = sb(x) \,^\frown\, \langle reg(r) \rangle \qquad \{skip\}$$
$$read_{PSO}(x,r) \;\widehat{=}\; (sb(x) = \langle\rangle \wedge readM_{TSO}(x,r)) \vee \quad \{rd(x,r)\}$$
$$(sb(x) \neq \varnothing \wedge readL_{PSO}(x,r)) \qquad \{r := n\}$$
$$(\text{where } n = lst_{PSO}(x,sb))$$
$$readM_{PSO} \;\widehat{=}\; reg' = reg[r \mapsto mem(x)]$$
$$readL_{PSO} \;\widehat{=}\; reg' = reg[r \mapsto lst_{PSO}(sb(x))]$$
$$fence_{PSO} \;\widehat{=}\; \forall v \in Var : sb(v) = \langle\rangle \qquad \{skip\}$$
$$flush_{PSO} \;\widehat{=}\; \exists x, n : sb(x) = \langle n \rangle \,^\frown\, sb'(x) \qquad \{wr(x,n)\}$$
$$\wedge\, mem' = mem[x \mapsto n]$$

where $lst_{PSO}(x,sb) = last(sb(x))$.

Fig. 3. Memory model PSO (Color figure online)

cache values of global variables. Such store buffers take different forms: in case of TSO it is a sequence of pairs (variable,value); in case of PSO it is a mapping from variables to sequence of values; in case of SC the store buffer is not existing (which we model by a set which is always empty). In the semantics, the store buffer is represented by sb.

We describe the semantics of program operations by logical formulae over sb, reg and mem. In this, primed variables are used to denote the state after execution of the operation. A formula like $(x = 0) \wedge (reg'(r_1) = 4)$ for instance describes the fact that currently x has to be 0 and in the next state the register r_1 has the value 4 (and all other registers stay the same). A state s for a process consists of a valuation of the variables pc (the program counter), sb and reg. We write $s \models p$ for a formula p to say that p holds true in s.

Definition 2. *A memory model* $MM = (type, init, read, write, flush, fence)$ *consists of*

– *the* type *of the store buffer, and*
– *formulae for initialization, read, write, flush and fence operations ranging over* mem, sb *and* reg.

Figures 2 and 3 give the types as well as operation formulae of TSO and PSO; Fig. 4 that of SC. In curly brackets (blue) we give the label of an operation (see below). We assume all registers and variables to initially have value 0.

The semantics of programs is given by assigning to every statement stm as semantics a predicate according to the given memory model, i.e., we fix $[stm]_{MM}$. Figure 5 then defines the semantics for sequential programs, parameterized in the memory model. We define $Ops(P)$ to be the set of all such predicates plus an operation $[\ell : flush]_{MM}$ for each program location $\ell \in \mathcal{L}$. Thus, $Ops(P)$ is the set of all *operations* of the program P. We assume that variables that are not mentioned by the predicates keep their value, e.g., not mentioning mem implicitly states $mem' = mem$.

Given the operations of a sequential program P, we can derive a *local* transition system of P. This local transition system abstracts from its environment (i.e., other processes running concurrently) in that it assumes arbitrary states of the global memory (which could be produced by other processes). We call this an *open* semantics.

$$
\begin{aligned}
type_{SC} &\mathrel{\hat{=}} 2^{Var} \\
init_{SC} &\mathrel{\hat{=}} sb = \varnothing \\
write_{SC}(x, n) &\mathrel{\hat{=}} mem' = mem[x \mapsto n] && \{wr(x, n)\} \\
write_{SC}(x, r) &\mathrel{\hat{=}} mem' = mem[x \mapsto reg(r)] && \{wr(x, reg(r))\} \\
read_{SC}(x, r) &\mathrel{\hat{=}} reg' = reg[r \mapsto mem(x)] && \{rd(x, r)\} \\
fence_{SC} &= true && \{skip\} \\
flush_{SC} &= false && \{skip\}
\end{aligned}
$$

Fig. 4. Memory model SC (Color figure online)

$$[\![\ell : stm]\!]_{MM} = (pc = \ell \wedge stm_{MM} \wedge pc' = succ(\ell))$$
$$[\![\ell : r := expr]\!]_{MM} = (pc = \ell \wedge r' = expr \wedge pc' = succ(\ell))$$
$$[\![\ell : flush]\!]_{MM} = flush_{MM}$$
$$[\![\ell : if\ (bexpr)\ then\ P_1\ else\ P_2\ fi]\!]_{MM} = (pc = \ell \wedge bexpr \wedge pc' = succ_T(\ell))$$
$$\vee\ (pc = \ell \wedge \neg bexpr \wedge pc' = succ_F(\ell))$$
$$[\![\ell : while\ (bexpr)\ do\ P\ od]\!]_{MM} = (pc = \ell \wedge bexpr \wedge pc' = succ_T(\ell))$$
$$\vee\ (pc = \ell \wedge \neg bexpr \wedge pc' = succ_F(\ell))$$
$$[\![\ell : skip)]\!]_{MM} = (pc = \ell \wedge pc' = suc(\ell))$$
$$[\![\ell : goto\ \ell')]\!]_{MM} = (pc = \ell \wedge pc' = \ell')$$

where $stm \in \{read(x, r), write(x, r), write(x, n), fence\}$.

Fig. 5. Semantics of program statements wrt. memory model MM

Transitions in the transition system will be labelled. The labels reflect the *effect of the operation as observed by the environment*. Therefore, a write to a store buffer gets a label *skip* whereas a flush operation gets a label *wr*. Memory model specific labels are given in Figs. 2, 3 and 4 behind the semantics of operations (enclosed in curly brackets, in blue). The labels for memory model independent operations are as follows: the label of *if bexpr then* ... is $reg(bexpr)$ (*reg* current valuation of registers), and similar for *while*, the label of $r := expr$ is $r := reg(expr)$ and that of *goto* simply *skip*. The set of all such labels is called *Lab*, and the label of an operation *op* is $label(op)$.

Definition 3. *The local transition system of a sequential program P on memory model MM, $lts_{MM}(P) = (S, \rightarrow, S_0)$, consists of*

- *a set of states $S = \{(pc, sb, reg) \mid pc \in \mathcal{L}, sb \in type_{MM}, reg \in (Reg \rightarrow \mathbb{Z})\}$,*
- *a set of initial states $S_0 = \{s \in S \mid s \models init_{MM} \wedge s \models (pc = \ell_0)\}$,*
- *a set of transitions $\rightarrow \subseteq S \times Lab \times S$ such that for $s = (pc, sb, reg)$ and $s' = (pc', sb', reg')$, we have $s \xrightarrow{lab} s'$ iff $\exists op \in Ops(P), \exists mem, mem' : ((s, mem), (s', mem')) \models op$ and $label(op) = lab$. For such transitions, we use the notation $s \xrightarrow{lab}_{mem,mem'} s'$.*

Processes typically run in parallel with other processes. The semantics for parallel compositions of processes is now a *closed* semantics already incorporating all relevant components. We just define it for two processes here; a generalisation to larger numbers of components is straightforward. The initial global state mem_0 assigns 0 to all global variables.

Definition 4. *Let P_j, $j \in \{1,2\}$, be two sequential programs and let $(S_j, \rightarrow_j, S_{0,j})$, be their process local (i.e., open) labelled transitions systems for memory model MM.*

The closed MM semantics of $P_1 \parallel P_2$, $lts_{MM}(P_1 \parallel P_2)$, is the labelled transition system $(S, \rightarrow, S_0)$ with $S \subseteq \{(mem, s_1, s_2) \mid s_1 \in S_1, s_2 \in S_2\}$, $S_0 = \{(mem_0, s_{0,1},$

$s_{0,2} \mid s_{0,j} \in S_{0,j}\}$, and $s = (mem, s_1, s_2) \xrightarrow{lab} s' = (mem', s'_1, s'_2)$ when $(s_1 \xrightarrow{lab}_{mem,mem'} s'_1 \wedge s_2 = s'_2)$ or $(s_2 \xrightarrow{lab}_{mem,mem'} s'_2 \wedge s_1 = s'_1)$.

Due to the open semantics for processes, we are thus able to give a *compositional* semantics for parallel composition. This is key to our transformation which operates on single processes.

Ultimately, we will be interested in comparing the weak memory model semantics of one program with the SC semantics of another. Our notion of equality is based on bisimulation equivalence [34]. Our definition of bisimulation compares transition systems with respect to their *labels* on transitions as well as their local *states*.

Definition 5. *Let* $T_1 = (S, \rightarrow_1, S_0)$ *be an* MM_1 *and* $T_2 = (Q, \rightarrow_2, Q_0)$ *an* MM_2 *transition system.*

Transition systems T_1 *and* T_2 *are locally bisimilar,* $T_1 \sim_\ell T_2$, *if there is a bisimulation relation* $\mathcal{R} \subseteq S \times Q$ *such that the following holds:*

1. *Local state equality:*
   $\forall (s, q) \in \mathcal{R}, s = (pc_1, sb_1, reg_1), q = (pc_2, sb_2, reg_2), \forall r \in Reg: reg_1(r) = reg_2(r)$.
2. *Matching on initial states:*
   $\forall s_0 \in S \, \exists q_0 \in Q_0$ *s.t.* $(s_0, q_0) \in \mathcal{R}$, *and reversely* $\forall q_0 \in Q_0 \, \exists s_0 \in S_0$ *s.t.* $(s_0, q_0) \in \mathcal{R}$.
3. *Mutual simulation of steps:*
   if $(s_1, q_1) \in \mathcal{R}$ *and* $s_1 \xrightarrow{lab} s_2$, *then* $\exists q_2$ *such that* $q_1 \xrightarrow{lab} q_2$ *and* $(s_2, q_2) \in \mathcal{R}$, *and vice versa, if* $(s_1, q_1) \in \mathcal{R}$ *and* $q_1 \xrightarrow{lab} q_2$, *then* $\exists s_2$ *such that* $s_1 \xrightarrow{lab} s_2$ *and* $(s_2, q_2) \in \mathcal{R}$.

Similarly, one can define *global bisimilarity* for the closed semantics of a parallel composition, in addition requiring equality of shared memory *mem*. We use the notation $\sim_g$ to denote global bisimilarity. This lets us state our first result: Local bisimilarity of processes implies global bisimilarity of their parallel compositions.

Theorem 1. *Let* P_1, P'_1, P_2, P'_2 *be sequential programs such that* $lts_{MM_1}(P_j) \sim_\ell lts_{MM_2}(P'_j)$, $j = 1, 2$. *Then*

$$lts_{MM_1}(P_1 \| P_2) \sim_g lts_{MM_2}(P'_1 \| P'_2).$$

Proof idea: Proofs of bisimilarity proceed by giving a relation and showing this relation to fulfill the properties of bisimilarity. Due to lack of space, we only give the relation here. Let $\mathcal{R}_j$, $j = 1, 2$, be the relations showing local bisimilarity of $lts_{MM_1}(P_j)$ and $lts_{MM_2}(P'_j)$. Out of this we construct a global bisimulation relation which is

$$\mathcal{R} := \{((mem, s_1, s_2), (mem, q_1, q_2)) \mid (s_j, q_j) \in \mathcal{R}_j, j = 1, 2\}.$$

4   Program Transformation

The basic principle behind our verification technique is to transform every sequential program P into a program P' such that $lts(P)$ for a weak memory model is locally bisimilar to $lts(P')$ for SC. The construction of P' proceeds by symbolic execution of P, and out of the thus constructed symbolic states generation of P'. The symbolic execution tracks - besides the operations being executed and the program locations reached - store buffer contents *only*, and only in a symbolic form. The symbolic form stores variable names together with either values of $\mathbb{Z}$ (in case a constant was used in the *write*), or register *names* (in case a register was used). A symbolic store buffer content for TSO might thus for instance look like this: $\langle (x, 3), (y, r_1), (x, r_2), (z, 5) \rangle$. The symbolic execution thereby generates a symbolic reachability graph, called *store-buffer graph*. Edges in the graph will get labels as well, however, only symbolic ones. We refer to these symbolic labels as the *name* of an operation (*Names*). For memory model specific operations, this is simply the name used for the operation (e.g., *flush*) with the exception that *read* is split into *readM* and *readL* (according to the semantics); for the other operations it is the (unevaluated) boolean condition *bexpr* or its negation (in case of if and while), or simply *goto*. We use names instead of the labels of the semantics here since we still need to see the operation which is executed.

Definition 6. *A store-buffer (or sb-)graph $G = (V, E, v_0)$ of a memory model MM consists of a set of nodes $V \subseteq \mathcal{L} \times type_{MM}$, edges $E \subseteq V \times Names \times V$ and initial node $v_0 \in V$ with $v_0 = (\ell_0, sb_0)$ with $sb_0 \models init_{MM}$.*

The store-buffer graph for a program P is constructed by a form of symbolic execution, executing program operations step by step without constructing the concrete states of registers.

Definition 7. *Let P be a sequential program. The sb-graph of P wrt. a memory model MM, $sg_{MM}(P)$, is inductively defined as follows:*

1. $v_0 := (pc, sb)$ *with* $sb \models init_{MM} \wedge pc = \ell_0$,
2. *if* $(pc, sb) \in V$, *we add a node* (pc', sb') *and an edge* $(pc, sb) \xrightarrow{name} (pc', sb')$
   if $\exists op \in Ops(P)$ *such that*
   - $(pc, pc') \models op$,
   - $(sb, sb') \models sym(op)$ *and*
   - $name = name(op)$.
   Here, $sym(op) = op$ except for $write(x, r)$ which is $sb' = sb^\frown\langle (x, r) \rangle$ for TSO and $sb'(x) = sb(x)^\frown\langle r \rangle$ for PSO.

Note that a store-buffer graph for memory model SC is simply a control-flow graph. Figures 6 and 7 show the store buffer graphs of process 1 of Fig. 1(c). Here, we directly see the difference between TSO and PSO: whereas for TSO, the value of x will always leave the store buffer before the value of y, this is different in PSO where both orders are possible.

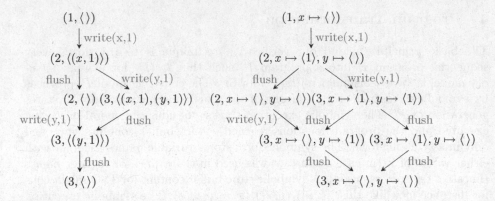

Fig. 6. TSO store-buffer graph of process 1 of Fig. 1(c)

Fig. 7. PSO store-buffer graph of process 1 of Fig. 1(c)

Note also that store-buffer graphs need not necessarily be finite. They are infinite if a program has loops with write operations, but no fences in order to enforce flushing of store buffer content. Since finiteness of the store buffer graph is a prerequisite to our technique, we state one restriction on the class of programs considered: all loops have to be *fenced* or *write-free*. A *loop* is a sequence $\ell_1, \ell_2, ..., \ell_n$ in the transition system such that $n > 1$ and $\ell_1 = \ell_n$. A loop is *write-free* if none of the operations executed in the loop is a write. A loop is *fenced*, if at least one of the operations on the loop is a fence.

We furthermore assume that all sequential programs are in *SSA-form* (static single assignment [17]), meaning that all the registers are (statically) assigned to only once. We furthermore assume that registers are never used before defined. Both, this and the SSA-form is guaranteed by modern compilers, e.g., the LLVM-framework[3] which we use for our approach.

Proposition 1. *Let P be a process program in which every loop is fenced or write-free. Then $sg_{MM}(P)$ is finite for every $MM \in \{TSO, PSO, SC\}$.*

In the generation of the SC-program out of an sb-graph, we transform every edge of the graph into an operation (predicate). In this, a flush operation (TSO or PSO) in the sb-graph, flushing a symbolic store buffer contents (either $(x, r) \in sb$ or $r \in sb(x)$, r being a register name), becomes a $write_{sc}(x, r)$ operation. For this to be sound (w.r.t. the intended equivalence of old and new program), we need to make sure that the contents of register r at a flush is still the same as the one at the time of writing the pair r into the (symbolic) store buffer. This is not the case when a write to a register is reaching a use of the register (a so-called *wd-chain*) *without* a fence operation in between. Unfenced wd-chains can be removed by introducing new auxiliary registers (see [43]) and our tool WEAK2SC automatically does so.

---

[3] http://www.llvm.org

The generation of the SC-program now proceeds by defining the operations of the new program (instead of program text[4]). Essentially, every edge in the sb-graph gives rise to one new operation, where the nodes of the graph act as new location labels. The sb-graph is thus the control flow graph of the new program.

Definition 8. *Let $G = (V, E, v_0)$ be an sb-graph of a program P on a memory model MM. The sequential SC program P' of G, $prog(G)$, is given by the new initial location $\ell_0 := v_0$ and the set of operations $Ops(P')$ defined as follows: for every edge $(\ell, sb) \xrightarrow{lab} (\ell', sb')$ we construct an operation*

$$(pc = (\ell, sb) \land op \land pc' = (\ell', sb'))$$

with

$$
op = \begin{cases}
skip & if\, lab \in \{fence, write(x, r), \\
 & \qquad\qquad write(x, n)\} \\
read_{SC}(x, r) & if\, lab = readM(x, r) \\
write_{SC}(x, r) & if\, lab = flush \\
 & \qquad \land\, flushed_{MM}(x, r, sb, sb') \\
r := r_{src} & if\, lab = readL(x, r) \\
 & \qquad \land\, r_{src} = lst_{MM}(x, sb) \\
lab & else
\end{cases}
$$

where $flushed_{TSO}(x, r, sb, sb') = (sb = \langle(x, r)\rangle^\frown sb')$ and $flushed_{PSO}(x, r, sb, sb') = (sb(x) = \langle r \rangle^\frown sb'(x)$.

Note that the transformation to SC does not necessarily yield a deterministic program again. This is for instance the case for the store buffers in Figs. 6 and 7. Out of both graphs, we generate a *nondeterministic* program. This does, however, not constitute a problem when the verification tool to be used allows for nondeterminism.

The transformation of a program on memory model MM into its SC form (weak to SC) is finally defined as

$$w2sc(P, MM) \,\widehat{=}\, prog(sg_{MM}(P))$$

which lets us state our main theorem.

Theorem 2. *Let P be a program with fenced or write-free loops only and with no unfenced wd-chains and $MM \in \{PSO, TSO\}$ a memory model. Then*

$$lts_{MM}(P) \sim_\ell lts_{SC}(w2sc(P, MM)).$$

Proof idea: Again, we only state the bisimulation relation here. The relation contains pairs (s_1, s_2), both consisting of tuples (pc_i, sb_i, reg_i), $i = 1, 2$. While s_1 can have a non-empty store buffer contents sb_1, s_2 (the SC state) always has an empty store buffer. The program counter value pc_2 consists of two parts:

---

[4] Input to Spin can then be generated from a set of operations.

a location of the original program and a symbolic store buffer contents (out of the sb-graph). The correspondence between s_1 and s_2 with respect to store buffer contents states that a concretisation of the symbolic content (in the pc of the SC state) yields the contents in the weak memory model. More formally, we define two functions $conc_{MM}(reg, sb)$ taking register values reg and a symbolic store buffer content sb as argument and returning a concrete store buffer content for a memory model $MM \in \{PSO, TSO\}$ as

$$conc_{TSO}(reg, \langle\rangle) = \langle\rangle$$
$$conc_{TSO}(reg, \langle(x,n)\rangle^\frown sb) = \langle(x,n)\rangle^\frown conc_{TSO}(reg, sb)$$
$$conc_{TSO}(reg, \langle(x,r)\rangle^\frown sb) = \langle(x, reg(r))\rangle^\frown conc_{TSO}(reg, sb)$$
$$conc_{PSO}(reg, sb)(v) = conc_{TSO}(reg, sb(v))$$

Here, $n \in \mathbb{Z}$ and $r \in Reg$. With this concretisation at hand, the bisimimulation relation for memory model MM is

$$\begin{aligned}\mathcal{R}_{MM} := \{(s_1, s_2) \mid s_i = (pc_i, sb_i, reg_i), i = 1, 2, \\ \wedge\ sb_2 = \varnothing \wedge pc_1 = first(pc_2) \\ \wedge\ \forall r \in Reg : reg_1(r) = reg_2(r) \\ \wedge\ conc_{MM}(reg_2, second(pc_2)) = sb_1\}\end{aligned}$$

5   Weak2SC – Tool

We automated our transformation in a tool framework called WEAK2SC. Figure 8 gives an overview of the verification process. The tool currently produces two sorts of outputs: code for the model checker SPIN [23] for verification of finite state concurrent programs, and input to the interactive prover KIV [22] for proving correctness of infinite state programs, in particular parameterized algorithms. We refrain from showing the KIV output here as the main difference to Promela is syntactical. We refer to [41] (which we extend and automate) for more details on the KIV output.

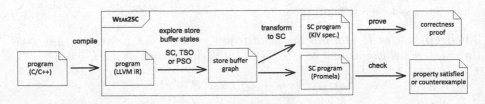

Fig. 8. Usage of WEAK2SC in the verification process

LLVM IR: We start with the C or C++ code of an algorithm that has to be compiled to intermediate representation (LLVM IR) and thus enables reasoning

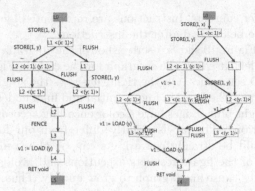

```
1:   @y = global i32* null, align 4
2:   @x = global i32* null, align 4

3:   define void @p0() nounwind {
4:   entry:
5:       store i32 1, i32* @x, align 4
6:       store i32 1, i32* @y, align 4
7:       fence seq_cst
8:       %1 = load i32* @y, align 4
9:       ret void
10: }
```

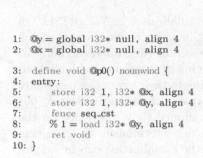

Fig. 9. Simple LLVM IR code with two writes followed by a fence and a read instruction.

Fig. 10. Store buffer graph visualization (PSO semantics) for the LLVM IR code in 9 (left with fence, right without fence); transition color encoding: writes (red), reads (green), flush (brown), fence (blue), local instructions (grey). (Color figure online)

about the algorithm on a low-level, i.e., single processor instructions like reads and writes. The intermediate representation is used for a variety of analyzes and optimizations by the compiler. Hence, we can decide whether we want to deal with the code before or after compiler optimization. The intermediate representation uses symbolic registers and addresses instead of explicit ones. Furthermore, it provides type information and preserves variable names from the original C/C++ program. All of this helps with understanding the low-level version of a program, which is crucial, since all further reasoning must be done low-level.

An example of the LLVM IR syntax is shown in Fig. 9. It shows a simple program, a method @p0 with two writes to two shared pointers (@x and @y declared globally) followed by fence instructions and a read of variable @y. Note that the pointers need to be initialized before they are used, which is usually done in the main method (not shown). Global variables in LLVM IR are prefixed with an '@'. All other variables are prefixed with an '%'.

SB-Graph: WEAK2SC parses the LLVM IR code of the compiled program and constructs an sb-graph for each of its methods. The semantics (SC, TSO or PSO) can be chosen upfront by the user. For TSO and PSO, WEAK2SC checks whether the program contains loops with writes but no fences. If it finds any, the user receives a warning as such loops cannot be represented by a finite store buffer graph and the tool would not terminate. Furthermore, WEAK2SC checks for wd-chains (see Sect. 4) in case of TSO or PSO and, if necessary, removes them.

By choosing a particular memory model, the underlying strategy for creation of the store buffer graph is determined. By choosing SC as the target memory model, WEAK2SC generates simply a control flow graph because SC does not have store buffers. The sb-graph is created by symbolically executing the program and memorizing all visited combinations of locations and symbolic store buffer contents. Instructions are represented by edges while nodes represent the state before and after the instruction.

Figure 10 shows a screenshot of the sb-graph visualization in WEAK2SC. On the left, it shows the sb-graph for the code in Fig. 9, which results from choosing PSO semantics. On the right, it shows the sb-graph for the same program, but without the fence in line 7. Initial and final nodes in the sb-graph are highlighted in light blue. Edges have a color encoding according to their semantics (see figure caption). The labels slightly differ from our formal definitions in Sect. 4, but should be straightforward. As can be observed immediately, the fence on the left of the figure restricts executions and avoids nondeterminism while missing fences cause an sb-graph to grow quickly. Thus, the visualization can be helpful for finding a good placement of fences or just serve as a simple view of possible executions. Note that we assume fences at invocation and return of a method, and thus return statements seem to act like a fence in the sb-graph. However, we can circumvent this assumption by unfolding different methods to one larger method and then create the sb-graph for it. By this, reorderings across method boundaries are included.

Promela Output: From the sb-graph, WEAK2SC generates a new SC-program. Generally, the new program can be the input to any verification tool that provides SC semantics and allows for non-determinism in programs. Figure 11 shows the output produced by WEAK2SC for the left sb-graph in Fig. 10.

The generated program has an explicit modelling of memory as an array of fixed but arbitrary size MEM_SIZE. A counter variable memUse always points to the next free cell of the memory and is incremented whenever memory is allocated. Memory is never freed, but can be allocated through the alloca statement, which increments memUse and checks whether MEM_SIZE was chosen sufficiently and throws an error, otherwise. When space is allocated, we do not differentiate between different types, i.e., each value (bit or integer) requires one entry in the memory array. This simplifies reasoning and sometimes also allows us to drop some value conversion instructions from the original program, e.g., casts from integer to bit and vice versa. For globally defined variables like x and y in the example, memory is allocated in the init process. All other variables represent registers and therefore are declared locally. In the example, register %1 from the LLVM code is represented by $v1$ in the Promela model.

WEAK2SC generates an inline statement for each method (p0 in the example), which can be used in different process definitions, e.g., two or more processes running the same method concurrently. Each inline statement starts with variable declarations for registers followed by the transition encoding. Each transition encoding starts at a certain label, performs some computation or non in case of skip, and then jumps via goto to the next label. Take for instance the transition A02y: memory[x] = 1; goto A02; in Fig. 11. It starts at A02y, writes value 1 to the memory address pointed to by x and then goes to the location A02. If more than one transition can be taken, then the choice is modeled by a nondeterministic if-statement. All labels are unique and prefixed by "A" if they belong to the first method, "B" for the second and so on. We use a compact

```
short memory[MEM_SIZE];
short memUse = 1;    //next free cell
short y = null;
short x = null;
...
inline p0(){
    short v1;
    goto A00;
    A00: goto A01x;
    A01x:
        if
        :: goto A02xy;
        :: memory[x] = 1; goto A01;
        fi;
    A02xy:
        if
        :: memory[x] = 1; goto A02y;
        :: memory[y] = 1; goto A02x;
        fi;
    A01: goto A02y;
    A02y: memory[y] = 1; goto A02;
    A02x: memory[x] = 1; goto A02;
    A02: goto A03;
    A03: v1 = memory[y]; goto A04;
    A04: goto AEnd;
    AEnd: skip;
}
...
init{atomic{
    alloca(1, y);
    alloca(1, x);
    ... }}
```

Algorithm	tso2sc	pso2sc
Dekker [21]	uwl	uwl
Dekker (TSO)	✓	✓
Peterson [36]	×	×
Peterson (TSO)	✓	×
Peterson (PSO)	✓	✓
Lamport bakery [28]	×	×
Lamport bakery (TSO)	✓	×
Lamport bakery (PSO)	✓	✓
Szymanski [39]	×	×
Szymanski (TSO)	✓	✓
fib_bench [1]	✓	✓
Arora queue [7]	×	×
Arora queue (TSO)	✓	×
Arora queue (PSO)	✓	✓
Treiber stack [42]	✓	×
Treiber stack (PSO)	✓	✓
TML [18]	✓	×
TML (PSO)	✓	✓

Fig. 11. Excerpt of generated program model (Promela) for the store buffer graph in Fig. 10.

Fig. 12. Verification results for the transformed programs (tso2sc, pso2sc). Brackets state the memory model for which a program was fenced.

encoding of labels (location numbers and store buffer content, if there is any). Each method has an "End" label used as the target label of return statements.

The generated model provides process definition stubs that have to be filled with calls of the methods/inline statements. The init definition defines the scenario for the state space exploration, i.e., a set of processes to run and initialization of variables if necessary. In SPIN, the properties to be verified can be either assertions or LTL formulae [37].

Implementation: WEAK2SC supports a subset of LLVM IR. In particular, it supports single word operations, e.g., 32/64 bit reads/writes but no multi-word operations. If unsupported operations are detected, WEAK2SC adds annotations in the generated programs, in order to draw a developer's attention to them. The generation of the sb-graph and of the new program are fully automated and the output is usually produced almost instantly.s

WEAK2SC is implemented as a plug-in for the Eclipse IDE[5]. It has a built-in parser for the LLVM-IR language, which is defined in the XText parser generator

---

[5] http://www.eclipse.org

framework. All internal models are based on the Eclipse Modelling Framework (EMF) and allow for easy customization or extension. All program models are generated by a template based model transformation (Acceleo framework). Support for a new target language can be added straightforwardly. It requires only a set of templates for the representation of the store buffer graph in the new target language. We provide a repository[6] for WEAK2SC with the latest files.

6   Experiments

With our automatic tool at hand, we carried out a number of experiments. The experiments aimed at answering the following research questions:

– **RQ1.** Is WEAK2SC able to find bugs in concurrent algorithms which are due to weak memory models?
– **RQ2.** Does our tool cover a sufficiently large range of concurrent programs?
– **RQ3.** How does WEAK2SC compare to other approaches in terms of verification time and state space?

Experimental Setup: In order to answer the above questions, we conducted experiments with a set of programs ranging from mutual exclusion algorithms to concurrent data structures (like the Treiber stack [42]) and software transactional memory algorithms (the Transactional Mutex Lock (TML) of [18]). All of these algorithms are defined for arbitrary many processes running concurrently and calling methods provided by these algorithms. Thus, for every experiment we first of all defined different usage scenarios (initial state, number of processes and their method calls). All of our experiments were conducted with SPIN 6.2.3 on a Linux virtual machine with 3 GB memory and 2 cores dedicated to it (Intel Core i5 M540, 2.53 GHz).

RQ1: In order to answer RQ1, we defined a correctness property for every scenario (in the form of temporal logic formulas or assertions in the code). Note that all of the algorithms are correct under SC. It turned out that almost all examples are incorrect on at least PSO. In that case, we inspected the code and added fences in order to fix the error. As a result, every example comes in up to 3 versions, original, fenced for TSO and fenced for PSO (in brackets after algorithm name). Table 1 shows the verification result for each algorithm. The columns "tso2sc" and "pso2sc" show the results for the programs transformed for TSO and PSO memory model, respectively. An entry × means violation and ✓ holding of the property. Some of the errors found for weak memory are actually new, e.g., TML has so far not been verified for PSO. For placing appropriate fences, we made extensive use of the sb-graph visualization.

RQ2: For answering question RQ2, we chose our case studies as to represent a wide range of highly concurrent programs (mutexes, concurrent data structures, STMs). In particular, we wanted to use algorithms which are vulnerable to weak

---

[6] https://github.com/oleg82upb/lina4wm-tools

memory errors. This is typically the case for concurrent data structures and STMs since – for performance reasons – these are intentionally designed with data races. For race-free programs it would be sufficient to verify the SC version. Out of our examples, we were only unable to construct the sb-graph due to an unfenced writing loop (uwl) for the unfenced version of Dekker [21]. The fenced version (where the fence is known to be needed anyway) was checkable. Furthermore, it turned out that the supported subset of LLVM IR is fully sufficient for all such algorithms. In fact, whenever we ran over a feature in an algorithm not covered so far, we slightly extended our subset.

RQ3: Research question RQ3 was the most difficult to answer. A lot of verification tools for weak memory models either do not support C, or use programs and property specifications syntactically streamlined for the SV-COMP competition. Thus, we could not find a set of benchmarks working for all tools. However, as the examples studied here are mainly the same, we give some remarks on the runtimes taken from publications and a comparison to ours at the end of this paragraph.

In order to gain proper data for performance comparison, we compared our approach against those using *operational* memory models for TSO and PSO [4, 35, 38, 40]. Operational models explicitly define store buffers and simulate their behaviour during program verification. In order to rule out influences on the results because of the usage of different tools, we implemented an operational model ourselves within SPIN. We took the operational memory model for TSO provided by [40] and adapted their technique to create an operational model for PSO. To this end, we extended WEAK2SC in order to generate program models for Promela that can be combined with any of the given operational models (SC, TSO or PSO) interchangeably. For the extension, we defined another set of code templates for Promela in Acceleo.

In order to determine the impact on the verification performance, we took the programs from the earlier experiments (in case of bugs, the fenced version) and measured the full exploration time and the number of explored states. For the mutex algorithms, it was a single scenario with two processes each. For the Treiber stack [42], the queue of Arora et al. [7] and the TML [18] we conducted results for several scenarios, varying in the number of processes and method calls per process. Table 1 shows the exploration performance results. The columns tso2sc and pso2sc again represent the results for the transformed programs; tso and pso are the columns for the explicit operational model. Column #i states the number of instructions per method in the LLVM IR code, whereas #n states the number of nodes in the generated sb-graph and therefore roughly indicates how much behavior is added due to the underlying weak memory. The numbers are "/"-separated where different or asymmetric methods are present. Column #s states the number of states explored and t the time in seconds that was required for the full exploration.

The scenarios for the mutex algorithms each consist of 2 processes trying to acquire the lock concurrently. For the non-mutex algorithms, the scenario is given in each row of the table. It is a parallel composition of a series of method

Table 1. Verification results for full state space exploration error: #i are lines of LLVM IR instructions ("/" separated for each method); #n number of nodes in the sb-graph ("/" separated for each sb-graph); #s the number of states explored t is the time in seconds.

Algorithm	#i	tso		tso2sc			pso		pso2sc		
		#s	t	#s	t	#n	#s	t	#s	t	#n
Dekker (TSO)	33	2232	0.01	1208	0.01	56	2368	0.02	1441	≈0	60
Peterson (PSO)	24	2149	0.01	1408	≈0	30	2149	0.02	1408	≈0	30
Lamp. B. (PSO)	49	12.2 k	0.04	7525	0.01	59	12.2 k	0.07	7525	0.01	59
Szymanski (TSO)	32/35	22.5 k	0.12	11.7 k	0.02	59/68	22.5 k	0.17	14.7 k	0.03	70/82
PGSQL [4]	11	1818	0.01	1817	0.01	26	7.2 k	0.14	7547	0.02	42
Burns (PSO) [14]	3/19/11	737	≈0	448	≈0	4/30/12	737	≈0	458	≈0	4/30/12
Fib_bench [1]	21	5 M	12.2	1.8 M	2.1	67	4.5 M	24.4	1.8 M	2.52	67
Arora Q. (PSO)											
uuuoouo∥sss	9/15/30	329.3 k	1.08	147.2 k	0.3	13/18/49	339.9 k	1.83	160.3 k	0.31	13/18/52
uououo∥ss∥ss	9/15/30	24.5 M	134	7.90 M	21.8	13/18/49	24.1 M	258	8.1 M	18.9	13/18/52
Treiber St.											
uouo∥uouo	13/16	215.3 k	0.74	132.2 k	0.34	16/29	218.6 k	1.53	143.7 k	0.31	16/31
uuuooo∥ ooouuu	13/16	2.6 M	11.2	1.8 M	4.94	16/29	2.7 M	22	2 M	4.89	16/31
TML (PSO)											
bwrc ∥ bwrc	30	2204	0.01	1426	0.01	55	2251	0.01	1578	≈0	59
bwrrc ∥ bwrrc	37	3739	0.01	1298	≈0	86	8176	0.07	5943	0.02	111
$\|_{i=1}^2$ (bwc$_i$ ∥ brc$_i$)	22/23	22.1 M	174	7.17 M	28.3	33/38	23.4 M	393	7,9 M	33.3	42/58

calls by each process, where u (resp. o) denotes the push (resp. pop) method of the queue and stack implementations. In case of the queue, an s denotes a steal method from non-owning process as it is a work-stealing queue. The TML algorithm provides methods begin (b), write (w), read (r) and commit (c). Since these would be rather short methods, we composed them into larger transactions.

Our results suggest that the transformation-based approach provides a significant performance improvement over the use of an operational memory model in most cases. All experiments show a roughly similar sized or a smaller state space for the transformed programs than for their operational counterpart (up to a factor of ca. 3). One reason for this is certainly the encoding of the operational memory, which requires lots of auxiliary variables to model the store buffer and which is not necessary[7] in the transformed programs. From our experiments, we can also observe a speedup of up to a magnitude and the results suggest that the speedup grows with the number of concurrent processes (see verification times for Arora queue and TML).

For a comparison with approaches other than the operational modelling, we could only look at publications. In [2], experiments were conducted with other tools like CBMC [5], goto-instrument [4], or Nidhugg [2], all of which take weak memory semantics into account during verification and which were used to verify the same set of mutex algorithms as we did. While CBMC and goto-instrument took seconds or in some cases even minutes to get results, only Nidhugg

---

[7] With the exception of max. 1 auxiliary variable per write in a loop.

performed comparably well to WEAK2SC. However, a direct comparison is not really fair as they verified slightly different implementations with different property instrumentation.

7   Conclusion

Verification of concurrent algorithms under weak memory, such as TSO and PSO, is difficult and only few tools have a built-in support for it. We presented a practical reduction approach from TSO or PSO to SC and its implementation in WEAK2SC. The reduction enables verification with standard tools for concurrent programs. The transformation exploits the fact that most of the possible reorderings under weak memory can be computed statically. Our approach uses these to generate a new program. Because we need finite sb-graphs, our approach is restricted to the class of programs that have no unfenced writing loops. Our results show that WEAK2SC is practically useful for finding errors within a short amount of time. As future work we plan for performance optimizations by developing more compact representations of sb-graphs.

Related work. In the last years, several approaches were proposed in order to deal with software verification under the influence of weak memory models, ranging from theoretical results to practical techniques. We comment on the latter.

A number of approaches (e.g., [4,8,15,20,31]) like us propose reduction techniques, which allow for a reuse of verification techniques developed for SC. In spirit closest to us is [8]. They propose a translation from a program run under TSO to SC, assuming a so called *age bound* on the time an entry can stay in the store buffer. Their approach needs a large number of auxiliary variables in the generated program. While this gives an equivalent program on SC (like in our case), [20] on the other hand generates an SC program overapproximating the behaviour of the TSO or PSO program.

Another class of approaches uses partial order (PO) representations of program executions and PO reduction techniques [2,5,9], as to cope with the large number of interleavings generated by concurrent programs plus the reorderings. From looking at the published runtimes of these approaches, we conjecture our tool to compare favourably well to these.

A further type of approaches apply underapproximating techniques like testing, bounded model checking or monitoring [5,11–13]. This is in particular often applied for automatic fence placement [10,27].

Acknowledgements. We thank Annika Mütze, Monika Wedel and Alexander Hetzer for their help with the implementation of WEAK2SC.

References

1. SV-Competition Benchmarks, April 2016. https://github.com/dbeyer/sv-benchmarks
2. Abdulla, P.A., Aronis, S., Atig, M.F., Jonsson, B., Leonardsson, C., Sagonas, K.: Stateless model checking for TSO and PSO. In: Baier, C., Tinelli, C. (eds.) TACAS 2015. LNCS, vol. 9035, pp. 353–367. Springer, Heidelberg (2015). doi:10.1007/978-3-662-46681-0_28
3. Abdulla, P.A., Atig, M.F., Chen, Y.-F., Leonardsson, C., Rezine, A.: Counter-example guided fence insertion under TSO. In: Flanagan, C., König, B. (eds.) TACAS 2012. LNCS, vol. 7214, pp. 204–219. Springer, Heidelberg (2012). doi:10.1007/978-3-642-28756-5_15
4. Alglave, J., Kroening, D., Nimal, V., Tautschnig, M.: Software verification for weak memory via program transformation. In: Felleisen, M., Gardner, P. (eds.) ESOP 2013. LNCS, vol. 7792, pp. 512–532. Springer, Heidelberg (2013). doi:10.1007/978-3-642-37036-6_28
5. Alglave, J., Kroening, D., Tautschnig, M.: Partial orders for efficient bounded model checking of concurrent software. In: Sharygina, N., Veith, H. (eds.) CAV 2013. LNCS, vol. 8044, pp. 141–157. Springer, Heidelberg (2013). doi:10.1007/978-3-642-39799-8_9
6. Alglave, J., Maranget, L., Tautschnig, M.: Herding cats: modelling, simulation, testing, and data mining for weak memory. ACM Trans. Program. Lang. Syst. 36(2), 7:1–7:74 (2014)
7. Arora, N.S., Blumofe, R.D., Plaxton, C.G.: Thread scheduling for multipro-grammed multiprocessors. In: Proceedings of the Tenth Annual ACM Symposium on Parallel Algorithms and Architectures, SPAA 1998, pp. 119–129. ACM, New York (1998)
8. Atig, M.F., Bouajjani, A., Parlato, G.: Getting rid of store-buffers in TSO analysis. In: Gopalakrishnan, G., Qadeer, S. (eds.) CAV 2011. LNCS, vol. 6806, pp. 99–115. Springer, Heidelberg (2011). doi:10.1007/978-3-642-22110-1_9
9. Bouajjani, A., Calin, G., Derevenetc, E., Meyer, R.: Lazy TSO reachability. In: Egyed, A., Schaefer, I. (eds.) FASE 2015. LNCS, vol. 9033, pp. 267–282. Springer, Heidelberg (2015). doi:10.1007/978-3-662-46675-9_18
10. Burckhardt, S., Alur, R., Martin, M.M.K.: Checkfence: checking consistency of concurrent data types on relaxed memory models. In: Ferrante, J., McKinley, K.S. (eds.) Proceedings of the ACM SIGPLAN 2007 Conference on Programming Language Design and Implementation, pp. 12–21. ACM (2007)
11. Burckhardt, S., Musuvathi, M.: Effective program verification for relaxed memory models. In: Gupta, A., Malik, S. (eds.) CAV 2008. LNCS, vol. 5123, pp. 107–120. Springer, Heidelberg (2008). doi:10.1007/978-3-540-70545-1_12
12. Burnim, J., Sen, K., Stergiou, C.: Sound and complete monitoring of sequential consistency for relaxed memory models. In: Abdulla, P.A., Leino, K.R.M. (eds.) TACAS 2011. LNCS, vol. 6605, pp. 11–25. Springer, Heidelberg (2011). doi:10.1007/978-3-642-19835-9_3
13. Burnim, J., Sen, K., Stergiou, C.: Testing concurrent programs on relaxed memory models. In: Dwyer, M.B., Tip, F. (eds.) ISSTA 2011, pp. 122–132. ACM (2011)
14. Burns, J., Lynch, N.A.: Mutual exclusion using indivisible reads and writes. In: 18th Allerton Conference on Communication, Control, and Computing, pp. 833–842 (1980)

15. Cohen, E., Schirmer, B.: From Total Store Order to Sequential Consistency: a practical reduction theorem. In: Kaufmann, M., Paulson, L.C. (eds.) ITP 2010. LNCS, vol. 6172, pp. 403–418. Springer, Heidelberg (2010). doi:10.1007/ 978-3-642-14052-5_28

16. I. Corporate SPARC International. The SPARC architecture manual: version 8. Prentice-Hall, Inc., Upper Saddle River, NJ, USA (1992)

17. Cytron, R., Ferrante, J., Rosen, B.K., Wegman, M.N., Zadeck, F.K.: Efficiently computing static single assignment form and the control dependence graph. ACM Trans. Program. Lang. Syst. 13(4), 451–490 (1991)

18. Dalessandro, L., Dice, D., Scott, M., Shavit, N., Spear, M.: Transactional mutex locks. In: D'Ambra, P., Guarracino, M., Talia, D. (eds.) Euro-Par 2010. LNCS, vol. 6272, pp. 2–13. Springer, Heidelberg (2010). doi:10.1007/978-3-642-15291-7_2

19. Dan, A.M., Meshman, Y., Vechev, M., Yahav, E.: Predicate abstraction for relaxed memory models. In: Logozzo, F., Fähndrich, M. (eds.) SAS 2013. LNCS, vol. 7935, pp. 84–104. Springer, Heidelberg (2013). doi:10.1007/978-3-642-38856-9_7

20. Dan, A., Meshman, Y., Vechev, M., Yahav, E.: Effective abstractions for verification under relaxed memory models. In: D'Souza, D., Lal, A., Larsen, K.G. (eds.) VMCAI 2015. LNCS, vol. 8931, pp. 449–466. Springer, Heidelberg (2015). doi:10. 1007/978-3-662-46081-8_25

21. Dijkstra, E.W.: Cooperating sequential processes. In: Genuys, F. (ed.) Programming Languages: NATO Advanced Study Institute, pp. 43–112. Academic Press, New York (1968)

22. Ernst, G., Pfähler, J., Schellhorn, G., Haneberg, D., Reif, W.: KIV - overview and VerifyThis competition. Softw. Tools Techn. Transfer 17(6), 1–18 (2014)

23. Holzmann, G.J.: The SPIN Model Checker - Primer and Reference Manual. Addison-Wesley, Reading (2004)

24. Intel, Santa Clara, CA, USA. Intel 64 and IA-32 Architectures Software Developer's Manual Volume 3A: System Programming Guide, Part 1, May 2012

25. Inverso, O., Nguyen, T.L., Fischer, B., La Torre, S., Parlato, G.: Lazy-cseq: a context-bounded model checking tool for multi-threaded C-programs. In: Cohen, M.B., Grunske, L., Whalen, M. (eds.) 30th IEEE/ACM International Conference on Automated Software Engineering, ASE 2015, Lincoln, NE, USA, pp. 807–812. IEEE, 9–13 November 2015

26. Kroening, D., Tautschnig, M.: CBMC – C bounded model checker. In: Ábrahám, E., Havelund, K. (eds.) TACAS 2014. LNCS, vol. 8413, pp. 389–391. Springer, Heidelberg (2014). doi:10.1007/978-3-642-54862-8_26

27. Kuperstein, M., Vechev, M.T., Yahav, E.: Automatic inference of memory fences. SIGACT News 43(2), 108–123 (2012)

28. Lamport, L.: A new solution of Dijkstra's concurrent programming problem. Commun. ACM 17(8), 453–455 (1974)

29. Lamport, L.: How to make a multiprocessor computer that correctly executes multiprocess programs. IEEE Trans. Comput. 28(9), 690–691 (1979)

30. Lattner, C., Adve, V.S.: LLVM: a compilation framework for lifelong program analysis & transformation. In: 2nd IEEE/ACM International Symposium on Code Generation and Optimization (CGO 2004), San Jose, CA, USA, pp. 75–88. IEEE Computer Society, 20–24 March 2004

31. Linden, A., Wolper, P.: An automata-based symbolic approach for verifying programs on relaxed memory models. In: Pol, J., Weber, M. (eds.) SPIN 2010. LNCS, vol. 6349, pp. 212–226. Springer, Heidelberg (2010). doi:10.1007/ 978-3-642-16164-3_16

32. Mador-Haim, S., Maranget, L., Sarkar, S., Memarian, K., Alglave, J., Owens, S., Alur, R., Martin, M.M.K., Sewell, P., Williams, D.: An axiomatic memory model for POWER multiprocessors. In: Madhusudan, P., Seshia, S.A. (eds.) CAV 2012. LNCS, vol. 7358, pp. 495–512. Springer, Heidelberg (2012). doi:10.1007/978-3-642-31424-7_36

33. Meshman, Y., Rinetzky, N., Yahav, E.: Pattern-based synthesis of synchronization for the C++ memory model. In: Kaivola, R., Wahl, T. (eds.) Formal Methods in Computer-Aided Design, FMCAD 2015, Austin, Texas, USA, pp. 120–127. IEEE, 27–30 September 2015

34. Milner, R. (ed.): A Calculus of Communicating Systems. LNCS, vol. 92. Springer, Heidelberg (1980)

35. Park, S., Dill, D.L.: An executable specification, analyzer and verifier for RMO (Relaxed Memory Order). In: SPAA, pp. 34–41 (1995)

36. Peterson, G.: Myths about the mutual exclusion problem. Inf. Process. Lett. **12**(3), 115–116 (1981)

37. Pnueli, A.: The temporal logic of programs. In: 18th Annual Symposium on Foundations of Computer Science, Providence, Rhode Island, USA, pp. 46–57. IEEE Computer Society, 31 October–1 November 1977

38. Sewell, P., Sarkar, S., Owens, S., Nardelli, F.Z., Myreen, M.O.: x86-TSO: a rigorous and usable programmer's model for x86 multiprocessors. Commun. ACM **53**(7), 89–97 (2010)

39. Szymanski, B.K.: A simple solution to Lamport's concurrent programming problem with linear wait. In: Proceedings of the 2nd International Conference on Supercomputing, ICS 1988, pp. 621–626. ACM, New York (1988)

40. Travkin, O., Mütze, A., Wehrheim, H.: SPIN as a linearizability checker under weak memory models. In: Bertacco, V., Legay, A. (eds.) HVC 2013. LNCS, vol. 8244, pp. 311–326. Springer, Heidelberg (2013). doi:10.1007/978-3-319-03077-7_21

41. Travkin, O., Wehrheim, H.: Handling TSO in mechanized linearizability proofs. In: Yahav, E. (ed.) HVC 2014. LNCS, vol. 8855, pp. 132–147. Springer, Heidelberg (2014). doi:10.1007/978-3-319-13338-6_11

42. Treiber, R.K.: Systems programming: coping with parallelism. Technical report RJ 5118, IBM Almaden Res. Ctr. (1986)

43. Wehrheim, H., Travkin, O.: TSO to SC via symbolic execution. In: Piterman, N. (ed.) HVC 2015. LNCS, vol. 9434, pp. 104–119. Springer, Heidelberg (2015). doi:10.1007/978-3-319-26287-1_7

44. Yang, Y., Gopalakrishnan, G., Lindstrom, G.: UMM: an operational memory model specification framework with integrated model checking capability. Concurrency Comput. Pract. Experience **17**(5–6), 465–487 (2005)

Petri Nets and Semilinear Sets
(Extended Abstract)

Hsu-Chun Yen[✉]

Department of Electrical Engineering, National Taiwan University,
Taipei 106, Taiwan, Republic of China
yen@cc.ee.ntu.edu.tw

Abstract. Semilinear sets play a key role in many areas of computer science, in particular, in theoretical computer science, as they are characterizable by Presburger Arithmetic (a decidable theory). The reachability set of a Petri net is not semilinear in general. There are, however, a wide variety of subclasses of Petri nets enjoying semilinear reachability sets, and such results as well as analytical techniques developed around them contribute to important milestones historically in the analysis of Petri nets. In this talk, we first give a brief survey on results related to Petri nets with semilinear reachability sets. We then focus on a technique capable of unifying many existing semilinear Petri nets in a coherent way. The unified strategy also leads to various new semilinearity results for Petri nets. Finally, we shall also briefly touch upon the notion of *almost semilinear sets* which witnesses some recent advances towards the general Petri net reachability problem.

Petri nets (or, equivalently, *vector addition systems*) represent one of the most popular formalisms for specifying, modeling, and analyzing concurrent systems. In spite of their popularity, many interesting problems concerning Petri nets are either undecidable or of very high complexity. For instance, the reachability problem is known to be decidable [13] (see also [6]) and exponential-space-hard [12]. (The reader is referred to [11] for an improved upper bound.) Historically, before the work of [13], a number of attempts were made to investigate the problem for restricted classes of Petri nets, in hope of gaining more insights and developing new tools in order to conquer the general Petri net reachability problem. A common feature of those attempts is that decidability of reachability for those restricted classes of Petri nets was built upon their reachability sets being *semilinear*. As semilinear sets precisely correspond to the those characterized by Presburger Arithmetic (a decidable theory), decidability of the reachability problem follows immediately.

Formally speaking, a *Petri net* (PN, for short) is a 3-tuple (P, T, φ), where P is a finite set of *places*, T is a finite set of *transitions*, and φ is a *flow function* $\varphi : (P \times T) \cup (T \times P) \to N$. A *marking* is a mapping $\mu : P \to N$, specifying a PN's *configuration*. (μ assigns tokens to each place of the PN.) A transition $t \in T$ is *enabled* at a marking μ iff $\forall p \in P$, $\varphi(p, t) \leq \mu(p)$. If a transition t is enabled, it may *fire* by removing $\varphi(p, t)$ tokens from each input place p and

© Springer International Publishing AG 2016
A. Sampaio and F. Wang (Eds.): ICTAC 2016, LNCS 9965, pp. 25–29, 2016.
DOI: 10.1007/978-3-319-46750-4_2

putting $\varphi(t, p')$ tokens in each output place p'. We then write $\mu \xrightarrow{t} \mu'$, where $\mu'(p) = \mu(p) - \varphi(p, t) + \varphi(t, p), \forall p \in P$. A sequence of transitions $\sigma = t_1...t_n$ is a *firing sequence* from μ_0 iff $\mu_0 \xrightarrow{t_1} \mu_1 \xrightarrow{t_2} \cdots \xrightarrow{t_n} \mu_n$ for some markings $\mu_1,...,\mu_n$. (We also write '$\mu_0 \xrightarrow{\sigma} \mu_n$' or '$\mu_0 \xrightarrow{\sigma}$' if μ_n is irrelevant.) Given a PN $\mathcal{P} = (P, T, \varphi)$, its *reachability set* w.r.t. the initial marking μ_0 is $R(\mathcal{P}, \mu_0) = \{\mu \mid \mu_0 \xrightarrow{\sigma} \mu,$ for some $\sigma \in T^*\}$. The *reachability relation* of $\mathcal{P}$ is $R(\mathcal{P}) = \{(\mu_0, \mu) \mid \mu_0 \xrightarrow{\sigma} \mu,$ for some $\sigma \in T^*\}$.

A subset L of N^k is a *linear set* if there exist vectors $v_0, v_1, \ldots, v_t$ in N^k such that $L = \{v \mid v = v_0 + m_1 v_1 + \cdots + m_t v_t, \ m_i \in N\}$. The vectors v_0 (referred to as the *constant vector*) and $v_1, v_2, \ldots, v_t$ (referred to as the *periods*) are called the *generators* of the linear set L. A set $SL \subseteq N^k$ is *semilinear* if it is a finite union of linear sets, i.e., $SL = \bigcup_{1 \le i \le d} \mathcal{L}_i$, where $\mathcal{L}_i$ ($\subseteq N^k$) is a linear set. It is worthy of noting that semilinear sets are exactly those that can be expressed by *Presburger Arithmetic* (i.e., first order theory over natural numbers with addition), which is a decidable theory.

A PN is said to be *semilinear* if has a semilinear reachability set. In addition to the trivial example of finite PNs, the following are notable classes of semilinear PNs, including PNs of dimension 5 [4], conflict-free [7], persistent [7], normal [15], sinkless [15], weakly persistent [14], cyclic [1], communication-free PNs [2,5], and several others (see [16] for more). It is also known that checking whether a PN has a semilinear reachability set is decidable [3]. In view of the above, a natural question to ask is to identity, if at all possible, the key behind the exhibition of semilinear reachability sets for such a wide variety of restricted PN classes, while their restrictions are imposed on the PN model either structurally or behaviorally. We are able to answer the question affirmatively to a certain extent. In what follows, we give a sketch for the idea behind our unified strategy. The idea was originally reported in [17]. As we shall explain later, for each of considered PNs, any reachable marking is witnessed by somewhat of a *canonical computation* which will be elaborated later. Furthermore, such canonical computations can be divided into a finite number of groups, each of which has a finite number of "minimal computations" associated with a finite number of "positive loops." As one might expect, such minimal computations and positive loops exactly correspond to the constant vectors and periods, respectively, of a semilinear set. It is worth pointing out that the implication of our approach is two-fold. First, we are able to explain in a unified way a variety of semilinearity results reported in the literature. Second, perhaps more importantly, our approach yields new results in the following aspects:

(i) new semilinearity results for additional subclasses of PNs,
(ii) unified complexity and decidability results for problems including reachability, model checking, etc.

Given an $\alpha = r_1 \cdots r_{d-1} \in T^*$ and an initial marking μ_0, a computation of the form

$$\pi: \ \mu_0 \xrightarrow{\sigma_0} \mu_1 \xrightarrow{r_1} \bar{\mu}_1 \xrightarrow{\sigma_1} \mu_2 \xrightarrow{r_2} \cdots \xrightarrow{r_{d-1}} \bar{\mu}_{d-1} \xrightarrow{\sigma_{d-1}} \mu_d,$$

where $\mu_i, \bar{\mu}_j \in N^k$, and $\sigma_r \in T^*$ $(0 \leq i \leq d, 1 \leq j \leq d-1, 0 \leq r \leq d)$, is called an α-*computation*. We write $cv(\pi) = (\mu_1, ..., \mu_d)$. Suppose $\delta_i \in T^*, 1 \leq i \leq d$, are transition sequences such that $\Delta(\delta_i) \geq 0$ and $(\mu_i + \sum_{j=1}^{i-1} \Delta(\delta_j)) \xrightarrow{\delta_i}$, then following the *monotonicity* property of PNs,

$$\pi' : \mu_0 \xrightarrow{\sigma_0 \delta_1} \mu_1' \xrightarrow{r_1} \bar{\mu}_1' \xrightarrow{\sigma_1 \delta_2} \mu_2' \xrightarrow{r_2} \cdots \xrightarrow{r_{d-1}} \bar{\mu}_{d-1}' \xrightarrow{\sigma_{d-1} \delta_d} \mu_d'$$

remains a valid PN computation. In fact, we have $\mu_0 \xrightarrow{\sigma_0 (\delta_1)^+ r_1 \sigma_1 (\delta_2)^+ r_2 \cdots r_{d-1} \sigma_{d-1} (\delta_d)^+}$, meaning that $\delta_1, ..., \delta_d$ constitute "pumpable loops". In view of the above and if we write $cv(\pi) \xrightarrow{(\delta_1, \cdots, \delta_d)} cv(\pi')$, clearly "$\Rightarrow$" is transitive as $v \xrightarrow{(\alpha_1, \cdots, \alpha_d)} v'$ and $v' \xrightarrow{(\delta_1, \cdots, \delta_d)} v''$ imply $v \xrightarrow{(\alpha_1 \delta_1, \cdots, \alpha_d \delta_d)} v''$, where $v, v', v'' \in (N^k)^d, k = |P|$.

It turns out that the following properties are satisfied by several interesting subclasses of PNs all of which have semilinear reachability sets. With respect to an $\alpha \in T^d$,

(1) there is a finite set of transition sequences $F \subseteq T^*$ with nonnegative displacements (i.e., $\forall \gamma \in F, \Delta(\gamma) \geq 0$) such that if $(\mu_1, \cdots, \mu_d) \xrightarrow{(\delta_1, \cdots, \delta_d)} (\mu_1', \cdots, \mu_d')$ in some α-computations, then $\delta_i = \gamma_1^i \cdots \gamma_{h_i}^i$, for some h_i where $\gamma_j^i \in F$ (i.e., δ_i can be decomposed into $\gamma_1^i \cdots \gamma_{h_i}^i$), and

(2) the number of "minimal" α-computations is finite.

Intuitively, (2) ensures the availability of a finite set of constant vectors of a semilinear set, while (1) allows us to construct a finite set of periods based on those $\Delta(\gamma), \gamma \in F$.

A PN $\mathcal{P} = (P, T, \varphi)$ with initial marking μ_0 is said to be *computationally decomposable* (or simply *decomposable*) if every reachable marking $\mu \in R(\mathcal{P}, \mu_0)$ is witnessed by an α-computation ($\alpha \in T^*$) which meets Conditions (1) and (2) above. $\mathcal{P}$ is called *globally decomposable* if $\mathcal{P}$ is decomposable for every initial marking $\mu_0 \in N^k$. Let $RR_\alpha(\mathcal{P}, \mu_0) = \{cv(\pi) \mid \pi$ is an α-computation from μ_0 for some $\alpha \in T^*\}$, and $RR_\alpha(\mathcal{P}) = \{(\mu_0, cv(\pi)) \mid \pi$ is an α-computation from μ_0 for some $\alpha \in T^*\}$. We are able to show that if a PN is decomposable (resp., globally decomposable) then $RR_\alpha(\mathcal{P}, \mu_0)$ (resp., $RR_\alpha(\mathcal{P})$) is semilinear.

Among various subclasses of PNs, conflict-free, persistent, normal, sinkless, weakly persistent, cyclic, and communication-free PNs can be shown to be decomposable. Furthermore, each of the above classes of PNs also enjoys a nice property that

(3) there exists a finite set $\{\alpha_1, ..., \alpha_r\} \subseteq T^*$ (for some r) such that every reachable marking of the PN is witnessed by an α_i-computation, for some $1 \leq i \leq r$.

As a result, our unified strategy shows $R(\mathcal{P}, \mu_0)$ of a PN $\mathcal{P}$ with initial marking μ_0 for each of the above subclasses to be semilinear. Furthermore, a stronger result shows that conflict-free and normal PNs are globally decomposable; hence, their reachability relations $R(\mathcal{P})$ are always semilinear.

For semilinear PNs, a deeper question to ask is: *What is the size of its semilinear representation?* An answer to the above question is key to the complexity analysis of various problems concerning such semilinear PNs. To this end, we are able to incorporate another ingredient into our unified strategy, yielding size bounds for the semilinear representations of the reachability sets. Consider a computation $\mu \xrightarrow{\sigma} \mu'$. Suppose $T = \{t_1, ..., t_h\}$. For a transition sequence $\sigma \in T^*$, let $PK(\sigma) = (\#_\sigma(t_1), ..., \#_\sigma(t_h))$ be an h-dimensional vector of nonnegative integers, representing the so-called *Parikh map* of σ. The i-th coordinate denotes the number of occurrences of t_i in σ. In addition to Conditions (1)-(3) above, if the following is also known for a PN:

(4) a function $f(\mu)$ which bounds the size of each of the minimal elements of
    $ER(\mu) = \{(PK(\sigma), \mu') \mid \mu \xrightarrow{\sigma} \mu'\}$ (i.e., the so-called *extended reachability set*),

then we are able to come up with a bound for the size of the semilinear representation of a PN's reachability set.

Semilinearity for PNs is also related to the concept of the so-called *flatness*. A PN is said to be *flat* if there exist some words $\sigma_1, ..., \sigma_r \in T^*$ such that every reachable marking μ is witnessed by a computation $\mu_0 \xrightarrow{\sigma} \mu$ with $\sigma \in \sigma_1^* \cdots \sigma_r^*$, i.e., it has a witnessing sequence of transitions belonging to a bounded language. It is not hard to see the reachability set of a flat PN to be semilinear, and as shown in [10], a variety of known PN classes are indeed flat. In a recent article [9], flatness is shown to be not only sufficient but also necessary for a PN to be semilinear. We shall compare flat PNs with the aforementioned decomposable PNs.

Finally, we also briefly touch upon recent advances for the general PN reachability problem in which the notion of *(almost) semilinearity* is essential in yielding a simpler decidability proof [8] in comparison with that of [6,13].

References

1. Araki, T., Kasami, T.: Decidability problems on the strong connectivity of Petri net reachability sets. Theor. Comput. Sci. **4**, 99–119 (1977)
2. Esparza, J.: Petri nets, commutative grammars and basic parallel processes. Fundamenta Informaticae **30**, 24–41 (1997)
3. Hauschildt, D.: Semilinearity of the Reachability Set is Decidable for Petri Nets. Technical report FBI-HH-B-146/90, University of Hamburg (1990)
4. Hopcroft, J., Pansiot, J.: On the reachability problem for 5-dimensional vector addition systems. Theor. Comput. Sci. **8**, 135–159 (1979)
5. Huynh, D.: Commutative grammars: the complexity of uniform word problems. Inf. Control **57**, 21–39 (1983)
6. Kosaraju, R.: Decidability of reachability in vector addition systems. In: 14th ACM Symposium on Theory of Computing, pp. 267–280 (1982)
7. Landweber, L., Robertson, E.: Properties of conflict-free and persistent Petri nets. JACM **25**, 352–364 (1978)
8. Leroux, J.: The general vector addition system reachability problem by presburger inductive invariants. In: LICS 2009, pp. 4–13. IEEE Computer Society (2009)

9. Leroux, J.: Presburger vector addition systems. In: LICS 2013, pp. 23–32. IEEE Computer Society (2013)
10. Leroux, J., Sutre, G.: Flat counter automata almost everywhere!. In: Peled, D.A., Tsay, Y.-K. (eds.) ATVA 2005. LNCS, vol. 3707, pp. 489–503. Springer, Heidelberg (2005). doi:10.1007/11562948_36
11. Leroux, J., Schmitz, S.: Demystifying reachability in vector addition systems. In: LICS 2015, pp. 56–67. IEEE Computer Society (2015)
12. Lipton, R.: The Reachability Problem Requires Exponential Space. Technical report 62, Yale University, Dept. of CS, January 1976
13. Mayr, E.: An algorithm for the general Petri net reachability problem. In: 13th ACM Symposium on Theory of Computing, pp. 238–246 (1981)
14. Yamasaki, H.: On weak persistency of Petri nets. Inf. Process. Lett. **13**, 94–97 (1981)
15. Yamasaki, H.: Normal Petri nets. Theor. Comput. Sci. **31**, 307–315 (1984)
16. Yen, H.: Introduction to Petri net theory. In: Esik, Z., Martin-Vide, C., Mitrana, V. (eds.) Recent Advances in Formal Languages and Applications. Studies in Computational Intelligence, vol. 25, pp. 343–373. Springer, Heidelberg (2006)
17. Yen, H.: Path decomposition and semilinearity of Petri nets. Int. J. Found. Comput. Sci. **20**(4), 581–596 (2009)

Program Verification

Termination of Single-Path Polynomial Loop Programs

Yi Li[(⊠)]

Chongqing Key Laboratory of Automated Reasoning and Cognition,
CIGIT, CAS, Chongqing, China
zm_liyi@163.com

Abstract. Termination analysis of polynomial programs plays a very important role in applications of safety critical software. In this paper, we investigate the termination problem of single-path polynomial loop programs (SPLPs) over the reals. For such a loop program, we first assume that the set characterized by its loop guards is closed, bounded and connected. And then, we give some conditions and prove that under such conditions, the termination of single-path loop programs is decidable over the reals.

1 Introduction

Termination analysis of loop programs is endowed with a great importance for software correctness. The popular method for termination analysis is based on the synthesis of ranking functions. Several methods have been presented in [1, 4–7, 11–15, 17, 22] on the synthesis of ranking functions. Also, the complexity of the linear ranking function problem for linear loops is discussed in [2–5].

For example, In 2001, Colón and Sipma [13] synthesized linear ranking functions (LRFs for short) to prove loop termination by the theory of polyhedra. For single-path linear loops, Podelski and Rybalcheko [22] first proposed a complete and efficient method for synthesizing LRFs based on linear programming when program variables range over the reals and rationals in 2004. Their method is dependent on Farkas' lemma which provides a technique to extract hidden constraints from a system of linear inequalities. Bradley et al. [6,7] extended the work presented in [13] and showed how to synthesize lexicographic LRFs with linear supporting invariants over multi-path linear constraint loops in 2005. In [12], Chen et al. gave a technique to generate non-linear ranking functions for polynomial programs by solving semi-algebraic systems. Cook et al. [14] described an automatic method for finding sound under-approximations to weakest preconditions to termination.

In 2012, [11] characterized a method to generate proofs of universal termination for linear simple loops based on the synthesis of disjunctive ranking relations. Their method is a generalization of the method given in [22]. In [17], a method was proposed by Ganty and Genaim to partition the transition relations, which can be applied to conditional termination analysis. Bagnara et al. [1] analysed termination of single-path linear constraint loops by the existence of eventual LRFs,

© Springer International Publishing AG 2016
A. Sampaio and F. Wang (Eds.): ICTAC 2016, LNCS 9965, pp. 33–50, 2016.
DOI: 10.1007/978-3-319-46750-4_3

where the eventual LRFs are linear functions that become ranking functions after a finite unrolling of the loop. In 2013, Cook et al. [15] presented a method for proving termination by Ramsey-based termination arguments instead of lexicographic termination arguments. For lasso programs, Heizmann et al. suggested a series of techniques to synthesize termination arguments in [18–20].

It is well known that the termination of loop programs is undecidable, even for the class of linear programs [25]. Existence of ranking function is only a sufficient (but not necessary) condition on the termination of a program. That is, it is easy to construct programs that terminate, but have no ranking functions. In contrast to the above methods for synthesizing ranking functions, [9,25] tried to detect decidable subclasses. In [25], Tiwari proves that the termination of a class of single-path loops with linear guards and assignments is decidable, providing a decision procedure via constructive proofs. Braverman [9] generalized the work of Tiwari, and showed that termination of a simple of class linear loops over the integer is decidable. Xia et al. [26] gave the NZM (Non-Zero Minimum) condition under which the termination problem of loops with linear updates and nonlinear polynomial loop conditions is decidable over the reals. In addition, there are some other methods for determining termination problem of loop programs. For instance, in [8] Bradley et al. applied finite difference trees to prove termination of multipath loops with polynomial guards and assignments. In [21], Liu et al. analyzed the termination problems for multi-path polynomial programs with equational loop guards and established sufficient conditions for termination and nontermination.

In this paper, we focus on the termination of single-path polynomial loop programs having the following form

$$P: \textbf{While } \mathbf{x} \in \Omega \textbf{ do}$$
$$\{\mathbf{x} := F(\mathbf{x})\} \tag{1}$$
$$\textbf{endwhile}$$

where Ω is a closed, bounded and connected subset in $\mathbb{R}^n$, defined by a set of polynomial equations and polynomial inequalities, and $F(\mathbf{x}) : \mathbb{R}^n \to \mathbb{R}^n$, is a polynomial mapping, i.e., $F(\mathbf{x}) = (f_1(\mathbf{x}), ..., f_n(\mathbf{x}))^T$ and $f_i(\mathbf{x})$ is a polynomial in $\mathbf{x}$, for $i = 1, ..., n$. For convenience, we say that Program P is defined by Ω and $F(\mathbf{x})$, i.e., $P \triangleq P(\Omega, F(\mathbf{x}))$. We say that Program P is non-terminating over the reals, if there exists a point $\mathbf{x}^* \in \mathbb{R}^n$ such that $F^k(\mathbf{x}^*) \in \Omega$ for any $k \geq 0$. If such $\mathbf{x}^*$ does not exist, then we say Program P is terminating over the reals.

In contrast to the existing methods mentioned above, for Program P, in this paper we give some conditions such that under such conditions the termination of P can be equivalently reduced to the computation of fixed points of $F(\mathbf{x})$. That is, if such conditions are satisfied, then P is nonterminating if and only if $F(\mathbf{x})$ has at least one fixed point in Ω. Otherwise, the termination of P remains unknown. In particular, Groebner basis technique is introduced, which sometimes can reduce a given polynomial mapping to another one with simpler structure. This helps us to further analyze the termination of P, when $F(\mathbf{x})$ has complex structure. Since the computation of fixed points of F can be equivalently reduced to semi-algebraic systems solving, in this paper, for convenience, we utilize the symbolic computation tool RegularChains [10] in Maple to solve such systems.

The rest of the paper is organized as follows. Section 2 introduces some basic notion and background information regarding ranking functions, semi-algebraic systems and Groebner basis. In Sect. 3, we give some proper conditions and prove that if such the conditions hold, then the termination of Program P is decidable over the reals. Moreover, some examples are given to illustrate our methods. Section 4 concludes the paper.

2   Preliminaries

In the section, some basic notion on ranking functions, semi-algebraic systems and Groebner basis will be introduced first.

2.1   Semi-Algebraic Systems

Let $\mathbb{R}$ be the field of real numbers. A semi-algebraic system is a set of equations, inequations and inequalities given by polynomials. And the coefficients of those polynomials are all real numbers. Let $\mathbf{v} = (v_1, ..., v_d)^T \in \mathbb{R}^d$, $\mathbf{x} = (x_1, ..., x_n)^T \in \mathbb{R}^n$. Next, we give the definition of semi-algebraic systems (SASs for short).

Definition 1 *(Semi-algebraic systems). A semi-algebraic system is a conjunctive polynomial formula of the following form*

$$\begin{cases} p_1(\mathbf{v}, \mathbf{x}) = 0, ..., p_r(\mathbf{v}, \mathbf{x}) = 0, \\ g_1(\mathbf{v}, \mathbf{x}) \geq 0, ..., g_k(\mathbf{v}, \mathbf{x}) \geq 0, \\ g_{k+1}(\mathbf{v}, \mathbf{x}) > 0, ..., g_t(\mathbf{v}, \mathbf{x}) > 0, \\ h_1(\mathbf{v}, \mathbf{x}) \neq 0, ..., h_m(\mathbf{v}, \mathbf{x}) \neq 0, \end{cases} \tag{2}$$

where $r > 1$, $t \geq k \geq 0$, $m \geq 0$ and all p_i's, g_i's and h_i's are polynomials in $\mathbb{R}[\mathbf{v}, \mathbf{x}] \setminus \mathbb{R}$. An semi-algebraic system is called parametric if $d \neq 0$, otherwise constant, where d is the dimension of $\mathbf{v}$.

The RegularChains library offers a set of tools for manipulating semi-algebraic systems. For instance, given a parametric SAS, RegularChains provides commands for getting necessary and sufficient conditions on the parameters under which the system has real solutions.

2.2   Ranking Functions

As a popular method for determining program termination, synthesis of ranking functions has received extensive attention in these years. We recall the definition of ranking functions, as follows.

Definition 2 *(Ranking functions for single-path polynomial loop programs).* *Given a single-path polynomial loop program P, we say $\rho(\mathbf{x})$ is a ranking function for P, if the following formula is true over the reals,*

$$\forall \mathbf{x}, \mathbf{x}', \big(\mathbf{x} \in \Omega \wedge \mathbf{x}' = F(\mathbf{x}) \Rightarrow \rho(\mathbf{x}) \geq 0 \wedge \rho(\mathbf{x}) - \rho(\mathbf{x}') \geq 1\big). \tag{3}$$

Note that the decrease by 1 in (3) can be replaced by any positive number δ. It is well known that the existence of ranking functions for P implies that Program P is terminating. Formula (3) holds if and only if the following two Formulae hold,

$$\forall \mathbf{x}, \mathbf{x}', \big(\mathbf{x} \in \Omega \wedge \mathbf{x}' = F(\mathbf{x}) \wedge \rho(\mathbf{x}) \geq 0\big), \tag{4}$$

$$\forall \mathbf{x}, \mathbf{x}', \big(\mathbf{x} \in \Omega \wedge \mathbf{x}' = F(\mathbf{x}) \wedge \rho(\mathbf{x}) - \rho(\mathbf{x}') \geq 1\big), \tag{5}$$

We now take an example to illustrate how to synthesize ranking functions by means of RegularChains.

Example 1. Consider the below single-path polynomial program.

$$P_1 : \textbf{While } x^2 + y^2 \leq 1 \textbf{ do}$$
$$\{x := x + 4y + 3; y := 3y + 1\} \tag{6}$$
$$\textbf{endwhile}$$

First, predefine a ranking function template $\rho(x, y) = ax + by + c$. we next will utilize the tool RegularChains to find a, b and c such that $\rho(x, y)$ meets Formula (4) and (5). Invoking the following commands in RegularChains,

$with(RegularChains);$

$with(SemiAlgebraicSetTools);$

$T_1 := \&E([x, y, x', y']), (x' = x + 4y + 3) \& and(y' = 3y + 1) \& and(x^2 + y^2 <= 1)$
$\qquad \& implies(ax + by \geq 0);$

$T_2 := \&E([x, y, x', y']), (x' = x + 4y + 3) \& and(y' = 3y + 1) \& and(x^2 + y^2 <= 1)$
$\qquad \& implies(a(x - x') + b(y - y') \geq 1);$

we get

$$(b \neq 0 \wedge \sqrt{a^2 + b^2} \leq c) \vee (b = 0 \wedge a \leq 0 \wedge -a \leq c) \vee (b = 0 \wedge - < a \wedge a \leq c),$$

and

$$0 \leq b \wedge -b \leq a \wedge a \leq -\frac{3}{7}b.$$

Thus, taking $a = -1, b = 2, c = 5$, we obtain a linear ranking function $\rho(x, y) = -x + 2y + 5$.

2.3   Polynomial Ideal and Groebner Basis

Let $\alpha = (a_1, ..., a_n)^T$ and let $\mathbf{x}^\alpha = x_1^{a_1} x_2^{a_2} \cdots x_n^{a_n}$. Let $f = \sum_\alpha c_\alpha \mathbf{x}^\alpha$ be a polynomial in $\mathbb{R}[\mathbf{x}]$. We call c_α the coefficient of the monomial $\mathbf{x}^\alpha$. The monomials $\mathbf{x}^\alpha$'s in f can also be ordered in terms of monomial orderings, such as lexicographic order, graded lexicographic order and graded reverse lex order. We state here the following well-known results on Polynomial ideal and Groebner basis briefly.

Definition 3. *Let $\mathbb{R}[\mathbf{x}]$ be a polynomial ring. A subset $I \subset \mathbb{R}[\mathbf{x}]$ is a polynomial ideal if it satisfies:*

(i) $0 \in I$.
(ii) If $f, g \in I$, then $f + g \in I$.
(iii) If $f \in I$ and $h \in \mathbb{R}[\mathbf{x}]$, then $h \cdot f \in I$.

Definition 4. *Let $f_1, ...f_s$ be polynomials in $\mathbb{R}[\mathbf{x}]$. Set*

$$\langle f_1, ..., f_s \rangle = \{ \sum_{i=1}^s h_i f_i : h_1, ..., h_s \in \mathbb{R}[\mathbf{x}] \}.$$

Then $\langle f_1, ..., f_s \rangle$ is a polynomial ideal of $\mathbb{R}[\mathbf{x}]$. We call $\langle f_1, ..., f_s \rangle$ the ideal generated by $f_1, ..., f_s$.

Definition 5. *Let k be a field, and let $f_1, ...f_s$ be polynomials in $k[\mathbf{x}]$. Then we set*

$$\mathbf{V}(f_1, ..., f_s) = \{ (a_1, ..., a_n) \in k^n : f_i(a_1, ..., a_n) = 0 \text{ for all } 1 \le i \le s \}.$$

We call $\mathbf{V}(f_1, ..., f_s)$ the affine variety defined by $f_1, ..., f_s$.

Definition 6. *Let $V \subset k^n$ be an affine variety. Set*

$$\mathbf{I}(V) = \{ f \in k[\mathbf{x}] : f(a_1, ..., a_n) = 0 \text{ for all } (a_1, ..., a_n) \in V \}.$$

Then $\mathbf{I}(V)$ is an ideal and we call it the ideal of V.

Definition 7. *Let I be an ideal. We will denote by $\mathbf{V}(I)$ the set*

$$\mathbf{V}(I) = \{ (a_1, ..., a_n) \in k^n : f(a_1, ..., a_n) = 0 \text{ for all } f \in I \}.$$

The following proposition follows from Hilbert's Basis Theorem.

Proposition 1. *With the above notion. $\mathbf{V}(I)$ is an affine variety. In particular, if $I = \langle f_1, ..., f_s \rangle$, then $\mathbf{V}(I) = \mathbf{V}(f_1, ..., f_s)$.*

The above proposition shows that even though a nonzero ideal I always contains infinitely many different polynomials, the set $\mathbf{V}(I)$ can still be defined by a finite set of polynomial equations. In particular, if $I_1 = I_2$, then $\mathbf{V}(I_1) = \mathbf{V}(I_2)$.

Definition 8. *Fix a monomial order. A finite subset $G = \{g_1, ..., g_s\}$ of an ideal I is said to be Groebner basis if*

$$\langle LT(g_1), ..., LT(g_s)\rangle = \langle LT(I)\rangle,$$

where $LT(I)$ is the set of leading terms of elements of I, and $\langle LT(I)\rangle$ is the ideal generated by the elements of $LT(I)$.

It has been proven that every ideal I other than $\{0\}$ has a Groebner basis, and any Groebner basis $G = \{g_1, ..., g_s\}$ for I is a basis having good properties. That is, $I = \langle G\rangle$. For convenience, we call G the affine Groebner basis of I, if g_i's are all affine. The computation of Groebner Basis has been implemented in Maple. For example, let $I = \langle f_1, f_2\rangle = \langle 3xy - 1, x^2 + 5y - x\rangle$. Invoking the following command,

$$Basis([f_1, f_2], plex(x, y), output = extended),$$

we can get

$$[45y^3 - 3y + 1, 15y^2 + x - 1], [[-3xy + 3y - 1, 9y^2], [1 - x, 3y]].$$

The first list is a Groebner basis of I, i.e., $G = [45y^3 - 3y + 1, 15y^2 + x - 1]$. If let

$$M_G = \begin{pmatrix} -3xy + 3y - 1, & 9y^2 \\ 1 - x, & 3y \end{pmatrix}, \qquad H = (f_1, f_2),$$

then we have $G = M_G \cdot H$. Since $I = \langle G\rangle$, every element of I can also be expressed by elements of G. For the example, invoking the commands in Maple,

- NormalForm(f1, G, plex(x, y), 'Q1')

- NormalForm(f2, G, plex(x, y), 'Q2'),

we obtain that $Q_1 = [-1, 3y], Q_2 = [5y, -15y^2 + x]$. And let

$$M_H = \begin{pmatrix} -1, & 3y \\ 5y, & -15y^2 + x \end{pmatrix}.$$

It is easy to check that $H = M_H \cdot G$.

3   Termination Analysis for SPLPs

In the section, we will give some conditions under which Program P as defined in (1) is nonterminating if and only if $F(\mathbf{x})$ has fixed points in Ω. We first give the following lemma, which enables us to build necessary and sufficient criteria for termination of several kinds of SPLPs.

Lemma 1. *Let Ω and F be defined as in (1). Let $P \triangleq P(\Omega, F(\mathbf{x}))$. If Program P is non-terminating over the reals, then for any continuous function $T(\mathbf{x})$, we have*

$$\Theta \cap \Omega \neq \emptyset,$$

where $\Theta = \{\mathbf{x} \in \mathbb{R}^n : T(\mathbf{x}) = T(F(\mathbf{x}))\}$.

Proof. The proof is simple. Assume that there exists a continuous function $T(\mathbf{x})$, such that $\Theta \cap \Omega = \emptyset$. There are two cases to consider.

(a) $\forall \mathbf{x} \in \Omega, T(\mathbf{x}) - T(F(\mathbf{x})) > 0$.
(b) $\forall \mathbf{x} \in \Omega, T(\mathbf{x}) - T(F(\mathbf{x})) < 0$.

Consider Case (a). Since Ω is a closed, bounded and connected set, and $T - T \circ F$ is continuous, it immediately follows that

$$\forall \mathbf{x} \in \Omega.(T(\mathbf{x}) - T(F(\mathbf{x})) \geq \delta_1 > 0 \wedge T(\mathbf{x}) \geq c_1),$$

for a certain positive number δ_1 and a certain constant c_1, by properties of continuous functions. It is not difficult to see that $\frac{1}{\delta_1}(T(\mathbf{x}) - c_1)$ is a ranking function for P, which implies that Program P is terminating. This contradicts the hypothesis that Program P is non-terminating. Similar analysis can also be applicable to Case (b). We just need to notice that if Case (b) occurs, then there must exist a certain positive number δ_2 and a certain constant c_2, such that

$$\forall \mathbf{x} \in \Omega.\big(-T(\mathbf{x}) - (-T(F(\mathbf{x}))) \geq \delta_2 > 0 \wedge -T(\mathbf{x}) \geq c_2 \big).$$

Hence, $\frac{1}{\delta_2}(-T(\mathbf{x}) - c_2)$ is a ranking function for P. The proof of the lemma is completed. $\qquad\square$

Following Lemma 1, we can get the below simple result.

Corollary 1. *Let Ω and F be defined as in (1). Give a Program $P \triangleq P(\Omega, F(\mathbf{x}))$. If there exists a continuous function $T(\mathbf{x})$, such that $\{\mathbf{x} \in \mathbb{R}^n : T(\mathbf{x}) = T(F(\mathbf{x}))\} \cap \Omega = \emptyset$, then Program P is terminating.*

It is not difficult to see that Corollary 1 presents a sufficient (but not necessary) criteria for Program P specified by $\Omega, F(\mathbf{x})$ to be terminating. In the following, we will establish necessary and sufficient condition under which the termination problem of Program P can be equivalently reduced to the problem of existence of fixed points of $F(\mathbf{x})$. To do this, we first introduce several useful lemmas as follows.

Lemma 2 *(separating hyperplane theorem [24]). Let C and D be two convex sets of $\mathbb{R}^n$, which do not intersect, i.e., $C \cap D = \emptyset$. Then there exist $a \neq 0$ and b such that $a^T \mathbf{x} \leq b$ for all $\mathbf{x} \in C$ and $a^T \mathbf{x} \geq b$ for all $\mathbf{x} \in D$.*

Lemma 2 tells us that if C and D are two disjoint nonempty convex subsets, then there exists the affine function $a^T \mathbf{x} - b$ that is nonpositive on C and nonnegative on D. The hyperplane $\{\mathbf{x} : a^T \mathbf{x} = b\}$ is called a separating hyperplane for C and D. Also, we say that the affine function $a^T \mathbf{x} - b$ strictly separates C and D as defined above, if $a^T \mathbf{x} < b$ for all $\mathbf{x} \in C$ and $a^T \mathbf{x} > b$ for all $\mathbf{x} \in D$. This is called strict separation of C and D. In general, disjoint convex sets need not be strictly separated by a hyperplane. However, the following lemma tells us that in the special case when C is a closed convex set and D is a single-point set, there indeed exists a hyperplane that strictly separates C and D.

Lemma 3 *(Strict separation of a point and a closed convex set [24]). Let C be a closed convex set and $\mathbf{x}_0 \notin C$. Then there exists a hyperplane that strictly separates $\mathbf{x}_0$ from C.*

Lemma 4 *([16]). Let $S \subseteq \mathbb{R}^n$ be a closed, bounded and connected set and let $\mathcal{H}$ be a polynomial mapping. Then, the image $\mathcal{H}(S)$ of S under the polynomial mapping $\mathcal{H}$ is still closed, bounded and connected.*

Given a polynomial mapping $F(\mathbf{x}) \in (\mathbb{R}[\mathbf{x}])^m$ and a vector $\alpha \in \mathbb{Z}_{\geq 0}^m$. Let $F(\mathbf{x})^\alpha = f_1^{a_1} f_2^{a_2} \cdots f_n^{a_n}$. Let $\mathcal{M}(\mathbf{x})$ be a vector of some monomials in $\mathbf{x}$. Let m be the number of elements in $\mathcal{M}$. Let

$$\mathcal{H}(\mathbf{x}) = \mathcal{M}(\mathbf{x}) - \mathcal{M}(F(\mathbf{x})).$$

Clearly, $\mathcal{H}(\mathbf{x})$ can be regarded as a polynomial mapping from k^n to k^m, where $k \in \{\mathbb{R}, \mathbb{C}\}$. For example, set $\mathbf{x} = (x_1, x_2)$, $F(\mathbf{x}) = (5x_1^2, x_1 - x_2 + 1)^T$ and $\mathcal{M}(\mathbf{x}) = (x_1, x_1 x_2)^T$. Then,

$$\mathcal{H}(\mathbf{x}) = (x_1 - 5x_1^2, x_1 x_2 - 5x_1^2(x_1 - x_2 + 1))^T$$

is a polynomial mapping from k^2 to k^2. In addition, for a given set Ω, define

$$\mathcal{H}(\Omega) = \{\mathcal{H}(\mathbf{x}) : \mathbf{x} \in \Omega\} \subseteq \mathbb{R}^m.$$

Let $\mathcal{U}(\mathbf{x}) = (u_1(\mathbf{x}), ..., u_s(\mathbf{x}))^T$ be a polynomial mapping. For convenience, we also use the same notation $\mathcal{U}(\mathbf{x})$ to denote a set of polynomials consisting of all the elements in polynomial mapping $\mathcal{U}(\mathbf{x})$. Define

$$V_{\mathbb{R}}(\mathcal{U}(\mathbf{x})) = \{\mathbf{x} \in \mathbb{R}^n : \mathcal{U}(\mathbf{x}) = 0\}$$
$$V_{\mathbb{C}}(\mathcal{U}(\mathbf{x})) = \{\mathbf{x} \in \mathbb{C}^n : \mathcal{U}(\mathbf{x}) = 0\}$$

to be the real algebraic variety and the complex algebraic variety defined by $\mathcal{U}(\mathbf{x}) = 0$, repectively.

Theorem 1. *Let Ω and F be defined as in (1). Given a Program $P \triangleq P(\Omega, F(\mathbf{x}))$. Let $\mathcal{M}(\mathbf{x})$ be a vector consisting of m monomials in $\mathbb{R}[\mathbf{x}]$. Define $\mathcal{H}(\mathbf{x}) = \mathcal{M}(\mathbf{x}) - \mathcal{M}(F(\mathbf{x}))$. If the following conditions are satisfied,*

(a) $V_{\mathbb{R}}(\mathcal{H}(\mathbf{x})) = V_{\mathbb{R}}(F(\mathbf{x}) - \mathbf{x})$,
(b) $\mathcal{H}(\Omega)$ is a convex set,

then, Program P is non-terminating over the reals if and only if $F(\mathbf{x})$ has at least one fixed point in Ω.

Proof. If $F(\mathbf{x})$ has one fixed point in Ω, then Program P does not terminate on its fixed point. Next, we will claim that if $F(\mathbf{x})$ has no fixed points in Ω, then Program P terminates. Since $F(\mathbf{x})$ has no fixed points in Ω, we know that $V_{\mathbb{R}}(F(\mathbf{x}) - \mathbf{x}) \cap \Omega = \emptyset$. It immediately follows that $V_{\mathbb{R}}(\mathcal{H}(\mathbf{x})) \cap \Omega = \emptyset$, since $V_{\mathbb{R}}(\mathcal{H}(\mathbf{x})) = V_{\mathbb{R}}(F(\mathbf{x}) - \mathbf{x})$. Therefore, for any $\mathbf{x} \in \Omega$, $\mathcal{H}(\mathbf{x}) \neq 0$. This implies that $\mathbf{0} \notin \mathcal{H}(\Omega)$, where $\mathbf{0} \in \mathbb{R}^m$, $\Omega \subseteq \mathbb{R}^n$ and

$$\mathcal{H}(\Omega) = \{\mathcal{H}(\mathbf{x}) : \mathbf{x} \in \Omega\} \subseteq \mathbb{R}^m.$$

Let $\mathbf{u} = \mathcal{H}(\mathbf{x})$. Since $\mathcal{H} : \mathbb{R}^n \to \mathbb{R}^m$ is a polynomial mapping and Ω is closed, bounded and connected, by Lemma 2, we know $\mathcal{H}(\Omega)$ is a closed, bounded and connected set. Also, by the hypothesis (b), we know that $\mathcal{H}(\Omega)$ is a convex set. Thus, $\mathcal{H}(\Omega)$ is a closed convex set. By Lemma 3, we know that in the space $\mathbb{R}^m$, there must exist a hyperplane $\mathbf{a}^T \cdot \mathbf{u} = \mathbf{b}$, which can strictly separate $\mathbf{0}$ from $\mathcal{H}(\Omega)$. That is to say, for any $\mathbf{u} \in \mathcal{H}(\Omega)$, $\mathbf{a}^T \cdot \mathbf{u} \neq \mathbf{b}$. Furthermore, since $\mathbf{0} \in \mathbb{R}^m$ is strictly separated from $\mathcal{H}(\Omega) \subseteq \mathbb{R}^m$ by the hyperplane $\mathbf{a}^T \cdot \mathbf{u} = \mathbf{b}$, it follows that the hyperplane $\mathbf{a}^T \cdot \mathbf{u} = \mathbf{0}$ must be disjoint from $\mathcal{H}(\Omega)$, which passes through the origin $\mathbf{0}$ and parallels to the hyperplane $\mathbf{a}^T \cdot \mathbf{u} = \mathbf{b}$. Therefore, since the hyperplane $\mathbf{a}^T \cdot \mathbf{u} = \mathbf{0}$ is disjoint from $\mathcal{H}(\Omega)$, we get that for any $\mathbf{u} \in \mathcal{H}(\Omega)$, $\mathbf{a}^T \cdot \mathbf{u} \neq \mathbf{0}$. This immediately implies that for any $\mathbf{x} \in \Omega$, we have $\mathbf{a}^T \cdot \mathcal{H}(\mathbf{x}) \neq \mathbf{0}$, since $\mathbf{u} = \mathcal{H}(\mathbf{x})$. Thus, for any $\mathbf{x} \in \Omega$, $\mathbf{a}^T \cdot (\mathcal{M}(\mathbf{x}) - \mathcal{M}(F(\mathbf{x}))) \neq \mathbf{0}$. Let $T(\mathbf{x}) = \mathbf{a}^T \cdot \mathcal{M}(\mathbf{x})$. By Corollary 1, Program P must terminate, since $\{\mathbf{x} \in \mathbb{R}^n : T(\mathbf{x}) = T(F(\mathbf{x}))\} \cap \Omega = \emptyset$. This completes the proof of the theorem. □

Remark 1. Let $S_{\mathcal{H}} = \{\mathbf{a} : \mathbf{a}^T \mathcal{H}(\mathbf{x}) \neq 0, \ for \ all \ \mathbf{x} \in \Omega\}$. By the proof of Theorem 1, we know that if the conditions (a) and (b) in the theorem are satisfied, then F has no fixed points in Ω implies that $S_{\mathcal{H}} \neq \emptyset$.

Example 2. Consider the termination of the below program.

$$P_2 : \textbf{While } 4 \leq x \leq 5 \wedge 1 \leq y \leq 2 \textbf{ do}$$
$$\{x := x; y := -xy + y^2 + 1\} \tag{7}$$
$$\textbf{endwhile}$$

Let $\Omega = \{(x, y) \in \mathbb{R}^2 : 4 \leq x \leq 5 \wedge 1 \leq y \leq 2\}$, $f_1(x, y) = x$ and $f_2(x, y) = y^2 - xy + 1$. Let $\mathcal{M}(\mathbf{x}) = (x, y, xy)^T$. Thus,

$$\mathcal{H}(\mathbf{x}) = \mathcal{M}(\mathbf{x}) - \mathcal{M}(F(\mathbf{x})) = (x - f_1, y - f_2, xy - f_1 f_2)^T.$$

Invoking the commands in RegularChains below,

```
/* to define the region Ω
c₁ := x >= 4; c₂ := x <= 5; c₃ := y >= 1; c₄ := y <= 2;
```

/* to describe the formula $\forall x \forall y ((x, y) \in V_{\mathbb{R}}(\mathcal{H}(x, y)) \Rightarrow V_{\mathbb{R}}(F(\mathbf{x}) - \mathbf{x}))$

```
q₁ := &A([x, y]), ((x − f₁ = 0)&and(y − f₂ = 0)&and(xy − f₁f₂ = 0))
        &implies((x − f₁ = 0)&and(y − f₂ = 0)));
```

/* to describe the formula $\forall x \forall y (V_{\mathbb{R}}(F(\mathbf{x}) - \mathbf{x}) \Rightarrow (x, y) \in V_{\mathbb{R}}(\mathcal{H}(x, y)))$

```
q₂ := &A([x, y]), ((x − f₁ = 0)&and(y − f₂ = 0))&implies((x − f₁ = 0)
        &and(y − f₂ = 0)&and(xy − f₁f₂ = 0));
```

/* to check if Formula q_1 is true

```
QuantifierElimination(q₁, output = rootof);
```

/* to check if Formula q_2 is true

```
QuantifierElimination(q₂, output = rootof);
```

we find that the conditions (a) in Theorem 1 is satisfied. Furthermore, invoking the commands as follows,

/ * to describe the relation between Ω and $\mathcal{H}(\Omega)$

$p_1 := \&E([x,y]), (x - f_1 = u_1)\&and(y - f_2 = u_2)\&and(xy - f_1 f_2 = u_3)$

$\&and(c_1)\&and(c_2)\&and(c_3)\&and(c_4);$

/ * to compute $\mathcal{H}(\Omega)$ by eliminating the quantified variables x and y from p_1

$Quantifier Elimination(p_1, output = rootof);$

we obtain that

$$u_1 = 0 \wedge u_3 = 0 \wedge 3 \le u_2 \le 7,$$

which defines $\mathcal{H}(\Omega)$. That is, $\mathcal{H}(\Omega) = \{\mathbf{u} = (u_1, u_2, u_3) \in \mathbb{R}^3 : u_1 = 0 \wedge u_3 = 0 \wedge 3 \le u_2 \le 7\}$. Clearly, $\mathcal{H}(\Omega)$ is convex. Thus, the condition (b) is satisfied. Therefore, by Theorem 1, Program P_2 is terminating, since $F(\mathbf{x})$ has no fixed points in Ω.

Corollary 2. *Let Ω and F be defined as in (1). Given a Program $P \triangleq P(\Omega, F(\mathbf{x}))$. Let $\mathcal{M}(\mathbf{x})$ be a vector consisting of some monomials in $\mathbf{x}$. Let $\mathcal{H}(\mathbf{x}) = \mathcal{M}(\mathbf{x}) - \mathcal{M}(F(\mathbf{x}))$. If*

(a) $V_{\mathbb{R}}(\mathcal{H}(\mathbf{x})) = V_{\mathbb{R}}(F(\mathbf{x}) - \mathbf{x})$,
(b) Ω is a convex set and $\mathcal{H}(\mathbf{x})$ is a convexity preserving mapping,

then, Program P is non-terminating over the reals if and only if $F(\mathbf{x})$ has at least one fixed point in Ω.

Proof. The proof is simple. We just need to notice that if condition (b) holds, then we have that $\mathcal{H}(\Omega)$ is a convex set. By Theorem 1, Program P is non-terminating if and only if $F(\mathbf{x})$ has at least one fixed point in Ω.   □

Corollary 3. *Let Ω and F be defined as in (1). Given a Program $P \triangleq P(\Omega, F(\mathbf{x}))$. If Ω is convex and $F(\mathbf{x})$ is an affine mapping, then, Program P is non-terminating over the reals if and only if $F(\mathbf{x})$ has at least one fixed point in Ω.*

Proof. Let $\mathcal{M}(\mathbf{x}) = (x, y)^T = \mathbf{x}$. Then $\mathcal{H}(\mathbf{x}) = \mathcal{M}(\mathbf{x}) - \mathcal{M}(F(\mathbf{x})) = \mathbf{x} - F(\mathbf{x})$. Clearly, $V_{\mathbb{R}}(\mathcal{H}(\mathbf{x})) = V_{\mathbb{R}}(F(\mathbf{x}) - \mathbf{x})$. Besides, since both $\mathcal{M}(\mathbf{x})$ and $F(\mathbf{x})$ are affine, $\mathcal{H}(\mathbf{x})$ is an affine mapping. Therefore, $\mathcal{H}(\Omega)$ is convex, since any affine mapping is a convexity preserving mapping and Ω is convex. By Theorem 1 or Corollary 2, Program P is non-terminating over the reals if and only if $F(\mathbf{x})$ has at least one fixed point in Ω.   □

Define

$$\widehat{F}(\mathbf{x}) = \begin{pmatrix} f_1(x_1) \\ f_2(x_2) \\ \vdots \\ f_i(x_i) \\ \vdots \\ f_n(x_n) \end{pmatrix}, \qquad \widehat{\Omega} = \{\mathbf{x} \in \mathbb{R}^n : a_i \le x_i \le b_i, for\ all\ i = 1, ..., n\}$$

where f_i's are all polynomials, and $\widehat{\Omega}$ is a set defined by box constraints.

Corollary 4. *With the above notion. Given a Program P specified by the above $\widehat{\Omega}$ and $\widehat{F}(\mathbf{x})$. Then Program P is non-terminating over the reals if and only if $\widehat{F}(\mathbf{x})$ has at least one fixed point in $\widehat{\Omega}$.*

Proof. Take $\mathcal{M}(\mathbf{x}) = \mathbf{x} = (x_1, ..., x_n)^T$. Then, $\mathcal{H}(\mathbf{x}) = \mathbf{x} - F(\mathbf{x})$. Clearly, $\mathbb{V}_R(\mathcal{H}(\mathbf{x})) = \mathbb{V}_R(\mathbf{x} - F(\mathbf{x}))$. Let $I_i = [a_i, b_i]$ be an interval. Since $f_i(x_i)$ is a polynomial in x_i and I_i is an interval, $f_i(I_i)$ is still an interval. Set $I_i^o = f_i(I_i)$. Then, $I_1^o \times I_2^o \times \cdots \times I_n^o$ is a hyperrectangle, which clearly defines $\mathcal{H}(\widehat{\Omega})$. That is, $\mathcal{H}(\widehat{\Omega}) = I_1^o \times I_2^o \times \cdots \times I_n^o$. It is very easy to see that $\mathcal{H}(\widehat{\Omega})$ is convex. Therefore, by Theorem 1, Program P is non-terminating over the reals if and only if $\widehat{F}(\mathbf{x})$ has at least one fixed point in $\widehat{\Omega}$. $\qquad\square$

In general, for given a closed, bounded, and connected set Ω, $\mathcal{H}(\Omega)$ is not necessarily a convex set. However, this requirement can be relaxed, when the number of program variables in Program P is 2.

Theorem 2. *Let $\mathbf{x} = (x_1, x_2)^T$. Given a Program P specified by a closed, bounded and connected set $\Omega \subseteq \mathbb{R}^2$ and a polynomial mapping $F(\mathbf{x})$. Let $\mathcal{M}(\mathbf{x}) = (\mathbf{x}^{\alpha_1}, \mathbf{x}^{\alpha_2})^T$, where $\alpha_1, \alpha_2 \in \mathbb{Z}_{\geq 0}^2$. Let $\mathcal{H}(\mathbf{x}) = \mathcal{M}(\mathbf{x}) - \mathcal{M}(F(\mathbf{x}))$. If the following conditions are satisfied,*

(a) $\mathbb{V}_R(\mathcal{H}(\mathbf{x})) = \mathbb{V}_R(\mathbf{x} - F(\mathbf{x}))$,
(b) For any two points $\mathbf{x}, \mathbf{y} \in \Omega$, and any $\lambda > 0$,

$$\mathcal{H}(\mathbf{x}) \neq -\lambda \cdot \mathcal{H}(\mathbf{y}),$$

then, Program P is non-terminating if and only if $F(\mathbf{x})$ has at least one fixed point in Ω.

Proof. Let $\mathbf{u} = \mathcal{H}(\mathbf{x})$. Sufficiency is clear, as the existence of fixed points in Ω of $F(\mathbf{x})$ implies that Program P does not terminate on such a fixed point. To see necessity, suppose that $F(\mathbf{x})$ has no fixed points in Ω. Therefore, for any $\mathbf{x} \in \Omega$, $\mathbf{x} - F(\mathbf{x}) \neq 0$. That is, $\mathbb{V}_R(\mathbf{x} - F(\mathbf{x})) \cap \Omega = \emptyset$. It directly follows that $\mathbb{V}_R(\mathcal{H}(\mathbf{x})) \cap \Omega = \emptyset$ by the hypothesis (a). Therefore, $\mathbf{0} \notin \mathcal{H}(\Omega)$. For any $\mathcal{H}(\mathbf{x}), \mathcal{H}(\mathbf{y})$ in $\mathcal{H}(\Omega)$, by the angle formula of two vectors, we define the angle of $\mathcal{H}(\mathbf{x}), \mathcal{H}(\mathbf{y})$ as

$$\theta = cos^{-1}\left(\frac{\mathcal{H}(\mathbf{x})\mathcal{H}(\mathbf{y})}{|\mathcal{H}(\mathbf{x})||\mathcal{H}(\mathbf{y})|}\right).$$

Clearly, $\theta \geq 0$. Let $g(\mathbf{x}, \mathbf{y}) = \frac{\mathcal{H}(\mathbf{x})\mathcal{H}(\mathbf{y})}{|\mathcal{H}(\mathbf{x})||\mathcal{H}(\mathbf{y})|}$. Since $\mathbf{0} \notin \mathcal{H}(\Omega)$, g is continuous on $\Omega \times \Omega$. And cos^{-1} is also continuous on $[-1, 1]$. Hence, the composition $cos^{-1} \circ g$ of cos^{-1} and g is continuous on $\Omega \times \Omega$. Since Ω is compact implies $\Omega \times \Omega$ is compact, continuous function $cos^{-1} \circ g$ has a maximum on $\Omega \times \Omega$, i.e., there exists θ^* such that for any $(\mathbf{x}, \mathbf{y}) \in \Omega \times \Omega$, we have $\theta \leq \theta^*$. Moreover, by the hypothesis that $\mathcal{H}(\mathbf{x}) \neq -\lambda \cdot \mathcal{H}(\mathbf{y})$ for any $\mathbf{x}, \mathbf{y} \in \Omega$ and any $\lambda > 0$, we get that the angle θ of any two vectors $\mathcal{H}(\mathbf{x}), \mathcal{H}(\mathbf{y}) \in \mathcal{H}(\Omega)$ cannot be π, i.e., $\theta \neq \pi$. We next further show that θ^* cannot be greater than π. Suppose that $\theta^* > \pi$. Then, since $0 \leq \theta \leq \theta^*$ and $cos^{-1} \circ g$ is a continuous function on $\Omega \times \Omega$, by properties of continuous function, there must exist $(\hat{\mathbf{x}}, \hat{\mathbf{y}}) \in \Omega \times \Omega$ such that $\theta = cos^{-1}(g(\hat{\mathbf{x}}, \hat{\mathbf{y}})) = \pi$. This clearly contradicts that $\theta \neq \pi$. Therefore, we have $\theta^* < \pi$. Because $\theta^* < \pi$, there exists a closed convex sector with vertex $\mathbf{0}$, whose angle is less than π, containing the set $\mathcal{H}(\Omega)$. Since the vertex $\mathbf{0}$ of this sector is not included in $\mathcal{H}(\Omega)$, i.e., $\mathbf{0} \notin \mathcal{H}(\Omega)$, there must exist a hyperplane $\mathbf{a}^T \cdot \mathbf{u} = \mathbf{a}^T \cdot \mathcal{H}(\mathbf{x}) = \mathbf{0}$ that intersects the sector only at the origin $\mathbf{0}$. Therefore, for any $\mathcal{H}(\mathbf{x}) \in \mathcal{H}(\Omega)$, $\mathbf{a}^T \cdot \mathcal{H}(\mathbf{x}) \neq \mathbf{0}$. It immediately follows by the definition of $\mathcal{H}(\Omega)$ that for any $\mathbf{x} \in \Omega$, $\mathbf{a}^T \cdot \mathcal{H}(\mathbf{x}) \neq \mathbf{0}$. Let $T(\mathbf{x}) = \mathbf{a}^T \cdot \mathcal{M}(\mathbf{x})$. By the above arguments, we get that $\{\mathbf{x} \in \mathbb{R}^2 : T(\mathbf{x}) = T(F(\mathbf{x}))\} \cap \Omega = \emptyset$. By Corollary 1, Program P is terminating over the reals. This completes the proof of the theorem. □

Next, we will introduce Groebner basis to analyze the termination of Program P. And the computations involved with Groebner basis and ideal will be done over $\mathbb{C}$. Given Program $P \triangleq P(F, \Omega)$ as defined in (1). Let $\mathcal{M}(\mathbf{x}) = (\mathbf{x}^{\alpha_1}, \mathbf{x}^{\alpha_2}, ..., \mathbf{x}^{\alpha_s})^T$. Let $\mathcal{H}(\mathbf{x}) = \mathcal{M}(\mathbf{x}) - \mathcal{M}(F(\mathbf{x})) = (h_1, ..., h_s)^T$ and let $G(\mathbf{x}) = (g_1, ..., g_\nu)^T$ be a Groebner basis for $\langle \mathcal{H}(\mathbf{x}) \rangle$. By the properties of Groebner basis, we have

$$\langle G(\mathbf{x}) \rangle = \langle \mathcal{H}(\mathbf{x}) \rangle \quad \text{and} \quad M(\mathbf{x}) \cdot G(\mathbf{x}) = \mathcal{H}(\mathbf{x}), \tag{8}$$

for a certain polynomial matrix $M(\mathbf{x}) \in (\mathbb{R}[\mathbf{x}])^{s \times \nu}$. For convenience, the notation $G(\mathbf{x})$ is also used to denote a polynomial mapping from k^n to k^ν. Define

$$S = \{v \in \mathbb{R}^\nu : v^T \cdot G(\mathbf{x}) \neq 0, for \; all \; \mathbf{x} \in \Omega\}. \tag{9}$$

Especially, if $G(\mathbf{x})$ is an affine Groebner basis and Ω is a bounded, closed convex polytope with finitely many vertices, i.e., $G(\mathbf{x}) = A\mathbf{x} + \mathbf{c}$, $\Omega = \{\mathbf{x} \in \mathbb{R}^n : B\mathbf{x} \geq \mathbf{b}\}$ and for all $\mathbf{x} \in \Omega$ there exists a positive number δ such that $|\mathbf{x}| \leq \delta$, then it can be shown that $S = \cup_{i=1}^t S_i$, where S_i is a convex polytope specified by semi-algebraic system $S_i = \{v \in \mathbb{R}^\nu : D_i v \geq 0 \wedge \mathbf{c}_i^T v > 0\}$, for $i = 1, ..., t$. Let $\triangleright = (\geq, >)^T$. S_i can be rewritten as

$$S_i = \{v \in \mathbb{R}^\nu : \tilde{D}_i v \triangleright 0\},$$

where $\tilde{D}_i = \begin{pmatrix} D_i \\ \mathbf{c}_i^T \end{pmatrix}$.

Theorem 3. *With the above notion. Let $\Omega \subseteq \mathbb{R}^n$ be a bounded, closed and convex polytope with finitely many vertices $\mathbf{x}_1, ..., \mathbf{x}_\mu$. Let S be defined as above. If $G(\mathbf{x})$ is a affine mapping, i.e., $G(\mathbf{x}) = A\mathbf{x} + \mathbf{c}$ for some constant matrix A and some constant vector $\mathbf{c}$, then, there exist $S_1, ..., S_t$, such that $S = \cup_{i=1}^t S_i$.*

Proof. By hypothesis, since Ω is a convex set, Ω is also a connected set. Hence, since $v^T G(\mathbf{x})$ is a continuous function on the bounded, closed and connected set Ω, by the properties of continuous functions, to check if $\forall \mathbf{x} \in \Omega \Rightarrow v^T \cdot G(\mathbf{x}) \neq 0$ is equivalent to check if

(1) $\forall \mathbf{x} \in \Omega, v^T \cdot G(\mathbf{x}) > 0$, or,

(2) $\forall \mathbf{x} \in \Omega, v^T \cdot G(\mathbf{x}) < 0$.

Denote by $T_{(1)}$ and $T_{(2)}$ the sets of the vectors v's satisfying the above (1) and (2), respectively. Clearly, $S = T_{(1)} \cup T_{(2)}$. We next show that $T_{(1)} = \cup_{i=1}^{t_1} S_i$. And similar analysis can be applied to $T_{(2)}$. Consider Formula (1). Since Ω is a bounded, closed and convex polytope with finitely many vertices $\mathbf{x}_1, ..., \mathbf{x}_\mu$, every point $\mathbf{x} \in \Omega$ is a convex combination of the vertices, i.e., $\mathbf{x} = \lambda_1 \mathbf{x}_1 + \lambda_2 \mathbf{x}_2 + \cdots + \lambda_\mu \mathbf{x}_\mu$ where $\sum_{i=1}^{\mu} \lambda_i = 1$ and $\lambda_i \geq 0$, $i = 1, ..., \mu$. Therefore, we have

$$\forall \mathbf{x} \in \Omega \Rightarrow v^T \cdot G(\mathbf{x}) > 0$$

is equivalent to

$$\forall \bar{\lambda}.(\wedge_{i=1}^{\mu} \lambda_i \geq 0 \wedge \sum_{i=1}^{\mu} \lambda_i = 1 \Rightarrow v^T \cdot (A(\sum_{i=1}^{\mu} \lambda_i \mathbf{x}_i) + \mathbf{c}) > 0) \qquad (10)$$

where $\bar{\lambda} = (\lambda_1, ..., \lambda_\mu)$ and v is regarded as parameter. Hence, eliminating the quantifiers λ_i's from (10), we get the desired $T_{(i)}$ that is a set of constraints only on v. Let $\mathbf{Obj}(\bar{\lambda}) = v^T \cdot (A(\sum_{i=1}^{\mu} \lambda_i \mathbf{x}_i) + \mathbf{c}) = \sum_{i=1}^{\mu} (v^T A \mathbf{x}_i) \lambda_i + v^T \mathbf{c}$. Clearly, Formula (10) can be converted to the following standard linear programming problem,

$$\text{minimize } \mathbf{Obj}(\bar{\lambda}) > 0$$

$$\text{subject to } \sum_{i=1}^{\mu} \lambda_i = 1, \qquad (11)$$

$$\lambda_i \geq 0.$$

The constraints $\{\sum_{i=1}^{\mu} \lambda_i = 1, \lambda_1 \geq 0, ..., \lambda_\mu \geq 0\}$ characterize a feasible region **Reg**. Also, it is not difficult to see that **Reg** is a simplex that has the vertices $e_1, ..., e_\mu$, where e_i denotes the vector with a 1 in the i-th coordinate and 0's elsewhere. It is well known that if the feasible region **Reg** is bounded, then the optimal solution is always one of the vertices of **Reg**. Therefore,

$$\text{minimize } \mathbf{Obj}(\bar{\lambda}) \triangleq \min(\{\mathbf{Obj}(e_i)\}_{i=1}^{\mu}),$$

where $\mathbf{Obj}(e_i) = v^T A \mathbf{x}_i + v^T \mathbf{c}$. Thus, to obtain $\mathbf{Obj}(\bar{\lambda})_{min}$ is equivalent to find the minimum value of $\{\mathbf{Obj}(e_1), ..., \mathbf{Obj}(e_\mu)\}$. Because v is regarded as parameter in (10) and (11), $\mathbf{Obj}(e_i)$'s are all linear homogenous polynomials in v with constant coefficients $(A\mathbf{x}_i + \mathbf{c})^T$. Therefore, to find the minimum value of $\{\mathbf{Obj}(e_i)\}_{i=1}^{\mu}$ and guarantee that its minimum value is positive, there will be μ cases to consider,

$$Ineq_{(1)}^i \triangleq \left(\bigwedge_{j \neq i} \mathbf{Obj}(e_j) \geq \mathbf{Obj}(e_i) \right) \wedge \mathbf{Obj}(e_i) > 0,$$

for $i = 1, ..., \mu$. Furthermore, since the inequalities in $Ineq_{(1)}^i$ are all linear homogenous polynomials in v, $Ineq_{(1)}^i$ can be rewritten as $Ineq_{(1)}^i \triangleq \widetilde{D}_i v \triangleright 0$. And let

$S_i = \{v \in \mathbb{R}^\nu : \ \widetilde{D}_i v \triangleright 0\}$ and set $t_1 = \mu$. Then we have $T_{(1)} = \cup_{i=1}^{t_1} S_i$. Consider Formula (2). Since $\forall \mathbf{x} \in \Omega \Rightarrow v^T \cdot G(\mathbf{x}) < 0$ is equivalent to $\forall \mathbf{x} \in \Omega \Rightarrow (-v)^T \cdot G(\mathbf{x}) > 0$, we can directly construct $Ineq_{(2)}^i$ by replacing v in $Ineq_{(1)}^i$ with $-v$, i.e., $Ineq_{(2)}^i \triangleq -\widetilde{D}_i v \triangleright 0$, for $i = 1, ..., \mu$. Let $S_i^- = S_{t_1+i} = \{v \in \mathbb{R}^\nu : \ -\widetilde{D}_i v \triangleright 0\}$ for $i = 1, ..., \mu$. Then we have $T_{(2)} = \cup_{i=1}^\mu S_{t_1+i}$. It immediately follows that $S = T_{(1)} \cup T_{(2)} = \cup_{i=1}^{2\mu} S_i$. This completes the proof of the theorem.   □

Remark 2. In fact, the proof of Theorem 3 proposes a method to directly construct the desired S, if $G(\mathbf{x})$ is affine and Ω is a bounded, closed and convex polytope with finitely many vertices.

Given two matrices $A, B \in \mathbb{R}^{m \times n}$, we say $A \geq B$, if $A_{ij} \geq B_{ij}$ for all $i = 1, ..., m, j = 1, ..., n$. It is not difficult to see that if $A \geq B$, then $Av \geq Bv$ for any non-negative vector v.

Let $F(\mathbf{x}) : \mathbb{R}^n \rightarrow \mathbb{R}^n$ be a polynomial mapping. Let $\mathcal{M}(\mathbf{x}) = (\mathbf{x}^{\alpha_1}, \mathbf{x}^{\alpha_2}, ..., \mathbf{x}^{\alpha_s})^T$ and let $\mathcal{H}(\mathbf{x}) = \mathcal{M}(\mathbf{x}) - \mathcal{M}(F(\mathbf{x})) = (h_1, ..., h_s)^T$. And let $G(\mathbf{x}) = (g_1, ..., g_\nu)^T = A\mathbf{x} + \mathbf{c}$ be an affine Groebner basis for $\langle \mathcal{H}(\mathbf{x}) \rangle$. Let $S = \{v \in \mathbb{R}^\nu : \ v^T \cdot G(\mathbf{x}) \neq 0, for \ all \ \mathbf{x} \in \Omega\}$. By Theorem 3, we have $S = \cup_{i=1}^t S_i$. Let $\Omega \subseteq \mathbb{R}^n$ be a bounded, closed and convex polytope with finitely many vertices $\mathbf{x}_1, ..., \mathbf{x}_\mu$. Let $M(\mathbf{x})$ be a polynomial matrix as defined in (8). We now establish the following two results.

Theorem 4. *With the above notion. Given Program $P \triangleq P(\Omega, F)$, where F, Ω are defined as above. If there exist nonempty set $S_i = \{v \in \mathbb{R}^\nu : \ \widetilde{D}_i v \triangleright 0\} \neq \emptyset$, $i \in \{1, ..., t\}$, and nonzero nonnegative vector $v^* \in \mathbb{R}^\nu$, such that*

$$\forall \mathbf{x} = \sum_{l=1}^\mu \lambda_l \mathbf{x}_l \in \Omega \Rightarrow \widetilde{D}_i M^T(\mathbf{x}) \geq \sum_{l=1}^\mu \lambda_l \widetilde{D}_i M^T(\mathbf{x}_l), \tag{12}$$

where $\lambda_l \geq 0$, $\sum_{l=1}^\mu \lambda_l = 1$ and

$$M^T(\mathbf{x}_1) v^*, M^T(\mathbf{x}_2) v^*, ..., M^T(\mathbf{x}_\mu) v^* \in S_i, \tag{13}$$

then, Program $P \triangleq P(\Omega, F)$ is terminating over the reals.

Proof. By the hypothesis, we have $S \neq \emptyset$ and

$$\forall \mathbf{x} = \sum_{l=1}^\mu \lambda_l \mathbf{x}_l \in \Omega \Rightarrow \widetilde{D}_i M^T(\mathbf{x}) v^* \geq \sum_{l=1}^\mu \lambda_l \widetilde{D}_i M^T(\mathbf{x}_l) v^*,$$

since v^* is a non-negative vector. And by (13), since for $l = 1, ..., \mu$, $M^T(\mathbf{x}_l) v^* \in S_i$, i.e., $\widetilde{D}_i M^T(\mathbf{x}_l) v^* \triangleright 0$, we have $\widetilde{D}_i M^T(\mathbf{x}) v^* \triangleright 0$, i.e., $M^T(\mathbf{x}) v^* \in S_i$. By the definition of S, we know that $(M^T(\mathbf{x}) v^*)^T \cdot G(\mathbf{x}) \neq 0$ for any $\mathbf{x} \in \Omega$, since $S_i \subseteq S$. That is, for all $\mathbf{x} \in \Omega$, $(M^T(\mathbf{x}) v^*)^T \cdot G(\mathbf{x}) = (v^*)^T M(\mathbf{x}) \cdot G(\mathbf{x}) \neq 0$. Because $M(\mathbf{x}) G(\mathbf{x}) = \mathcal{H}(\mathbf{x})$, we have for all $\mathbf{x} \in \Omega$,

$$(v^*)^T M(\mathbf{x}) \cdot G(\mathbf{x}) = (v^*)^T \mathcal{H}(\mathbf{x}) \neq 0.$$

Let $T(\mathbf{x}) = (v^*)^T \mathcal{M}(\mathbf{x})$. Then, we get $\{\mathbf{x} \in \mathbb{R}^n : T(\mathbf{x}) = T(F(\mathbf{x}))\} \cap \Omega = \emptyset$. It immediately follows that Program P is terminating by Corollary 1. This completes the proof of the theorem. □

Remark 3. Note that to check if Formula (12) holds is equivalent to check if

$$\forall \bar{\lambda}. (\bigwedge_{l=1}^{\mu} \lambda_l \geq 0 \wedge \sum_{l=1}^{\mu} \lambda_l = 1 \Rightarrow \widetilde{D}_i M^T (\sum_{l=1}^{\mu} \lambda_l \mathbf{x}_l) \geq \sum_{l=1}^{\mu} \lambda_l \widetilde{D}_i M^T (\mathbf{x}_l)), \quad (14)$$

since Ω is bounded convex polytope, and each point in Ω can be expressed as a convex combination of the vertices of Ω by the properties of bounded convex polytope. In addition, in terms of the definition of S_i, to check if Formula (13) holds is equivalent to check if the following semi-algebraic system,

$$v^* \geq 0 \wedge v^* \neq 0 \wedge \widetilde{D}_i M^T (\mathbf{x}_1) v^* \rhd 0 \wedge \widetilde{D}_i M^T (\mathbf{x}_2) v^* \rhd 0 \wedge \cdots \wedge \widetilde{D}_i M^T (\mathbf{x}_\mu) v^* \rhd 0. \quad (15)$$

has real solutions, where $\rhd = (\geq, >)^T$.

Theorem 5. *With the above notion. Given Program $P \triangleq P(\Omega, F)$, where F, Ω are defined as above. If the following conditions are satisfied,*

(a) $V_{\mathbb{R}}(\mathcal{H}(\mathbf{x})) = V_{\mathbb{R}}(F(\mathbf{x}) - \mathbf{x})$,
(b) $\mathcal{H}(\Omega) \subseteq G(\Omega)$,

then Program $P \triangleq P(\Omega, F)$ is non-terminating over the reals if and only if F has at least one fixed point in Ω.

Proof. If F has fixed points in Ω, then Program P is non-terminating. We next show that if F has no fixed points in Ω, then Program P must terminate. First, let $\mathbf{u} = G(\mathbf{x})$. And since $G(\mathbf{x})$ is an affine Groenber basis of $\langle \mathcal{H}(\mathbf{x}) \rangle$, by (8), we have $\langle \mathcal{H}(\mathbf{x}) \rangle = \langle G(\mathbf{x}) \rangle$ and $M(\mathbf{x})G(\mathbf{x}) = \mathcal{H}(\mathbf{x})$. This immediately implies that $V_{\mathbb{C}}(\mathcal{H}(\mathbf{x})) = V_{\mathbb{C}}(G(\mathbf{x}))$. Hence, $V_{\mathbb{R}}(\mathcal{H}(\mathbf{x})) = V_{\mathbb{R}}(G(\mathbf{x})) = V_{\mathbb{R}}(F(\mathbf{x}) - \mathbf{x})$, according to condition (a). Since F has no fixed points in Ω, we have $\mathbf{0} \notin G(\Omega)$, where $G(\Omega) = \{\mathbf{u} = G(\mathbf{x}) \in \mathbb{R}^\nu : \text{for all } \mathbf{x} \in \Omega\}$. In addition, since $G(\mathbf{x})$ is affine and Ω is a bounded, closed and convex polytope, $G(\Omega)$ is also bounded, closed and convex set. By Lemma 3 and the similar arguments presented in the proof of Theorem 1, we know that there must exist a hyperplane $v^T \cdot \mathbf{u} = \mathbf{b}$ strictly separates $\mathbf{0} \in \mathbb{R}^\nu$ from $G(\Omega) \subseteq \mathbb{R}^\nu$. That is, $v^T \cdot \mathbf{u} = \mathbf{b}$ does not intersect with $G(\Omega)$. This immediately indicates that the hyperplane $v^T \cdot \mathbf{u} = 0$ also does not intersect with $G(\Omega)$. That is, for any $\mathbf{u} \in G(\Omega)$, $v^T \cdot \mathbf{u} \neq 0$. Therefore, by the definition of $G(\Omega)$, we have for any $\mathbf{x} \in \Omega$, $v^T \cdot G(\mathbf{x}) \neq 0$. This suggests that $S = \cup_{i=1}^t S_i \neq \emptyset$. Furthermore, by condition (b), since $\mathcal{H}(\Omega) \subseteq G(\Omega)$, it immediately follows that $v^T \cdot \mathcal{H}(\Omega) \neq 0$ for any $v \in S$. This implies Program P is terminating. □

Example 3. Consider the termination of the below program.

$$P_3 : \textbf{While } 1 \leq x \leq 2 \wedge 1 \leq y \leq 2 \textbf{ do}$$
$$\{x := -5x - 12; y := 3y - x^2 - 1\} \quad (16)$$
$$\textbf{endwhile}$$

Let $f_1 = -5x - 12$ and $f_2 = 3y - x^2 - 1$. Define $\Omega = \{(x, y)^T \in \mathbb{R}^2 : 1 \leq x \leq 2 \wedge 1 \leq y \leq 2\}$. Set VP $= \{(1, 1), (2, 1), (1, 2), (2, 2)\}$ to be a set of all vertices of Ω. And let $\mathcal{M}(\mathbf{x}) = (x, y)^T$. Then $\mathcal{H}(\mathbf{x}) = \mathcal{M}(\mathbf{x}) - \mathcal{M}(F(\mathbf{x})) = (x - f_1, y - f_2)^T$. First, invoking the command $Basis$ in Maple, we get the Groebner basis of ideal $\langle \mathcal{H}(\mathbf{x}) \rangle$ and the corresponding transformation matrix $M(\mathbf{x})$,

$$G(\mathbf{x}) = (-5 + 2y, x + 2)^T,$$

$$M(\mathbf{x}) = \begin{pmatrix} 0, & 6 \\ -1, & x - 2 \end{pmatrix}.$$

Since each component in G is affine, $G(\mathbf{x})$ is an affine Groebner basis. We next check if the hypothesis in Theorem 4 holds. To do this, we first need to compute S. Let $G(\mathbf{x}) = (g_1(\mathbf{x}), g_2(\mathbf{x}))^T$ and $v = (v_1, v_2)^T$. To compute S is equivalent to eliminate quantified variables x, y from the following quantified formula:

$$\forall \mathbf{x} \in \Omega \Rightarrow v_1 \cdot g_1(\mathbf{x}) + v_2 \cdot g_1(\mathbf{x}) \neq 0. \tag{17}$$

This can be done easily, since $G(\mathbf{x})$ is affine. However, by the proof of Theorem 3 and Remark 2, we can directly construct S as follows,

$$S = \bigcup_{i=1}^{4} S_i \cup \bigcup_{i=1}^{4} S_i^- = \bigcup_{i=1}^{4} \{(v_1, v_2)^T : \; \widetilde{D}_i v \rhd 0\} \cup \bigcup_{i=1}^{4} \{(v_1, v_2)^T : \; \widetilde{D}_i(-v) \rhd 0\},$$

where $\rhd = (\geq, \geq, \geq, >)^T$ and

$$\widetilde{D}_1 = \begin{pmatrix} 1 & 0 \\ 0 & 2 \\ 1 & 2 \\ -1 & 7 \end{pmatrix}, \widetilde{D}_2 = \begin{pmatrix} -1 & 0 \\ -1 & 2 \\ 0 & 2 \\ 0 & 7 \end{pmatrix}, \widetilde{D}_3 = \begin{pmatrix} 0 & -2 \\ 1 & -2 \\ 1 & 0 \\ -1 & 9 \end{pmatrix}, \widetilde{D}_4 = \begin{pmatrix} -1 & -2 \\ 0 & -2 \\ -1 & 0 \\ 0 & 9 \end{pmatrix}. \tag{18}$$

Next, we will check if Formula (12) and Formula (13) hold. According to Remark 3, to check if Formula (12) and Formula (13) hold is equivalent to check if Formula (14) and Formula (15) hold. By computation, we find that when $i = 3$, both Formula (14) and Formula (15) hold. Therefore, by Theorem 4, Program P_3 must terminate over the reals.

4   Conclusion

We have analyzed the termination of single-path polynomial loop programs (SPLPs). Some conditions are given such that under such conditions the termination of this kind of loop programs over the reals can be equivalently reduced to computation of real fixed points. In other words, once such conditions are satisfied, an SPLP $P(\Omega, F)$ is not terminating over the reals if and only if F has at least one fixed point in Ω. Furthermore, such conditions can be described by quantified formulae. This enables us to apply the tools for real quantifier elimination, such as RegularChains, to automatically check if such conditions are satisfied.

Acknowledgments. The author would like to thank the anonymous reviewers for their helpful suggestions. This research is partially supported by the National Natural Science Foundation of China NNSFC (61572024, 61103110).

References

1. Bagnara, R., Mesnard, F.: Eventual linear ranking functions. In: Proceedings of the 15th Symposium on Principles and Practice of Declarative Programming, pp. 229–238. ACM, Madrid (2013)
2. Bagnara, R., Mesnard, F., Pescetti, A., Zaffanella, E.: A new look at the automatic synthesis of linear ranking functions. Inf. Comput. **215**, 47–67 (2012)
3. Ben-Amram, A.: The hardness of finding linear ranking functions for Lasso programs. Electron.Proc. Theor. Comput. Sci. **161**, 32–45 (2014)
4. Ben-Amram, A., Genaim, S.: On the linear ranking problem for integer linear-constraint loops. In: POPL 2013 Proceedings of the 40th Annual ACM SIGPLAN-SIGACT Symposium on Principles of Programming Languages, pp. 51–62. ACM, Rome (2013)
5. Ben-Amram, A., Genaim, S.: Ranking functions for linear-constraint loops. J. ACM **61**(4), 1–55 (2014)
6. Bradley, A.R., Manna, Z., Sipma, H.B.: Linear ranking with reachability. In: Etessami, K., Rajamani, S.K. (eds.) CAV 2005. LNCS, vol. 3576, pp. 491–504. Springer, Heidelberg (2005). doi:10.1007/11513988_48
7. Bradley, A.R., Manna, Z., Sipma, H.B.: The polyranking principle. In: Caires, L., Italiano, G.F., Monteiro, L., Palamidessi, C., Yung, M. (eds.) ICALP 2005. LNCS, vol. 3580, pp. 1349–1361. Springer, Heidelberg (2005). doi:10.1007/11523468_109
8. Bradley, A.R., Manna, Z., Sipma, H.B.: Termination of polynomial programs. In: Cousot, R. (ed.) VMCAI 2005. LNCS, vol. 3385, pp. 113–129. Springer, Heidelberg (2005). doi:10.1007/978-3-540-30579-8_8
9. Braverman, M.: Termination of integer linear programs. In: Ball, T., Jones, R.B. (eds.) CAV 2006. LNCS, vol. 4144, pp. 372–385. Springer, Heidelberg (2006). doi:10.1007/11817963_34
10. Chen, C., Maza, M.: Quantifier elimination by cylindrical algebraic decomposition based on regular chains. In: Proceedings of the 39th International Symposium on Symbolic and Algebraic Computation, pp. 91–98. ACM (2014)
11. Chen, H.Y., Flur, S., Mukhopadhyay, S.: Termination proofs for linear simple loops. In: Miné, A., Schmidt, D. (eds.) SAS 2012. LNCS, vol. 7460, pp. 422–438. Springer, Heidelberg (2012). doi:10.1007/978-3-642-33125-1_28
12. Chen, Y., Xia, B., Yang, L., Zhan, N., Zhou, C.: Discovering non-linear ranking functions by solving semi-algebraic systems. In: Jones, C.B., Liu, Z., Woodcock, J. (eds.) ICTAC 2007. LNCS, vol. 4711, pp. 34–49. Springer, Heidelberg (2007). doi:10.1007/978-3-540-75292-9_3
13. Colón, M.A., Sipma, H.B.: Synthesis of linear ranking functions. In: Margaria, T., Yi, W. (eds.) TACAS 2001. LNCS, vol. 2031, pp. 67–81. Springer, Heidelberg (2001). doi:10.1007/3-540-45319-9_6
14. Cook, B., Gulwani, S., Lev-Ami, T., Rybalchenko, A., Sagiv, M.: Proving conditional termination. In: Gupta, A., Malik, S. (eds.) CAV 2008. LNCS, vol. 5123, pp. 328–340. Springer, Heidelberg (2008). doi:10.1007/978-3-540-70545-1_32
15. Cook, B., See, A., Zuleger, F.: Ramsey vs. lexicographic termination proving. In: Piterman, N., Smolka, S.A. (eds.) TACAS 2013. LNCS, vol. 7795, pp. 47–61. Springer, Heidelberg (2013). doi:10.1007/978-3-642-36742-7_4

16. Duistermaat, J., Kolk, J.: Multidimensional Real Analysis. Cambridge University Press, Cambridge (2004)
17. Ganty, P., Genaim, S.: Proving termination starting from the end. In: Sharygina, N., Veith, H. (eds.) CAV 2013. LNCS, vol. 8044, pp. 397–412. Springer, Heidelberg (2013). doi:10.1007/978-3-642-39799-8_27
18. Heizmann, M., Hoenicke, J., Leike, J., Podelski, A.: Linear ranking for linear Lasso programs. In: Hung, D., Ogawa, M. (eds.) ATVA 2013. LNCS, vol. 8172, pp. 365–380. Springer, Heidelberg (2013). doi:10.1007/978-3-319-02444-8_26
19. Leike, J., Heizmann, M.: Ranking templates for linear loops. In: Ábrahám, E., Havelund, K. (eds.) TACAS 2014. LNCS, vol. 8413, pp. 172–186. Springer, Heidelberg (2014). doi:10.1007/978-3-642-54862-8_12
20. Leike, J., Tiwari, A.: Synthesis for polynomial Lasso programs. In: McMillan, K.L., Rival, X. (eds.) VMCAI 2014. LNCS, vol. 8318, pp. 434–452. Springer, Heidelberg (2014). doi:10.1007/978-3-642-54013-4_24
21. Liu, J., Xu, M., Zhan, N.J., Zhao, H.J.: Discovering non-terminating inputs for multi-path polynomial programs. J. Syst. Sci. Complex. **27**, 1284–1304 (2014)
22. Podelski, A., Rybalchenko, A.: A complete method for the synthesis of linear ranking functions. In: Steffen, B., Levi, G. (eds.) VMCAI 2004. LNCS, vol. 2937, pp. 239–251. Springer, Heidelberg (2004). doi:10.1007/978-3-540-24622-0_20
23. Sohn, K., Van Gelder, A.: Termination detection in logic programs using argument sizes (extended abstract). In: Proceedings of the Tenth ACM SIGACT- SIGMOD-SIGART Symposium on Principles of Database Systems, pp. 216-226. ACM, Association for Computing Machinery, Denver (1991)
24. Stephen, B., Lieven, V.: Convex Optimization. Cambridge University Press, New York (2004)
25. Tiwari, A.: Termination of linear programs. In: Alur, R., Peled, D.A. (eds.) CAV 2004. LNCS, vol. 3114, pp. 70–82. Springer, Heidelberg (2004). doi:10.1007/978-3-540-27813-9_6
26. Xia, B., Zhang, Z.: Termination of linear programs with nonlinear constraints. J. Symb. Comput. **45**(11), 1234–1249 (2010)

Relation-Algebraic Verification of Prim's Minimum Spanning Tree Algorithm

Walter Guttmann[✉]

Department of Computer Science and Software Engineering,
University of Canterbury, Christchurch, New Zealand
walter.guttmann@canterbury.ac.nz

Abstract. We formally prove the correctness of Prim's algorithm for computing minimum spanning trees. We introduce new generalisations of relation algebras and Kleene algebras, in which most of the proof can be carried out. Only a small part needs additional operations, for which we introduce a new algebraic structure. We instantiate these algebras by matrices over extended reals, which model the weighted graphs used in the algorithm. Many existing results from relation algebras and Kleene algebras generalise from the relation model to the weighted-graph model with no or small changes. The overall structure of the proof uses Hoare logic. All results are formally verified in Isabelle/HOL heavily using its integrated automated theorem provers.

1 Introduction

A well-known algorithm commonly attributed to Prim [43] – and independently discovered by Jarník [27] and Dijkstra [17] – computes a minimum spanning tree in a weighted undirected graph. It starts with an arbitrary root node, and constructs a tree by repeatedly adding an edge that has minimal weight among the edges connecting a node in the tree with a node not in the tree. The iteration stops when there is no such edge, at which stage the constructed tree is a minimum spanning tree of the component of the graph that contains the root (which is the whole graph if it is connected).

The aim of this paper is to demonstrate the applicability of relation-algebraic methods for verifying the correctness of algorithms on weighted graphs. Accordingly, we will use an implementation of Prim's algorithm close to the above abstraction level. Since its discovery many efficient implementations of this and other spanning tree algorithms have been developed; for example, see the two surveys [21,35]. These implementations typically rely on specific data structures, which can be introduced into a high-level algorithm by means of data refinement; for example, see [4]. We do not pursue this in the present paper.

Relation-algebraic methods have been used to develop algorithms for unweighted graphs; for example, see [4,5,7]. This works well because such a graph can be directly represented as a relation; an adjacency matrix is a Boolean matrix. Weighted graphs do not have a direct representation as a binary relation.

© Springer International Publishing AG 2016
A. Sampaio and F. Wang (Eds.): ICTAC 2016, LNCS 9965, pp. 51–68, 2016.
DOI: 10.1007/978-3-319-46750-4_4

Previous relational approaches to weighted graphs therefore use many-sorted representations such as an incidence matrix and a weight function. In this paper, we directly work with a matrix of weights.

In the context of fuzzy systems, relations have been generalised from Boolean matrices to matrices over the real interval $[0, 1]$ or over arbitrary complete distributive lattices [19]. The underlying idea is to extend qualitative to quantitative methods; see [41] for another instance based on automata. We propose to use matrices over lattices to model weighted graphs, in particular in graph algorithms. Previous work based on semirings and Kleene algebras deals well with path problems in graphs [20]. We combine these algebras with generalisations of relation algebras to tackle the minimum spanning tree problem.

Tarski's relation algebras [46], which capture Boolean matrices, have been generalised to Dedekind categories to algebraically capture fuzzy relations [30]; these categories are also known as locally complete division allegories [18]. In the present paper we introduce a new generalisation – Stone relation algebras – which maintains the signature of relation algebras and weakens the underlying Boolean algebra structure to Stone algebras. We show that matrices over extended reals are instances of Stone relation algebras and of Kleene algebras, and can be used to represent weighted graphs.

Most of the correctness proof of Prim's minimum spanning tree algorithm can be carried out in these general algebras. Therefore, most of our results hold for many instances, not just weighted graphs. A small part of the correctness proof uses operations beyond those available in relation algebras and in Kleene algebras, namely for summing edge weights and identifying minimal edges. We capture essential properties of these operations in a new algebraic structure.

With this approach we can apply well-developed methods and concepts of relation algebras and Kleene algebras to reason about weighted graphs in a new, more direct way. The contributions of this paper are:

- Stone relation algebras, a new algebraic structure that generalises relation algebras but maintains their signature. Many theorems of relation algebras already hold in these weaker algebras. Combined with Kleene algebras, they form a general yet expressive setting for most of the correctness proof of the minimum spanning tree algorithm.
- A new algebra that extends Stone-Kleene relation algebras by dedicated operations and axioms for finding minimal edges and for computing the total weight of a graph.
- Models of the above algebras, including weighted graphs represented by matrices over extended reals. This includes a formalisation of Conway's automata-based construction for the Kleene star of a matrix.
- A Hoare-logic correctness proof of Prim's minimum spanning tree algorithm entirely based on the above algebras.
- Isabelle/HOL theories that formally verify all of the above and all results in and about the algebras stated in the present paper. Proofs are omitted in this paper and can be found in the Isabelle/HOL theory files available at http://www.csse.canterbury.ac.nz/walter.guttmann/algebra/.

Section 2 recalls basic algebraic structures and introduces Stone relation algebras, which we use to represent weighted graphs. They are extended by the Kleene star operation to describe reachability in graphs in Sect. 3. Operations for summing weights and for finding their minimum are added in Sect. 4. In this setting, Sect. 5 presents the minimum spanning tree algorithm, details aspects of its correctness proof and shows how the proof uses the various algebras. Related work is discussed in Sect. 6.

2   Stone Relation Algebras

In this section we introduce Stone relation algebras, which generalise relation algebras so as to model not just Boolean matrices but matrices over arbitrary numbers required to represent weighted graphs. Each entry in such a matrix is taken from the set of real numbers $\mathbb{R}$ extended with a bottom element $\bot$ and a top element $\top$; let $\mathbb{R}' = \mathbb{R} \cup \{\bot, \top\}$. If the entry in row i and column j of the matrix is $\bot$, this means there is no edge from node i to node j. If the entry is a real number, there is an edge with that weight. An entry of $\top$ is used to record the presence of an edge without information about its weight; see below. The order $\leq$ and the operations max and min on $\mathbb{R}'$ are extended from reals so that $\bot$ is the $\leq$-least element and $\top$ is the $\leq$-greatest element. To work with extended reals (weights) and matrices of extended reals (weighted graphs) we use the following well-known algebraic structures [9,12,22].

Definition 1. A *bounded semilattice* is an algebraic structure $(S, \sqcup, \bot)$ where $\sqcup$ is associative, commutative and idempotent and has unit $\bot$:

$$x \sqcup (y \sqcup z) = (x \sqcup y) \sqcup z \qquad x \sqcup y = y \sqcup x \qquad x \sqcup x = x \qquad x \sqcup \bot = x$$

A *bounded distributive lattice* is an algebraic structure $(S, \sqcup, \sqcap, \bot, \top)$ where $(S, \sqcup, \bot)$ and $(S, \sqcap, \top)$ are bounded semilattices and the following distributivity and absorption axioms hold:

$$x \sqcup (y \sqcap z) = (x \sqcup y) \sqcap (x \sqcup z) \qquad x \sqcup (x \sqcap y) = x$$
$$x \sqcap (y \sqcup z) = (x \sqcap y) \sqcup (x \sqcap z) \qquad x \sqcap (x \sqcup y) = x$$

The *lattice order* is given by

$$x \leq y \Leftrightarrow x \sqcup y = y$$

A *distributive p-algebra* $(S, \sqcup, \sqcap, {}^{-}, \bot, \top)$ expands a bounded distributive lattice $(S, \sqcup, \sqcap, \bot, \top)$ with a pseudocomplement operation $^{-}$ satisfying the equivalence

$$x \sqcap y = \bot \Leftrightarrow x \leq \overline{y}$$

This means that $\overline{y}$ is the $\leq$-greatest element whose meet with y given by $\sqcap$ is $\bot$. A *Stone algebra* is a distributive p-algebra satisfying the equation

$$\overline{x} \sqcup \overline{\overline{x}} = \top$$

An element $x \in S$ is *regular* if $\overline{\overline{x}} = x$. A *Boolean algebra* is a Stone algebra whose elements are all regular.

Note that there is no obvious way to introduce a Boolean complement on $\mathbb{R}'$, which is why we use the weaker Stone algebras. We obtain the following consequences for Stone algebras; in particular, extended reals form a Stone algebra and so do matrices over extended reals. See [20] for similar matrix semirings and the max-min semiring of extended reals. The set of square matrices with indices from a set A and entries from a set S is denoted by $S^{A \times A}$. It represents a graph with node set A and edge weights taken from S.

Theorem 1. *Let* $(S, \sqcup, \sqcap, \bar{\ }, \bot, \top)$ *be a Stone algebra and let* A *be a set.*

1. *The regular elements of* S *form a Boolean algebra that is a subalgebra of* S [22].
2. $(S^{A \times A}, \sqcup, \sqcap, \bar{\ }, \bot, \top)$ *is a Stone algebra, where the operations* $\sqcup$, $\sqcap$, $\bar{\ }$, $\bot$, $\top$ *and the lattice order* $\leq$ *are lifted componentwise.*
3. $(\mathbb{R}', \max, \min, \bar{\ }, \bot, \top)$ *is a Stone algebra with*

$$\bar{x} = \begin{cases} \top \ \textit{if } x = \bot \\ \bot \ \textit{if } x \neq \bot \end{cases}$$

and the order $\leq$ *on* $\mathbb{R}'$ *as the lattice order.*

The regular elements of the Stone algebra $\mathbb{R}'$ are $\bot$ and $\top$. In particular, applying the pseudocomplement operation $\bar{\ }$ twice maps $\bot$ to itself and every other element to $\top$. Applying $\bar{\ }$ twice to a matrix over $\mathbb{R}'$, which represents a weighted graph, yields a matrix over $\{\bot, \top\}$ that represents the structure of the graph forgetting the weights. A related operation called the 'support' of a matrix is discussed in [33]; it works on matrices over natural numbers and maps 0 to 0 and each non-zero entry to 1. Relations are used to describe the 'shape' of a matrix of complex numbers in [16]; a shape represents a superset of the non-zero entries of a matrix, but an operator to obtain the non-zero entries is not discussed there.

The matrices over $\{\bot, \top\}$ are the regular elements of the matrix algebra and form a subalgebra of it. This situation, shown in Fig. 1, is analogous to that of partial identities – subsets of the identity relation used to represent

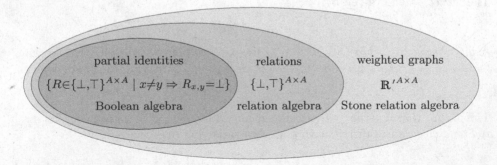

Fig. 1. Relations form a substructure of weighted graphs as partial identities form a substructure of relations

conditions in computations – which form a substructure of the encompassing relation algebra. In both cases, the substructure can be obtained as the image of a closure operation. The regular matrices are the image of the closure operation $\lambda x.\overline{\overline{x}}$ that is used in the correctness proof whenever only the structure of the graph is important, not the weights. The graph structure can be represented as a (Boolean) relation; in the context of fuzzy systems these are also called 'crisp' relations to distinguish them from fuzzy relations [19]. An operation to obtain the least crisp relation containing a given fuzzy relation is discussed in [47].

The order $\leq$ of Stone algebras allows us to compare edge weights. For matrices the comparison and all operations of Stone algebras work componentwise. These operations cannot be used to propagate information about edges through a graph. To combine information from edges between different pairs of nodes we add a relational structure with the operations of composition and converse. In unweighted graphs, they would be provided by relation algebras. To handle weighted graphs, we introduce the following generalisation.

Definition 2. A *Stone relation algebra* $(S, \sqcup, \sqcap, \cdot, ^-, ^\mathsf{T}, \bot, \top, 1)$ is a Stone algebra $(S, \sqcup, \sqcap, ^-, \bot, \top)$ with a composition $\cdot$ and a converse $^\mathsf{T}$ and a constant 1 satisfying Eqs. (1)–(10). We abbreviate $x \cdot y$ as xy and let composition have higher precedence than the operators $\sqcup$ and $\sqcap$. The axioms are:

$$(xy)z = x(yz) \tag{1}$$

$$1x = x \tag{2}$$

$$(x \sqcup y)z = xz \sqcup yz \tag{3}$$

$$(xy)^\mathsf{T} = y^\mathsf{T} x^\mathsf{T} \tag{4}$$

$$(x \sqcup y)^\mathsf{T} = x^\mathsf{T} \sqcup y^\mathsf{T} \tag{5}$$

$$x^{\mathsf{T}\mathsf{T}} = x \tag{6}$$

$$\bot x = \bot \tag{7}$$

$$xy \sqcap z \leq x(y \sqcap x^\mathsf{T} z) \tag{8}$$

$$\overline{\overline{xy}} = \overline{\overline{x}}\,\overline{\overline{y}} \tag{9}$$

$$\overline{\overline{1}} = 1 \tag{10}$$

An element $x \in S$ is a *vector* if $x\top = x$, *symmetric* if $x = x^\mathsf{T}$, *injective* if $xx^\mathsf{T} \leq 1$, *surjective* if $1 \leq x^\mathsf{T}x$ and *bijective* if x is injective and surjective. An element $x \in S$ is an *atom* if both $x\top$ and $x^\mathsf{T}\top$ are bijective. A *relation algebra* $(S, \sqcup, \sqcap, \cdot, ^-, ^\mathsf{T}, \bot, \top, 1)$ is a Stone relation algebra whose reduct $(S, \sqcup, \sqcap, ^-, \bot, \top)$ is a Boolean algebra.

We reuse the concise characterisations of vectors, atoms, symmetry, injectivity, surjectivity and bijectivity known from relation algebras [44]. In the instance of relations over a set A, a vector represents a subset of A and an atom represents a relation containing a single pair. Hence, in the graph model a vector describes a set of nodes – such as the ones visited in Prim's algorithm – and an atom describes an edge of the graph. Injectivity then means that two nodes cannot have the same successor, which is a property of trees.

Observe that relation algebras and Stone relation algebras have the same signature. The main difference between them is the weakening of the lattice structure from Boolean algebras to Stone algebras. In particular, the property

$$x^\mathsf{T}\overline{xy} \leq \overline{y} \tag{11}$$

holds in Stone relation algebras. Tarski's relation algebras require a Boolean algebra, axioms (1)–(6), and property (11) [34]. Axioms (7)–(10) follow in relation algebras.

Axiom (8) has been called 'Dedekind formula' or 'modular law' [8, 30]. Besides being typed, Dedekind categories require that composition has a left residual and that each Hom-set is a complete distributive lattice [29] and therefore a Heyting algebra, which is more restrictive than a Stone algebra. Rough relation algebras [13] weaken the lattice structure of relation algebras to double Stone algebras, which capture properties of rough sets. Axioms (9) and (10) state that regular elements are closed under composition and its unit.

Many results of relation algebras hold in Stone relation algebras directly or with small modifications. For example, $x \leq xx^\mathsf{T}x$, the complement of a vector is a vector, and composition with an injective element distributes over $\sqcap$ from the right. We also obtain the following variant of the so-called Schröder equivalence:

$$xy \leq \overline{z} \iff x^\mathsf{T}z \leq \overline{y}$$

The following result shows further consequences for Stone relation algebras. In particular, every Stone algebra can be extended to a Stone relation algebra, and the Stone relation algebra structure can be lifted to matrices by using the usual matrix composition (taking $\sqcup$ and $\cdot$ from the underlying Stone relation algebra as addition and multiplication, respectively).

Theorem 2. *1. The regular elements of a Stone relation algebra S form a relation algebra that is a subalgebra of S.*
2. Let $(S, \sqcup, \sqcap, \overline{}, \bot, \top)$ be a Stone algebra. Then $(S, \sqcup, \sqcap, \sqcap, \overline{}, \lambda x.x, \bot, \top, \top)$ is a Stone relation algebra with the identity function as converse.
3. Let $(S, \sqcup, \sqcap, \cdot, \overline{}, \mathsf{T}, \bot, \top, 1)$ be a Stone relation algebra and let A be a finite set. Then $(S^{A \times A}, \sqcup, \sqcap, \cdot, \overline{}, \mathsf{T}, \bot, \top, 1)$ is a Stone relation algebra, where the operations $\cdot$, T and 1 are defined by

$$(M \cdot N)_{i,j} = \bigsqcup\nolimits_{k \in A} M_{i,k} \cdot N_{k,j}$$
$$(M^\mathsf{T})_{i,j} = (M_{j,i})^\mathsf{T}$$
$$1_{i,j} = \begin{cases} 1 & \text{if } i = j \\ \bot & \text{if } i \neq j \end{cases}$$

Hence weighted graphs form a Stone relation algebra as follows: for weights the operations are $x \cdot y = \min\{x, y\}$ and $x^\mathsf{T} = x$ according to Theorem 2.2, and these operations are lifted to matrices as shown in Theorem 2.3. Because in this instance the converse operation of the underlying Stone relation algebra is the identity, the lifted converse operation only transposes the matrix. Thus for a finite set A, the set of matrices $\mathbb{R}'^{A \times A}$ is a Stone relation algebra with the following operations:

$$(M \sqcup N)_{i,j} = \max(M_{i,j}, N_{i,j})$$
$$(M \sqcap N)_{i,j} = \min(M_{i,j}, N_{i,j})$$
$$(M \cdot N)_{i,j} = \max_{k \in A} \min(M_{i,k}, N_{k,j})$$
$$\overline{M}_{i,j} = \overline{M_{i,j}}$$
$$M^{\mathsf{T}}_{i,j} = M_{j,i}$$
$$\bot_{i,j} = \bot$$
$$\top_{i,j} = \top$$
$$1_{i,j} = \begin{cases} \top & \text{if } i = j \\ \bot & \text{if } i \neq j \end{cases}$$

The order in this structure is $M \leq N \Leftrightarrow \forall i, j \in A : M_{i,j} \leq N_{i,j}$.

3 Stone-Kleene Relation Algebras

In this section, we combine Stone relation algebras with Kleene algebras [32] in order to obtain information about reachability in graphs. Kleene algebras are used to model finite iteration for regular languages and relations. In particular, they expand semirings by a unary operation – the Kleene star – which instantiates to the reflexive-transitive closure of relations. The properties of the Kleene star have been studied in [14] and we use the axiomatisation given in [32].

Definition 3. An *idempotent semiring* is an algebraic structure $(S, \sqcup, \cdot, \bot, 1)$ where $(S, \sqcup, \bot)$ is a bounded semilattice and $\cdot$ is associative, distributes over $\sqcup$ and has unit 1 and zero $\bot$:

$$x(y \sqcup z) = xy \sqcup xz \qquad x\bot = \bot \qquad x1 = x \qquad x(yz) = (xy)z$$
$$(x \sqcup y)z = xz \sqcup yz \qquad \bot x = \bot \qquad 1x = x$$

A *Kleene algebra* $(S, \sqcup, \cdot, {}^*, \bot, 1)$ is an idempotent semiring $(S, \sqcup, \cdot, \bot, 1)$ with an operation $*$ satisfying the unfold and induction axioms

$$1 \sqcup yy^* \leq y^* \qquad z \sqcup yx \leq x \Rightarrow y^*z \leq x$$
$$1 \sqcup y^*y \leq y^* \qquad z \sqcup xy \leq x \Rightarrow zy^* \leq x$$

A *Stone-Kleene relation algebra* is a structure $(S, \sqcup, \sqcap, \cdot, {}^-, {}^{\mathsf{T}}, {}^*, \bot, \top, 1)$ such that the reduct $(S, \sqcup, \sqcap, \cdot, {}^-, {}^{\mathsf{T}}, \bot, \top, 1)$ is a Stone relation algebra, the reduct $(S, \sqcup, \cdot, {}^*, \bot, 1)$ is a Kleene algebra and the following equation holds:

$$\overline{\overline{x^*}} = (\overline{\overline{x}})^* \tag{12}$$

An element $x \in S$ is *acyclic* if $xx^* \leq \overline{1}$ and x is a *forest* if x is injective and acyclic. A *Kleene relation algebra* $(S, \sqcup, \sqcap, \cdot, \overline{}, ^\mathsf{T}, ^*, \bot, \top, 1)$ is a Stone-Kleene relation algebra whose reduct $(S, \sqcup, \sqcap, \cdot, \overline{}, ^\mathsf{T}, \bot, \top, 1)$ is a relation algebra.

Axiom (12) states that regular elements are closed under the operation *. Many results of Kleene relation algebras hold in Stone-Kleene relation algebras directly or with small modifications. For example, $(xx^\mathsf{T})^* = 1 \sqcup xx^\mathsf{T}$ for each vector x, the operations converse and Kleene star commute, and

$$x^* x^{\mathsf{T}*} \sqcap x^\mathsf{T} x \leq 1$$

for each forest x. The latter follows using the cancellation property

$$xy \leq 1 \;\Rightarrow\; x^* y^* \leq x^* \sqcup y^*$$

which we have proved in Kleene algebras as part of the present verification work; such properties can also be interpreted in rewrite systems [45]. Proofs of the above properties – and other algebraic results and consequences stated in this paper – can be found in the Isabelle/HOL theory files mentioned in the introduction. The following result shows further consequences for Kleene algebras, Stone-Kleene relation algebras and Kleene relation algebras.

Theorem 3. *1. The regular elements of a Stone-Kleene relation algebra S form a Kleene relation algebra that is a subalgebra of S.*
2. Let $(S, \sqcup, \sqcap, \bot, \top)$ be a bounded distributive lattice. Then $(S, \sqcup, \sqcap, \lambda x.\top, \bot, \top)$ is a Kleene algebra with the constant $\top$ function as the star operation.
3. Let $(S, \sqcup, \sqcap, \overline{}, \bot, \top)$ be a Stone algebra.
   Then $(S, \sqcup, \sqcap, \sqcap, \overline{}, \overline{}, \lambda x.x, \lambda x.\top, \bot, \top, \top)$ is a Stone-Kleene relation algebra.
4. Let $(S, \sqcup, \sqcap, \cdot, \overline{}, ^\mathsf{T}, ^*, \bot, \top, 1)$ be a Stone-Kleene relation algebra and let A be a finite set. Then $(S^{A \times A}, \sqcup, \sqcap, \cdot, \overline{}, ^\mathsf{T}, ^*, \bot, \top, 1)$ is a Stone-Kleene relation algebra, where the operation * is defined recursively using Conway's automata-based construction [14]:

$$\begin{pmatrix} a & b \\ c & d \end{pmatrix}^* = \begin{pmatrix} e^* & a^* b f^* \\ d^* c e^* & f^* \end{pmatrix} \qquad \text{where} \qquad \begin{pmatrix} e \\ f \end{pmatrix} = \begin{pmatrix} a \sqcup b d^* c \\ d \sqcup c a^* b \end{pmatrix}$$

This shows the recursive case, which splits a matrix into smaller matrices. At termination, the Kleene star is applied to the entry of a one-element matrix.

In particular, this provides a formally verified proof of Conway's construction for the Kleene star of matrices, which is missing in existing Isabelle/HOL theories of Kleene algebras [3, Sect. 5.7].

As a consequence, weighted graphs form a Stone-Kleene relation algebra as follows: for weights the max-min lattice is extended with the Kleene star operation $x^* = \top$ according to Theorem 3.3, and the Kleene star is defined for matrices by Conway's construction shown in Theorem 3.4.

4   An Algebra for Minimising Weights

In this section we extend Stone-Kleene relation algebras by dedicated operations for the minimum spanning tree application. First, the algorithm needs to select an edge with minimal weight; this is done by the operation m. Second, the sum of edge weights needs to be minimised according to the specification; the sum is obtained by the operation s. Third, the axioms of s use the operation $+$ to add the weights of corresponding edges of two graphs. These operations are captured in the following algebraic structure.

Definition 4. An *M-algebra* $(S, \sqcup, \sqcap, \cdot, +, {}^-, {}^\mathsf{T}, *, s, m, \bot, \top, 1)$ is a Stone-Kleene relation algebra $(S, \sqcup, \sqcap, \cdot, {}^-, {}^\mathsf{T}, *, \bot, \top, 1)$ with an addition $+$, a summation s and a minimum selection m satisfying the following properties:

$$\overline{\overline{x}} = \overline{\overline{y}} \ \wedge \ x \le y \Rightarrow z + x \le z + y \tag{13}$$

$$x + s(\bot) = x \tag{14}$$

$$s(x) + s(y) = s(x \sqcup y) + s(x \sqcap y) \tag{15}$$

$$s(x^\mathsf{T}) = s(x) \tag{16}$$

$$x \ne \bot \Rightarrow s(y) \le \overline{\overline{s(x)}} \tag{17}$$

$$m(x) \le \overline{\overline{x}} \tag{18}$$

$$\overline{\overline{m(x)}} = m(x) \tag{19}$$

$$x \ne \bot \Rightarrow m(x) \text{ is an atom} \tag{20}$$

$$y \text{ is an atom} \ \wedge \ \overline{\overline{y}} = y \ \wedge \ y \sqcap x \ne \bot \Rightarrow s(m(x) \sqcap x) \le s(y \sqcap x) \tag{21}$$

Among the new operations, only m is used in the algorithm. The axioms have the following meaning:

(13) The operation $+$ is $\le$-isotone in its second argument as long as no new edges are introduced (this is required because edges may have negative weights).

(14) The empty graph adds no weight; the given axiom is weaker than the conjunction of $s(\bot) = \bot$ and $x + \bot = x$.

(15) This generalises the inclusion-exclusion principle to sets of numbers.

(16) Reversing edges does not change the sum of weights.

(17) The result of s is represented by a graph with one fixed edge.

(18) The minimal edge is contained in the graph.

(19) The result of m is represented as a relation.

(20) The result of m is just one edge, if the graph is not empty.

(21) Any edge y in the graph x weighs at least as much as $m(x)$; the operation s is used to compare the weights of edges between different nodes.

A precise definition of the operations $+$, s and m on weighted graphs is given in the following result, which shows that weighted graphs form an M-algebra.

Theorem 4. *Let A be a finite set. Let $\prec$ be a strict total order on A with least element h. Then the set of matrices $\mathbb{R}'^{A \times A}$ is an M-algebra with the following operations:*

$$(M + N)_{i,j} = M_{i,j} + N_{i,j} \tag{22}$$

$$s(M)_{i,j} = \begin{cases} \sum_{k,l \in A} M_{k,l} \ if \ i = j = h \\ \bot \qquad\qquad if \ i \neq h \vee j \neq h \end{cases} \tag{23}$$

$$m(M)_{i,j} = \begin{cases} \top \ if \ M_{i,j} \neq \bot \ \wedge \\ \quad \forall k, l \in A : (M_{k,l} \neq \bot \Rightarrow M_{i,j} \leq M_{k,l}) \wedge \\ \quad\qquad ((k \prec i \vee (k = i \wedge l \prec j)) \Rightarrow M_{i,j} \neq M_{k,l}) \\ \bot \ otherwise \end{cases} \tag{24}$$

The addition $+$ on $\mathbb{R}'$ used in (22) is defined by

$x + y =$	$y = \bot$	$y \in \mathbb{R}$	$y = \top$
$x = \bot$	$\bot$	y	$\top$
$x \in \mathbb{R}$	x	$x +_{\mathbb{R}} y$	$\top$
$x = \top$	$\top$	$\top$	$\top$

The finite summation $\sum$ on $\mathbb{R}'$ used in (23) is defined recursively using this binary addition, which is associative and commutative.

Equation (24) means that $m(M)_{i,j} = \top$ if (i, j) is the smallest pair (according to the lexicographic order based on $\prec$) such that $M_{i,j}$ is minimal among the weights different from $\bot$. The function s uses the entry in row h and column h to store the sum of the weights different from $\bot$.

5   Correctness of the Minimum Spanning Tree Algorithm

In this section we present a minimum spanning tree algorithm and prove its correctness. In particular, we show how the algebras introduced in the previous sections are used to reason about graph properties. The algorithm is shown in Fig. 2. It is a while-program with variables whose values range over an M-algebra S.

The input of the algorithm is a weighted graph $g \in S$ and a root node $r \in S$. The algorithm constructs a minimum spanning tree $t \in S$ and maintains a set of visited nodes v. Both r and v are represented as vectors. The algorithm starts with an empty tree t and the single visited node r. The expression $v\overline{v}^\mathsf{T} \sqcap g$ restricts g to the edges starting in v and ending outside of v. In each iteration an edge e is chosen with minimal weight among these edges; then e is added to t and the end node of e is added to v. When there are no edges from v to its complement set, the while-loop finishes and the output of the algorithm is t.

We show correctness of the algorithm relative to two assumptions:

1. The while-loop terminates. This follows since a new edge is added to the spanning tree in each iteration and the graph is finite. Such termination proofs can also be done algebraically [23], but this is not part of the present paper.

```
input g, r
t ← ⊥
v ← r
while vv̄ᵀ ⊓ g ≠ ⊥ do
    e ← m(vv̄ᵀ ⊓ g)
    t ← t ⊔ e
    v ← v ⊔ eᵀ⊤
end
output t
```

Fig. 2. A relational minimum spanning tree algorithm

2. There exists a minimum spanning tree. This follows since the number of spanning trees of a finite graph is finite. A proof of this is not part of the present paper, but could be based on cardinalities of relations [28].

We do not assume that the graph g is connected. As a consequence, the above algorithm will produce a minimum spanning tree of the component of g that contains r. In M-algebras, the nodes in this component are given by

$$c(g,r) = r^\mathsf{T}\overline{\overline{g}}{}^*$$

which is the converse of a vector that represents the set of nodes reachable from r in the graph g ignoring edge weights. It follows that for connected g the result is a minimum spanning tree of the whole graph. The correctness proof uses the following predicates.

Definition 5. Let S be an M-algebra and let $g, r, t, v \in S$. Then t is a *spanning tree* of g with root r if t is a forest, t is regular and

$$t \le c(g,r)^\mathsf{T} c(g,r) \sqcap \overline{\overline{g}} \tag{25}$$

$$c(g,r) \le r^\mathsf{T} t^* \tag{26}$$

Such a t is a *minimum spanning tree* of g with root r if, additionally,

$$s(t \sqcap g) \le s(u \sqcap g)$$

for each spanning tree u of g with root r. Next, the *precondition* requires that g is symmetric, that r is regular, injective and a vector, and that a minimum spanning tree of g with root r exists. Next, the *loop invariant* requires the precondition and $v^\mathsf{T} = r^\mathsf{T} t^*$ and that t is a spanning tree of $vv^\mathsf{T} \sqcap g$ with root r, and that $t \le w$ for some minimum spanning tree w of g with root r. Finally, the *postcondition* requires that t is a minimum spanning tree of g with root r.

By lattice properties and since $c(g,r)$ is the converse of a vector, inequality (25) is equivalent to the conjunction of $t \le \overline{\overline{g}}$ and $t \le c(g,r)^\mathsf{T}$ and $t \le c(g,r)$. The first of these inequalities states that all edges of t are contained in g (ignoring the weights). The second inequality states that each edge of t starts in a node in the component of g that contains r. The third inequality expresses the same for the end nodes of the edges of t.

Also inequality (26) is concerned with the component of g that contains r. It states that all nodes in this component are reachable from r using edges in t. Observe that $r^\mathsf{T} t^* = c(t,r)$ since t is regular, so together with $t \le \overline{\overline{g}}$ we obtain $c(g,r) = c(t,r)$ as a consequence.

Symmetry of g specifies that the graph is undirected. The properties of r in the precondition state that r represents a single node. Assumption 2 amounts to the existence of a minimum spanning tree in the precondition.

The verification conditions to establish the postcondition are automatically generated from the precondition and the loop invariant using Hoare logic. We use an implementation of Hoare logic that comes with Isabelle/HOL; see [36,37]. The generated conditions are predicates whose variables range over an M-algebra; all calculations take place in this algebra or its reducts. The high-level structure of the proof is standard; the difference here is that the whole argument is carried out in new algebraic structures that directly model weighted graphs.

Theorem 5. *Assume the precondition stated in Definition 5 holds. Then the postcondition stated there holds after the algorithm in Fig. 2 finishes.*

In the following we discuss several parts of the proof, which are carried out in different algebraic structures. Our aim is not completeness, but to show that many results used in the proof actually hold in more general settings. We focus on the preservation of the loop invariant for the current tree t and the current set of visited nodes v. Let $t' = t \sqcup e$ and $v' = v \sqcup e^\mathsf{T} \top$ be the values of these variables at the end of the body of the while-loop.

First, the proof involves showing that t' is a spanning tree of $v'v'^\mathsf{T} \sqcap g$ with root r, that is, of the subgraph of g restricted to nodes in v'. In particular, this requires that t' is injective. To this end, we use the following property given in [39] that also holds in Stone relation algebras.

Lemma 1. *Let S be a Stone relation algebra. Let $t, e \in S$ such that t and e are injective and $et^\mathsf{T} \le 1$. Then $t \sqcup e$ is injective.*

The assumptions of Lemma 1 are established as follows:

- Injectivity of t follows from the invariant.
- e is an atom by axiom (20), so $e\top$ is injective, whence e is injective.
- $et^\mathsf{T} = \bot \le 1$ follows by another general result of Stone relation algebras from $e \le v\overline{v}^\mathsf{T}$ and $t \le vv^\mathsf{T}$ and that v is a vector.

We also require that t' is contained in the subgraph of g restricted to the nodes in v'. For this we use the following result of Stone relation algebras.

Lemma 2. *Let S be a Stone relation algebra. Let $t, e, v, g \in S$ such that $t \le vv^\mathsf{T} \sqcap g$ and $e \le v\overline{v}^\mathsf{T} \sqcap g$. Then $t' \le v'v'^\mathsf{T} \sqcap g$ where $t' = t \sqcup e$ and $v' = v \sqcup e^\mathsf{T} \top$.*

Next, we also require that t' is acyclic. To show this, we use the following result of Stone-Kleene relation algebras.

Lemma 3. *Let S be a Stone-Kleene relation algebra. Let $t, e, v \in S$ such that t is acyclic, v is a vector and $e \le v\overline{v}^\mathsf{T}$ and $t \le vv^\mathsf{T}$. Then $t \sqcup e$ is acyclic.*

Note that this lemma does not require that t is a tree or that e contains just one edge. It is a much more general statement that can be used in reasoning about graphs in other contexts than the minimum spanning tree algorithm – in fact, it holds not only for weighted graphs but for any other instance of Stone-Kleene relation algebras. The same observation applies to the previous lemmas and many others used in the correctness proof.

Next, the invariant maintains that v is the set of nodes reachable from r in t, which is formulated as $v^\mathsf{T} = r^\mathsf{T} t^*$. To preserve this property, we use the following result of Stone-Kleene relation algebras.

Lemma 4. *Let S be a Stone-Kleene relation algebra. Let $t, e, r, v \in S$ such that v is a vector, $e \le v\overline{v}^\mathsf{T}$ and $et = \bot$ and $v^\mathsf{T} = r^\mathsf{T} t^*$. Then $v'^\mathsf{T} = r^\mathsf{T} t'^*$ where $t' = t \sqcup e$ and $v' = v \sqcup e^\mathsf{T} \mathsf{T}$.*

The assumption $et = \bot$ follows similarly to $et^\mathsf{T} = \bot$ for Lemma 1.

Finally, we discuss how to preserve the property that the currently constructed spanning tree t can be extended to a minimum spanning tree. The situation is shown in Fig. 3. Assuming that there is a minimum spanning tree w of g such that $t \le w$, we have to show that there is a minimum spanning tree w' of g such that $t' = t \sqcup e \le w'$ where $e = m(v\overline{v}^\mathsf{T} \sqcap g)$ is an edge of g with minimal weight going from a node in v to a node not in v. We do this by explicitly constructing the new minimum spanning tree w'. To this end, we need to find the edge f in w that crosses the cut from v to $\overline{v}$, and replace it with the edge e – this does not increase the weight due to minimality of e. An algebraic expression for the edge f is

$$f = w \sqcap v\overline{v}^\mathsf{T} \sqcap \mathsf{T} ew^{\mathsf{T}*}$$

The three terms on the right hand side enforce that f is in w, that f starts in v and ends in $\overline{v}$, and that there is a path in w from the end node of f to the end node of e. It can be shown algebraically that f is an atom, that is, that f represents the unique edge satisfying these conditions. An algebraic expression for the path p from the end of f to the end of e is

$$p = w \sqcap \overline{v}\overline{v}^\mathsf{T} \sqcap \mathsf{T} ew^{\mathsf{T}*}$$

The three terms on the right hand side enforce that the edges in p are in w, that they start and end in $\overline{v}$, and that there is a path in w from each of their end nodes to the end node of e. The required tree w' is then obtained by removing the edge f from w, turning around the path p, and inserting the edge e. An algebraic expression for w' is

$$w' = (w \sqcap \overline{f} \sqcup p) \sqcup p^\mathsf{T} \sqcup e$$

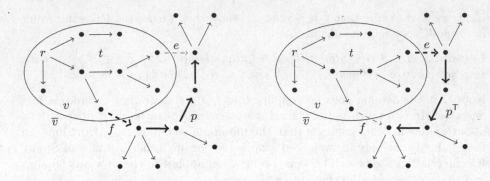

Fig. 3. Replacing the edge f in w (left) with the minimal edge e in w' (right) where t is the tree in the oval and v is the set of nodes in t

We then show that w' so defined is a minimum spanning tree of g with root r and that $t \sqcup e \leq w'$. In the following we focus on the part of this proof that shows $s(w' \sqcap g) \leq s(u \sqcap g)$ for each spanning tree u of g with root r. This follows by the calculation

$$s(w' \sqcap g) = s(w \sqcap \overline{f \sqcup p} \sqcap g) + s(p^\mathsf{T} \sqcap g) + s(e \sqcap g) \tag{27}$$

$$\leq s(w \sqcap \overline{f \sqcup p} \sqcap g) + s(p \sqcap g) + s(f \sqcap g) \tag{28}$$

$$= s(((w \sqcap \overline{f \sqcup p}) \sqcup p \sqcup f) \sqcap g) \tag{29}$$

$$= s(w \sqcap g) \tag{30}$$

$$\leq s(u \sqcap g) \tag{31}$$

Equation (27) holds by axioms (14) and (15) since $w \sqcap \overline{f \sqcup p}$ and p^T and e are pairwise disjoint (that is, their pairwise meet given by $\sqcap$ is $\bot$). A similar argument justifies Eq. (29). Axiom (21) is used to show $s(e \sqcap g) \leq s(f \sqcap g)$ in inequality (28). Axiom (16) is used to show that replacing p with p^T does not change the weight there. Equation (30) follows by a simple calculation, most of which takes place in Stone algebras. Finally, Eq. (31) holds since w is a minimum spanning tree of g with root r.

This is the main part of the overall proof where the operations and axioms of M-algebras are used. Most of the proof, however, can already be carried out in Stone-Kleene relation algebras or weaker structures as discussed above. We expect such results to be useful for reasoning about other graph algorithms.

6   Related Work

In this section we compare the present paper with related work on algorithms for minimum spanning trees. Often the correctness of such algorithms is argued informally with varying amounts of mathematical rigour and details; for example, see [15,31]. Our results are fully verified in Isabelle/HOL [38] based on formal definitions and models.

A formal derivation of Prim's minimum spanning tree algorithm in the B event-based framework using Atelier B is presented in [1]. The paper also discusses the role of refinement in this process, which is not part of the present paper. The B specification is based on sets and relations, uses an inductive definition of trees, and represents weights by functions, whence objects of several different sorts are involved.

Our formalisation is based on Stone-Kleene relation algebras, which generalise relation algebras and Kleene algebras, and can be instantiated directly by weight matrices. The generalisation is crucial as weight matrices do not support a Boolean complement; accordingly we do not use implementations of relation algebras such as [2,24]. Nevertheless we can build on well-developed relational concepts and methods for our new algebras – such as algebraic properties of trees – which are useful also in other contexts.

We mostly apply equational reasoning based on a single-sorted algebra. This is well supported by automated theorem provers and SMT solvers such as those integrated in Isabelle/HOL via the Sledgehammer tool [11,42] that we heavily use in the verification. Typically the tool can automatically find proofs of steps at a granularity comparable to manual equational reasoning found in papers. Automation works less well in some cases, for example, chains of inequalities, applications of isotone operations, and steps that introduce intermediate terms that occur on neither side of an equation. On the other hand, in some cases the tool can automatically prove a result that would take several manual steps.

A distributed algorithm for computing minimum spanning trees is verified using the theorem prover Nqthm in [25]. The specification is again based on sets and a weight function. The main focus of the paper is on the distributed aspects of the algorithm, which uses asynchronous messages and differs essentially from Prim's minimum spanning tree algorithm. The distributed algorithm is the topic of a number of other papers using a variety of formalisms including Petri nets and modal logic.

Relation algebras are used to derive spanning tree algorithms in [6]. The given proof is created manually and not verified using a theorem prover. It uses relations and, in absence of weighted matrices, an incidence matrix representation and a weight function in a setting with several different sorts.

Constraint-based semirings are used to formulate minimum spanning tree algorithms in [10]. These semirings abstract from the edge weights and represent graphs by sets of edges. The semiring structure is not lifted to the graph level, whereas we lift Stone algebras to Stone relation algebras – and similarly for Kleene algebras – and can therefore exploit the algebraic structure of graphs. Detailed proofs are not presented and there is no formal verification of results. The paper is mainly concerned with extending the algorithms to partially ordered edge weights, which is not part of the present paper.

Semirings with a pre-order, so-called dioids, are used to formulate various shortest-path problems in [20]. The corresponding algorithms are generalisations of methods for solving linear equations over these structures. Other approaches to path problems are based on Kleene algebras; for example, see [26], which also discusses many previous works in this tradition. Semirings and Kleene algebras are suitable for path problems as they capture the essential operations of choosing between alternatives, composing edges and building paths. It is not clear how to model the minimum spanning tree problem using Kleene algebras only.

Relational methods based on allegories are used for algorithm development in [8], but there relations mostly represent computations rather than the involved data. An extension to quantitative analysis is discussed in [40].

7   Conclusion

The generalisation of Boolean algebras to Stone algebras gives a promising way to extend correctness reasoning from unweighted to weighted graphs. When applied to relation algebras, many results continue to hold with no changes or small changes. In combination with Kleene algebras, we could carry out most of the correctness proof of Prim's minimum spanning tree algorithm.

A small part of the proof needed some additional operations; we captured a few key properties in the present paper, but the underlying structure should be studied further. To this end, we will look at variants of the minimum spanning tree algorithm and other graph algorithms. We will also consider the integration of termination proofs, complexity reasoning and combinatorial arguments using cardinalities of relations.

Using algebras for proving the correctness of programs is well supported by Isabelle/HOL. We have benefited from the existing verification condition generator for Hoare logic, from the structuring mechanisms that allow the development of hierarchies of algebras and their models, and heavily from the integrated automated theorem provers, SMT solvers and counterexample generators.

Acknowledgements. I thank the anonymous referees for helpful feedback including the suggestion to generalise Lemma 1 to its present form. I thank Rudolf Berghammer for discussions about the cardinality of relations and ways to generalise it. I thank Peter Höfner and Bernhard Möller for discussing alternative approaches to minimum spanning trees in Kleene algebras. I thank the participants of the 73rd meeting of IFIP WG 2.1, the 14th Logic and Computation Seminar of Kyushu University and the 2016 Workshop on Universal Structures in Mathematics and Computing for the opportunity to talk about this work and for their valuable feedback. The presentation at Kyushu University was part of a JSPS Invitation Fellowship for Research in Japan.

References

1. Abrial, J.-R., Cansell, D., Méry, D.: Formal derivation of spanning trees algorithms. In: Bert, D., Bowen, J.P., King, S., Waldén, M. (eds.) ZB 2003. LNCS, vol. 2651, pp. 457–476. Springer, Heidelberg (2003). doi:10.1007/3-540-44880-2_27
2. Armstrong, A., Foster, S., Struth, G., Weber, T.: Relation algebra. Archive of Formal Proofs (2016). First version (2014)
3. Armstrong, A., Gomes, V.B.F., Struth, G., Weber, T.: Kleene algebra. Archive of Formal Proofs (2016). First version (2013)

4. Berghammer, R., Fischer, S.: Combining relation algebra and data refinement to develop rectangle-based functional programs for reflexive-transitive closures. J. Log. Algebr. Methods Program. **84**(3), 341–358 (2015)
5. Berghammer, R., von Karger, B.: Relational semantics of functional programs. In: Brink, C., Kahl, W., Schmidt, G. (eds.) Relational Methods in Computer Science, chap. 8, pp. 115–130. Springer, Wien (1997)
6. Berghammer, R., von Karger, B., Wolf, A.: Relation-algebraic derivation of spanning tree algorithms. In: Jeuring, J. (ed.) MPC 1998. LNCS, vol. 1422, pp. 23–43. Springer, Heidelberg (1998). doi:10.1007/BFb0054283
7. Berghammer, R., Rusinowska, A., de Swart, H.: Computing tournament solutions using relation algebra and RelView. Eur. J. Oper. Res. **226**(3), 636–645 (2013)
8. Bird, R., de Moor, O.: Algebra of Programming. Prentice Hall, Englewood Cliffs (1997)
9. Birkhoff, G.: Lattice Theory. Colloquium Publications, vol. XXV, 3rd edn. American Mathematical Society, Providence (1967)
10. Bistarelli, S., Santini, F.: C-semiring frameworks for minimum spanning tree problems. In: Corradini, A., Montanari, U. (eds.) WADT 2008. LNCS, vol. 5486, pp. 56–70. Springer, Heidelberg (2009). doi:10.1007/978-3-642-03429-9_5
11. Blanchette, J.C., Böhme, S., Paulson, L.C.: Extending Sledgehammer with SMT solvers. In: Bjørner, N., Sofronie-Stokkermans, V. (eds.) CADE 2011. LNCS (LNAI), vol. 6803, pp. 116–130. Springer, Heidelberg (2011). doi:10.1007/978-3-642-22438-6_11
12. Blyth, T.S.: Lattices and Ordered Algebraic Structures. Springer, Berlin (2005)
13. Comer, S.D.: On connections between information systems, rough sets and algebraic logic. In: Rauszer, C. (ed.) Algebraic Methods in Logic and in Computer Science. Banach Center Publications, vol. 28, pp. 117–124. Institute of Mathematics, Polish Academy of Sciences (1993)
14. Conway, J.H.: Regular Algebra and Finite Machines. Chapman and Hall, London (1971)
15. Cormen, T.H., Leiserson, C.E., Rivest, R.L.: Introduction to Algorithms. MIT Press, Cambridge (1990)
16. Desharnais, J., Grinenko, A., Möller, B.: Relational style laws and constructs of linear algebra. J. Log. Algebr. Methods Program. **83**(2), 154–168 (2014)
17. Dijkstra, E.W.: A note on two problems in connexion with graphs. Numerische Mathematik **1**(1), 269–271 (1959)
18. Freyd, P.J., Ščedrov, A.: Categories, Allegories. North-Holland Mathematical Library, vol. 39. Elsevier Science Publishers (1990)
19. Goguen, J.A.: L-fuzzy sets. J. Math. Anal. Appl. **18**(1), 145–174 (1967)
20. Gondran, M., Minoux, M.: Graphs, Dioids and Semirings. Springer, Heidelberg (2008)
21. Graham, R.L., Hell, P.: On the history of the minimum spanning tree problem. Ann. Hist. Comput. **7**(1), 43–57 (1985)
22. Grätzer, G.: Lattice Theory: First Concepts and Distributive Lattices. W. H. Freeman and Co., San Francisco (1971)
23. Guttmann, W.: Algebras for correctness of sequential computations. Sci. Comput. Program. **85**(Part B), 224–240 (2014)
24. Guttmann, W., Struth, G., Weber, T.: A repository for Tarski-Kleene algebras. In: Höfner, P., McIver, A., Struth, G. (eds.) Automated Theory Engineering. CEUR Workshop Proceedings, vol. 760, pp. 30–39 (2011)
25. Hesselink, W.H.: The verified incremental design of a distributed spanning tree algorithm: extended abstract. Formal Aspects Comput. **11**(1), 45–55 (1999)

26. Höfner, P., Möller, B.: Dijkstra, Floyd and Warshall meet Kleene. Formal Aspects Comput. **24**(4), 459–476 (2012)
27. Jarník, V.: O jistém problému minimálním (Z dopisu panu O. Borůvkovi). Práce moravské přírodovědecké společnosti **6**(4), 57–63 (1930)
28. Kawahara, Y.: On the cardinality of relations. In: Schmidt, R.A. (ed.) RelMiCS/AKA 2006. LNCS, vol. 4136, pp. 251–265. Springer, Heidelberg (2006). doi:10.1007/11828563_17
29. Kawahara, Y., Furusawa, H.: Crispness in Dedekind categories. Bull. Inf. Cybern. **33**(1–2), 1–18 (2001)
30. Kawahara, Y., Furusawa, H., Mori, M.: Categorical representation theorems of fuzzy relations. Inf. Sci. **119**(3–4), 235–251 (1999)
31. Knuth, D.E.: The Art of Computer Programming. Fundamental Algorithms, vol. 1, 3rd edn. Addison-Wesley, Reading (1997)
32. Kozen, D.: A completeness theorem for Kleene algebras and the algebra of regular events. Inf. Comput. **110**(2), 366–390 (1994)
33. Macedo, H.D., Oliveira, J.N.: A linear algebra approach to OLAP. Formal Aspects Comput. **27**(2), 283–307 (2015)
34. Maddux, R.D.: Relation-algebraic semantics. Theor. Comput. Sci. **160**(1–2), 1–85 (1996)
35. Mareš, M.: The saga of minimum spanning trees. Comput. Sci. Rev. **2**(3), 165–221 (2008)
36. Nipkow, T.: Winskel is (almost) right: Towards a mechanized semantics textbook. Formal Aspects Comput. **10**(2), 171–186 (1998)
37. Nipkow, T.: Hoare logics in Isabelle/HOL. In: Schwichtenberg, H., Steinbrüggen, R. (eds.) Proof and System-Reliability, pp. 341–367. Kluwer Academic Publishers (2002)
38. Nipkow, T., Paulson, L.C., Wenzel, M.: Isabelle/HOL: A Proof Assistant for Higher-Order Logic. LNCS, vol. 2283. Springer, Heidelberg (2002)
39. Oliveira, J.N.: Extended static checking by calculation using the pointfree transform. In: Bove, A., Barbosa, L.S., Pardo, A., Pinto, J.S. (eds.) LerNet 2008. LNCS, vol. 5520, pp. 195–251. Springer, Heidelberg (2009). doi:10.1007/978-3-642-03153-3_5
40. Oliveira, J.N.: Towards a linear algebra of programming. Formal Aspects Comput. **24**(4), 433–458 (2012)
41. Oliveira, J.N.: Weighted automata as coalgebras in categories of matrices. Int. J. Found. Comput. Sci. **24**(6), 709–728 (2013)
42. Paulson, L.C., Blanchette, J.C.: Three years of experience with Sledgehammer, a practical link between automatic and interactive theorem provers. In: Sutcliffe, G., Ternovska, E., Schulz, S. (eds.) Proceedings of the 8th International Workshop on the Implementation of Logics, pp. 3–13 (2010)
43. Prim, R.C.: Shortest connection networks and some generalizations. Bell Syst. Tech. J. **36**(6), 1389–1401 (1957)
44. Schmidt, G., Ströhlein, T.: Relations and Graphs. Springer, Berlin (1993)
45. Struth, G.: Abstract abstract reduction. J. Log. Algebr. Program. **66**(2), 239–270 (2006)
46. Tarski, A.: On the calculus of relations. J. Symbol. Log. **6**(3), 73–89 (1941)
47. Winter, M.: A new algebraic approach to L-fuzzy relations convenient to study crispness. Inf. Sci. **139**(3–4), 233–252 (2001)

Certified Impossibility Results and Analyses in *Coq* of Some Randomised Distributed Algorithms

Allyx Fontaine[1](✉) and Akka Zemmari[2]

[1] Université de la Guyane, UMR ESPACE-DEV, Cayenne, France
allyx.fontaine@gmail.com
[2] Université de Bordeaux, LaBRI UMR CNRS 5800, Bordeaux, France
akka.zemmari@labri.fr

Abstract. Randomised algorithms are generally simple to formulate. However, their analysis can become very complex, especially in the field of distributed computing. In this paper, we formally model in *Coq* a class of randomised distributed algorithms. We develop some tools to help proving impossibility results about classical problems and analysing this class of algorithms. As case studies, we examine the handshake and maximal matching problems. We show how to use our tools to formally prove properties about algorithms solving those problems.

1 Introduction

Randomised distributed problems were studied intensively in the past few decades. Generally, randomised distributed algorithms are defined in a concise way. However, their analysis remains delicate and complex, which makes their proof difficult. Model checkers give an automatic way to check whether the results of the algorithms verify a certain specification, however they proceed exhaustively, leading to an explosion of space complexity. An alternative is to use proof assistants. They assist the user to prove properties and certify the proof at its end. The proof assistant *Coq* [Tea] is powerful to model and prove properties or impossibility results thanks to its higher order logic.

1.1 The Theoretical Model

There exists various models for distributed systems depending on the features we allow: message passing model, shared memory model, mobile robots model, etc. Here, we restrict our study to the *standard message passing model*. It consists of a point-to-point communication network described by a connected graph $G = (V, E)$, where the vertices V represent network processes and the edges E represent bidirectional communication channels. Processes communicate by message passing: a process sends a message to another by depositing the message in the corresponding channel.

© Springer International Publishing AG 2016
A. Sampaio and F. Wang (Eds.): ICTAC 2016, LNCS 9965, pp. 69–81, 2016.
DOI: 10.1007/978-3-319-46750-4_5

We assume the system is fully *synchronous*, namely, all processes start at the same time and time proceeds in synchronised rounds. A round of each process is composed of the following three steps. Firstly, it sends messages to neighbours; secondly, it receives messages from neighbours; thirdly, it performs some local computations. Note that we consider only reliable systems: no fault can occur on processes or communication links. This hypothesis is strong but it allows to analyse complexities that give a lower bound for systems based on weaker assumptions (and therefore more realistic).

The network G is *anonymous*: unique identities are not available to distinguish the processes. We do not assume any global knowledge of the network, not even its size or an upper bound on its size. The processes do not require any position or distance information. The anonymity hypothesis is often seen for privacy reasons. In addition, each process can be integrated in a large-scale network making it difficult or impossible to guarantee the uniqueness of identifiers.

Each process knows from which channel it receives or to which it sends a message, thus one supposes that the network is represented by a connected graph with a port numbering function defined as follows (where $\mathcal{N}_G(u)$ denotes the set of vertices of G adjacent to u): given a graph $G = (V, E)$, a *port numbering* function ϕ is a set of local functions $\{\phi_u \mid u \in V\}$ such that for each vertex $u \in V$, ϕ_u is a bijection between $\mathcal{N}_G(u)$ and the set of natural numbers between 1 and $|\mathcal{N}_G(u)|$.

A *probabilistic algorithm* makes some random choices based on some given probability distributions. A *distributed probabilistic algorithm* is a collection of local probabilistic algorithms. Since the network is anonymous, their local probabilistic algorithms are identical. We assume that choices of vertices are independent. A *Las Vegas algorithm* is a probabilistic algorithm which terminates with a positive probability (in general 1) and always produces a correct result.

1.2   Our Contribution

We provide a library to prove properties on (randomised) distributed algorithms. We use the *Coq* proof assistant, library *Alea* [APM09] and plugin *ssreflect* [GMT08]. We first define, in Sect. 2, the algorithm class of anonymous distributed algorithms according to the model previously described. We explain why our definitions are valid. Our main contribution is the tools we developed to enable the user to analyse anonymous distributed algorithms described in Sect. 3.

Section 4 illustrates how to use those tools by analysing solutions for two case studies: the handshake and maximal matching problems. A communication takes place only if the participant processors are waiting for the communication: this is termed handshake. A solution of the handshake problem gives a matching of the graph. A matching is a subset M of E such that no two edges of M have a common vertex. A matching M is said to be maximal if any edge of G is in M or has an extremity linked to an edge in M. First, we show that randomisation may be required to solve distributed problems in particular the handshake problem. Hence, we formally prove an impossibility result which is: "there is no deterministic algorithm in this class that solves the handshake problem".

This also proves that there is no deterministic algorithm that solves the (maximal) matching problem. Then we implement a solution of the handshake problem and we prove that this solution is correct. We analyse the handshake and the maximal matching problems by proving some probabilistic properties.

We believe this is the first work about formal proof of anonymous synchronous randomised distributed message passing algorithms. The examples, used as case studies, are certified. Lemmas and theorems, presented in frame in our paper, are denoted by their name in the *Coq* development available at [FZ].

1.3   Related Works

Distributed Algorithms. Several models are used to represent distributed algorithms. Models such as fault-tolerant [KNR12], population protocols [DM09], mobile robots [CRTU15] are also studied and certified with proof assistants.

C.T. Chou [Cho95] uses the *HOL* proof assistant to show the correctness of distributed algorithms, modelled by labelled transition systems. Specifications are expressed in terms of temporal logic, and their proofs of correctness use simulation proofs with the help of joint invariants.

The approach of D. Méry et al. [MMT11] is directly related to the design of *correct-by-construction* programs. From the formal specifications of the distributed algorithms, they use the refinement for controlling their correctness with *Event-B*.

The development Loco library [CF11] by P. Castéran and V. Filou consists of a set of libraries on labelled graphs and graph relabelling systems. It allows the user to specify tasks, and to prove the correctness of relabelling systems with reference to these tasks and also impossibility results.

Randomised Distributed Algorithms. In our model, we consider that the algorithm operates in rounds, applying a local algorithm to each vertex. This removes the non-determinism due to the latency. But several approaches take into account the dual paradigm of randomised distributed systems: probabilistic aspect and non-determinism due to the response time that changes from one processor to another. They require models with non-deterministic choice between several probability distributions. These choices can be made by a scheduler or an opponent. Equivalent models are following this idea: probabilistic automata [PS95], Markov decision processes [Der70]. To specify properties of randomised distributed algorithms, one can use the temporal logic with probabilistic operators and a threshold.

The *Model Checking* is a tool used to ensure system correction. However, used with probabilities, it leads to an explosion of space complexity. There are methods for reducing the explosion. A qualitative analysis of randomised distributed algorithms is feasible thanks to the model checker PRISM [KNP02]. M. Kwiatkowska et al. [KNS01] use *PRISM* model checker and *Cadence* proof assistant, to obtain automated proofs. Consensus protocol is proved for its non-probabilistic part with *Cadence* and for its probabilistic part with *PRISM*.

J. Hurd et al. [HMM05] formalise in higher order logic the language $pGCL$ used to reason on probabilistic choices or choices made by an adversary. They prove the mutual exclusion algorithm of Rabin: consider N processes, sometimes some of them need to access a critical zone; the algorithm consists in electing one of them. However they do not model the processor concurrently but use an interpretation consisting in reducing the number of processes to 1.

1.4   Preliminaries

Different evaluations of the same probabilistic expression lead to different values. To reason about such expression in a functional language, a solution consists in studying the distribution of this expression rather than its result. In *Alea*, a probabilistic expression $(e : \tau)$ is interpreted as a distribution whose type is $(\tau \to [0,1]) \to [0,1]$. This monadic type is denoted distr τ. We will use the notation $(\mu\ e)$ to represent the associated measure of expression e. Let Q be a property and let $\mathbb{1}_Q$ be its characteristic function. The probability that the result of the expression e satisfies Q is represented by $(\mu\ e)\ \mathbb{1}_Q$.

To construct monadic expressions, *Alea* provides the following functions. Munit a returns the Dirac distribution at point a. Mlet x = d1 in d2 evaluates d1, links the result to x and then evaluate d2 where d1 and d2 are random expressions (not necessarily of the same type). Random n returns a number between 0 and the natural number n with a uniform probability $1/(n + 1)$.

Most of proofs presented in this paper are based on both transformations [APM09]:

Lemma Munit_simpl: $\forall$ (P: τ) (f: τ $\to$ [0,1]), $(\mu$ (Munit P)) f = (f P).
Lemma Mlet_simpl: $\forall$ (P: distr τ) (Q: $\tau \to$ distr τ') (f: $\tau' \to$ [0,1]),

$(\mu\ ($ Mlet x = P in (Q x))) f = $(\mu$ P) (fun x $\Rightarrow$ $(\mu$ (Q x)) f).

2   Our Formal Model

We provide semantics and functions to express anonymous randomised distributed algorithms. They can be used by the user to define his/her own algorithms in the distributed systems we focus on. Once the algorithms are defined, the user can do tests by evaluating them, prove correctness and analyse them.

2.1   Formal Distributed Systems

Synchronous anonymous message passing model can be represented by a connected graph $G = (V, E)$ with a port numbering function ϕ. To encode the graph, we use an adjacency function Adj that, given two vertices, returns a boolean saying either they are connected or not. The edge that links two adjacent vertices v and w is denoted by $\{v, w\}$. We model the *port numbering function* $\phi : V \mapsto (seq\ V)$ as the ordered sequence of the neighbours of a vertex. For all v, $\phi(v) = [v_1, v_2]$ means that v has two neighbours: the first one is v_1 and the second one is v_2. Two axioms are required: the function ϕ only links adjacent vertices and does not contain duplicated vertices:

```
Hypothesis Hφ1 : ∀ v w, Adj v w = v ∈ (φ w).
Hypothesis Hφ2 : ∀ v, uniq (φ v).
```

Each process sends a message to its neighbour by putting it in the corresponding link. A port (pair of vertices) represents the link whereby a vertex put its message. We define P as the set of ports. Thus, if v sends a message to its ith neighbour, it sends its message by the port (v, w) where w is the ith element of the sequence $(\phi\ v)$. We model the exchange of messages, in a global way, by a *port labelling function* over the graph G. The set of labels over ports is denoted Ψ. A port labelling function $\psi : P \mapsto \Psi$ maps a port to its associated label. The state of each process is represented by a label $\lambda(v)$ associated to the corresponding vertex $v \in V$. Hence, each vertex has a status represented by a *vertex labelling function* $\lambda : V \mapsto \Lambda$ where Λ is the set of labels over the vertices. Consider $\sigma = (\lambda, \psi)$ the pair of labelling functions which maps a vertex (resp. a port) to its state. Type of such pair, the *global state* of the graph, will be denoted by State.

For instance, v_2 only distinguishes its four neighbours but it knows nothing about its identity or the ones of its neighbours. We can see, with a global view, that v_5 is the fourth neighbour of v_2 according to ϕ; the fact that v_2 sends a message m to its fourth neighbour consists in replacing the label of the port (v_2, v_5) by m.

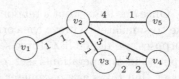

Graph supplied with a port numbering ϕ such that $(\phi\ v_1) = [v_2]$, $(\phi\ v_2) = [v_1, v_3, v_4, v_5]$, $(\phi\ v_3) = [v_2, v_4]$, $(\phi\ v_4) = [v_2, v_3]$, $(\phi\ v_5) = [v_2]$.

A processor sends (write a message on the corresponding port) and receives (reads the corresponding port) messages. We define the *writing (resp. reading) area* of a vertex v as the set of port labels it is able to update (resp. to read), that is port labels of the form (v, w) (resp. (w, v)) where w is a neighbour of v. From a global state σ, we define the *local view* of a vertex v as the triplet composed by its local state, the sequence of local states of the port in its writing and in its reading area given with the order of ϕ and extracted from σ. The local view of a vertex corresponds to the local information it owns. We define two local functions for a vertex to model received message from all neighbours (read) and sent messages to all neighbours (write):

- read:State × V → Λ × (seq Ψ) × (seq Ψ): consider σ and v, (read σ v) returns the local view of v in σ.
- write:State×V×Λ × (seq Ψ) → State: consider σ, v, λ, ψ, (write σ v λ ψ) returns the new global state obtained from the old one σ such that the local state of v is updated by λ and the one of its writing area by the sequence ψ.

2.2　Syntax and Semantics

Randomisation appears in local computations $(A \times (\text{seq } \Psi))$ made by a vertex. All of those local computations will create a random State. We define the inductive type for randomisation GR, that will be used to construct random local computations GR $(A \times (\text{seq } \Psi))$ and global states GR State. In Haskell [Has], monads are structures that represent computations and the way they can be combined. To express randomisation in *Coq* in a monadic form, we introduce three abstract operators Greturn, Gbind and Grandom.

```
Inductive GR (B:Type): Type :=
  | Greturn (b:B)
  | Gbind {A :Type}(a:GR A)(f : A → GR B)
  | Grandom (n:nat)(f : nat → GR B).
```

To improve the readability of the code, we use the following abbreviations. Let stmts be any statement block, n be an integer and f a function:

$$\text{Gbind } x \text{ (fun } v \Rightarrow \{<\text{stmts}>\}) \leftrightarrow \text{Glet } v = x \text{ in } \{<\text{stmts}>\}.$$
$$\text{Grandom n f} \leftrightarrow \text{Glet } x = (\text{random n}) \text{ in f}.$$

Once one has set out a randomised algorithm thanks to our syntax, one would like to simulate, to prove the correctness or to analyse this algorithm. For those purposes, we define [FZ] three semantics that interpret their input, a monad of type GR, in an operational, set, or distributional way. The operational semantic Opsem, that takes as a parameter a random number generator, is used to evaluate computations. The set monad Setsem is used to handle the set of transitional and final results of a randomised algorithm. We can then prove properties of correctness by reasoning on this set. The distributional monad Distsem is used to reason about distribution. We define it according to the monad of *Alea* by using the operators Munit, Mlet and Random [APM09].

2.3　Randomised Distributed Algorithms

We model a distributed algorithm by local algorithms executed by each processes during a round. We represent local algorithms by rewriting rules: from its local view, a vertex v can rewrite its own state and its writing area by applying a local computation of type GLocT. A round GRound is the state obtained from the application of a local computation to all vertices. Note that the update of the global state is not made concurrently but sequentially (see Sect. 3.1). Let LCs be a sequence of local computations, then a step GStep corresponds to the application of rounds taking successively as input the local computations of LCs. The execution of an algorithm with at most n steps is modelled by (GMC n LCs s init) where s is the list of vertices and init is the initial global state.

```
Definition GLocT := Λ → (seq Ψ) → (seq Ψ) → GR(Λ*seq Ψ).
Fixpoint GRound (s:seq V)(res: State)(LCs:GLocT):GR State:=
  match s with       |nil ⇒ Greturn res
    |v::t ⇒Glet s=(GRound t res LCs) in Glet p=(LCs (read res v)) in
           Greturn (write s v p)
  end.
Fixpoint GStep (LCs:seq GLocT)(s:seq V)(res:State):GR State:=
  match LCs with      | nil ⇒ Greturn res
    |a1::a2 ⇒ Glet y = (GRound s res a1) in (GStep a2 s y)
  end.
Fixpoint GMC(n:nat) (LCs:seq GLocT) (s:seq V)(init:State):GR State:=
  match n with        |0 ⇒ Greturn init
    | S m ⇒ Glet y = (GStep LCs s init) in (GMC m LCs s y)
  end.
```

According to the semantic, the result of the distributed algorithm (of type GR State) is either a possible global state that can be obtained from the algorithm with a random number generator (operational semantic); the set of all global states that the algorithm can produce (set semantic); or the distribution of global states resulting from the algorithm (distributional semantic).

3   General Results

In this section, we only use the distributional semantic. To ensure readability, let e: GR B be an randomised expression of type B, then instead of writing (Distsem (Greturn e)), we write (Dreturn E). Similarly, we introduce new functions beginning with D (instead of F) as distributional: Dlet, DRound, etc. Let LC be a local computation and LCs be a sequence of local computations.

3.1   Validity of Our Model

The algorithm simulates the sending of messages by updating the state σ for each vertex using the deterministic functions read and write. As the writing areas are pairwise disjoint (relabellings do not overlap), two calls of write, each applied to a different vertex, permute.

```
Lemma write_comm: ∀ v w, v ≠ w →
  (write (write σ w c₂) v c₁) = (write (write σ v c₁) w c₂).
```

The global function depends formally on the implementation of the sequence of vertices (enum V). It describes sequentially the simulation of the application of the local function simultaneously on all the vertices. In fact, our system is distributed, this implies that several vertices can relabel their writing area at the same time. However, to reason over such algorithm, we want it to be sequential. Then we have to show that the order of application of the local algorithm is not relevant. This property is ensured thanks to the permutability of function write.

Thus, the result will be the same than the one obtained if vertices would execute this algorithm at the same time.

Lemma DRoundCommute3: Let σ be a global state of G and LC be a discrete local computation. Let lv be a sequence of vertices of G. Let lv' be a permutation of lv, then:

DRound lv σ LC = DRound lv' σ LC.

3.2 Tools to Prove Properties on Algorithms of Our Model

Composition. A way to prove properties on function DRound is to proceed by induction on the sequence of vertices, especially if we want to prove a property about a vertex v. The method is based on the decomposition of the randomised expression into the measure of one vertex and the measure for the remaining. As an example, we can prove properties such as the termination with probability one as stated below.

Lemma DRound_total: $\forall \; \sigma \; s$, $(\forall \; v$, Term (LC (read v))) $\rightarrow$
$(\mu \; (\text{DRound } s \; \sigma \; \text{LC})) \; \mathbb{1} = 1$.

Non-null Probability. Probability that an event occurs in a randomised algorithm is not null if there is a possible execution of the algorithm whereby this event is verified. Therefore, it suffices to highlight a witness.

Lemma proba_not_null: Let A be a randomised algorithm and E an event. Let t be a witness, if $(\mu \; A) \; \mathbb{1}_{.=t} > 0$ and $(E \; t)$ then $(\mu \; A) \; \mathbb{1}_{(E \; t)} > 0$.

Termination. A randomised distributed algorithm repeats a step until a certain property is verified by the labelling graph. In general, this property is that all the vertices stop to interact with each other, *i.e.* until all vertices are inactive. This lead us to consider the algorithm with a property of termination TermB.

```
Fix DLV (sV: seq V)(σ: State) (LCs: seq DLocT) (TermB: State → bool)
  : distr (State) :=     if (TermB σ) then Dreturn σ
    else Dlet r = (DStep LCs sV σ) in DLV sV r LCs TermB
```

In *Coq* we need to highlight a variant which decrements at each round in order to prove the termination. However there exists some algorithms which terminate with probability 1 but in which some executions could possibly be infinite. To deal with this kind of programs, there is, in *Alea*, a tool to handle limits of sequences of distributions. Hence, when a recursive function is introduced, we interpret it as a fix point and then compute the least upper bound.

Lemma termglobal: For all randomised update of a global state to another
rd : State → distr State, for all global state σ, for all ended property **TermB**,
for all variant (**cardTermB: State → nat**), for all real c between 0 and 1 and for
all state property (**PR:State→bool**), if:

1. $\forall$ s, Term (rd s)
2. $\forall$ s, cardTermB s = 0 → TermB s = true
3. $0 < c$
4. $\forall$ s, 0<cardTermB s → PR s → $c \leq \mu$ (rd s) $\mathbb{1}_{(cardTermB \, . < cardTermB \, s)}$
5. $\forall$ s, PR s → μ (rd s) $\mathbb{1}_{(cardTermB \, s < cardTermB \, .)} = 0$
6. $\forall$ s f, PR s → μ (rd s) $\mathbb{1}_{PR. \, \wedge \, f.} = \mu$ (rd s) $\mathbb{1}_f$
7. PR σ

then: **Term (fglobal rd TermB σ)**.

From Lemma **termglobal**, we obtain Lemma **DPLV_total**: function **DLV** terminates by taking as input the state update (**DStep LCs (enum V)**). Thus, to prove
that a Las Vegas algorithm terminates with probability 1, it suffices to show
that the probability for a certain variant (such that, if it is null, it implies the
termination) to decrement is non-null and to increase is null after a step (**DStep
LCs (enum V)**). The property **PR** specify the global states with a property always
true: the two last hypotheses mean that it has no impact on the probability
computations and that it is verified by the initial state.

4 Applications

As a case study, we focus on the Handshake problem. We first show that we
require randomisation, we define a randomised solution and we prove its correctness. Then, we analyse this solution. As a generalisation of this problem, we
analyse a solution to the maximal matching problem.

4.1 Correctness of an Handshake Solution

Handshake Specification. We specify an handshake solution as a structure
hsAlgo containing three components.

- **HsR** is the local computation sequence (that each node executes in successive
  rounds).
- **HsP** is the local handshake function (each node knows if it is in handshake).
- **HsI** is the initial state.

  Hypotheses required by the above components are the following.

- **HsI1**: the initial state is *consistent*, *i.e.*, for each v, if v is in handshake with
  one of its neighbours (say w), then w is also in handshake with v.
- **HsI2** : the initial state is *uniform*, *i.e.* each vertex has the same label and each
  port also.
- **HsP1** : the global handshake function (obtained from **HsP**) applied to a vertex
  v returns numbers lesser than the degree of v.
- **HsRind**: consistency is preserved by a step of the algorithm.

We assume an important hypothesis on the graph: it must contain at least an edge, otherwise no handshake can occur. The aim of this algorithm is to realise handshakes (`hsRealisation:`Λ Ψ `(A: hsAlgo` Λ Ψ`)`), *i.e.*, for any graph, there is an execution in which one reachable state contains a handshake.

Impossibility Result. We have seen that the difference between deterministic algorithms and randomised algorithms is the use of `random`. We show the interest of randomised algorithms by proving that there is no deterministic algorithm that solves the handshake problem for any graph. We define a property `Adet` (`1:seq GLocT`) verifying that all the computational rules are deterministic.

```
Lemma NotReal:∀ Λ Ψ(A:hsAlgo Λ Ψ), Adet(HsR A)→∼(hsRealisation A).
```

The Randomised Algorithm. The algorithm `randHSLoc` is defined as follows: each vertex v chooses uniformly at random one of its neighbours $c(v)$, sends 1 to $c(v)$ and 0 to the others. There is a handshake between v and $c(v)$ if v receives 1 from $c(v)$. The function (`randSendChosen` n l) returns a boolean sequence of size $|l|$ where each component takes the value 0 except the nth.

```
Definition randHSLoc (λ:Λ) (ψ_out ψ_in:seq Ψ) : GR (Λ × seq Ψ) :=
  match |ψ_in| with |0 ⇒ Greturn (None, nil)        (*isolated vertex*)
    |S n ⇒ Glet k=(random n) in Greturn(None,randSendChosen(k+1) ψ_in)
  end.
```

Correctness of the Randomised Algorithm. To define formally our algorithm, we first define the components. The rule sequence `randHsR` corresponds to a single local rule `randHSLoc`. The function `randHsP` returns `None` if the vertex is not in a handshake or `Some i` if the vertex is in handshake with its ith neighbour. The initial state `randHsI` is the one where all labels are valued at `None` and all the labels of the ports at 0. We prove the hypotheses: consistency of the initial state (`randHSI1`), uniformity of the initial state (`randHsI2`), domain of the handshake function (`randHsP1`) and stability of consistency by a computation step (`randHsRind`). We then build the algorithm: `randhs`. We prove that the only hypothesis that differs is the determinism: `NonADet`. We finally prove that there exists at least an execution of the algorithm that realises a handshake `Real`.

```
Definition randhs : (hsAlgo Λ Ψ) :=
   (Build_hsAlgo randHsI1 randHsI2 randHsP1 randHsRind).
Lemma NonADet: ∼ Adet (HsR randhs).
Lemma Real : hsRealisation randhs.
```

4.2 The Handshake Algorithm in *Coq*

The local computation we consider here (`DHSLoc`) is similar to `randHSLoc` except that we applied it only on active vertices, that is on the active subgraph. Active vertices are required to construct the maximal matching of the next section. We denote by `DHS` the global algorithm based on the local algorithm `DHSLoc`.

```
Definition DHSLoc (λ:Λ) (ψ_out ψ_in: seq Ψ): dist (Λ*seq Ψ) :=
  if (active λ) then
    match (numberActive ψ_in) with
      |0 ⇒ Dreturn (Some |ψ_out|, nseq |ψ_out| false)
      |S n ⇒ Dlet k = (Random n) in
                  Dreturn (λ, sendChosen k.+1 ψ_in)
    end
  else Dreturn (λ, ψ_out).

Definition DHSRound (sV: seq V) (σ:State) := DRound sV σ DHSLoc.
```

Composition. As the order is irrelevant, sV can be rewritten into $v :: (sV \setminus v)$, where sV is the sequence of vertices in the graph. We can apply composition technique to prove the lemma below where $P(v, w)$ is the property *"v chooses w"*.

```
Lemma DHS_degv_global: ∀ G σ {v,w}, (μ (DHS sV σ )) 𝕀_{P(v,w)} = 1/d(v).
```

Analysis of the Success. Let $\mathcal{HS}(e)$ denote the event "there is a handshake on the edge e". We define $\mathcal{H}(e)$ as the characteristic function of $\mathcal{HS}(e)$. The goal of our establishment of a model is to write the formal proof of results from [MSZ03]. To be able to prove this theorem, the general results were used, as for example Lemma proba_not_null.

```
Theorem DHS_deg: ∀ sV σ, μ (DHS (enum V) σ) (∃ e, H(e)) ≥ 1 − e^{−1/2}.
```

4.3   The Maximal Matching Algorithm

Here is the definition of the maximal matching algorithm. We show that this algorithm terminates with probability 1. This algorithm consists in iterating the handshake algorithm (DMMLoc2) only by considering the active vertices where vertices in handshake becomes inactive (DMMLoc1). At the beginning, every vertex is active. At the end, every vertex is inactive (termB).

```
Definition DMMLoc1 (λ:Λ) (ψ_out ψ_in:seq Ψ) : dist (Λ × seq Ψ) :=
  if (active λ) then
    if (agreed ψ_out ψ_in) then
         Dreturn ( Some (index true ψ_out) , ψ_out)
    else Dreturn (None, map (fun x ⇒ true) ψ_out)
  else Dreturn (λ, ψ_out).

Definition DMMLoc2 (λ:Λ) (ψ_out ψ_in:seq Ψ) : dist (Λ × seq Ψ) :=
  DHSLoc λ ψ_out ψ_in.

Definition termB (f: State) : bool :=
  [∀ v, active (f.1 v)].

Definition DMMLV (sV: seq V) (σ: State) :=
  DLV sV σ (DMMLoc1::DMMLoc2::nil) termB.
```

The general lemma `DPLV_total` (see Lemma `termglobal`) implies that this algorithm terminates with probability 1.

Theorem `DMMLV_term`: $\forall\ \sigma,\ \mu\ (DMMLV\ (enum\ V)\ \sigma)\ \mathbb{I} = 1$.

Proof. To prove this lemma, we used the general result `termglobal`. We first show that the probability to have a handshake during a round is strictly positive which means that the number of active vertices decrements with a non null probability. Hence as a variant we take the number of active vertices. The property always true `PR` in our labelling is that every active vertex sends 1 to all of its neighbour and every inactive vertex sends 0. We then prove the 7 hypotheses of Lemma `termglobal`.

5   Conclusion

We develop on this paper tools to reason about (randomised) distributed algorithms in anonymous networks. We prove negative results but also we prove properties over randomised algorithms which solve handshake and maximal matching problems. More particularly, for the handshake problem, we analyse the probability of at least a handshake in a round. We then iterate this algorithm to construct a maximal matching. We prove that this algorithm terminates with probability 1.

Many of the techniques used in this paper can be applied to analyse solutions for other similar problems like symmetry break, local election algorithms and distributed computing of maximal independent sets. One of the future work consists in proving properties about time complexity by providing tools to handle the number of rounds.

Acknowledgement. The authors are grateful to P. Castéran who follows this work all along. We particularly thank him for his first proof in *Coq* of the impossibility result stated in Sect. 4.1 and for the development of the semantics that is the base of their development. They also thank C. Paulin-Mohring and A. Mahboubi for their help using *Alea* and *ssreflect* respectively.

References

[APM09] Audebaud, P., Paulin-Mohring, C.: Proofs of randomized algorithms in Coq. Sci. Comput. Program. **74**(8), 568–589 (2009)

[CF11] Castéran, P., Filou, V.: Tasks, types and tactics for local computation systems. Studia Informatica Universalis **9**(1), 39–86 (2011)

[Cho95] Chou, C.T.: Mechanical verification of distributed algorithms in higher-order logic. Comput. J. **38**(2), 152–161 (1995)

[CRTU15] Courtieu, P., Rieg, L., Tixeuil, S., Urbain, X.: Impossibility of gathering, a certification. Inf. Process. Lett. **115**(3), 447–452 (2015)

[Der70] Derman, C.: Finite State Markovian Decision Processes. Mathematics in Science and Engineering. Academic Press, Orlando (1970)

[DM09]   Deng, Y., Monin, J.F.: Verifying self-stabilizing population protocols with Coq. In: TASE, pp. 201–208 (2009)

[FZ]   Fontaine, A., Zemmari, A.: RDA: a Coq Library on Randomised Distributed Algorithms. http://www.allyxfontaine.com/RDA

[GMT08] Gonthier, G., Mahboubi, A., Tassi, E.: A Small Scale Reflection Extension for the Coq system. Rapport de recherche RR-6455, INRIA (2008)

[Has]   http://www.haskell.org/haskellwiki/monad

[HMM05] Hurd, J., McIver, A., Morgan, C.: Probabilistic guarded commands mechanized in ol. Electr. Notes Theor. Comput. Sci. **112**, 95–111 (2005)

[KNP02] Kwiatkowska, M.Z., Norman, G., Parker, D., Prism: probabilistic symbolic model checker. In: Computer Performance Evaluation/TOOLS, pp. 200–204 (2002)

[KNR12] Küfner, P., Nestmann, U., Rickmann, C.: Formal verification of distributed algorithms. In: Baeten, J.C.M., Ball, T., Boer, F.S. (eds.) TCS 2012. LNCS, vol. 7604, pp. 209–224. Springer, Heidelberg (2012). doi:10.1007/978-3-642-33475-7_15

[KNS01] Kwiatkowska, M.Z., Norman, G., Segala, R.: Automated verification of a randomized distributed consensus protocol using cadence SMV and PRISM. In: Berry, G., Comon, H., Finkel, A. (eds.) CAV 2001. LNCS, vol. 2102, pp. 194–206. Springer, Heidelberg (2001)

[MMT11] Méry, D., Mosbah, M., Tounsi, M.: Refinement-based verification of local synchronization algorithms. In: Butler, M., Schulte, W. (eds.) FM 2011. LNCS, vol. 6664, pp. 338–352. Springer, Heidelberg (2011). doi:10.1007/978-3-642-21437-0_26

[MSZ03] Métivier, Y., Saheb, N., Zemmari, A.: Analysis of a randomized rendezvous algorithm. Inf. Comput. **184**(1), 109–128 (2003)

[PS95]   Pogosyants, A., Segala, R.: Formal verification of timed properties for randomized distributed algorithms. In: PODC, pp. 174–183 (1995)

[Tea]   "Coq Development Team". The Coq Proof Assistant Reference Manual. coq.inria.fr

Calculating Statically Maximum Log Memory Used by Multi-threaded Transactional Programs

Anh-Hoang Truong[1], Ngoc-Khai Nguyen[2](✉), Dang Van Hung[1],
and Duc-Hanh Dang[1]

[1] VNU University of Engineering and Technology, Hanoi, Vietnam
[2] Hanoi University of Natural Resources and Environment, Hanoi, Vietnam
nnkhai@hunre.edu.vn

Abstract. During the execution of multi-threaded and transactional programs, when new threads are created or new transactions are started, memory areas called logs are implicitly allocated to store copies of shared variables so that the threads can independently manipulate these variables. It is not easy to manually calculate the peak of memory allocated for logs when programs have arbitrary mixes of nested transactions and new thread creations. We develop a static analysis to compute the amount of memory used by logs in the worst execution scenarios of the programs. We prove the soundness of our analysis and we show a prototype tool to infer the memory bound.

Keywords: Memory bound · Transactional memory · Static analysis

1 Introduction

We address the problem of determining the memory bound of transactional programs at compile time to ensure that they can run smoothly without out of memory errors. To describe the problem more precisely, we use a core language in which transactional and multi-threading statements are based on [12] and other features are generalized so that transactional programs in other imperative languages can be translated to our language for the memory estimation problem. The key features we borrow from [12] allow programmers to mix creating new threads and opening new transactions.

When one transaction is nested in another, we called the former parent transaction, and the latter child one. A child transaction must commit before its parent does. When a transaction is started, a memory area called *log* is allocated for storing a copy of shared variables. A transaction that started but has not committed yet is called an *open* transaction. Inside open transactions, the programmers can also create new threads. A new thread in this case will make a copy of transaction logs of its parent thread. When a parent thread commits a

This research is funded by Vietnam National Foundation for Science and Technology Development (NAFOSTED) under grant number 102.03-2014.23.

A. Sampaio and F. Wang (Eds.): ICTAC 2016, LNCS 9965, pp. 82–99, 2016.
DOI: 10.1007/978-3-319-46750-4_6

transaction, all the child threads that are created inside the parent transaction must join the commit with their parent. This kind of commits is called *joint commit*, and the time when these commits occur is *joint commit point*. Joint commits act as implicit synchronizations of parallel threads. If a transaction has no child threads, the commit is a normal (local) commit. Both types of commits release the memory allocated for the logs. Now, we can formulate the problem as follows. Given the size of transaction logs in the program, compute the maximal memory requirement for the whole program.

In our previous studies [13, 15, 16], we built type systems to count the maximum number of logs that can coexist at runtime. This number gives us raw information about the memory used by transaction logs. To infer precisely the maximal amount of memory that transaction logs may use, we need information about the size of each log. Therefore, in this work we extend the start transaction statement in our previous work to contain this information. But this does not mean the programmers have to annotate this size information as it can be synthesized by identifying shared variables of transactions. Then, we develop a type system to estimate the maximum memory that transaction logs may require. It turns out that the ideas of type structures in our previous work can be reused, but the type semantics and typing rules are novel and different from those in our previous works. The type system with its soundness proofs and a prototype tool are our main contributions in this work.

Estimating resource usage in general and estimating memory resource in particular has always been an active research problem. In [17], Wegbereiter gave methods to analyze the complexity of Lisp programs by using recursive function. Hughes and Pareto [11] introduce a strict, first-order functional language with a type system such that well-typed programs run within the space specified by the programmer. Hofmann and Jost [10] compute the linear bounds on heap space for a first-order functional language. Later, they use a type system to calculate the heap space bound as a function of input for an object oriented language. Wei-Ngan Chin *et al.* [8] studied memory usages of object-oriented programs. In [7] the authors statically compute upper bounds of resource consumption of a method using a non-linear function of method's parameters. The bounds are not precise and their work is not type-based. Braberman *et al.* [4, 6] calculate symbolic approximation of memory bounds for Java programs. In [5] the authors propose type systems for component languages with parallel composition but the threads run independently. Albert *et al.* have many works in resource estimation for programs. In [3], they compute the heap consumption of a program as a function of its data size. In [1, 2], they studied the problem in the context of distributed and concurrent programs. In [14], Pham *et al.* proposes a fast algorithm to statically find the upper bounds of heap memory for a class of JavaCard programs. In [9], Jan Hoffmann and Zhong Shao also use type system to estimate resource usage of parallel programs but for a functional language.

The works mentioned above focus only on sequential or functional language. The language that we study here is different as it is a multi-threaded and nested transactional language with complex and implicit synchronization. The type system that we develop in this work is significantly different from the ones in our previous work even though it looks similar. Note that compared to our previous work, the semantics of several type elements are completely different.

The rest of the paper is structured as follows. In the next section we informally explain the problem and the approach via a motivating example. Section 3 introduces the formal syntax and operational semantics of the calculus. Section 4 presents a new type system. The soundness of the analysis is represented in Sect. 5. A prototype tool with its main algorithm to compute memory bound is described in Sect. 6. Section 7 concludes and outlines our future work.

2 Motivating Example

We use the sample program in Listing 1.1, borrowed from [15], to explain the problem and our approach. Note that this example focuses on the core of the language. Real programs will have many other constructs of programming languages such as procedures, method calls, message passing, variables, and other computation primitives. These programs can be converted to equivalent programs, w.r.t. transactional and multi-threading behaviours, in our core language.

In this code snippet, the statements onacid and commit are for starting and closing a transaction [12]. The statement spawn is for creating a thread with the code represented by the parameters of the statement. The onacid statement in our previous works has no parameters, but in this work it is associated with a number to denote the size of the memory needed to allocate to the log of the transaction at runtime.

Listing 1.1. A nested multi-threaded program.

```
1   onacid(1); //thread 0
2     onacid(2);
3       spawn(onacid(4);commit;commit;commit);//thread 1
4       onacid(3);
5         spawn(onacid(5);commit;commit;commit;commit);//thread 2
6       commit;
7       onacid(6);commit;
8     commit;
9   onacid(7);commit;
10  commit
```

The behavior of this program is depicted in Fig. 1. The starting transaction statement onacid and ending transaction statement commit are denoted by [and] in the figure, respectively. The statement spawn creates a new thread running in parallel with its parent thread and is described by the horizontal lines. The new thread duplicates the logs of the parent thread for storing a copy of the value of variables of the parent thread so that it can manipulate these variables independently.

In this example, when spawning thread 1, thread 0 has opened two transactions, so thread 1 makes two copies of thread 0's logs and hence on line 3 the parameter of spawn contains the last two commits to close them. These commits must be synchronized with the commits in lines 8 and 10 of the thread 0, and form a so-called joint commit. Joint commits are described by the rectangular dotted line in Fig. 1. The right-hand edges of the boxes mark these synchronizations. The left-hand edges are the corresponding open transactions that the joint commits must jointly close.

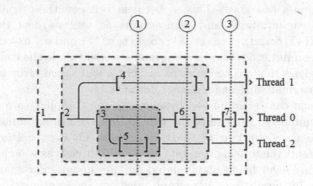

Fig. 1. Threads dependencies and join commits.

We now try to manually calculate the maximum memory used by logs to answer the question: How much more memory is required by the software transactional memory mechanism?

- At the point ①: The total memory used for logs, denoted by m_1, is the sum of:
  - log memory for the first two and the fourth transactions of thread 0: $1 + 2 + 3 = 6$,
  - log memory for thread 1: $(1 + 2) + 4 = 7$, since thread 1 clones two logs of its parent thread,
  - log memory for thread 2: $(1 + 2 + 3) + 5 = 11$ since thread 2 clones three logs of its parent thread.
  So $m_1 = 6 + 7 + 11 = 24$.
- At the point ②: The total memory used for logs, denoted by m_2, is the sum of:
  - log memory for the first three transactions of thread 0: $1 + 2 + 6 = 9$,
  - log memory for thread 1: $(1 + 2) + 4 = 7$ as above,
  - log memory for thread 2: $(1 + 2) = 3$, since thread 2 now has only two logs copied from its parent thread.
  So $m_2 = 9 + 7 + 3 = 19$.
- At the point ③: Similarly, we have $m_3 = 8 + 1 + 1 = 10$.

So in the worst case, the maximum memory allocated to logs is $\max(m_1, m_2, m_3) = 24$ units.

Note that our language was inspired and abstracted from [12] as we focus on transactional and multi-threaded features to estimate the additional memory, which is implicitly allocated by the implementation the language at runtime. The additional commits that a child thread has, e.g., the last two commits in `spawn(onacid(4);commit;commit;commit);`, make the language harder to use but this gives programmers more power to control when the commits can start. This is because between these commits, there can be other computations and transactions as well as new threads, e.g., `spawn(onacid(4);commit;commit;e';commit;e'');` and `e, e''` may be some other lengthy computations and the programmers want to do the commits before them. The analysis that we will present in Sect. 4 will signal error for programs that do not have matching `onacids` and `commits`.

The language can be simplified by allowing the compiler to automatically insert these commits to the end of the thread, but then programmers will have less control on the behaviour of the transactions. Or even better, a smarter compiler can insert the missing commits for a child threads as soon as the shared variables are no longer being manipulated by the threads. For example, if `e''` in the example in the previous paragraph does not access any shared variables, then the compiler can insert a commit before `e''`. In both situations, the type system that we present here can compute the worst scenario of log memory that the runtime required.

So, there is a trade-off in the language design. Our language design here can cover the other cases where commits are automatically inserted to the child threads by the compiler. Last, here the language does not have loops because we believe that spawning threads and creating transactions are expensive operations and one usually puts these constructs outside loops with unknown number of iterations. For loops with a fixed number of iterations, the sequential composition can encode them.

3   Transactional Language

3.1   Syntax

Figure 2 gives the syntax of our language, called TM (transactional memory). In the first line, *program* P can be *empty*, notation 0, or a composition of parallel threads $P \parallel P$. $p(e)$ denotes a thread with *identifier* p executing term e. For *term* e, we assume the language has a set of atomic statements A, ranged over by α. **onacid**(n) and **commit** are statements for starting and committing a transaction. Parameter n in **onacid**(n) represents the number of memory units allocated when opening the new transaction. Note that in reality n can be synthesized by the compiler based on the sizes of shared variables in the scope of the transaction. That means programmers do not have to annotate this size information. $e_1; e_2$ denotes sequencing of statements and $e_1 + e_2$ denotes branching. The last statement **spawn**(e) is for creating a new thread executing e.

$$P ::= 0 \quad | \quad P \parallel P \quad | \quad p(e)$$
$$e ::= \alpha \quad | \quad \mathbf{onacid}(n) \quad | \quad \mathbf{commit} \quad |$$
$$e_1; e_2 \quad | \quad e_1 + e_2 \quad | \quad \mathbf{spawn}(e)$$

Fig. 2. TM syntax

3.2 Dynamic Semantics

The (global) run-time environment is structured as a collection of local environments. Each local environment is a sequence of logs with their sizes. We formally define the local and global environments as follows.

Definition 1 (Local environment). *A* local environment E *is a finite sequence of* log id*'s and their size:* $l_1:n_1; \ldots; l_k:n_k$. *The environment with no element is called the* empty environment, *denoted by* ϵ.

For an environment $E = l_1:n_1; \ldots; l_k:n_k$, we denote $[\![E]\!] = \sum_{i=1}^{k} n_i$ the number of memory units used in E, and $|E| = k$ the number of elements in E.

Definition 2 (Global environment). *A* global environment Γ *is a collection of* thread id*'s and their local environments,* $\Gamma = \{p_1:E_1, \ldots, p_k:E_k\}$.

The log memory used by $\Gamma = \{p_1:E_1, \ldots, p_k:E_k\}$, denoted by $[\![\Gamma]\!]$, is defined by: $[\![\Gamma]\!] = \sum_{i=1}^{k} [\![E_i]\!]$.

For a global environment Γ and a set P of threads, we call the pair Γ, P a *state*. We have a special state *error* for stuck states—the states at which no other transition rules can be applied. The dynamic semantics is defined by *transition rules* between states of the form $\Gamma, P \Rightarrow \Gamma', P'$ or $\Gamma, P \Rightarrow error$ in Table 1.

Table 1. TM dynamic semantics

$$\frac{p' \, fresh \quad spawn(p, p', \Gamma) = \Gamma'}{\Gamma, P \parallel p(\mathbf{spawn}(e_1); e_2) \Rightarrow \Gamma', P \parallel p(e_2) \parallel p'(e_1)} \text{ S-SPAWN}$$

$$\frac{l \, fresh \quad start(l:n, p, \Gamma) = \Gamma'}{\Gamma, P \parallel p(\mathbf{onacid}(n); e) \Rightarrow \Gamma', P \parallel p(e)} \text{ S-TRANS}$$

$$\frac{intranse(\Gamma, l:n) = \mathbf{p} = \{p_1, .., p_k\} \quad commit(\mathbf{p}, \Gamma) = \Gamma'}{\Gamma, P \parallel \coprod_1^k p_i(\mathbf{commit}; e_i) \Rightarrow \Gamma', P \parallel \coprod_1^k p_i(e_i)} \text{ S-COMM}$$

$$\frac{i = 1, 2}{\Gamma, P \parallel p(e_1 + e_2) \Rightarrow \Gamma, P \parallel p(e_i)} \text{ S-COND} \qquad \frac{}{\Gamma, P \parallel p(\alpha; e) \Rightarrow \Gamma, P \parallel p(e)} \text{ S-SKIP}$$

$$\frac{\Gamma = \Gamma' \cup \{p : E\} \quad |E| = 0}{\Gamma, P \parallel p(\mathbf{commit}; e) \Rightarrow error} \text{ S-ERROR-C} \qquad \frac{\Gamma = \Gamma' \cup \{p : E\} \quad |E| > 0}{\Gamma, P \parallel p() \Rightarrow error} \text{ S-ERROR-O}$$

Table 1 uses some auxiliary functions described as follows. Note that the function names are from [12] and congruence rules are applied for processes: $P \parallel P' \equiv P' \parallel P$, $P \parallel (P' \parallel P'') \equiv (P \parallel P') \parallel P''$ and $P \parallel 0 \equiv P$.

- In the rule S-SPAWN, the function $spawn(p, p', \Gamma)$ adds to Γ a new element with thread id p' and a local environment cloned from the local environment of p. Formally, suppose $\Gamma = \{p : E\} \cup \Gamma''$ and $spawn(p, p', \Gamma) = \Gamma'$, then $\Gamma' = \Gamma \cup \{p' : E'\}$ where $E' = E$.
- In the rule S-TRANS, the function $start(l : n, p, \Gamma)$ creates one more log with the label l and with the size n units of memory at the end of the local environment of p_i. If $start(l{:}n, p_i, \Gamma) = \Gamma'$ where $\Gamma = \{p_1 : E_1, \ldots, p_i : E_i, \ldots, p_k : E_k\}$ and l is a fresh label, then $\Gamma' = \{p_1 : E_1, \ldots, p_i : E'_i, \ldots, p_k : E_k\}$, where $E'_i = E_i; l : n$.
- In the rule S-COMM, the function $intranse(\Gamma, l : n)$ returns a set of all threads, denoted by $\mathbf{p}$, in Γ whose local environments contain log id l and this log id is the last element of the local environments. That is $intranse(\Gamma, l{:}n) = \mathbf{p} = \{p_1, .., p_k\}$ then:
  - for all $i \in \{1..k\}$, p_i has the form $E'_i; l : n$.
  - for all $p' : E' \in \Gamma$ such that $p' \notin \{p_1, .., p_k\}$ we have E' does not contain log id l.
- Also in the rule S-SPAWN, the function $commit(\mathbf{p}, \Gamma)$ removes the last log id in the local environments of all threads in $\mathbf{p}$. That is, suppose $intranse(\Gamma, l{:}n) = \mathbf{p}$ and $commit(\mathbf{p}, \Gamma) = \Gamma'$, then for all $p' : E' \in \Gamma'$, if $p' \in \mathbf{p}$, then $p' : (E'; l : n) \in \Gamma$. Otherwise, $p' : E' \in \Gamma$.

Note that function $spawn$ copies the labels of the parent thread's environment to the local environment of the new thread and the function $intranse$ finds these labels to identify threads that need synchronization in a joint commit.

The rules in Table 1 have the following meanings:

- The rule S-SPAWN says that a new thread is created with the statement **spawn**. The statement **spawn**(e_1) creates a new thread p' executing e_1 in parallel with its parent thread p, and changes the environment from Γ to Γ'.
- The rule S-TRANS is for the cases where thread p creates a new transaction with the statement **onacid**. A new transaction with label l is created, and changes the environment from Γ to environment Γ'.
- The rule S-COMM is for committing a transaction. In this rule $\coprod_1^k p_i(E_i)$ stands for $p_1(e_1) \parallel .. \parallel ..p_k(e_k)$. If the current transaction of thread p is l, then all threads in the transaction l have to joint commit when transaction l commits.
- The rule $S - COND$ is to select one of the two branches e_1 or e_2 to continue.
- The rule $S - SKIP$ is for other computation statements of the language, which we assume they do not interfere with our multi-threading and transactional semantics, so we can skip them.
- The rules S-ERROR-C and S-ERROR-O are used in cases where there are mismatches in starting and committing transactions. For instance, `onacid;spawn(commit;commit);commit` has a mismatch in the second commit in `spawn(commit;commit)`. $p()$ in S-ERROR-O means there is missing commit(s) in the program.

4 Type System

The main purpose of our type system is to identify the maximum log memory that a TM program may require. The type of a term in our system is computed from what we call sequences of *tagged numbers*, which is an abstract representation of the term's transactional behavior w.r.t. log memory.

4.1 Types

Inspired from our previous works [15], our types are finite sequences over the set of so called *tagged numbers*. A tagged number is a pair of a *tag* and a non-negative natural number $\mathbb{N}^+$. We use four tags, or signs, $\{+, -, \neg, \sharp\}$ for denoting opening, commit, joint commit and accumulated maximum of memory used by logs, respectively. The set of all tagged number is denoted by $^T\mathbb{N}$. So $^T\mathbb{N} = \{ {}^+n, {}^-n, {}^\sharp n, {}^\neg n \mid n \in \mathbb{N}^+ \}$. The meanings of these tag numbers is described below.

- The tag number ^+n says that the open transaction has a log whose size is n units of memory. Note that this semantics is different from ones in our previous works where it denotes the number of consecutive onacids.
- The tag number ^-n means there are n consecutive commits statements,
- The tag number $^\neg n$ means there are n threads that require synchronization at a joint commit,
- The tag number $^\sharp n$ says the current maximum of memory units used by the term is n.

To help the readers better understand types, we give the following type examples. onacid(2) has type $^+2$, commit has type $^-1$, onacid(2);commit has type $^+2\,^-1$. Later, we will explain how this type can be converted to its equivalent form: $^\sharp 2$, by matching and combining $+$ and $-$ elements. For the sequential composition statement onacid(1); onacid(2); commit; commit, its type is $^+1\,^+2\,^-1\,^-1$ or its equivalent form $^+1\,^\sharp 2\,^-1$ or $^\sharp 3$. For spawn(onacid(4);commit;commit;commit), its type is $^+4\,^-1\,^-1\,^-1$ and can be simplified to $^\sharp 4\,^\neg 1\,^\neg 1$ by matching and combining $+$ and $-$ elements and identifying joint commits elements. Note that we do not combine the two consecutive $^\neg$. Instead, we will match and combine with a suitable $^\neg$ of some other term that will be executed in another thread. $^\sharp 2\,^\sharp 4$ and $^\sharp 4\,^\sharp 3$ can be converted to its equivalent type $^\sharp 4$ since they all reflect that the maximum units of memory used is 4.

We will develop rules to associate a sequence of tagged numbers with a term in TM. During computation, a tag with zero (e.g. $^+0$, $^-0$, etc.) may be produced but it has no effect to the semantics of the sequence so we will automatically discard it when it appears. To simplify the presentation we also automatically insert $^\sharp 0$ element whenever needed.

In the following, let s range over $^T\mathbb{N}$, $^T\bar{\mathbb{N}}$ be the set of all sequences of tagged numbers, S range over $^T\bar{\mathbb{N}}$ and let $m, n, l, ..$ range over $\mathbb{N}$. The empty sequence is denoted by ϵ as usual. For a sequence S we denote by $|S|$ the length of S, and write $S(i)$ for the ith element of S. For a tagged number s, we denote $\mathrm{tag}(s)$ the tag of s, and $|s|$ the natural number of s (i.e. $s = {}^{\mathrm{tag}(s)}|s|$). For a sequence $S \in {}^T\bar{\mathbb{N}}$, we write $\mathrm{tag}(S)$ for the sequence of the tags of the elements of S and $\{S\}$ for the set of tags appearing in S. Note that $\mathrm{tag}(s_1 \dots s_k) = \mathrm{tag}(s_1) \dots \mathrm{tag}(s_k)$. We also write $\mathrm{tag}(s) \in S$ instead of $\mathrm{tag}(s) \in \{S\}$ for simplicity.

The set ${}^T\bar{\mathbb{N}}$ can be partitioned into equivalence classes such that all elements in the same class represent the same transactional behavior, and for each class we use the most compact sequence as the representative for the class and we call it *canonical* element.

Definition 3 (Canonical sequence). *A sequence S is* canonical *if $\mathrm{tag}(S)$ does not contain '$--$', '$\sharp\sharp$', '$+-$', '$+\sharp-$', '$+\neg$' or '$+\sharp\neg$' and $|S(i)| > 0$ for all i.*

The intuition here is that we can always simplify/shorten a sequence S without changing its interpretation. The seq function below reduces a sequence in ${}^T\bar{\mathbb{N}}$ to a canonical one. Note the pattern '$+-$' does not appear on the left, but we can insert $\sharp 0$ to apply the function. The last two patterns, '$+\neg$' and '$+\sharp\neg$', will be handled by the function jc later in Definition 8.

Definition 4 (Simplification). *Function* seq *is defined recursively as follows:*

$$\mathrm{seq}(S) = S \text{ when } S \text{ is canonical}$$
$$\mathrm{seq}(S\,{}^\sharp m\,{}^\sharp n\,S') = \mathrm{seq}(S\,{}^\sharp\max(m,n)\,S')$$
$$\mathrm{seq}(S\,{}^- m\,{}^- n\,S') = \mathrm{seq}(S\,{}^-(m+n)\,S')$$
$$\mathrm{seq}(S\,{}^+ k\,{}^\sharp l\,{}^- n\,S') = \mathrm{seq}(S\,{}^\sharp(l+k)\,{}^-(n-1)\,S')$$

In this definition, the second and the third lines are for simplifying the representation. The last line is for local commits—the commits that do not synchronize with other threads.

As illustrated by Fig. 1, threads are synchronized by joint commits (dotted rectangles). So these joint commits split a thread into so-called *segments* and only some segments can run in parallel. For instance, in the running example, onacid(5) on line 5 cannot run in parallel with onacid(6) on line 7.

With our type given to a term e, segments can be identified by examining the type of e in **spawn**(e) for extra $-$ or $\neg$. For example, in **spawn**$(e_1); e_2$, if the canonical sequence of e_1 has $-$ or $\neg$; then the thread of e_1 must be synchronized with its parent which is the thread of e_2. Function merge in Definition 6 is used in these situations, but to define it we need some auxiliary functions:

For $S \in {}^T\bar{\mathbb{N}}$ and for a tag $sig \in \{+, -, \neg, \sharp\}$, we introduce the function $first(S, sig)$ that returns the smallest index i such that $\mathrm{tag}(S(i)) = sig$. If no such element exists, the function returns 0. A commit can be a local commit or, implicitly, a joint commit. At first, we presume all commits to be local commits. Then, when we discover that there is no local transaction starting statement (i.e. onacid) to match with a local commit, that commit should be a joint commit. The following function performs that job and converts a canonical sequence that has no $+$ element to a so-called *joint sequence*.

Definition 5 (Join). *Let $S = s_1 \ldots s_k$ be a canonical sequence such that $+ \notin \{S\}$ and assume $i = first(S, -)$. Then, function $join(S)$ recursively replaces $-$ in S by $\neg$ as follows:*

$$join(S) = S \qquad\qquad\qquad\qquad\qquad\qquad\qquad if\ i = 0$$
$$join(S) = s_1..s_{i-1}\ \neg 1\ join(\ \neg(|s_i| - 1)s_{i+1}..s_k) \qquad otherwise$$

Note that in Definition 5 the canonical sequence S contains only $\sharp$ elements interleaved with $-$ or $\neg$ elements. After applying the join function, we get joint sequences. These joint sequences contain only $\sharp$ elements interleaved with $\neg$ elements.

A joint sequence is used to type a term inside a **spawn** or a term in the main thread. The joint sequences are merged together in the following definition:

Definition 6 (Merge). *Let S_1 and S_2 be joint sequences such that the number of $\neg$ elements in S_1 and S_2 are the same (can be zero). The merge function is defined recursively as:*

$$merge(\sharp m_1, \sharp m_2) = \sharp(m_1 + m_2)$$
$$merge(\sharp m_1\ \neg n_1\ S_1', \sharp m_2\ \neg n_2\ S_2') = \sharp(m_1 + m_2)\ \neg(n_1 + n_2)\ merge(S_1', S_2')$$

The definition is well-formed, since S_1, S_2 are joint sequences so they have only $\sharp$ and $\neg$ elements. In addition, the number of $\sharp$s are the same in the assumption of the definition. So we can insert $\sharp 0$ to make the two sequences match over the defined patterns. Note that for the merge function is used for terms like **spawn**$(e_1); e_2$, in which we compute the type for e_1, then apply the join function to obtain a joint sequence—the type of **spawn**(e_1). Then, we need to compute a matching joint sequence from e_2 to merge with the joint sequence of the type of **spawn**(e_1).

We need one more function, which we use to type terms of the form $e_1 + e_2$. For these terms, we require that the external transactional behaviors of e_1 and e_2 are the same, i.e., when removing all the elements with the tag $\sharp$ from them, the remaining sequences are identical. Let S_1 and S_2 be such two sequences. Then, they can always be written as $S_i = \sharp m_i \ *n\ S_i'$, $i = 1, 2$, $* = \{+, -, \neg\}$, where S_1' and S_2' in turn have the same transactional behaviors. On this condition for S_1 and S_2, we define the choice operator as follows:

Definition 7 (Choice). *Let S_1 and S_2 be two sequences such that if we remove all $\sharp$ elements from them, then the remaining two sequences are identical. The alt function is recursively defined as:*

$$alt(\sharp m_1, \sharp m_2) = \sharp \max(m_1, m_2)$$
$$alt(\sharp m_1\ *n\ S_1', \sharp m_2\ *n\ S_2') = \sharp \max(m_1, m_2)\ *n\ alt(S_1', S_2')$$

4.2 Typing Rules

The language of types T is defined by the following syntax:

$$T = S \mid S^\rho$$

The second kind of type S^ρ is used for term **spawn**(e) as it needs to synchronizes with their parent thread if there is any joint commit. The treatments of two cases are different, so we denote $kind(T)$ the kind of T, which can be empty (normal) or ρ depending on which case T is.

The type environment encodes the transaction context for the term being typed. The typing judgment is of the form:

$$n \vdash e : T$$

where $n \in \mathbb{N}$ is the type environment. When n is negative, it means e uses n units of memory for its logs when executing e. When n is positive, it means e can free n units of memory of some log.

Table 2. Typing rules

$$\frac{}{-n \vdash \mathbf{onacid}(n) : \,^+n} \text{ T-ONACID} \qquad \frac{n \in \mathbb{N}^+}{n \vdash \mathbf{commit} : \,^-1} \text{ T-COMMIT}$$

$$\frac{n \vdash e : S}{n \vdash \mathbf{spawn}(e) : \mathrm{join}(S)^\rho} \text{ T-SPAWN} \qquad \frac{n \vdash e : S}{n \vdash e : \mathrm{join}(S)^\rho} \text{ T-PREP}$$

$$\frac{n_i \vdash e_i : S_i \quad i = 1,2 \quad S = \mathrm{seq}(S_1 S_2)}{n_1 + n_2 \vdash e_1; e_2 : S} \text{ T-SEQ}$$

$$\frac{n_1 \vdash e_1 : S_1 \quad n_2 \vdash e_2 : S_2^\rho \quad S = \mathrm{jc}(S_1, S_2)}{n_1 + n_2 \vdash e_1; e_2 : S} \text{ T-JC}$$

$$\frac{n \vdash e_i : S_i^\rho \quad i = 1,2 \quad S = \mathrm{merge}(S_1, S_2)}{n \vdash e_1; e_2 : S^\rho} \text{ T-MERGE}$$

$$\frac{n \vdash e_i : T_i \quad i = 1,2 \quad kind(T_1) = kind(T_2) \quad T_i = S_i^{kind(T_i)}}{n \vdash e_1 + e_2 : \mathrm{alt}(S_1, S_2)^{kind(S_1)}} \text{ T-COND}$$

The typing rules for our calculus are shown in Table 2. Note that we do not have rules for typing α as we assume they do not interfere with multi-threading and transactional semantics, so we can remove them when typing the programs. We assume that in these rules functions seq, jc, merge, alt are applicable, i.e., their arguments satisfy the conditions of the functions. The rule T-SPAWN converts S to the joint sequence and marks the new type by ρ so that we can merge with its parent in T-MERGE. The rule T-PREP allows us to make a matching type for the e in T-MERGE. The remaining rules are straightforward except for the rule T-JC in which we need the new function jc (in Definition 8). The rule T-JC handles the joint commit between the threads running in parallel. The last $+$ element in S_1, say ^+n, will be matched with the first $\neg$ element in S_2, say ^-l (Fig. 3). But after ^+n, there can be a $\sharp$ element, say $^\sharp n'$, so the local peak of memory units used by the term having type $^+n\,^\sharp n'$ is $n + n'$. Before ^-l there can be a $^\sharp l'$, so when we do the joint commit of terms having type ^-l with its starting transaction having type ^+n the type of the segment will be $l' + l * n$. After combining ^+n from S_1 and ^-l from S_2 we can simplify the new sequences and repeat the join commits of jc. Thus, the function jc is defined as follows:

Definition 8 (Joint commit). *Function* jc *is defined recursively as follows:*

$$\text{jc}(S_1' {}^{+n}\,{}^{\sharp}n'\,,\, {}^{\sharp}l'\, \neg l\, S_2') = \text{jc}(\text{seq}(S_1'\, {}^{\sharp}(n + n'))\,,\, \text{seq}(\, {}^{\sharp}(l' + l * n)\, S_2')) \ \text{if}\ l > 0$$

$$\text{jc}(\, {}^{\sharp}n'\,,\, {}^{\sharp}l'\, S_2') = \text{seq}(\, {}^{\sharp}\max(n', l')\, S_2') \ \text{otherwise}$$

Note that in this definition of jc the pattern matching in the first line has higher priority than one in the second line.

As our type reflects the behavior of a term, so the type of a well-typed program contains only a sequence of single ${}^{\sharp}n$ element where n is the maximum number of units of memory used when implementing the program.

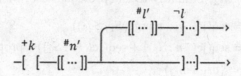

Fig. 3. Joint commit parallel threads

Definition 9 (Well-typed). *A term e is* well-typed *if there exists a type derivation for e such that $0 \vdash e : {}^{\sharp}n$ for some n.*

A typing judgment has a crucial property for our correctness proofs. It states that the typing environment combined with the type of its term always produces a 'well-formed' structure.

Theorem 1 (Type judgment property). *If $n \vdash e : T$ and $n \geqslant 0$, then* $\text{sim}(\, {}^{+}n\,, T) = {}^{\sharp}m$ *for some m (i.e. $\text{sim}(\, {}^{+}n\,, T)$ has the form of single element with tag $\sharp$) and $m \geqslant n$ where $\text{sim}(T_1, T_2) = \text{seq}(\text{jc}(S_1, S_2))$ with S_1, S_2 is T_1, T_2 without ρ.*

Proof. By induction on the typing rules in Table 2.

- The case T-ONACID does not apply as $n < 0$.
- The case T-COMMIT we have $\text{seq}(\text{jc}(\, {}^{+}n\,, \neg 1)) = {}^{\sharp}n$ so $m = n$.
- For T-SEQ, by induction hypotheses (IH) we have $\text{seq}(\text{jc}(\, {}^{+}n_i, S_i)) = \text{seq}(\, {}^{+}n_i\, S_i) = {}^{\sharp}m_i$ with $i = 1, 2$ since S_i have no $\neg$ elements. We need to prove that $\text{sim}(\, {}^{+}(n_1 + n_2)\,, S) = {}^{\sharp}m$ and $m \geqslant m_1 + m_2$. We have

$$\begin{aligned}
\text{sim}(\, {}^{+}(n_1 + n_2)\,, S) &= \text{seq}(\text{jc}(\, {}^{+}(n_1 + n_2)\,, \text{seq}(S_1 S_2))) \\
&= \text{seq}(\, {}^{+}n_2\, {}^{+}n_1\, S_1 S_2) && \neg \notin S_1 S_2 \\
&= \text{seq}(\, {}^{+}n_2\, (\, {}^{+}n_1\, S_1) S_2) && \text{Definition 4} \\
&= \text{seq}(\, {}^{+}n_2\, {}^{\sharp}m_1\, S_2) && \text{IH} \\
&= {}^{\sharp}(m_1 + m_2) && \text{IH, Definition 4}
\end{aligned}$$

- For T-JC, by induction hypotheses we have $\text{seq}(^+n_1\, S_1) = {}^\sharp m_1$ and $\text{seq}(\text{jc}(^+n_2, S_2)) = {}^\sharp m_2$. Similarly to the previous case, we have

$$
\begin{aligned}
\text{sim}(^+(n_1+n_2), S) &= \text{seq}(\text{jc}(^+(n_1+n_2), \text{jc}(S_1,S_2))) && S = \text{jc}(S_1,S_2)\\
&= \text{seq}(\text{jc}(^+(n_1+n_2)\, S_1, S_2)) && \text{jc}\\
&= \text{seq}(\text{jc}(^+n_2\, {}^\sharp m_1, S_2)) && \text{IH}\\
&= {}^\sharp\max(n_2+m_1, m_2) && \text{jc, IH}\\
&\geqslant {}^\sharp\max(n_2+n_1) && m_1 \geqslant n_1 \text{ by IH}
\end{aligned}
$$

- For T-MERGE, by induction hypotheses we have $\text{seq}(\text{jc}(^+n_1, S_1)) = {}^\sharp m_1$ and $\text{seq}(\text{jc}(^+n_2, S_2)) = {}^\sharp m_2$. Similarly to the previous case, we have

$$
\begin{aligned}
\text{seq}(\text{jc}(^+n, S)) &= \text{seq}(\text{jc}(^+n, \text{merge}(S_1,S_2))) && S = \text{merge}(S_1,S_2)\\
&= \text{seq}(\text{jc}(^+n, S_1)) + \text{seq}(\text{jc}(^+n, S_2)) && \text{properties of } S_1, S_2\\
&= {}^\sharp(m_1+m_2) && \text{IH}
\end{aligned}
$$

- For the remaining rules, the lemma holds by the induction hypotheses.   □

Typing the running example program. Let us try to make a type derivation for the program in Listing 1.1. We denote e_m^l for the part of the program from line l to line m. First, using T-SEQ, T-ONACID, T-COMMIT we have:

$$
6 \vdash \mathbf{onacid}(5); \mathbf{commit}; \mathbf{commit}; \mathbf{commit}; \mathbf{commit} : {}^\sharp 5 \ {}^{\neg}1 \ {}^{\neg}1 \ {}^{\neg}1 \tag{1}
$$

Then, by applying the rule T-SPAWN, we have:

$$
6 \vdash e_5^5 : ({}^\sharp 5 \ {}^{\neg}1 \ {}^{\neg}1 \ {}^{\neg}1)^\rho \tag{2}
$$

Now, we want to use T-MERGE and we need a term such that its type matches the type of e_5^5. We find that e_{10}^6 satisfies this condition since its type can be derived using T-SEQ, T-ONACID, T-COMMIT as follows:

$$
6 \vdash e_{10}^6 : {}^{\neg}1 \ {}^\natural 6 \ {}^{\neg}1 \ {}^\natural 7 \ {}^{\neg}1
$$

By applying T-PREP, we have a matching type with (2). So we can apply T-MERGE to get the type for e_{10}^5 as follows:

$$
6 \vdash e_{10}^5 : ({}^\sharp 5 \ {}^{\neg}2 \ {}^\natural 6 \ {}^{\neg}2 \ {}^\natural 7 \ {}^{\neg}2)^\rho
$$

With $-3 \vdash e_4^4 : {}^+3$, we can apply T-JC to get the type of e_{10}^4 as follows:

$$
3 \vdash e_{10}^4 : {}^\natural 11 \ {}^{\neg}2 \ {}^\natural 7 \ {}^{\neg}2 \tag{3}
$$

as

$$
\begin{aligned}
\text{jc}({}^+3, {}^\sharp 5 \ {}^{\neg}2 \ {}^\natural 6 \ {}^{\neg}2 \ {}^\natural 7 \ {}^{\neg}2) &= \text{jc}(\text{seq}({}^\natural 3), \text{seq}({}^\sharp(5+3*2) \ {}^\natural 6 \ {}^{\neg}2 \ {}^\natural 7 \ {}^{\neg}2))\\
&= \text{jc}({}^\natural 3, {}^\natural 11 \ {}^{\neg}2 \ {}^\natural 7 \ {}^{\neg}2) = \text{seq}({}^\natural 11 \ {}^{\neg}2 \ {}^\natural 7 \ {}^{\neg}2) = {}^\natural 11 \ {}^{\neg}2 \ {}^\natural 7 \ {}^{\neg}2
\end{aligned}
$$

Similarly, we can calculate the type for the term on line 3: $3 \vdash e_3^3$: $(^{\natural}4 \ ^{-}1 \ ^{-}1)^{\rho}$. This type matches (3) so we can apply T-MERGE and get the type for e_{10}^3 as follows:

$$3 \vdash e_{10}^3 : {}^{\natural}15 \ ^{-}3 \ ^{\flat}7 \ ^{-}3$$

Type for line 1 to line 2 is: $-3 \vdash e_2^1 : {}^{+}1 \ ^{+}2$. Apply T-JC for e_2^1 and e_{10}^3 we get:

$$0 \vdash e_{10}^1 : {}^{\flat}24$$

since

$$\mathrm{jc}({}^{+}1 \ ^{+}2, \ {}^{\natural}15 \ ^{-}3 \ ^{\flat}7 \ ^{-}3) = \mathrm{jc}(\mathrm{seq}({}^{+}1 \ ^{\flat}2), \mathrm{seq}({}^{\flat}21 \ ^{\flat}7 \ ^{-}3))$$
$$= \mathrm{jc}({}^{+}1 \ ^{\flat}2, \ {}^{\flat}21 \ ^{-}3) = \mathrm{jc}(\mathrm{seq}({}^{\flat}3), \mathrm{seq}({}^{\flat}24)) = \mathrm{jc}({}^{\flat}3, \ {}^{\flat}24) = {}^{\flat}24$$

The program is well-typed and the maximum memory that it needs in this case is 24 units. In the next section, we will show the soundness of the type system.

5 Correctness

To prove the correctness of our type system, we need to show that a well-typed program does not use more memory than the amount expressed in its type. Let our well-typed program be e and its type is ${}^{\natural}n$. We need to show that when executing e according to the semantics in Sect. 3, the total number of units of memory used for logs by the program in the global environment is always smaller than or equal to n.

A state is a pair Γ, P where $\Gamma = \{p_1 : E_1, \ldots, p_k : E_k\}$ and $P = \coprod_1^k p_i(e_i)$. We say Γ satisfies P, notation $\Gamma \models P$, if there exist $S_1, \ldots, S_k$ such that $[\![E_i]\!] \vdash e_i : S_i$ for all $i = 1, \ldots, k$. For a component i, E_i represents the number of units of memory that have been created or copied in thread p_i, and S_i represents the number of units of memory that will be created when executing e_i. Therefore, total memory used by thread p_i is expressed by $\mathrm{sim}({}^{+}[\![E_i]\!], S_i)$, where the sim function is defined in Theorem 1. We will show that $\mathrm{sim}({}^{+}[\![E_i]\!], S) = {}^{\natural}n$ for some n. We denote this value n as $[\![E_i, S_i]\!]$. Then, the total memory of logs of a program state, included in Γ and the potential logs that will be created when executing the remaining program, denoted by $[\![\Gamma, P]\!]$, and is defined by:

$$[\![\Gamma, P]\!] = \sum_{i=1}^{k} [\![E_i, S_i]\!]$$

Since $[\![\Gamma, P]\!]$ represents the maximum number of units *from* the current state and $[\![\Gamma]\!]$ is the number of units *in* the current state, we have the following lemma.

Lemma 1. *If $\Gamma \models P$, then $[\![\Gamma, P]\!] \geqslant [\![\Gamma]\!]$.*

Proof. By the definition of $[\![\Gamma, P]\!]$ and $[\![\Gamma]\!]$, we only need to show $[\![E_i, S_i]\!] \geqslant [\![E_i]\!]$ for all i. This follows from Theorem 1. □

Lemma 2 (Subject reduction). *If $\Gamma \models P$ and $\Gamma, P \Rightarrow \Gamma', P'$, then $\Gamma' \models P'$ and $[\![\Gamma, P]\!] \geqslant [\![\Gamma', P']\!]$.*

Proof (Sketch). The proof is done by checking one by one all the semantics rules in Table 1. For each rule, we need to prove two parts: (i) $\Gamma' \models P'$ and (ii) $[\![\Gamma, P]\!] \geqslant [\![\Gamma', P']\!]$.

- For S-SPAWN, by the assumption we have $\Gamma \models P$ and $\{p : E\} \in \Gamma$ and $P = P_1 \parallel p(\mathbf{spawn}(e_1); e_2)$ and $[\![E]\!] \vdash \mathbf{spawn}(e_1); e_2 : S$ for some P_1, S. By the definition of function *spawn* we have $\Gamma' = \Gamma \cup \{p' : E\}$ and by the rule S-SPAWN, we have $P' = P_1 \parallel p'(e_1) \parallel p(e_2)$.
    - For (i), by definition of $\models$ we only need to prove that $[\![E]\!] \vdash e_1 : S_1$ and $[\![E]\!] \vdash e_2 : S_2$ for some S_1, S_2 because P' differs from P only in the terms e_1 and e_2 of p' and p.
    Since $[\![E]\!] \vdash \mathbf{spawn}(e_1); e_2 : S$ and $\mathbf{spawn}(e_1); e_2$ can only be typed by $T - MERGE$, we have $[\![E]\!] \vdash \mathbf{spawn}(e_1) : S_1'$ and $[\![E]\!] \vdash e_2 : S_2$ for some S_1' and S_2. By $T - SPAWN$ we have $[\![E]\!] \vdash e_1 : S_1$ where $S_1' = \mathrm{join}(S_1)^\rho$. So (i) holds and we also have $S = \mathrm{merge}(S_1, S_2)$.
    - For (ii), first we denote $n_i = \mathrm{seq}(\mathrm{jc}(^+[\![E]\!], S_i))$ with $i = 1, 2$. We have:

$$
\begin{aligned}
&[\![\Gamma, P]\!] - [\![\Gamma', P']\!] \\
&= [\![E, S]\!] - ([\![E, S_1]\!] + [\![E, S_2]\!]) & \text{def. of } [\![.]\!] \\
&= \mathrm{seq}(\mathrm{jc}(^+[\![E]\!], S)) - (n_1 + n_2))) & \text{def. of } [\![.]\!] \\
&= \mathrm{seq}(\mathrm{jc}(^+[\![E]\!], \mathrm{merge}(S_1, S_2))) - (n_1 + n_2) & S = \mathrm{merge}(S_1, S_2) \\
&= (n_1 + n_2) - (n_1 + n_2) & \text{properties of } S_1, S_2 \\
&= 0
\end{aligned}
$$

    So (ii) holds.
- The remaining cases can be proved similarly but due to lack of space we omit here[1]

Now we come to the correctness property of our type system. A well-typed program will not use more units of memory than the one stated in its type.

Theorem 2 (Correctness). *Suppose $0 \vdash e : {}^\natural n$ and $p_1 : \epsilon, p_1(e) \Rightarrow^* \Gamma, P$, then $[\![\Gamma]\!] \leq n$.*

Proof. For the starting environment we have: $[\![p_1 : \epsilon, p_1(e)]\!] = \mathrm{sim}(0, {}^\natural n) = {}^\natural n$. So from Lemma 2 and Theorem 1, the theorem holds by induction on the length of transitions. □

---

[1] Full proof version avaiable at https://github.com/truonganhhoang/tm-infer/blob/master/tm-full.pdf.

6   Type Inference

We have implemented a prototype tool[2] in F# language that can check and infer types for well-typed programs. Listing 1.3 is the main `infer` function, which is similar to the algorithm in [15], that takes a `term` and an 'environment' `headseq` in line 3. The differences are in the implementation of other functions such as seq, jc and merge. Listing 1.2 shows the simplified implementation of seq in Definition 4 where in line 1 T.P, T.M, T.X, T.J denote tags $+, -, \sharp, \neg$, respectively. The function finds the patterns defined in Definition 4 and simplifies them.

A program or term is encoded as a list of branches and leaves. The algorithm travels the program from the first statement to the last statement and makes recursive calls when hitting **spawn** statements. When we reach the end of `term` in line 5 we need to compact the result type by calling the seq function (Definition 4). Otherwise, we check if the next statement x is a branch or a leaf. If it is a leaf we just update the `headseq` with the new leaf (line 10) and then repeat the inference process. In case x is a branch (line 12), we infer its term `br` and merge it with the remaining part `xs`. The merged type is combined with the head 'environment' to produce the final result.

We tested our tool on several examples. The code contains automated tests and all test cases are passed, i.e., actual results are equal to our expected ones.

Listing 1.2. seq function used in `infer`

```
1    type T = P = 0 | M = 1 | X = 2 | J = 3
2    let rec seq (lst : TagSeq) : TagSeq =
3      match lst with
4      | [] -> []
5      | (_,0)::xs -> seq xs
6      | (T.X,1)::(T.X,m)::xs -> seq ((T.X,max 1 m)::xs)
7      | (T.M,1)::(T.M,m)::xs -> seq ((T.M,1+m)::xs)
8      | (T.P,1)::(T.M,m)::xs -> seq ((T.X,1)::(T.M,m-1)::xs)
9      | (T.P,1)::(T.X,n)::(T.M,m)::xs -> seq ((T.X,1+n)::(T.M,m-1)
10         ::xs)
11     | x::xs -> x::(seq xs)
```

Listing 1.3. Main type inference algorithm

```
1    type TagNum = T * int
2    type Tree =  | Branch of Tree list | Leaf of TagNum
3    let rec infer (term: Tree list) (headseq: TagNum list) =
4      match term with
5      | [] -> seq headseq (* simplifies the result *)
6      | x::xs ->
7        match x with
8        | Leaf tagnum ->
9          (* expand the head part *)
10         let new_head = seq (List.append headseq [tagnum]) in
11         infer xs new_head
```

---

[2] Available at https://github.com/truonganhhoang/tm-infer.

```
12      | Branch br -> (* a new thread *)
13         (* infer the child and parent tail *)
14         let child = join (infer br []) in
15         let parent = join (infer xs []) in
16         (* merge the child and the parent tail *)
17         let tailseq = seq (merge child parent) in
18         (* join commit with the head *)
19         jc headseq tailseq
```

7 Conclusion

We have presented a generalized language whose main features are a mixing of multi-threading and nested transactions. A key new feature in the language is that it contains size information of transaction logs, which in practice can be automatically synthesized by identifying shared variables in the transactions. Then, based on the size information, we can infer statically the maximum memory units needed for transaction logs. Although the language is not easy to use directly as it is designed to give control power to programmers on the behavior of transactions, the analysis we presented can be applied to popular transactional languages as they are special cases of our presented language. The type system of our analysis looks similar to the ones in our previous works, but the semantics of type elements and typing rules are novel and the maximum memory obtained from well-typed programs is of practical value.

We are extending our tool to take real world transactional programs as input and produce the worst execution scenarios of the program where maximum log memory are used.

References

1. Albert, E., Arenas, P., Fernández, J.C., Genaim, S., Gómez-Zamalloa, M., Puebla, G., Román-Díez, G.: Object-sensitive cost analysis for concurrent objects. Softw. Test. Verification Reliab. 25(3), 218–271 (2015)
2. Albert, E., Correas, J., Román-Díez, G.: Peak cost analysis of distributed systems. In: Müller-Olm, M., Seidl, H. (eds.) SAS 2014. LNCS, vol. 8723, pp. 18–33. Springer, Heidelberg (2014). doi:10.1007/978-3-319-10936-7_2
3. Albert, E., Genaim, S., Gomez-Zamalloa, M.: Heap space analysis for Java bytecode. In: Proceedings of the 6th International Symposium on Memory Management, ISMM 2007, pp. 105–116. ACM, New York (2007)
4. Aspinall, D., Atkey, R., MacKenzie, K., Sannella, D.: Symbolic and analytic techniques for resource analysis of java bytecode. In: Wirsing, M., Hofmann, M., Rauschmayer, A. (eds.) TGC 2010. LNCS, vol. 6084, pp. 1–22. Springer, Heidelberg (2010). doi:10.1007/978-3-642-15640-3_1
5. Bezem, M., Hovland, D., Truong, H.: A type system for counting instances of software components. Theoret. Comput. Sci. 458, 29–48 (2012)
6. Braberman, V., Garbervetsky, D., Hym, S., Yovine, S.: Summary-based inference of quantitative bounds of live heap objects. Sci. Comput. Program. Part A 92, 56–84 (2014). Special issue on Bytecode 2012

7. Braberman, V.A., Garbervetsky, D., Yovine, S.: A static analysis for synthesizing parametric specifications of dynamic memory consumption. J. Object Technol. **5**(5), 31–58 (2006)
8. Chin, W.-N., Nguyen, H.H., Qin, S., Rinard, M.: Memory usage verification for OO programs. In: Hankin, C., Siveroni, I. (eds.) SAS 2005. LNCS, vol. 3672, pp. 70–86. Springer, Heidelberg (2005). doi:10.1007/11547662_7
9. Hoffmann, J., Shao, Z.: Automatic static cost analysis for parallel programs. In: Vitek, J. (ed.) ESOP 2015. LNCS, vol. 9032, pp. 132–157. Springer, Heidelberg (2015). doi:10.1007/978-3-662-46669-8_6
10. Hofmann, M., Jost, S.: Static prediction of heap space usage for first-order functional programs. vol. 38, pp. 185–197. ACM, New York, January 2003
11. Hughes, J., Pareto, L.: Recursion and dynamic data-structures in bounded space: towards embedded ML programming. SIGPLAN Not. **34**(9), 70–81 (1999)
12. Jagannathan, S., Vitek, J., Welc, A., Hosking, A.: A transactional object calculus. Sci. Comput. Program. **57**(2), 164–186 (2005)
13. Mai Thuong Tran, T., Steffen, M., Truong, H.: Compositional static analysis for implicit join synchronization in a transactional setting. In: Hierons, R.M., Merayo, M.G., Bravetti, M. (eds.) SEFM 2013. LNCS, vol. 8137, pp. 212 228. Springer, Heidelberg (2013). doi:10.1007/978-3-642-40561-7_15
14. Pham, T.-H., Truong, A.-H., Truong, N.-T., Chin, W.-N.: A fast algorithm to compute heap memory bounds of Java Card applets. In: Cerone, A., Gruner, S. (eds.) Sixth IEEE International Conference on Software Engineering and Formal Methods, SEFM 2008, Cape Town, South Africa, 10–14 November 2008, pp. 259–267. IEEE Computer Society (2008)
15. Truong, A.-H., Hung, D., Dang, D.-H., Vu, X.-T.: A type system for counting logs of multi-threaded nested transactional programs. In: Bjørner, N., Prasad, S., Parida, L. (eds.) ICDCIT 2016. LNCS, vol. 9581, pp. 157–168. Springer, Heidelberg (2016). doi:10.1007/978-3-319-28034-9_21
16. Vu, X.-T., Mai Thuong Tran, T., Truong, A.-H., Steffen, M.: A type system for finding upper resource bounds of multi-threaded programs with nested transactions. In: Symposium on Information and Communication Technology 2012, SoICT 2012, Halong City, Quang Ninh, Viet Nam, 23–24 August 2012, pp. 21–30 (2012)
17. Wegbreit, B.: Mechanical program analysis. Commun. ACM **18**(9), 528–539 (1975)

Design, Synthesis and Testing

Synthesis of Petri Nets with Whole-Place Operations and Localities

Jetty Kleijn[1], Maciej Koutny[2], and Marta Pietkiewicz-Koutny[2]([✉])

[1] LIACS, Leiden University, P.O. Box 9512, 2300 RA Leiden, The Netherlands
[2] School of Computing Science, Newcastle University,
Newcastle upon Tyne NE1 7RU, UK
marta.koutny@ncl.ac.uk

Abstract. Synthesising systems from behavioural specifications is an attractive way of constructing implementations which are correct-by-design and thus requiring no costly validation efforts. In this paper, systems are modelled by Petri nets and the behavioural specifications are provided in the form of step transition systems, where arcs are labelled by multisets of executed actions. We focus on the problem of synthesising Petri nets with whole-place operations and localities (WPOL-nets), which are a class of Petri nets powerful enough to express a wide range of system behaviours, including inhibition of actions, resetting of local states, and locally maximal executions.

The synthesis problem was solved for several specific net classes and later a general approach was developed within the framework of τ-nets. In this paper, we follow the synthesis techniques introduced for τ-nets that are based on the notion of a region of a transition system, which we suitably adapt to work for WPOL-nets.

Keywords: Concurrency · Theory of regions · Transition system · Synthesis problem · Petri net · Step semantics · Locality · Whole-place operations net

1 Introduction

The starting point of a scientific investigation that aims at describing and analysing a dynamic system or an experiment is very often a record of a series of observations as depicted, for example, by a graph like that in Fig. 1(a). The observation graph captures important information about the system, e.g., the fact that it can be in three different states in which the quantity of some crucial resource ξ has been measured to be equal to 2, 1, or 0 units. Other relevant information is that the moves between these three states result from executions of three distinct actions: A, B, and C. Moreover, these actions can sometimes be performed simultaneously (for example, B and C), as well as individually (for example, A).

Suppose now that one would like to construct a formal system model matching the observations depicted by the graph in Fig. 1(a). Such a model could then

© Springer International Publishing AG 2016
A. Sampaio and F. Wang (Eds.): ICTAC 2016, LNCS 9965, pp. 103–120, 2016.
DOI: 10.1007/978-3-319-46750-4_7

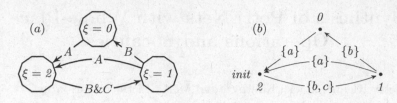

Fig. 1. A record of real-life observations of a system (a); and its step transition system representation (b).

be used for further analyses of the real-life system using suitable techniques and tools. Since the observation graph conveys a mix of state and action information, a natural way of proceeding might be to develop a Petri net model, as Petri nets deal explicitly with both state and action based issues and are able to express different relationships between actions and/or states: causality, simultaneity, and competing for resources.

To construct a Petri net model for the observation graph in Fig. 1(a), we first convert it into a slightly more formal representation in terms of a transition system as shown in Fig. 1(b) where the actions A, B, and C are respectively represented by net-transitions a, b, and c, the arcs are labelled by sets of executed net-transitions, and the nodes are labelled by integers representing the volume of the crucial resource ξ. Moreover, one node is designated as the initial state. A key idea is that the quantity of the crucial resource can be represented by a specific place (local state) p_1 in a Petri net model to be constructed, and the overall aim of the synthesis process is to build a Petri net whose reachability graph is isomorphic to the graph in Fig. 1(b), and the tokens assigned to place p_1 in different markings (global states) are as specified by the integers labelling the nodes.

It is natural to aim at a model as simple as possible, and so one might attempt to synthesise from Fig. 1(b) a Place/Transition net (PT-net). since these are the simplest Petri net model allowing one to represent integer-valued quantities. However, such an attempt would fail, as the transition system in Fig. 1(b) does not represent the behaviour of any PT-net. The first reason is that to be so it should have contained two more arcs, labelled by $\{b\}$ and $\{c\}$, outgoing from the initial state. Another problem is that it contains two $\{a\}$-labelled arcs coming to the same (initial) state from two distinct states. Since PT-nets are backward-deterministic, they would never produce this kind of behaviour.

The Petri net model we will use to construct a suitable formal model for behavioural descriptions like that in Fig. 1(b), will be nets with whole-place operations (i.e., the weight of an arc may depend on the current total number of tokens in a subset of places) and localities (WPOL-nets). Grouping net-transitions in different localities and introducing an execution semantics that allows only maximal multisets of enabled net-transitions to 'fire' within a given locality helps to overcome the first problem mentioned above. Allowing the weights of connections between places and transitions to depend on the current marking and, in consequence, introducing *whole-place operations* addresses the second problem concerning the backward non-deterministic behaviour.

The synthesis of a WPOL-net from a transition system specification will be based on the notion of a *region* of a transition system [2,3,10] suitably adapted to WPOL-nets and their locally maximal execution semantics, a special kind of *step firing policy* (see [7,13]). This paper shows for the first time how to synthesise a net, whose execution depends dynamically on the current marking (distribution of 'resources'), under an additional constraint in the form of a step firing policy.

Synthesising systems from behavioural specifications is an attractive way of constructing implementations which are correct-by-design and thus requiring no costly validation efforts. The synthesis problem was solved for many specific classes of nets, e.g., [4,5,8,14–17]. Later, a general approach was developed within the framework of τ-nets that takes a *net-type* as a parameter [3]. In this paper, we focus on the problem of synthesising WPOL-nets from behavioural specifications provided by step transition systems. WPOL-nets are nets with whole-place operations (WPO-nets) extended with transition localities. WPO-nets in turn are derived from transfer/reset nets [9] and *affine* nets [11], extending PT-nets with whole-place operations [1]. A solution to the synthesis problem for WPO-nets was outlined in [12], and we use some of the ideas introduced there in this paper, at the same time dealing with the additional constraint of the locally maximal execution semantics.

The paper is organised as follows. The next section recalls some basic notions concerning transition systems, PT-nets, and τ-nets. Section 3 introduces WPO-nets and WPOL-nets, and Sects. 4 and 5 present a solution to the synthesis problem for WPOL-nets, treating them as a special kind of τ-nets. The paper ends with a brief conclusion that outlines some directions for future work.

2   Preliminaries

An *abelian monoid* is a set $\mathbb{S}$ with a commutative and associative binary operation $+$, and an identity element $\mathbf{0}$. The result of composing n copies of $s \in \mathbb{S}$ is denoted by $n \cdot s$ and so $\mathbf{0} = 0 \cdot s$. Two examples of abelian monoids are: (i) $\mathbb{S}_{PT} = \mathbb{N} \times \mathbb{N}$, where $\mathbb{N}$ are all non-negative integers, with the pointwise arithmetic addition operation and $\mathbf{0} = (0,0)$ and (ii) the free abelian monoid $\langle T \rangle$ generated by a set T. $\mathbb{S}_{PT}$ will represent (weighted) arcs between places and transitions in PT-nets, whereas $\langle T \rangle$ will represent *steps* (multisets of transitions) of nets with transition set T. The free abelian monoid $\langle T \rangle$ can be seen as the set of all finite multisets over T, e.g., $aab = aba = baa = \{a, a, b\}$. We use $\alpha, \beta, \gamma, \ldots$ to range over the elements of $\langle T \rangle$. For $t \in T$ and $\alpha \in \langle T \rangle$, $\alpha(t)$ denotes the multiplicity of t in α, and so $\alpha = \sum_{t \in T} \alpha(t) \cdot t$. Then $t \in \alpha$ whenever $\alpha(t) > 0$, and $\alpha \leq \beta$ whenever $\alpha(t) \leq \beta(t)$ for all $t \in T$. The size of α is $|\alpha| = \sum_{t \in T} \alpha(t)$.

Transition systems. A *(deterministic) transition system* $\langle Q, \mathbb{S}, \delta \rangle$ over an abelian monoid $\mathbb{S}$ consists of a set of *states* Q and a partial *transition function*[1] δ :

---

[1] Transition functions are not related to (Petri) net-transitions.

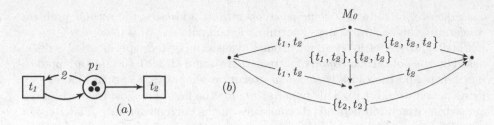

Fig. 2. A PT-net (*a*); and its concurrent reachability graph (*b*).

$Q \times \mathbb{S} \to Q$ such that $\delta(q, \mathbf{0}) = q$ for all $q \in Q$. An *initialised* transition system $\langle Q, \mathbb{S}, \delta, q_0 \rangle$ is a transition system with an *initial* state $q_0 \in Q$ such that each state $q \in Q$ is *reachable*, i.e., there are $s_1, \ldots, s_n$ and $q_1, \ldots, q_n = q$ ($n \geq 0$) with $\delta(q_{i-1}, s_i) = q_i$, for $1 \leq i \leq n$. For every state q of a transition system TS, we denote by $enb_{TS}(q)$ the set of all s which are *enabled* at q, i.e., $\delta(q, s)$ is defined. TS is *bounded* if $enb_{TS}(q)$ is finite for every state q of TS. Moreover, such a TS is *finite* if it has finitely many states. In diagrams, $\mathbf{0}$-labelled arcs are omitted and singleton steps written without brackets.

Initialised transition systems $\mathcal{T}$ over free abelian monoids — called *step transition systems* or *concurrent reachability graphs* — represent behaviours of Petri nets. *Net-types* are non-initialised transition systems τ over abelian monoids used to define various classes of nets.

Let $\mathcal{T} = \langle Q, \langle T \rangle, \delta, q_0 \rangle$ and $\mathcal{T}' = \langle Q', \langle T \rangle, \delta', q_0' \rangle$ be step transition systems. $\mathcal{T}$ and $\mathcal{T}'$ are *isomorphic*, $\mathcal{T} \cong \mathcal{T}'$, if there is a bijection f with $f(q_0) = q_0'$ and $\delta(q, \alpha) = q' \Leftrightarrow \delta'(f(q), \alpha) = f(q')$, for all $q, q' \in Q$ and $\alpha \in \langle T \rangle$.

Place/Transition nets. A *Place/Transition net* (PT-net, for short) is a tuple $N = \langle P, T, W, M_0 \rangle$, where P and T are disjoint sets of *places* and *transitions*, $W : (P \times T) \cup (T \times P) \to \mathbb{N}$ is a *weight function*, and M_0 is an *initial marking* belonging to the set of *markings* defined as mappings from P to $\mathbb{N}$. We use the standard conventions concerning the graphical representation of PT-nets, as illustrated in Fig. 2(*a*).

For all $p \in P$ and $\alpha \in \langle T \rangle$, we denote $W(p, \alpha) = \sum_{t \in T} \alpha(t) \cdot W(p, t)$ and $W(\alpha, p) = \sum_{t \in T} \alpha(t) \cdot W(t, p)$. Then a *step* $\alpha \in \langle T \rangle$ is *enabled* and may be *fired* at a marking M if, for every $p \in P$, $M(p) \geq W(p, \alpha)$. We denote this by $\alpha \in enb_N(M)$. *Firing* such a step leads to the marking M', for every $p \in P$ defined by $M'(p) = M(p) - W(p, \alpha) + W(\alpha, p)$. We denote this by $M[\alpha\rangle M'$. The *concurrent reachability graph* $CRG(N)$ of N is the step transition system formed by firing inductively from M_0 all possible enabled steps, i.e., $CRG(N) = \langle [M_0\rangle, \langle T \rangle, \delta, M_0 \rangle$ where

$$[M_0\rangle = \{M_n \mid \exists \alpha_1, \ldots, \alpha_n \, \exists M_1, \ldots M_{n-1} \, \forall 1 \leq i \leq n \; : \; M_{i-1}[\alpha_i\rangle M_i\}$$

is the set of *reachable* markings and $\delta(M, \alpha) = M'$ *iff* $M[\alpha\rangle M'$. Figure 2(b) shows the concurrent reachability graph of the PT-net in Fig. 2(*a*).

Petri nets defined by net-types. A net-type $\tau = \langle \mathcal{Q}, \mathbb{S}, \Delta \rangle$ is a parameter in the definition of τ-*nets*. It specifies the values (markings) that can be stored in places ($\mathcal{Q}$), the operations and tests (inscriptions on the arcs) that a net-transition may perform on these values ($\mathbb{S}$), and the enabling condition and the newly generated values for steps of transitions (Δ).

A τ-*net* is a tuple $N = \langle P, T, F, M_0 \rangle$, where P and T are respectively disjoint sets of places and transitions, $F : (P \times T) \rightarrow \mathbb{S}$ is a *flow mapping*, and M_0 is an *initial marking* belonging to the set of *markings* defined as mappings from P to $\mathcal{Q}$. N is *finite* if both P and T are finite.

For all $p \in P$ and $\alpha \in \langle T \rangle$, we denote $F(p, \alpha) = \sum_{t \in T} \alpha(t) \cdot F(p, t)$. Then a step $\alpha \in \langle T \rangle$ is *enabled* at a marking M if, for every $p \in P$, $F(p, \alpha) \in enb_\tau(M(p))$. We denote this by $\alpha \in enb_N(M)$. *Firing* such a step produces the marking M', for every $p \in P$ defined by $M'(p) = \Delta(M(p), F(p, \alpha))$. We denote this by $M[\alpha\rangle M'$, and then define the *concurrent reachability graph* $CRG(N)$ of N as the step transition system formed by firing inductively from M_0 all possible enabled steps.

As in [3,7], it is possible to encode a PT-net $N = \langle P, T, W, M_0 \rangle$ as a τ-net without affecting its concurrent reachability graph, It is enough to take $F(p, t) = (W(p, t), W(t, p))$. Thus $F(p, t) = (i, o)$ means that i is the weight of the arc from p to t, and o the weight of the arc in the opposite direction. With this encoding, N becomes a τ_{PT}-net where $\tau_{PT} = \langle \mathbb{N}, \mathbb{S}_{PT}, \Delta_{PT} \rangle$ is an infinite net-type over $\mathbb{S}_{PT}$ defined earlier, with Δ_{PT} given by $\Delta_{PT}(n, (i, o)) = n - i + o$ provided that $n \geq i$ (see Fig. 5(a)).

3   Nets with Whole-Place Operations

Assuming an ordering of places, markings can be represented as vectors. The i-th component of a vector $\mathbf{x}$ is denoted by $\mathbf{x}^{(i)}$. For $\mathbf{x} = (x_1, \ldots, x_n)$ and $\mathbf{y} = (y_1, \ldots, y_n)$, $(\mathbf{x}, 1) = (x_1, \ldots, x_n, 1)$ and $\mathbf{x} \otimes \mathbf{y} = x_1 \cdot y_1 + \cdots + x_n \cdot y_n$. Moreover, $\otimes$ will also denote the multiplication of two-dimensional arrays.

A *net with whole-place operations* (WPO-net) is a tuple $N = \langle P, T, W, \mathbf{m}_0 \rangle$, where $P = \{p_1, \ldots, p_n\}$ is a finite set of ordered *places*, T is a finite set of *transitions* disjoint with P, $W : (P \times T) \cup (T \times P) \rightarrow \mathbb{N}^{n+1}$ is a *whole-place* weight function, and $\mathbf{m}_0$ is an *initial marking* belonging to the set $\mathbb{N}^n$ of *markings*.

For $p \in P$ and $\alpha \in \langle T \rangle$, $W(p, \alpha) = \sum_{t \in T} \alpha(t) \cdot W(p, t)$ and $W(\alpha, p) = \sum_{t \in T} \alpha(t) \cdot W(t, p)$. Then α is *enabled* at a marking $\mathbf{m}$ if, for every $p \in P$,

$$\mathbf{m}(p) \geq (\mathbf{m}, 1) \otimes W(p, \alpha). \tag{1}$$

We denote this by $\alpha \in enb_N(\mathbf{m})$. An enabled α can be *fired* leading to a new marking such that, for every $p \in P$,

$$\mathbf{m}'(p) = \mathbf{m}(p) + (\mathbf{m}, 1) \otimes (W(\alpha, p) - W(p, \alpha)). \tag{2}$$

We denote this by $\mathbf{m}[\alpha\rangle\mathbf{m}'$, and define the *concurrent reachability graph* $CRG(N)$ of N as one built by firing inductively from $\mathbf{m}_0$ all possible enabled steps.

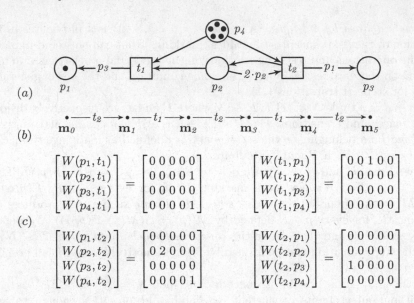

Fig. 3. A WPO-net generating the first six Fibonacci numbers (a); its concurrent reachability graph (b); and the weight function (c).

It is convenient to specify the weight function using arc annotations which are linear expressions involving the p_i's. For example, if $n = 3$ then $W(p_3, t) = (3, 0, 1, 5)$ can be written down as $3 \cdot p_1 + p_3 + 5$. A place p_j $(1 \leq j \leq n)$ is a *whole-place* if $W(p, t)^{(j)} > 0$ or $W(t, p)^{(j)} > 0$, for some $p \in P$ and $t \in T$. In such a case we also write $p_j \rightsquigarrow p$. Note that it may happen that $p = p_j$; for example, if $W(p_1, t) = p_1$. This is useful, e.g., for simulating inhibitor arcs (see $W(p_2, t_2) = 2 \cdot p_2$ in Fig. 3(a)). In diagrams, arcs with '0' annotation are dropped, and '1' annotations are not shown.

Figure 3 shows a modified example, taken from [9], of a WPO-net for the generation of the first six Fibonacci numbers. Its markings are as follows: $\mathbf{m_0} = (\mathbf{1}, 0, 0, 5)$, $\mathbf{m_1} = (1, \mathbf{1}, 1, 4)$, $\mathbf{m_2} = (\mathbf{2}, 0, 1, 3)$, $\mathbf{m_3} = (2, 1, \mathbf{3}, 2)$, $\mathbf{m_4} = (\mathbf{5}, 0, 3, 1)$, and $\mathbf{m_5} = (5, 1, \mathbf{8}, 0)$. Hence, the markings of places p_1 and p_3, in alternation, represent the first six Fibonacci numbers (written above in bold). As $W(t_1, p_1)^{(3)} = 1 > 0$, $W(t_2, p_3)^{(1)} = 1 > 0$ and $W(p_2, t_2)^{(2)} = 2 > 0$, the net has three whole-places, p_1, p_2 and p_3 with $p_3 \rightsquigarrow p_1$, $p_1 \rightsquigarrow p_3$ and $p_2 \rightsquigarrow p_2$. Moreover, p_4, acting as a simple counting place, is a *non-whole-place*.

A WPO-*net with localities* (or WPOL-net) is a tuple $N = \langle P, T, W, \mathbf{m_0}, \ell \rangle$ such that $N' = \langle P, T, W, \mathbf{m_0} \rangle$ is a WPO-net, and $\ell : T \rightarrow \{1, 2, \dots, l\}$, where $l \geq 1$, is the *locality mapping* of N and $\{1, 2, \dots, l\}$ are the *localities* of N. In diagrams, nodes representing transitions assigned the same locality are shaded in the same way, as illustrated in Fig. 4(a) for transitions b and c. Finally N inherits the notations introduced for N'.

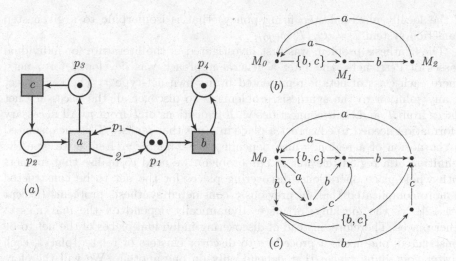

Fig. 4. A WPOL-net (*a*); its concurrent reachability graph (*b*); and the concurrent reachability graph of the underlying WPO-net (*c*).

WPOL-nets are executed under the *locally maximal* step firing policy. A step $\alpha \in \langle T \rangle$ is *resource enabled* at a marking $\mathbf{m}$ if, for every $p \in P$, the inequality (1) is satisfied. Such a step is then *control enabled* if there is no $t \in T$ such that there exists a transition $t' \in \alpha$ with $\ell(t) = \ell(t')$ and the step $t + \alpha$ is resource enabled at $\mathbf{m}$. A control enabled step α can be then fired leading to the marking $\mathbf{m}'$, for every $p \in P$ given by the formula (2).

In general (see [7]), a step firing policy is given by a *control disabled steps* mapping $cds : 2^{\langle T \rangle} \to 2^{\langle T \rangle \setminus \{\mathbf{0}\}}$ that, for a set of resource enabled steps at some reachable marking, returns the set of steps disabled by this policy at that marking. For the locally maximal step firing policy this mapping will be denoted by cds_{lmax} and we will identify this policy with its cds_{lmax} mapping:[2]

$$cds_{lmax}(X) = \{\alpha \in X \setminus \{\mathbf{0}\} \mid \exists \beta \in X : \ell(\beta) \subseteq \ell(\alpha) \wedge \alpha \leq \beta \wedge \alpha \neq \beta\}.$$

Step firing policies are a means of controlling and constraining the potentially huge number of execution paths generated by a concurrent system. The concurrent reachability graph of a net executed under a step firing policy contains only the control enabled steps (see Fig. 4(*b*, *c*)).

4   Synthesis of WPOL-nets

The net synthesis problem we consider here aims to devise a procedure which constructs a WPOL-net with a concurrent reachability graph (reflecting the use

---

[2] Control disabled steps mappings are defined in [7] in the context of τ-nets, and this is how cds_{lmax} will be used in Sect. 4.

of the locally maximal step firing policy) that is isomorphic to a given step transition system $\mathcal{T} = \langle Q, \langle T \rangle, \delta, q_0 \rangle$.

The synthesis problem was first investigated in the literature for individual classes of Petri nets, and later a general approach was developed for τ-nets, where each class of nets is represented by its own net-type τ. The key aspect of any solution to the synthesis problems is to discover all the necessary net places from $\mathcal{T}$ and their connections with transitions of T from τ. All necessary information needed to construct a place in a net that realises $\mathcal{T}$ is encapsulated in the notion of a region, which depends on parameter τ. Before we give the definition of a region relevant to our problem, we need to realise that for nets with whole-place operations, discovering places for the net to be constructed is more complicated than in previously considered synthesis problems (except for [12]), as the markings of places dynamically depend on the markings of other places. Therefore, instead of discovering individual places of the net to be constructed, one needs a procedure to discover clusters of related places, each cluster containing places that depend only on one another. We will therefore re-define WPOL-nets as nets containing clusters of — at most k — related places (k-WPOL-nets) and express them as τ-nets, so that we can synthesise them as τ-nets, using the general framework of net synthesis theory.

4.1   k-WPOL-nets and Their Net-Type

WPOL-nets allow arc weights to depend on the current marking of all places. This may be too generous, e.g., in the case of systems where places are distributed among different neighbourhoods, forming the scopes where their markings can influence the token game. One way of capturing this is to restrict the number of places which can influence arc weights.

A k-restricted WPOL-net (k-WPOL-net, $k \geq 1$) is a WPOL-net N for which there is a partition $P_1 \uplus \cdots \uplus P_r$ of the set of places such that each P_i has at most k places and, for all $p \in P_i$ and $p' \notin P_i$, $p \not\rightarrow p' \not\rightarrow p$. In other words, the places can be partitioned into clusters of bounded size so that there is no exchange of whole-place marking information between different clusters.

Although k-WPOL-nets (as well as WPOL-nets) are not τ-nets in the sense of the original definition as the change of a marking of a place does not only depend on its marking and the connections to the transitions, they still fit the ideas behind the definition of τ-nets. All we need to do is to define a suitably extended net-type capturing the behaviour of sets of k places rather than the behaviour of single places. More precisely, for each $k \geq 1$, the k-WPOL-net-type is a transition system[3]:

$$\tau^k = \langle \mathbb{N}^k, (\mathbb{N}^{k+1})^k \times (\mathbb{N}^{k+1})^k, \Delta^k \rangle$$

where

$$\Delta^k : \mathbb{N}^k \times ((\mathbb{N}^{k+1})^k \times (\mathbb{N}^{k+1})^k) \rightarrow \mathbb{N}^k$$

---

[3] As will be explained later, the same net-type can be defined for a given kind of nets to be executed without any specific policy or with some policy. Therefore, we can re-use here the τ_{wpo}^k net-type introduced in [12], which coincides with τ^k.

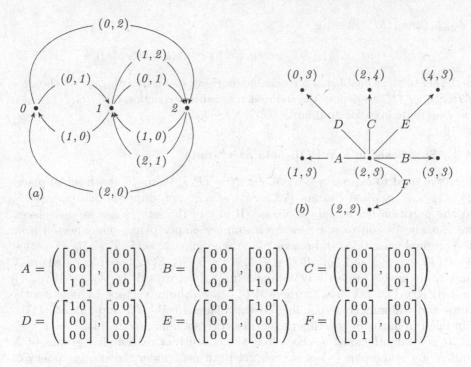

Fig. 5. Fragments of two infinite net-types: τ_{PT} (a); and τ^2 (b).

is a partial function such that $\Delta^k(\mathbf{x}, (X, Y))$ is defined if $\mathbf{x} \geq (\mathbf{x}, 1) \otimes X$ and, if that is the case,

$$\Delta^k(\mathbf{x}, (X, Y)) = \mathbf{x} + (\mathbf{x}, 1) \otimes (Y - X).$$

Note that here we treat tuples of vectors in $(\mathbb{N}^{k+1})^k$ as $(k+1) \times k$ arrays.

Having defined a net-type τ^k, a τ^k-*net* is a tuple $N = \langle \mathcal{P}, T, F, M_0, \ell \rangle$, where $\mathcal{P} = \{P_1, \ldots, P_r\}$ is a set of disjoint sets of implicitly ordered places comprising exactly k places each, T is a set of transitions being different from the places in the sets of $\mathcal{P}$, $F : (\mathcal{P} \times T) \to (\mathbb{N}^{k+1})^k \times (\mathbb{N}^{k+1})^k$ is a *flow mapping*, M_0 is an *initial marking* belonging to the set of *markings* defined as mappings from $\mathcal{P}$ to $\mathbb{N}^k$, and ℓ is a locality mapping for the transitions in T.

For all $P_i \in \mathcal{P}$ and $\alpha \in \langle T \rangle$, we set $F(P_i, \alpha) = \sum_{t \in T} \alpha(t) \cdot F(P_i, t)$. Then a step $\alpha \in \langle T \rangle$ is *resource enabled* at a marking M if, for every $P_i \in \mathcal{P}$, $F(P_i, \alpha) \in enb_{\tau^k}(M(P_i))$. We denote this by $\alpha \in enb_N(M)$. *Firing* such a step (for now we ignore the firing policy) produces the marking M', for every $P_i \in \mathcal{P}$, defined by $M'(P_i) = \Delta^k(M(P_i), F(P_i, \alpha))$. We denote this by $M[\alpha\rangle M'$, and then define the *concurrent reachability graph* $CRG(N)$ of N as the step transition system formed by firing inductively from M_0 all possible enabled steps.

However, we want to execute N under the locally maximal step firing policy. The related control disabled steps mapping cds_{lmax}, when applied to N, would control disable at each marking M all the resource enabled steps that belong to

$cds_{lmax}(enb_N(M))$. That is,

$$enb_{N,cds_{lmax}}(M) = enb_N(M) \setminus cds_{lmax}(enb_N(M)) \tag{3}$$

is the set of steps enabled at a reachable marking M under cds_{lmax}. We then use $CRG_{cds_{lmax}}(N)$ to denote the induced reachable restriction of $CRG(N)$, which may be finite even for an infinite $CRG(N)$.

4.2   Synthesising k-WPOL-nets as τ^k-nets

First we need to express a k-WPOL-net $N = \langle P, T, W, \mathbf{m}_0, \ell \rangle$, with set of places $P = \{p_1, \ldots, p_n\}$ and clusters $P_1, \ldots, P_r$, as a τ^k-net. Suppose that each set P_i in the partition has exactly k places. (If any of the sets P_i has $m < k$ places, we can always add to it $k - m$ fresh dummy empty places disconnected from the original transitions and places.) We then define $\widehat{N} = \langle \mathcal{P}, T, F, M_0, \ell \rangle$ so that $\mathcal{P} = \{P_1, \ldots, P_r\}$ and, for all $P_i \in \mathcal{P}$ and $t \in T$: (i) $F(P_i, t) = (X, Y)$ where X and Y are arrays respectively obtained from the arrays $[W(p_1, t), \ldots, W(p_n, t)]$ and $[W(t, p_1), \ldots, W(t, p_n)]$, where $W(\cdot, \cdot)$ are column vectors, by deleting the rows and columns corresponding to the places in $P \setminus P_i$; and (ii) $M_0(P_i)$ is obtained from $\mathbf{m}_0$ by deleting the entries corresponding to the places in $P \setminus P_i$.

It is straightforward to check that the concurrent reachability graphs of N and $\widehat{N}$ are isomorphic (when we execute both nets under the cds_{lmax} policy or ignore the policy in both nets). Conversely, one can transform any τ^k-net into an equivalent k-WPOL-net, and trivially each WPOL-net is a $|P|$-WPOL-net.

We can turn the WPO-net of Fig. 3(a) into a WPOL-net with locality mapping ℓ such that $\ell(t_1) = 1$ and $\ell(t_2) = 2$. The result can be represented as a τ^2-net $\widehat{N} = \langle \{P_1, P_2\}, \{t_1, t_2\}, F, M_0, \ell \rangle$, where $P_1 = \{p_1, p_3\}$ and $P_2 = \{p_2, p_4\}$, $M_0(P_1) = (1, 0)$, $M_0(P_2) = (0, 5)$ and:

$$F(P_1, t_1) = \left(\begin{bmatrix} 0 & 0 \\ 0 & 0 \\ 0 & 0 \end{bmatrix}, \begin{bmatrix} 0 & 0 \\ 1 & 0 \\ 0 & 0 \end{bmatrix} \right) \qquad F(P_1, t_2) = \left(\begin{bmatrix} 0 & 0 \\ 0 & 0 \\ 0 & 0 \end{bmatrix}, \begin{bmatrix} 0 & 1 \\ 0 & 0 \\ 0 & 0 \end{bmatrix} \right)$$

$$F(P_2, t_1) = \left(\begin{bmatrix} 0 & 0 \\ 0 & 0 \\ 1 & 1 \end{bmatrix}, \begin{bmatrix} 0 & 0 \\ 0 & 0 \\ 0 & 0 \end{bmatrix} \right) \qquad F(P_2, t_2) = \left(\begin{bmatrix} 2 & 0 \\ 0 & 0 \\ 0 & 1 \end{bmatrix}, \begin{bmatrix} 0 & 0 \\ 0 & 0 \\ 1 & 0 \end{bmatrix} \right)$$

The above discussion implies that k-WPOL-net synthesis can be reduced to the following two problems of τ^k-net synthesis.

Problem 1 (feasibility). *Let $\mathcal{T} = \langle Q, \langle T \rangle, \delta, q_0 \rangle$ be a bounded step transition system, k be a positive integer, and ℓ be a locality mapping for $\mathcal{T}$.*
Provide necessary and sufficient conditions for $\mathcal{T}$ to be realised by some τ^k-net, $\widehat{N}$, executed under the cds_{lmax} policy defined by ℓ ($\mathcal{T} \cong CRG_{cds_{lmax}}(\widehat{N})$).

Problem 2 (effective construction). *Let $\mathcal{T} = \langle Q, \langle T \rangle, \delta, q_0 \rangle$ be a finite step transition system, k be a positive integer, and ℓ be a locality mapping for $\mathcal{T}$.*

Decide whether there is a finite τ^k-net realising $\mathcal{T}$ when executed under the cds_{lmax} policy defined by ℓ. Moreover, if the answer is positive construct such a τ^k-net.

To address Problem 1, we define a τ^k-region of $\mathcal{T} = \langle Q, \langle T \rangle, \delta, q_0 \rangle$ as a pair:

$$\langle \sigma : Q \to \mathbb{N}^k, \eta : T \to (\mathbb{N}^{k+1})^k \times (\mathbb{N}^{k+1})^k \rangle$$

such that, for all $q \in Q$ and $\alpha \in enb_{\mathcal{T}}(q)$,

$$\eta(\alpha) \in enb_{\tau^k}(\sigma(q)) \text{ and } \Delta^k(\sigma(q), \eta(\alpha)) = \sigma(\delta(q, \alpha)),$$

where $\eta(\alpha) = \sum_{t \in T} \alpha(t) \cdot \eta(t)$. Moreover, for every state q of Q, we denote by $enb_{\mathcal{T}, \tau^k}(q)$ the set of all steps α such that $\eta(\alpha) \in enb_{\tau^k}(\sigma(q))$, for all τ^k-regions $\langle \sigma, \eta \rangle$ of $\mathcal{T}$. Hence for every state q of $\mathcal{T}$, we have

$$enb_{\mathcal{T}}(q) \subseteq enb_{\mathcal{T}, \tau^k}(q). \tag{4}$$

In the context of the synthesis problem, a τ^k-region represents a cluster of places whose local states (in τ^k) are consistent with the global states (in $\mathcal{T}$). Then, to deliver a realisation of $\mathcal{T}$, one needs to find *enough* τ^k-regions to construct a τ^k-net $\widehat{N}$ satisfying $\mathcal{T} \cong CRG_{cds_{lmax}}(\widehat{N})$. The need for the existence of such τ^k-regions is dictated by the following two *regional axioms*:

Axiom 1 (state separation). *For any pair of states $q \neq r$ of $\mathcal{T}$, there is a τ^k-region $\langle \sigma, \eta \rangle$ of $\mathcal{T}$ such that $\sigma(q) \neq \sigma(r)$.*

Axiom 2 (forward closure). *For every state q of $\mathcal{T}$, $enb_{\mathcal{T}}(q) = enb_{\mathcal{T}, \tau^k}(q) \setminus cds_{lmax}(enb_{\mathcal{T}, \tau^k}(q))$.*

The above axioms provide a full characterisation of realisable transition systems. The first axiom links the states of $\mathcal{T}$ with markings of the net to be constructed, making sure that a difference between two states of $\mathcal{T}$ is reflected in a different number of tokens held in the two markings of the net representing the said states. The second axiom means that, for every state q and every step α in $\langle T \rangle \setminus enb_{\mathcal{T}}(q)$, we have either of the following:

1. There is a τ^k-region $\langle \sigma, \eta \rangle$ of $\mathcal{T}$ such that $\eta(\alpha) \notin enb_{\tau^k}(\sigma(q))$ (the step α is not *region enabled*) or
2. $\alpha \in cds_{lmax}(enb_{\mathcal{T}, \tau^k}(q))$ (the step α is not *control enabled*, meaning that it is rejected by the cds_{lmax} policy).

Note that when a τ^k-net under cds_{lmax} realises $\mathcal{T}$, every cluster of places of the net still determines a corresponding τ^k-region of the transition system, without taking cds_{lmax} into account. This is why the same kind of regions would be used if we are asked to synthesise a WPO-net (rather than a WPOL-net).

Before we prove the main result of the paper that gives the solution to Problem 1, we need two auxiliary results. The first one presents an important property enjoyed by control disabled steps mappings, and in particular by cds_{lmax}.

Proposition 1. *Let X be a finite set of resource enabled steps at some reachable marking of some τ^k-net and Y be its subset ($Y \subseteq X$). Then:*

$$X \setminus cds_{lmax}(X) \subseteq Y \implies cds_{lmax}(X) \cap Y \subseteq cds_{lmax}(Y).$$

Proof. Let $\alpha \in cds_{lmax}(X) \cap Y$. We need to show that $\alpha \in cds_{lmax}(Y)$. From $\alpha \in cds_{lmax}(X)$ it follows that there is $\beta \in X$ such that $\ell(\beta) \subseteq \ell(\alpha)$ and $\alpha < \beta$. We now consider two cases:

Case 1: $\beta \in Y$. Then $\alpha \in cds_{lmax}(Y)$.

Case 2: $\beta \in X \setminus Y$. Then, by $X \setminus cds_{lmax}(X) \subseteq Y$, we have that $\beta \in cds_{lmax}(X)$. Hence, there is $\gamma \in X$ such that $\ell(\gamma) \subseteq \ell(\beta)$ and $\beta < \gamma$. If $\gamma \in Y$ we can continue as in case 1, with γ replacing β and obtain $\alpha \in cds_{lmax}(Y)$ due to the transitivity of $\subseteq$ and $<$. Otherwise, we continue as in case 2 with γ replacing β and so $\gamma \in cds_{lmax}(X)$. Then we can repeat the same argument. Now, because X is a finite set, one must find sooner or later in this iteration some step $\phi \in Y$ such that case 1 holds with ϕ replacing β, and so $\alpha \in cds_{lmax}(Y)$. $\square$

The second auxiliary result associates a region of a step transition system $\mathcal{T}$ with a particular cluster of places of the net to be synthesised from $\mathcal{T}$. The mappings σ and η hold all the information about the associated cluster of places, their connections to transitions in the net and their markings for every state of the net. In fact, for the mapping σ, if we know η, it is enough to know its value for the initial state q_0 to uniquely compute the values for the remaining states of $\mathcal{T}$.

Proposition 2. *Let $\mathcal{T} \cong CRG_{cds_{lmax}}(\widehat{N})$ for a τ^k-net $\widehat{N} = \langle \mathcal{P}, T, F, M_0, \ell \rangle$. Then, for each cluster $P_i \in \mathcal{P}$ ($i = 1, \ldots, r$), there is exactly one τ^k-region $\langle \sigma, \eta \rangle$ of $\mathcal{T}$ such that $\sigma(q_0) = M_0(P_i)$ and $\eta(\alpha) = F(P_i, \alpha)$ for all steps $\alpha \in \langle T \rangle$.*

Proof. All step transition systems we consider in this paper are deterministic. Observe that both δ and Δ^k are functions rather than relations. Also observe that $\mathcal{T}$ is reachable (i.e., each of its states is reachable from the initial one). Hence, $\sigma(q_0)$ and $\eta : \langle T \rangle \to (\mathbb{N}^{k+1})^k \times (\mathbb{N}^{k+1})^k$ determine at most one map $\sigma : Q \to \mathbb{N}^k$ such that $\Delta^k(\sigma(q), \eta(\alpha)) = \sigma(\delta(q, \alpha))$ whenever $\alpha \in enb_{\mathcal{T}}(q)$, and therefore they determine at most one τ^k-region of $\mathcal{T}$.

We now define the map σ. Let $P_i \in \mathcal{P}$ ($i = 1, \ldots, r$). By assumption $\mathcal{T} \cong CRG_{cds_{lmax}}(\widehat{N})$, and $CRG_{cds_{lmax}}(\widehat{N})$ is a sub-graph of $CRG(\widehat{N})$. Let $\sigma : Q \to \mathbb{N}^k$ be defined as follows: $\sigma(q) = f(q)(P_i)$, where $f(q)$ is the image of q through the isomorphism $\cong$ ($f(q)$ is a marking of $\widehat{N}$). Then, for every $\alpha \in enb_{\mathcal{T}}(q)$, we have, from $\mathcal{T} \cong CRG_{cds_{lmax}}(\widehat{N})$, that α is resource enabled at $f(q)$ in $\widehat{N}$, and hence $F(P_i, \alpha) \in enb_{\tau^k}(f(q)(P_i))$ and the marking of P_i after α is fired is $f(\delta(q, \alpha))(P_i) = \Delta^k(f(q)(P_i), F(P_i, \alpha))$. Therefore, we have, for σ defined as above and $\eta(\alpha) = F(P_i, \alpha)$ (as stated in the assumptions), that $\eta(\alpha) \in enb_{\tau^k}(\sigma(q))$ and $\sigma(\delta(q, \alpha)) = \Delta^k(\sigma(q), \eta(\alpha))$. Hence $\langle \sigma, \eta \rangle$, with σ defined as above, is a τ^k-region of $\mathcal{T}$ associated with P_i. Also, $\sigma(q_0) = f(q_0)(P_i) = M_0(P_i)$ as $\cong$ is an isomorphism preserving the initial states. Therefore, the result holds. $\square$

Theorem 1. *Let $\mathcal{T} = \langle Q, \langle T \rangle, \delta, q_0 \rangle$ be a bounded step transition system and cds_{lmax} be the locally maximal step firing policy associated with a locality mapping ℓ defined for $\mathcal{T}$.*

Then $\mathcal{T}$ can be realised by a τ^k-net ($k \geq 1$) under cds_{lmax} iff Axioms 1 and 2 are satisfied.

Proof. ($\Longrightarrow$) $\mathcal{T}$ can be realised by the τ^k-net $\widehat{N}$ under cds_{lmax}. That means that $\mathcal{T} \cong CRG_{cds_{lmax}}(\widehat{N})$. Let $f : Q \to (\mathcal{P} \to \mathbb{N}^k)$ be a bijection linking the states of $\mathcal{T}$ with the reachable markings of $\widehat{N}$. First, we prove that:

$$enb_{\mathcal{T},\tau^k}(q) \subseteq enb_{\widehat{N}}(f(q)). \tag{5}$$

Let $\alpha \notin enb_{\widehat{N}}(f(q))$. Then there is a cluster $P_i \in \mathcal{P}$ ($1 \leq i \leq r$) in $\widehat{N}$ such that $F(P_i, \alpha) \notin enb_{\tau^k}(f(q)(P_i))$. Let $\langle \sigma, \eta \rangle$ be the τ^k-region of $\mathcal{T}$ induced by P_i according to Proposition 2. Then $\sigma(q) = f(q)(P_i)$ and $\eta(\alpha) = F(P_i, \alpha)$. Hence, $\eta(\alpha) \notin enb_{\tau^k}(\sigma(q))$ and so $\alpha \notin enb_{\mathcal{T},\tau^k}(q)$.

To show Axiom 1 let $q \neq r$ in Q. As $\mathcal{T} \cong CRG_{cds_{lmax}}(\widehat{N})$, we have $f(q) \neq f(r)$, and therefore $f(q)(P_i) \neq f(r)(P_i)$, for some $1 \leq i \leq r$. Let $\langle \sigma, \eta \rangle$ be the τ^k-region of $\mathcal{T}$ induced by P_i according to Proposition 2. Then $\sigma(q) = f(q)(P_i) \neq f(r)(P_i) = \sigma(r)$. Hence, $\sigma(q) \neq \sigma(r)$.

To show Axiom 2, we first show that, for all $\alpha \in \langle T \rangle$ and $q \in Q$:

$$\alpha \notin enb_{\mathcal{T}}(q) \implies \alpha \notin enb_{\mathcal{T},\tau^k}(q) \setminus cds_{lmax}(enb_{\mathcal{T},\tau^k}(q)). \tag{6}$$

Let $q \in Q$ and $\alpha \notin enb_{\mathcal{T}}(q)$. From (3) and $\mathcal{T} \cong CRG_{cds_{lmax}}(\widehat{N})$, either:

(i) $\alpha \notin enb_{\widehat{N}}(f(q))$ or
(ii) $\alpha \in enb_{\widehat{N}}(f(q)) \cap cds_{lmax}(enb_{\widehat{N}}(f(q)))$.

If (i) holds then, by (5), we have $\alpha \notin enb_{\mathcal{T},\tau^k}(q)$ and so (6) holds. In (ii) two cases are possible. If $\alpha \notin enb_{\mathcal{T},\tau^k}(q)$ we have (6); otherwise $\alpha \in enb_{\mathcal{T},\tau^k}(q)$ and we set the following: $X = enb_{\widehat{N}}(f(q))$ and $Y = enb_{\mathcal{T},\tau^k}(q)$. By (5), we have $Y \subseteq X$. Moreover, by (3 and 4) and $\mathcal{T} \cong CRG_{cds_{lmax}}(\widehat{N})$, we have $X \setminus cds_{lmax}(X) \subseteq Y$. Hence, by Proposition 1 and the fact that $\mathcal{T}$ is bounded, $\alpha \in cds_{lmax}(X) \cap Y \subseteq cds_{lmax}(enb_{\mathcal{T},\tau^k}(q))$, and so (6) holds.

To finish the proof of Axiom 2, we show that, for all $q \in Q$:

$$enb_{\mathcal{T}}(q) \subseteq enb_{\mathcal{T},\tau^k}(q) \setminus cds_{lmax}(enb_{\mathcal{T},\tau^k}(q)). \tag{7}$$

By isomorphism $\mathcal{T} \cong CRG_{cds_{lmax}}(\widehat{N})$ and (3), we have $enb_{\mathcal{T}}(q) = enb_{\widehat{N}}(f(q)) \setminus cds_{lmax}(enb_{\widehat{N}}(f(q)))$. Hence $enb_{\mathcal{T}}(q) \cap cds_{lmax}(enb_{\widehat{N}}(f(q))) = \varnothing$. Thus, by (5) and $cds_{lmax}(Y) \subseteq cds_{lmax}(X)$ (for $Y \subseteq X$), $enb_{\mathcal{T}}(q) \cap cds_{lmax}(enb_{\mathcal{T},\tau^k}(q)) = \varnothing$. Moreover, by (4), which always holds, we can conclude that (7) holds.

($\Longleftarrow$) Let $\mathcal{R}$ be the set of all τ^k-regions of $\mathcal{T}$. Let $\widehat{N} = \langle \mathcal{P}, T, F, M_0, \ell \rangle$ be a τ^k-net defined as follows: $\mathcal{P} = \mathcal{R}$, $M_0(P_i) = \sigma(q_0)$ and $F(P_i, t) = \eta(t)$ for any $P_i = \langle \sigma, \eta \rangle \in \mathcal{P}$ and $t \in T$. We will show that if $\mathcal{T}$ satisfies Axioms 1 and 2 then $\mathcal{T} \cong CRG_{cds_{lmax}}(\widehat{N})$.

We denote by $\mathcal{RM}_{cds_{lmax}}$ the set of all reachable markings in $CRG_{cds_{lmax}}(\widehat{N})$ and by $M \xrightarrow{\alpha} M'$ the directed arcs in this graph. We now define a relation $\sim \subseteq Q \times \mathcal{RM}_{cds_{lmax}}$ as the smallest relation that includes $q_0 \sim M_0$ and such that

$$q \sim M, \delta(q,\alpha) = q' \text{ and } M \xrightarrow{\alpha} M' \text{ implies } q' \sim M'.$$

We prove first that $\sim$ is a partial bijection between Q and $\mathcal{RM}_{cds_{lmax}}$. By construction of $\widehat{N}$, $M_0(P_i) = \sigma(q_0)$ for every $P_i = \langle \sigma, \eta \rangle$ of $\widehat{N}$. Now let $q \sim M$ with $\delta(q,\alpha) = q'$ and $M \xrightarrow{\alpha} M'$, and assume for the sake of an induction that $M(P_i) = \sigma(q)$ for every $P_i = \langle \sigma, \eta \rangle$ of $\widehat{N}$. As $\langle \sigma, \eta \rangle$ is a τ^k-region of $\mathcal{T}$, $\sigma(\delta(q,\alpha)) = \Delta^k(\sigma(q), \eta(\alpha))$. As $P_i = \langle \sigma, \eta \rangle$ is a cluster of places in $\widehat{N}$ and $F(P_i, t) = \eta(t)$ for all $t \in T$ by construction of $\widehat{N}$, we have $\sigma(\delta(q,\alpha)) = \Delta^k(M(P_i), F(P_i,\alpha))$. From $M \xrightarrow{\alpha} M'$, we have $M'(P_i) = \Delta^k(M(P_i), F(P_i,\alpha))$. As a result, $M'(P_i) = \sigma(\delta(q,\alpha)) = \sigma(q')$ and we have $q' \sim M'$. So, $q \sim M$ implies $M(P_i) = \sigma(q)$ for all $P_i = \langle \sigma, \eta \rangle$ of $\widehat{N}$. Furthermore, from Axiom 1, $q \neq r$ implies $\sigma(q) \neq \sigma(r)$ for some τ^k-region $\langle \sigma, \eta \rangle$ of $\mathcal{T}$. Therefore, the relation $\sim$ is a partial bijection between Q and $\mathcal{RM}_{cds_{lmax}}$.

Next, we show that the following implication is satisfied:

$$q \sim M \implies enb_{\mathcal{T},\tau^k}(q) = enb_{\widehat{N}}(M). \tag{8}$$

Let $\alpha \in enb_{\mathcal{T},\tau^k}(q)$. This means that $\eta(\alpha) \in enb_{\tau^k}(\sigma(q))$, for all τ^k-regions $\langle \sigma, \eta \rangle$ of $\mathcal{T}$. It was shown above that, for every cluster of places $P_i = \langle \sigma, \eta \rangle$ of $\widehat{N}$, $M(P_i) = \sigma(q)$, where $q \sim M$. Furthermore, by construction of $\widehat{N}$, $F(P_i, t) = \eta(t)$, for all $t \in T$, and $P_i = \langle \sigma, \eta \rangle$. Hence, $\eta(\alpha) = F(P_i, \alpha)$. Therefore, $F(P_i, \alpha) \in enb_{\tau^k}(M(P_i))$, for every cluster of places P_i of $\widehat{N}$. This in turn means that α is resource enabled at M in $\widehat{N}$: $\alpha \in enb_{\widehat{N}}(M)$.

To show the reverse inclusion, let $\alpha \in enb_{\widehat{N}}(M)$. Then, by the fact that α is resource enabled at M, in $\widehat{N}$, we have $F(P_i, \alpha) \in enb_{\tau^k}(M(P_i))$, for every cluster P_i of $\widehat{N}$. From the construction of $\widehat{N}$, it follows that $F(P_i, t) = \eta(t)$ for all $t \in T$ and $P_i = \langle \sigma, \eta \rangle$, hence $\eta(\alpha) \in enb_{\tau^k}(M(P_i))$. For every cluster $P_i = \langle \sigma, \eta \rangle$ of $\widehat{N}$, $M(P_i) = \sigma(q)$ when $q \sim M$. So, $\eta(\alpha) \in enb_{\tau^k}(\sigma(q))$ for every τ^k-region of $\mathcal{T}$. Hence, $\alpha \in enb_{\mathcal{T},\tau^k}(q)$.

We now observe that $q \sim M$ implies $enb_{\mathcal{T}}(q) = enb_{\widehat{N}, cds_{lmax}}(M)$, which follows from (8), Axiom 2, and (3). Hence $\sim$ is a bijection between Q and $\mathcal{RM}_{cds_{lmax}}$, and so $\mathcal{T} \cong CRG_{cds_{lmax}}(\widehat{N})$. $\qquad\square$

To solve Problem 2 using the feasibility result provided by Theorem 1 one needs to find an effective representation of the τ^k-regions of $\mathcal{T}$. Similarly as in [12], one can define a system of equations and inequalities encoding the conditions that must be satisfied by τ^k-regions. Let $Q = \{q_0, q_1, \ldots, q_m\}$ and $T = \{t_1, \ldots, t_n\}$. The encoding employs the following variables:

- $\mathbf{x}_0, \mathbf{x}_1, \ldots, \mathbf{x}_m$ are k-vectors of non-negative integer variables which encode the mapping σ; and

– $\mathbf{X}_1, \ldots, \mathbf{X}_n$ and $\mathbf{Y}_1, \ldots, \mathbf{Y}_n$ are $(k+1) \times k$ arrays of non-negative integer variables which encode the mapping η.

We then define the homogeneous system $\mathcal{S}_{\mathcal{T}}$:

$$\begin{cases} \mathbf{x}_s - (\mathbf{x}_s, 1) \otimes \sum_{i=1}^n \alpha(t_i) \cdot \mathbf{X}_i \geq \mathbf{0} & \text{for all } \delta(q_s, \alpha) = q_r \\ \mathbf{x}_r - \mathbf{x}_s - (\mathbf{x}_s, 1) \otimes \sum_{i=1}^n \alpha(t_i) \cdot (\mathbf{Y}_i - \mathbf{X}_i) = \mathbf{0} & \text{in } \mathcal{T}. \end{cases} \quad (9)$$

Then the non-negative integer solutions of $\mathcal{S}_{\mathcal{T}}$ are in a one-to-one correspondence with the τ^k-regions of $\mathcal{T}$. Therefore, Axioms 1 and 2 can be checked using the solutions of $\mathcal{S}_{\mathcal{T}}$.

In the case of PT-net synthesis, a similar procedure has been shown to be effective since the homogeneous system considered there was linear and one could always find a sufficiently representative finite basis for all the solutions. Here, however, the situation is much harder as the system $\mathcal{S}_{\mathcal{T}}$ is quadratic. In practice, one would often want to impose bounds on the allowed range of the whole-place coefficients used in arc annotations. Then Problem 2 has a solution since one could replace $\mathcal{S}_{\mathcal{T}}$ by finitely many linear systems that can be dealt with using the techniques developed for PT-nets. However, one can consider a modified version of Problem 2 without bounding the whole-place coefficients and still obtain a solution, as described in the next section.

5   Synthesis with Known Whole-Places

We will now outline how one can develop a fully satisfactory procedure for synthesis problems like that discussed in the introduction.

Problem 3 (effective construction with known whole-places). *Let $\mathcal{T} = \langle Q, \langle T \rangle, \delta, q_0 \rangle$ be a finite step transition system, m be a positive integer, and κ be a mapping assigning tuples in $\mathbb{N}^m$ to Q. Decide whether there is a WPOL-net N with implicitly ordered places $p_1, \ldots, p_m, \ldots, p_n$ realising $\mathcal{T}$ such that:*

1. *each whole-place p_i of N satisfies $i \leq m$, and*
2. *for every state $q \in Q$, it is the case that $\kappa(q) = (\mu(q)^{(1)}, \ldots, \mu(q)^{(m)})$, where μ is a bijection from Q to the reachable markings of N establishing the isomorphism between $\mathcal{T}$ and the concurrent reachability graph of N.*

Moreover, if the answer is positive, construct such a WPOL-net N.

Figure 1(b) defines an instance of the above problem with $m = 1$. We will now describe how the above problem can be solved using results from the last section.

Since $\mathcal{T}$ is finite, there are only finitely many semantically distinct ways in which one can assign localities to the transitions in T. We can explore them all one-by-one, and below we assume that ℓ is a *fixed* locality mapping for T. Note that for the the example in Fig. 1(b), we must have $\ell(b) = \ell(c)$ since otherwise the initial state would have to enable also a step α with $b \in \alpha$ and $c \notin \alpha$. Hence here one only needs to consider two locality assignments.

We next discuss the coefficients on the arcs adjacent to $p_1, \ldots, p_m$. Suppose first that $i, j \leq m$ and $W(p_i, t) = v_1 \cdot p_1 + \cdots + v_m \cdot p_m + v_0$ in a net solving Problem 3, and μ is a corresponding bijection. We consider two cases:

– $\kappa(q)^{(j)} > 0$, for some $\delta(q, \alpha) = q'$ with $t \in \alpha$. Then, since α is enabled at $\mu(q)$, it must be the case that $\kappa(q)^{(i)} \geq v_j \cdot \kappa(q)^{(j)} \cdot \alpha(t)$, and so

$$v_j \leq \min \left\{ \frac{\kappa(q)^{(i)}}{\kappa(q)^{(j)} \cdot \alpha(t)} \;\middle|\; \delta(q, \alpha) = q' \; and \; t \in \alpha \right\}.$$

Hence, the range of possible values for v_j is finite.

– $\kappa(q)^{(j)} = 0$, for each $\delta(q, \alpha) = q'$ with $t \in \alpha$. Then we can assume $v_j = 1 + \max\{\kappa(q)^{(i)} \mid q \in Q\}$. This does not 'contradict' any of the arcs in T and, at the same time, ensures a maximal disabling power of coefficient v_j.

Suppose next that $i, j \leq m$ and $W(t, p_i) = v_1 \cdot p_1 + \cdots + v_m \cdot p_m + v_0$. We again consider two cases:

– $\kappa(q)^{(j)} > 0$, for some $\delta(q, \alpha) = q'$ with $t \in \alpha$. Then, since executing α at $\mu(q)$ leads to $\mu(q')$, it must be the case that $\kappa(q')^{(i)} \geq v_j \cdot \kappa(q)^{(j)} \cdot \alpha(t)$, and so

$$v_j \leq \min \left\{ \frac{\kappa(q')^{(i)}}{\kappa(q)^{(j)} \cdot \alpha(t)} \;\middle|\; \delta(q, \alpha) = q' \; and \; t \in \alpha \right\}.$$

Hence, the range of possible values for v_j is again finite.

– $\kappa(q)^{(j)} = 0$, for each $\delta(q, \alpha) = q'$ with $t \in \alpha$. Then we set $v_j = 0$.

Note that for the example in Fig. 1(b), all coefficients v_j satisfy $0 \leq v_j \leq 1$. Moreover, as $\{b, c\}$ is an enabled step, it is not possible to have both $W(p_1, b) = p_1 + v$ and $W(p_1, c) = p_1 + w$.

As a result, we need to take into account only finitely many assignments of values to the whole-place coefficients of arcs between the transitions in T and $p_1, \ldots, p_m$. We can consider them one-by-one and, after filtering out those inconsistent with κ, carry out independent searches for a solution. Therefore, below we assume that such whole-place coefficients are *fixed*, and proceed further unless the net constructed so far is a solution (the initial marking is $\kappa(q_0)$).

Having fixed transition localities and whole-place coefficients involving the potential whole-places, we can proceed with the main part of the decision procedure, i.e., the construction of additional non-whole-places that can use $p_1, \ldots, p_m$ in their arc annotations.

First, we derive the system $\mathcal{S}_T$ as in (9) with $k = m + 1$, implicitly assuming that the first m components correspond to $p_1, \ldots, p_m$, and the k-th component corresponds to a generic non-whole-place p being constructed. We then delete all equations and inequalities which concern $p_1, \ldots, p_m$, i.e., those beginning with $\mathbf{x}_s^{(i)}$, for $1 \leq i \leq m$. We finally replace by concrete values all those variables which are 'fixed' by the mapping κ, and the fact that p must be a non-whole-place. The homogeneous system $\mathcal{S}_T'$ obtained in this way is *linear*.

Assume some arbitrary ordering of the variables of $\mathcal{S}_T'$. Using the results from [6], one can find a finite set $\mathbf{p}^1, \ldots, \mathbf{p}^r$ of non-negative integer solutions of $\mathcal{S}_T'$ such that each non-negative integer solution $\mathbf{p}$ of $\mathcal{S}_T'$ is a linear combination $\mathbf{p} = \sum_{l=1}^r a_l \cdot \mathbf{p}^l$ with non-negative rational coefficients a_l. For every non-negative integer solution $\mathbf{p}$ of $\mathcal{S}_T'$, let $\psi(\mathbf{p})$ be a corresponding τ^k-region.

The $\mathbf{p}^l$'s are *fixed* and some of them turned into new places if Problem 3 has a solution under the fixed localities and coefficients. This, in turn, is the case if we can verify Axioms 1 and 2. Clearly, if $r = 0$ then the problem is not feasible for the current fixed parameters. Otherwise, we proceed as follows.

To check state separation (Axiom 1), let q_i and q_j be a pair of distinct states of $\mathcal{T}$. If $\kappa(q_i) \neq \kappa(q_j)$, then we are done. Suppose then that $\kappa(q_i) = \kappa(q_j)$, and ρ is a τ^k-region separating q_i and q_j. Then there is a solution $\mathbf{p} = \sum_{l=1}^{r} a_l \cdot \mathbf{p}^l$ such that $\rho = \psi(\mathbf{p})$. This means that $\mathbf{p}$ assigns different values to q_i and q_j. Hence, since the coefficients a_l are non-negative, there must be $\mathbf{p}^l$ which also assigns different values to q_i and q_j. Therefore, $\psi(\mathbf{p}^l)$ separates q_i and q_j. We therefore only need to check the $\mathbf{p}^l$'s in order to establish the separation of q_i and q_j. If a suitable $\mathbf{p}^l$ is found, we add a non-whole-place p corresponding to the last place of $\psi(\mathbf{p}^l)$ to the net being constructed.

Checking forward closure (Axiom 2) is carried out for each state q_i, and considers steps $\alpha \in \langle T \rangle$ that are not enabled at q_i in $\mathcal{T}$. Moreover, one does not need to consider $\alpha \neq \mathbf{0}$ in the following cases:

- α is already disabled by the whole-places, or $|\alpha| > max$, where max is the maximum size of steps labelling arcs in $\mathcal{T}$. In the latter case, one can always add a standard PT-net place which is connected with each transition by an incoming and outgoing arc of weight 1, and is initially marked with max tokens. Such a non-whole-place disables all steps with more than max transitions, and does not disable any other steps.
- There is $\beta \neq \alpha$ enabled at q_i such that $\ell(\beta) \subseteq \ell(\alpha)$ and $\alpha \leq \beta$.

In all other cases, α is not τ^k-region enabled at q_i iff $\psi(\mathbf{p})$ disables α, for some solution $\mathbf{p} = \sum_{l=1}^{r} a_l \cdot \mathbf{p}^l$. Hence, since the coefficients a_l are non-negative, α is not τ^k-region enabled at q_i iff there is $\mathbf{p}^l$ such that $\psi(\mathbf{p}^l)$ disables α. We therefore only need to check the $\mathbf{p}^l$'s in order to establish the disabling of α. If a suitable $\mathbf{p}^l$ is found, we add a non-whole-place p corresponding to the last place of $\psi(\mathbf{p}^l)$ to the net being constructed.

Finally, if one can validate all cases of state separation and forward closure, the resulting net is a solution to Problem 3, and otherwise there is no solution.

6   Conclusions

Among the possible directions for future work, we single out two challenges. The first one is the development of a synthesis approach for WPO-nets executed under more general step firing policies, e.g., those based on linear rewards of steps, where the reward for firing a single transition is either fixed or it depends on the current net marking [7]. The second task, more specific to k-WPOL-nets, is to investigate the relationship between the locality mapping and the grouping of the places into clusters.

Acknowledgements. We would like to thank the anonymous reviewers for useful comments and suggestions.

References

1. Abdulla, P.A., Delzanno, G., Van Begin, L.: A language-based comparison of extensions of Petri nets with and without whole-place operations. In: Dediu, A.H., Ionescu, A.M., Martín-Vide, C. (eds.) LATA 2009. LNCS, vol. 5457, pp. 71–82. Springer, Heidelberg (2009). doi:10.1007/978-3-642-00982-2_6
2. Badouel, E., Bernardinello, L., Darondeau, P.: Petri Net Synthesis. Texts in Theoretical Computer Science. An EATCS Series. Springer, Heidelberg (2015)
3. Badouel, E., Darondeau, P.: Theory of regions. In: Reisig, W., Rozenberg, G. (eds.) ACPN 1996. LNCS, vol. 1491, pp. 529–586. Springer, Heidelberg (1998). doi:10.1007/3-540-65306-6_22
4. Bernardinello, L., De Michelis, G., Petruni, K., Vigna, S.: On the synchronic structure of transition systems. In: Desel, J. (ed.) Structures in Concurrency Theory. Workshops in Computing, pp. 69–84. Springer, London (1995)
5. Busi, N., Pinna, G.M.: Synthesis of nets with inhibitor arcs. In: Mazurkiewicz, A., Winkowski, J. (eds.) CONCUR 1997. LNCS, vol. 1243, pp. 151–165. Springer, Heidelberg (1997). doi:10.1007/3-540-63141-0_11
6. Chernikova, N.: Algorithm for finding a general formula for the non-negative solutions of a system of linear inequalities. USSR Comput. Math. Math. Phys. **5**, 228–233 (1965)
7. Darondeau, P., Koutny, M., Pietkiewicz-Koutny, M., Yakovlev, A.: Synthesis of nets with step firing policies. Fundam. Informaticae **94**, 275–303 (2009)
8. Desel, J., Reisig, W.: The synthesis problem of Petri nets. Acta Informatica **33**, 297–315 (1996)
9. Dufourd, C., Finkel, A., Schnoebelen, P.: Reset nets between decidability and undecidability. In: Larsen, K.G., Skyum, S., Winskel, G. (eds.) ICALP 1998. LNCS, vol. 1443, pp. 103–115. Springer, Heidelberg (1998). doi:10.1007/BFb0055044
10. Ehrenfeucht, A., Rozenberg, G.: Partial 2-structures; Part I: basic notions and the representation problem, and Part II: state spaces of concurrent systems. Acta Informatica **27**, 315–368 (1990)
11. Finkel, A., McKenzie, P., Picaronny, C.: A well-structured framework for analysing Petri net extensions. Inf. Comput. **195**, 1–29 (2004)
12. Kleijn, J., Koutny, M., Pietkiewicz-Koutny, M., Rozenberg, G.: Applying Regions. Theoret. Comput. Sci. (2016)
13. Koutny, M., Pietkiewicz-Koutny, M.: Synthesis of Petri nets with localities. Sci. Ann. Comp. Sci. **19**, 1–23 (2009)
14. Mukund, M.: Petri nets and step transition systems. Int. J. Found. Comput. Sci. **3**, 443–478 (1992)
15. Nielsen, M., Rozenberg, G., Thiagarajan, P.S.: Elementary transition systems. Theoret. Comput. Sci. **96**, 3–33 (1992)
16. Pietkiewicz-Koutny, M.: Transition systems of elementary net systems with inhibitor arcs. In: Azéma, P., Balbo, G. (eds.) ICATPN 1997. LNCS, vol. 1248, pp. 310–327. Springer, Heidelberg (1997). doi:10.1007/3-540-63139-9_43
17. Schmitt, V.: Flip-flop nets. In: Puech, C., Reischuk, R. (eds.) STACS 1996. LNCS, vol. 1046, pp. 517–528. Springer, Heidelberg (1996). doi:10.1007/3-540-60922-9_42

Schedulers and Finishers: On Generating the Behaviours of an Event Structure

Annabelle McIver[1], Tahiry Rabehaja[1(✉)], and Georg Struth[2]

[1] Department of Computing, Macquarie University, Sydney, Australia
tahiry.rabehaja@mq.edu.au
[2] Department of Computer Science, The University of Sheffield, Sheffield, UK

Abstract. It is well known that every trace of a transition system can be generated using a scheduler. However, this basic completeness result does not hold in event structure models. The reason for this failure is that, according to its standard definition, a scheduler chooses which action to schedule and, at the same time, finishes the one scheduled last. Thus, scheduled events will never be able to overlap. We propose to separate scheduling from finishing and introduce the dual notion of finishers which, together with schedulers, are enough to regain completeness back. We then investigate all possible interactions between schedulers and finishers, concluding that simple alternating interactions are enough to express complex ones. Finally, we show how finishers can be used to filter behaviours to the extent to which they capture intrinsic system characteristics.

1 Introduction

Formal software analysis is principally based on the meaning given to computations. Often this semantics is defined as the set of behaviours a program can perform. For instance, the sequential behaviours of a labelled transition system are given by traces. In general, these behaviours are generated by schedulers. For labelled transition systems, schedulers are complete in the sense that each and every trace of the system can be generated by a scheduler. This completeness, however, fails if we are to model truly-concurrent behaviours using event structures. This paper introduces the concept of *finishers* which complement schedulers by providing a complete technique for the generation of all the behaviours of an arbitrary event structure.

A trace belonging to the language of a labelled transition system systematically records a totally ordered sequence of actions that are performed sequentially over time. In other words, a new action, to be appended to the end of the trace, cannot start unless the last action in that trace has terminated. The total order between the actions captures exactly how a sequential (or interleaved) system behaves. However, there are cases where we need to model situations with

This research was supported by an iMQRES from Macquarie University, the ARC Discovery Grant DP1092464 and the EPSRC Grant EP/J003727/1.

A. Sampaio and F. Wang (Eds.): ICTAC 2016, LNCS 9965, pp. 121–138, 2016.
DOI: 10.1007/978-3-319-46750-4_8

overlapping actions [13], or parallel executions with inter-process communication [8]. Totally ordering the actions fails to capture these situations faithfully and the most natural solution is to weaken the total ordering of actions into a partial ordering of events [3,8,9,17,20,23]. Thus, the behaviours of an event structure are encoded as labelled partially ordered sets, or *lposets* for short, where comparable events must occur in the given order and incomparable ones are concurrent. These concurrent events may happen in any order (interleaving) or they may overlap (true-concurrency).

Every trace of a labelled transition system can be generated by a scheduler. Intuitively, the scheduler walks through the transition system and resolves all choices by selecting one of the next available actions based on the execution history. The same technique can be defined for event structures but it is not complete because such a scheduler forces sequential dependencies, specified by the order in which events are scheduled. In other words, the scheduler does two different jobs in one go: it determines which events have occurred and which are scheduled to happen. By assigning the first task to a different entity, which we call *finishers*, we are able to schedule an event without terminating the actions of all previously scheduled events.

In this paper, the sole role of a scheduler is to choose an event that is available or *enabled*, given the current history of the computation encoded as a lposet. Once an event is scheduled, its associated action is considered to be ready to run or has started to execute but *not* yet terminated. Termination is the job of a finisher. Intuitively, a finisher looks at a lposet corresponding to the scheduled events ordered with causal dependencies, and decides which part of that lposet can be safely terminated. Thus, a finisher has at least two basic properties: finished events must have been scheduled sometime in the past and they must remain finished as the computation progresses. Through this dichotomy, we show that each and every behaviour of an event structure results from the interaction between a scheduler and a finisher.

The main contributions of this paper are listed below.

1. We give an insight into the basic nature of schedulers and introduce finishers to account for the dual counterpart of scheduling (Sect. 3.2).
2. We show that schedulers and finishers provide a complete technique to generate each and every behaviour of an event structure (Theorem 1). This technique gives a novel operational perspective at the dynamics of event structures.
3. We show that all complex interactions between schedulers and finishers can be obtained from simple alternating interactions (Theorem 2).
4. We introduce how to use finishers to filter behaviours that satisfy intrinsic characteristics such as safety or feasibility properties (Sect. 6).

This paper is organised as follows. Section 2 gives a summary of the important notions related to event structures. Section 3 introduces schedulers and finishers whose alternating interactions are elaborated in Sect. 4 to generate all the behaviours of an event structure. Section 5 shows that arbitrary interactions between schedulers and finishers can be expressed using the simpler alternating interactions. Finally, Sect. 6 exposes how to use finishers to express intrinsic properties of true-concurrent systems, and Sect. 7 discusses works that are related to our approach. Most of the results of this paper are proved in the appendix and proof sketches are added to achieve a smooth reading flow.

2   Bundle Event Structure

Event structures come in many variations such as stable and prime event structures [23], bundle event structure [9], configuration structures [20] and there are even quantitative extensions [7,11,18,21]. In this paper, we use Langerak's bundle event structure [9] to model concurrent computations. Bundle event structures form a sound model of Hoare et al.'s Concurrent Kleene Algebra and thus provide a compositional algebraic program verification framework, which is describe in our previous work [10,11,18]. Notice, however, that the notion of scheduler and finisher (and their fundamental properties) can be readily applied to stable and prime event structures as well as configuration structures.

A bundle event structure has *events* as its fundamental objects. Intuitively, an event is the occurrence of an action at a certain moment in time. Thus, an action can be repeated, but each of its occurrences is associated with a unique event. For instance, the act of writing a bit into a register is an action that repeats over time and each occurrence of the writing is a particular event.

Events are (partially) ordered by a *causality relation* which we denote by $\mapsto$: if an event e' causally depends on e, written $\{e\} \mapsto e'$, then e must happen before e' can happen. For instance, the events of writing a bit and reading it from the register are causally dependent.

It is possible for an event e'' to depend on two mutually conflicting events, that is, the event e'' may occur only after either of the events e or e' has happened. In this case, we write $\{e, e'\} \mapsto e''$ and the relationship between e and e' is called *conflict*, written $e \# e'$, because both events cannot occur simultaneously.

A set x that contains mutually conflicting events is called a *bundle set*, that is, for every $e, e' \in x$ such that $e \neq e'$, we have $e \# e'$. We will also write $x \# x$ when x is a bundle set.

Definition 1. *A bundle event structure, or simply an event structure, is a tuple* $(E, \#, \mapsto, \lambda)$, *such that E is a set of events,* the conflict relation $\# \subseteq E \times E$ *is an irreflexive and symmetric binary relation on E; the relation* $\mapsto \subseteq \mathbb{P}E \times E$ *is a bundle relation, that is, for any event $e \in E$, $x \mapsto e$ implies that x is a bundle set; and the map* $\lambda : E \to \Sigma$ *is a labelling function which associates actions from an alphabet Σ to events.*

In the sequel, we fix an event structure $\mathcal{E}$ with E as its set of events.

2.1   Trace and Configuration

The maximal behaviours of a bundle event structure can be expressed using three equivalent techniques, namely: event traces, configurations, and lposets. The following two definitions are from Langerak's work [9].

Definition 2. *A finite sequence of events $e_1 e_2 \cdots e_n$ from E is called an* event trace *if for every $i \geq 1$, $e_i \notin \{e_1, \ldots, e_{i-1}\}$, e_i is not in conflict with any of the events $e_1, \ldots, e_{i-1}$, and for every bundle relation $x \mapsto e_i$, there exists $j < i$ such that $e_j \in x$. The set of all traces of $\mathcal{E}$ is denoted by $\mathcal{T}(\mathcal{E})$ or simply $\mathcal{T}$ if no confusion may arise.*

Intuitively, an event trace is a conflict-free sequence of events such that every event in the trace is enabled from all previous events.

Given a trace $\alpha = e_1 e_2 \cdots e_n$, we denote $\leq_\alpha$ the ordering of the events in that trace, that is, $e_1 \leq_\alpha e_2 \leq_\alpha e_3 \cdots e_{n-1} \leq_\alpha e_n$. A configuration is obtained by forgetting the order of the trace α.

Definition 3. *A* configuration *is a subset $x \subseteq E$ such that $x = \{e_1, \ldots, e_n\}$ for some event trace $e_1 \cdots e_n$ referred to as a* linearization of x. *The set of all configurations of $\mathcal{E}$ is denoted by $\mathcal{C}(\mathcal{E})$ or simply $\mathcal{C}$ if no confusion may arise.*

Example 1. Consider the bundle event structure $\mathcal{E}$ depicted in Fig. 1 where the actions are left implicit since the behaviours of an event structure is provided by the configurations as per Definition 3. The events e_1 (resp. e_3) and e_2 can occur concurrently. The configurations of $\mathcal{E}$ are $\emptyset, \{e_1\}, \{e_2\}, \{e_1, e_2\}, \{e_1, e_3\}, \{e_1, e_2, e_3\}$ and $\{e_1, e_2, e_3, e_4\}$ $\qquad\qquad\square$

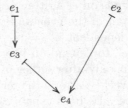

The events e_1 (resp. e_3) and e_2 are concurrent and the event e_4 requires both events e_3 and e_2 to have occurred before it can happen. Thus, it requires two causality relations $\{e_2\} \mapsto e_4$ and $\{e_3\} \mapsto e_4$.

Fig. 1. An example of event structure.

2.2 Labelled Partially Ordered Set

A *labelled partially ordered set* (lposet) is constructed from a configuration by recovering the coarsest partial order on events. More generally, a lposet is a tuple $(x, \leq, \lambda)$ such that $\leq$ is a partial order on x and $\lambda : x \to \Sigma$ is a labelling of the events in x with actions from Σ. Recall, from Szpilrajn's theorem, that a partial order is the intersection of its linearizations [1]. If x is a configuration of a bundle event structure $\mathcal{E}$, then the lposet generated by x is defined by $(x, \leq_x, \lambda_x)$ where

$$\leq_x = \bigcap_{\alpha \text{ linearization of } x} \leq_\alpha$$

and λ is the restriction of the labelling function of $\mathcal{E}$ to x. We refer to this order as the *canonical order* of x. The canonical order $\leq_x$ is the coarsest causal order that the events in x must obey. It is dictated by the bundle relation of $\mathcal{E}$ and all other behaviour, such as traces, must at least contain it. The set of lposets of $\mathcal{E}$ is denoted by $\mathcal{L}(\mathcal{E})$ or simply $\mathcal{L}$ if no confusion may arise. In the sequel, we do not distinguish between a configuration x and its associated canonical lposet $(x, \leq_x, \lambda_x)$ as long as the event structure $\mathcal{E}$ is provided by the context.

Example 2. The configuration $\{e_1, e_2, e_3\}$ of the event structure depicted in Fig. 1 has exactly three linearisations: $e_2 e_1 e_3$, $e_1 e_2 e_3$ and $e_1 e_3 e_2$. Since e_1 occurs before e_3 in all three traces and e_2 occurs before as well as after e_1 and e_3 in some traces, we deduce that the canonical order $\leq_{\{e_1, e_2, e_3\}}$ is the reflexive closure of $\{(e_1, e_3)\}$ which is a partial order on the set $\{e_1, e_2, e_3\}$.   □

In this paper, a lposet u is also denoted by $(\mathbf{set}(u), \leq_u, \lambda_u)$ so that we can refer explicitly to the set of events $\mathbf{set}(u)$, order relation $\leq_u$ or labelling function λ_u respectively. A lposet u *implements* another lposet v (or u is *subsumed* by v) if there exists a label-preserving monotonic bijection $f{:}\mathbf{set}(v){\rightarrow}\mathbf{set}(u)$. If u implements v then we write $u \sqsubseteq_s v$ (s stands for subsumption [3]). Intuitively, if u implements v, then the order in which actions are executed in v are respected in u. The lposet u may specify some extra causal dependency but concurrent events in u must also be concurrent in v. If we assume that concurrency generates nondeterminism, then we can deduce that u is more deterministic than v which is compatible with the notion of program refinement [14]. Two lposets u, v are *equivalent* if $u \sqsubseteq_s v$ and $v \sqsubseteq_s u$. For finite lposets, equivalence is the same as isomorphism. Two lposets u and v are isomorphic if there exists a label-preserving bijection $f{:}\mathbf{set}(v){\rightarrow}\mathbf{set}(u)$ such that f and f^{-1} are monotonic. Following Pratt and Gischer [3,17], we do not make any distinction between equivalent lposets .

3   Behaviours, Schedulers and Finishers

In the previous section, we have seen that every event structure $\mathcal{E}$ is associated with a set of lposets $\mathcal{L}$. When the concurrent computation modelled by $\mathcal{E}$ runs, we can observe a lposet $v \in \mathcal{L}$ or some lposet u which implements v. The reason is that we allow concurrent events to happen in any order or to overlap. Thus, the behaviours of the event structure $\mathcal{E}$ are members of the downward-closed set

$$\downarrow\mathcal{L} = \{u \mid \exists v \in \mathcal{L} \cdot u \sqsubseteq_s v\},$$

the set of all lposets subsumed by some element of $\mathcal{L}$. In [3], Gischer has shown that the inclusion of these downward-closed sets provides the order for a true-concurrent model of Hoare et al.'s Concurrent Kleene Algebra [6]. That soundness result justifies our use of $\downarrow\mathcal{L}$ to express the behaviours of an event structure.

3.1   Prefix Relation on lposet

Computational progress in a sequential system is expressed by using prefixes. Prefixes are approximations of a computation and, as time passes, the trace is getting longer meaning that the approximation becomes more precise [4]. This notion is extended to the concurrent setting by defining the prefixes of lposets.

Similarly to traces, it is clear that a prefix v of a lposet u must be a restriction of u. This is however not enough because when v "progresses" into u (i.e. the approximation gets better), no new event can occur in-between events of v. In other words, every newly occurring events of u must happen "after" v. This property is provided by the $\leq_u$-downward-closure of v as a sub-lposet of u and is formalised by Property 1 below. In particular, it ensures that if v is a prefix of u and $u \in \mathcal{L}$, then $v \in \mathcal{L}$.

Definition 4. *A lposet v is a prefix of a lposet u, written $v \trianglelefteq u$, iff* $\mathbf{set}(v) \subseteq \mathbf{set}(u)$, $\lambda_v = \lambda_u \cap (\mathbf{set}(v) \times \Sigma)$, $\leq_v = \leq_u \cap (\mathbf{set}(v) \times \mathbf{set}(v))$ *and*

$$e \leq_v e' \wedge e' \in \mathbf{set}(u) \Rightarrow e \in \mathbf{set}(u) \wedge e \leq_u e'. \tag{1}$$

We have shown elsewhere [10] that $\trianglelefteq$ is a partial order. For configurations and canonical lposets, $\trianglelefteq$ is equivalent to $\subseteq$.

Proposition 1. *If $x, y \in \mathcal{C}$ and $x \subseteq y$, then $(x, \leq_x, \lambda_x) \trianglelefteq (y, \leq_y, \lambda_y)$.*

Proof. The proof consists of showing that if $x \subseteq y$ then every linearization of x is the restriction of some linearization of y. Conversely, every linearization of y extends into a linearization of x. The full proof is given in Appendix A.   □

3.2   Scheduling and Finishing Events

In this paper, we argue that schedulers are but one portion of the technique that guarantees the generation of every lposet of a given event structure. For instance, in Fig. 1, no scheduler will be able to generate the lposet $(x, \leq_x \cup \{(e_2, e_3)\}, \lambda_x)$ where $x = \{e_1, e_2, e_3\}$ because the causality relation $e_2 \leq e_3$ requires that e_2 has *finished occurring* before e_3 is scheduled to occur (see Example 3 for more details). Therefore, we will introduce the dual notion of *finisher* in Definition 6 to account for finished events.

Definition 5. *A scheduler on the BES $\mathcal{E}$ is a partial function $\sigma \colon \downarrow\mathcal{L} \rightarrow E$ such that for every lposet $u \in \downarrow\mathcal{L}$ in the domain of σ, we have:*

i. $\sigma(u) \notin \mathbf{set}(u)$, and
ii. $(\mathbf{set}(u) \cup \{\sigma(u)\}, \leq_u \cup \leq_{\mathbf{set}(u) \cup \{\sigma(u)\}}, \lambda_{\mathbf{set}(u) \cup \{\sigma(u)\}})$ is in $\downarrow\mathcal{L}$,

where $\leq_{\mathbf{set}(u) \cup \{\sigma(u)\}}$ is the canonical order associated with the configuration $\mathbf{set}(u) \cup \{\sigma(u)\}$ and $\lambda_{\mathbf{set}(u) \cup \{\sigma(u)\}}$ is the restriction of the labelling function λ to $\mathbf{set}(u) \cup \{\sigma(u)\}$. We write $\sigma[u]$ the lposet given in (ii).

A scheduler will always schedule a fresh event that is enabled (Definition 5 (i)). However, it is not forced to do so and may stop scheduling after a certain amount of time. Iterating from the empty lposet, a scheduler σ generates a sequence of finite lposets $\emptyset \trianglelefteq \sigma[\emptyset] \trianglelefteq \sigma[\sigma[\emptyset]] \trianglelefteq \cdots$ (Definition 5 (ii)).

As discussed above, the scheduler's task is limited to scheduling enabled events and it is up to the finisher to decide when events are terminated.

Definition 6. *A finisher is a total function* $\varphi{:}{\downarrow}\mathcal{L}{\to}{\downarrow}\mathcal{L}$ *such that:*

i. $\varphi(u) \trianglelefteq u$, *for every lposet* $u{\in}{\downarrow}\mathcal{L}$, *and*
ii. φ *is* $\trianglelefteq$-*monotonic.*

Intuitively, a lposet u can be thought of as a set of scheduled events ordered by causal dependencies. The lposet $\varphi(u)$ captures the set of events in u that have terminated (Definition 6 (i)). Thus, all events scheduled after this "point of time" will causally depend on the events in $\varphi(u)$. The monotonicity property (Definition 6 (ii)) ensures that, as new events are scheduled, terminated ones cannot be unfinished.

Observe that the identity **id** on ${\downarrow}\mathcal{L}$ gives a special example of a finisher. This finisher enforces that every scheduled event will be finished instantaneously. When the identity finisher **id** is used, concurrency reduces to interleaving.

Example 3. Let $\mathcal{E}$ be the event structure outlined in Fig. 1. We define a scheduler σ on $\mathcal{E}$ such that $\sigma(\emptyset) = e_1$, $\sigma(\{e_1\}) = e_2$ and $\sigma(\{e_1, e_2\}) = e_3$, where each set in the argument of σ should be read as the canonical lposet associated to the respective configuration. This scheduler schedules e_2 before e_3 but that does not mean that e_2 will happen before e_3 because that order is not enforced by the bundle relation of $\mathcal{E}$. The resulting chain of (canonical) lposets is

$$\emptyset \trianglelefteq \{e_1\} \trianglelefteq \{e_1, e_2\} \trianglelefteq \{e_1, e_2, e_3\} \ .$$

To enforce the behaviour that e_3 happens after e_2 has occurred, we need a finisher that satisfies $\{e_2\} \trianglelefteq \varphi(\{e_1, e_2\})$[1]. An example of such a finisher is $\varphi(\emptyset) = \emptyset, \varphi(\{e_1\}) = \emptyset, \varphi(\{e_1, e_2\}) = \{e_1, e_2\}, \varphi(\{e_1, e_2, e_3\}) = \{e_1, e_2, e_3\}$, and $\varphi(\{e_1, e_2, e_3, e_4\}) = \{e_1, e_2, e_3, e_4\}$. □

4   Generating Lposets from Schedulers and Finishers

The dynamics of an event structure is obtained through the interactions between pairs of schedulers and finishers. The *state* of the event structure $\mathcal{E}$ is described by a tuple $(u, v){\in}{\downarrow}\mathcal{L}^2$ such that $v \trianglelefteq u$[2]. Intuitively, u is the scheduled lposet while v describes all "finished" events (the order in which these events occurred is constrained by the order of v).

Example 4. The pair $(\{e_1, e_2, e_3\}, \{e_1, e_2\})$ of canonical lposets is a state of the bundle event structure $\mathcal{E}$ of Fig. 1. Intuitively, it says that the events e_1, e_2 and e_3 have been scheduled, and events e_1 and e_2 have occurred. □

Let us define how a scheduler operates on the states of an event structure. Let σ be a scheduler on $\mathcal{E}$, we define $\overline{\sigma}{:}{\downarrow}\mathcal{L}^2{\to}{\downarrow}\mathcal{L}^2$ such that $\overline{\sigma}(u, v) = (u', v)$ where

$$\mathbf{set}(u') = \mathbf{set}(u) \cup \{\sigma(u)\}$$
$$\leq_{u'} = \leq_u \cup \leq_{\mathbf{set}(u')} \cup (\mathbf{set}(v) \times \{\sigma(u)\})$$
$$\lambda_{u'} = \lambda_{\mathbf{set}(u')}$$

---

[1] More generally, we want $\{e_2\} \trianglelefteq \varphi(u)$ whenever e_3 is enabled at u and e_2 occurs in u.
[2] ${\downarrow}\mathcal{L}^2$ abbreviates $({\downarrow}\mathcal{L}) \times ({\downarrow}\mathcal{L})$.

Intuitively, $e \leq_{u'} e'$ holds in the new lposet u' if either:

- $e \leq_u e'$: e and e' have been scheduled and the ordering holds in u, or
- $e \leq_{\mathsf{set}(u')} e'$: $e' = \sigma(u)$ is the newly scheduled event and the order $e \leq_{u'} e'$ is enforced by the transitive closure of the bundle relation of $\mathcal{E}$, or
- $e \in \mathsf{set}(v)$ has already happened and $e' = \sigma(u)$ is the newly scheduled event.

In the sequel, we denote the lposet u' in this construction by $\sigma[u \leftarrow v]$, that is, $\overline{\sigma}(u, v) = (\sigma[u \leftarrow v], v)$. The intuition behind this notation is that the newly scheduled event $\sigma(u)$ causally depends on the finished events of v. Notice that if $v \trianglelefteq u$ then $u \trianglelefteq \sigma[u \leftarrow v]$ and in particular $v \trianglelefteq \sigma[u \leftarrow v]$. Thus if (u, v) is a state of $\mathcal{E}$ then $\overline{\sigma}(u, v) = (\sigma[u \leftarrow v], v)$ is also a state of $\mathcal{E}$.

Similarly, every finisher φ generates a map $\overline{\varphi}: \downarrow \mathcal{L}^2 \to \downarrow \mathcal{L}^2$ such that $\overline{\varphi}(u, v) = (u, \varphi(u))$. It is clear from this definition and Definition 6 that if (u, v) is a state of $\mathcal{E}$, i.e. $v \trianglelefteq u$, then $\overline{\varphi}(u, v)$ is also a state of $\mathcal{E}$.

Since $\overline{\sigma}$ and $\overline{\varphi}$ preserves the states of $\mathcal{E}$, it also follows that for every state (u, v), we can construct a increasing chain of states by alternatively applying $\overline{\sigma}$ and $\overline{\varphi}$ (until σ is undefined in which case the chain stops at the next finishing operation). That is,

$$(u, v) \trianglelefteq \overline{\sigma}(u, v) \trianglelefteq \overline{\varphi}(\overline{\sigma}(u, v)) \trianglelefteq \overline{\sigma}(\overline{\varphi}(\overline{\sigma}(u, v))) \trianglelefteq \cdots , \tag{2}$$

where $(u, v) \trianglelefteq (u', v')$ means $u \trianglelefteq u'$ and $v \trianglelefteq v'$. When the initial state is $(\emptyset, \emptyset)$, this chain is called the *resolution* of the event structure $\mathcal{E}$ *wrt* σ and φ, and is denoted $(\sigma\varphi)^*$. The remainder of this section is devoted to showing that such a chain is powerful enough to generate each and every lposet in $\downarrow\mathcal{L}$.

Let us denote by $(u_0, v_0) \trianglelefteq (u_1, v_1) \trianglelefteq \cdots$, where $u_0 = v_0 = \emptyset$, the states involved in Sequence (2) above. We write $\sup(\sigma\varphi)^* = (\cup_i u_i, \cup_i v_i)$ where the union of lposets is defined componentwise — $u \cup v = (\mathsf{set}(u) \cup \mathsf{set}(v), \leq_u \cup \leq_v , \lambda_u \cup \lambda_v)$. This union is well defined. Firstly, each labelling function λ_i of u_i is the restriction of the labelling function of the event structure on $\mathsf{set}(u_i)$. Hence the union of the λ_i will again be a function. Secondly, since the sequence is increasing, the union of the order relations will again be a partial order on $\cup_i \mathsf{set}(u_i)$. Thus, $\sup(\sigma\varphi)^*$ is a lposet.

Example 5. Reconsider the event structure given in Fig. 1. The scheduler σ and finisher φ of Example 3 generate the resolution illustrated in Fig. 2a. In the lposet $\sup(\sigma\varphi)^*$, the events satisfy $e_1 \leq e_3$ and $e_2 \leq e_3$, allowing e_1 and e_2 to be concurrent. By contrast, Fig. 2b generates the events trace $e_1 e_2 e_3$.   □

The following proposition establishes the properties of each and every state that occurs in the resolution $(\sigma\varphi)^*$.

Proposition 2. *Let σ be a scheduler, φ be a finisher on $\mathcal{E}$ and (u, v) be a state in $(\sigma\varphi)^*$. The implication*

$$\forall e, e' \in \mathsf{set}(u) : e \leq_u e' \Rightarrow e \leq_{\mathsf{set}(u)} e' \lor e \in \mathsf{set}(v) \tag{3}$$

holds, where $\leq_{\mathsf{set}(u)}$ is the canonical order of the configuration $\mathsf{set}(u)$.

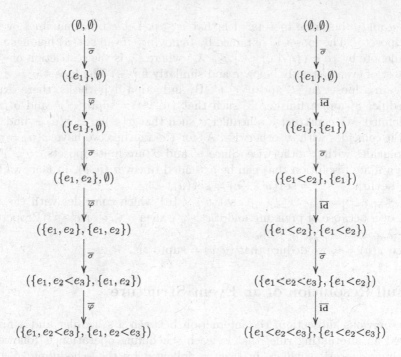

(a) Resolution $(\sigma\varphi)^*$.   (b) Resolution $(\sigma\mathbf{id})^*$.

These two figures show a comparison of the resolutions of an event structure *wrt* a single scheduler σ and two different finishers φ and $\mathbf{id}$. The explicit use of $<$ in the sets of events shows the causal dependencies resulting from the interaction of the scheduler and finishers. The other dependencies can be inferred from the bundle relation.

Fig. 2. Two examples of resolutions.

Proof. By simple induction on the reachability of (u,v) from $(\emptyset, \emptyset)$.   □

The following theorem implies that Property 3 is also sufficient for a pair (u,v), satisfying $v \trianglelefteq u$, to be a state in some resolution. In particular, if $u \in \downarrow\mathcal{L}$, then Theorem 1 ensures the existence of a pair (σ, φ) of a scheduler and finisher such that $\sup(\sigma\varphi)^* = (u,u)^3$. This is the completeness result we sought.

Theorem 1. *Let* $(u,v) \downarrow \mathcal{L}^2$ *such that* $v \trianglelefteq u$. *If* u *and* v *satisfy Property 3, then there exists a scheduler* σ *and a finisher* φ *such that* $(u,v) = \sup(\sigma\varphi)^*$.

Proof. We reason by induction on the size of $\mathbf{set}(u)$:

– the empty pair $(\emptyset, \emptyset)$ is obtained from the scheduler that is undefined everywhere and the finisher $\mathbf{id}$.

---

[3] In fact, every partial order can be generated using Theorem 1 when all pairs of events in the underlying event structure are concurrent.

– Let u and v be finite lposets such that $v \trianglelefteq u$. Let e be a maximal event in the lposet u. The lposet u' obtained by removing e from $\mathbf{set}(u)$ belongs to $\downarrow\mathcal{L}$. We denote by $\downarrow e = (\{e' \mid e' < e\}, \leq_e, \lambda_e)$ where $\leq_e$ is the restriction of $\leq_u$ on the set of events strictly below e and similarly for λ_e. The lposet $v' = v \cap \downarrow e^4$ is again a lposet in $\downarrow\mathcal{L}$ and $v' \trianglelefteq u'$. By induction hypothesis, there exists a scheduler σ' and a finisher φ' such that $(u', v') = \sup(\sigma'\varphi')^*$ and $\sigma'(u')$ is undefined. We construct a scheduler σ such that $\sigma(u') = e$, $\sigma(u)$ is undefined and it coincides with σ' otherwise. As for the finisher, we have $\varphi(u) = v$ and it coincides with φ' otherwise. Since u' and v' are finite lposets, $(\sigma'\varphi')^*$ will give a finite resolution that can be extended to cover (u, v). In fact, we have

- $\mathbf{set}(\sigma[u' \leftarrow v']) = \mathbf{set}(u') \cup \{e\} = \mathbf{set}(u)$,
- $\leq_{\sigma[u' \leftarrow v']} = \leq_{u'} \cup \leq_{\mathbf{set}(u)} \cup \mathbf{set}(v') \times \{e\}$ which coincides with the order of u because of prefixing and if $e' \leq_u e$ then $e' \leq_u e$ or $e' \in v'$ (Property 3).
- $\lambda_{\sigma[u' \leftarrow v']} = \lambda_{\mathbf{set}(u)} = \lambda_u$.

Since $\varphi(u) = v$, we deduce that $(u, v) = \sup(\sigma\varphi)^*$. $\square$

5   Full Resolution of an Event Structure

In the previous subsection, the interaction between a scheduler and a finisher followed an alternating rule, that is, each scheduling operation is followed by a finishing operation which, in turn, is followed by the scheduling of a new event, and so on. In general, these two operations can happen in any order but the most important characteristic is that scheduled events causally depend on finished events.

The goal of this subsection is to prove that simpler form of resolution is enough to generate all possible interactions between a scheduler and a finisher. Firstly, let us establish some key properties of the successive composition of the functions $\overline{\sigma}$ and $\overline{\varphi}$.

Proposition 3. *For every natural numbers $m, n \in \mathbb{N}$ and $(u, v) \in \downarrow\mathcal{L}^2$, if $n \geq 1$ then $\overline{\sigma}^m \circ \overline{\varphi}^n(u, v) = \overline{\sigma}^m \circ \overline{\varphi}(u, v)$*

Proof. $\overline{\varphi}$ is idempotent. $\square$

Thus, for $n \geq 1$, we can reduce the expression $(\overline{\sigma}^m \circ \overline{\varphi}^n)^k(u, v)$ into $(\overline{\sigma}^m \circ \overline{\varphi})^k(u, v)$.

Next we show that every given state (u, v), such that $v \trianglelefteq \varphi(u)$, is a prefix of all subsequent states as expected. This ensures that $(\sigma^m \circ \varphi)^n$ satisfies basic progress requirement properties [4].

Proposition 4. *Given σ and φ, if $v \trianglelefteq \varphi(u)$ then*

$$(u, v) \trianglelefteq (\overline{\sigma}^m \circ \overline{\varphi})^n(u, v)$$

for every $m, n \in \mathbb{N}$.

---

[4] The intersection of two lposets is $(x, \leq_x, \lambda_x) \cap (y, \leq_y, \lambda_y) = (x \cap y, \leq_x \cap \leq_y, \lambda_x \cap \lambda_y)$.

Proof. It suffices to show that $(u, v) \trianglelefteq \overline{\sigma}(u, v)$ and $(u, v) \trianglelefteq \overline{\varphi}(u, v)$, which are clear from the definition of $\overline{\sigma}$ and $\overline{\varphi}$. The result, with arbitrary m and n, follows by simple inductions and the transitivity of $\trianglelefteq$.   □

Lastly, we show that the property $v \trianglelefteq \varphi(u)$ is an invariant for every state (u, v) generated from schedulers and finishers. Therefore, if v was finished when u was scheduled then v remains finished after any subsequent scheduling and finishing applied to $\mathcal{E}$ from u.

Proposition 5. *For every* $(u, v) \in \downarrow\mathcal{L}^2$ *such that* $v \trianglelefteq u$, *if* $v \trianglelefteq \varphi(u)$ *then* $v_m^n \trianglelefteq \varphi(u_m^n)$ *where* $(u_m^n, v_m^n) = (\overline{\sigma}^m \circ \overline{\varphi})^n (u, v)$ *and* $m, n \in \mathbb{N}$.

Proof. The proof is by induction on n. The full proof is given in Appendix B. □

We are now ready to introduce the notion of the full resolution of an event structure $\mathcal{E}$ *wrt* some given scheduler and finisher.

Definition 7. *The full interaction of a scheduler* σ *and a finisher* φ *is the directed graph*

$$(\downarrow\mathcal{L}^2, \{((u, v), \overline{\varphi}(u, v)), ((u, v), \overline{\sigma}(u, v)) \mid (u, v) \in \downarrow\mathcal{L}^2\}).$$

The subgraph composed of nodes that are reachable with a finite path from $(\emptyset, \emptyset)$ *is denoted by* $(\sigma * \varphi)^*$ *and is called the* full resolution *of* $\mathcal{E}$ *wrt* σ *and* φ. *Note that we remove self loops in the full resolution graph.*[5]

We start by showing that $(\sigma * \varphi)^*$ is a directed acyclic graph.

Proposition 6. *For scheduler* σ *and finisher* φ *of an event structure* $\mathcal{E}$, *the graph* $(\sigma * \varphi)^*$ *is acyclic.*

Proof. Assume that $(\sigma * \varphi)^*$ has a cycle that is not a self-loop. Since $\overline{\varphi}$ is idempotent, that cycle needs to contain at least one application of $\overline{\sigma}$. Moreover, if there is such a cycle, then it contains a state (u, v) such that u is exactly the same as the first component of the state obtained after a finite application of σ and φ. But σ strictly increases the left lposet of an arbitrary pair, which makes it impossible to find such a state (u, v).   □

The following proposition shows that every partial function defined on an increasing chain of lposets can always be extended into a finisher. This extension is not necessarily unique but it allows finishers to be defined on a chain of $\downarrow\mathcal{L}$ rather than on the whole set $\downarrow\mathcal{L}$.

Proposition 7. *If* $f : \downarrow\mathcal{L} \rightarrow \downarrow\mathcal{L}$ *is a partial function defined on an increasing sequence of lposets* $\emptyset = u_0 \trianglelefteq u_1 \trianglelefteq \cdots$ *and satisfies the two properties of a finisher, i.e.* $u_i \trianglelefteq f(u_i)$ *for every* i *and* f *is monotonic, then there exists a finisher* φ *(i.e. totally defined) such that* $\varphi(u_i) = f(u_i)$ *for every* i.

---

[5] These self loops are mainly due to idempotency of finishers.

Proof. It suffices to prove that the extension

$$\varphi(u) = \begin{cases} f(u_i) & \text{if there is a maximal } i \text{ such that } u_i \trianglelefteq u \\ u & \text{if } u_i \trianglelefteq u \text{ for every } i \\ \emptyset & \text{otherwise} \end{cases}$$

is indeed a finisher. The full proof is given in Appendix C. □

We finally show that every full resolution is the union of (alternating) resolutions associated with each and every path in the directed acyclic graph.

Theorem 2. *For every scheduler σ and finisher φ there exists a (countable) family of finishers $\varphi_0, \varphi_1, \ldots$ such that the full resolution $(\sigma*\varphi)^*$ is the union of the family of resolutions $(\sigma\varphi_0)^*, (\sigma\varphi_1)^*, \ldots$.*

Proof. Every expression $(\sigma^m \circ \varphi)^n$ generates a path π in the full resolution graph. Each path provides a partial function f_π that satisfies the premises of Proposition 7. Thus, each f_π can be extended into a finisher φ_π which, together with σ, generates a resolution. We conclude that $(\sigma*\varphi)^*$ is the union of the $(\sigma\varphi_\pi)^*$s. The details of the proof are given in Appendix D. □

Example 6. The full resolution of the event structure given in Fig. 1 *wrt* the scheduler and finisher of Example 3 is depicted in Fig. 3. The full resolution possesses two suprema which are attained by following two different resolutions. □

6   Using Finishers for Behaviour Filtering

It is sometimes useful to study the behaviours of an event structure that have certain characteristic properties. For instance, let us consider a concurrent 1-bit register with *write* and *read* actions. It would be beneficial to confine to *feasible* behaviours such as the action of reading 0 can only occur after or at the same time as the writing 0. Such a property is easily expressed using our notion of finisher: if w (resp. r) is the event carrying the write-0 (resp. read-0) action, then we are only interested in finishers that satisfy $\varphi(\{r<w\}) = \emptyset$.

More general intrinsic properties of the underlying system are encoded by shrinking the codomain of finishers. If $A \subseteq \downarrow\mathcal{L}$ denotes a subset of lposets that have the desired characteristic properties, then every resolution with respect to finishers of type $\downarrow\mathcal{L} \to A$ will contain states whose second component (finished lposet) is in A. If the set A is characterised by a safety property then finishers of type $\downarrow\mathcal{L} \to A$ will only finish safe behaviours.

It should be noted that the set A must, at least, have some basic properties such as prefix closure. Further investigation is required to elaborate the properties of such sets and their practical applications. We leave this for future work.

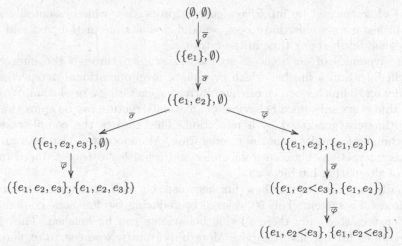

In the full resolution, the supremum depends on the path followed which is given by how the scheduler and finisher interact. The branching occurs at the state $(\{e_1, e_2\}, \emptyset)$ where the choice between scheduling e_3 and finishing $\{e_1, e_2\}$ dictates the causality dependency between e_2 and e_3.

Fig. 3. The full resolution of an event structure.

7 Related Works

The notion of scheduler for event structures has been applied to transform such structures into transition systems rather than applied in the generation of the behaviours themselves [7,9,22]. The resulting transition systems can be analysed using standard techniques from the well-developed theories of sequential process communications [5,12,19,22]. This transformation, however, renders the much-desired modelling of truly-concurrent behaviours worthless since concurrency reduces back to interleaving.

To the best of our knowledge, this paper contains the first attempt to understand true-concurrency within the operational perspective provided by schedulers. Note however that other event structure analysis techniques exist such as the transformation described above [7,9,23], the correspondence with Petri nets [15,20,23], the algebraic analysis in the style of process and Kleene algebras [7,10,18] and the notions of simulations for event structures [2,11,16]. However, most of these techniques assume in advance that the set of behaviours of the underlying event structure has already been computed or provided as part of the specification. In this paper, we show how to obtain these behaviours in the first place, using the elementary operations of schedulers and finishers.

8 Conclusion

This paper introduced the notion of finishers by naturally capturing tasks related to terminated actions in event structures. Finishers are quite simple in nature and are characterised by intuitive algebraic properties, which essentially state that finished events must have been scheduled sometime in the past and that they remain finished any time in the future.

The dynamics of an event structure was studied through the interaction of a scheduler and a finisher which provide a novel operational perspective. A scheduler and a finisher can be run alternatively, generating a resolution. We have shown that every subsumed behaviour of an event structure can be approximated and, ultimately, generated by a resolution. This ensures the completeness of our technique for finite behaviour generation. Moreover, we have proven that complex interactions between a scheduler and a finisher can be reduced to the union of alternating interactions.

Finally, we have shown how finishers can be used to express distinctive behaviours of a system. This is achieved by reducing the finishers' co-domains, which ensures that only these specific behaviours can be finished. This, however, requires further investigation. More importantly, we need to explore the connection of this work to seemingly related techniques, such as Interval Logics.

A   Proof of Proposition 1

Let $\mathcal{E}$ be an event structure and $\alpha \in \mathcal{T}$ be an event trace. We write $\overline{\alpha}$ for the set of events occurring in α, which is a configuration of $\mathcal{E}$.

Lemma 1. *Let $\alpha \in \mathcal{T}$ and $x \in \mathcal{C}$ such that $x \subseteq \overline{\alpha}$, then the restriction $\alpha|_x$ of α to events in x is an event trace.*

Proof. Let $\alpha = e_1 e_2 \cdots e_n$, $x \in \mathcal{C}$ and write $\alpha|_x = e_{i_1} e_{i_2} \cdots e_{i_m}$. Let us show that $\alpha|_x$ is an event trace of $\mathcal{E}$. Let $e_{i_k} \in x$ and $z \mapsto e_{i_k}$ be a bundle of $\mathcal{E}$. Since α is an event trace, there exists an event e_j such that $e_j \in z$ and $j < i_k$. Since x is a configuration and $e_{i_k} \in x$, there exists $e_{i_l} \in z$ and $l < k$. By definition, the bundle set z contains mutually conflicting events only and since $\overline{\alpha}$ is conflict free, $e_{i_l} = e_j$. That is, $z \cap \{e_{i_1}, \ldots, e_{i_{k-1}}\} \neq \emptyset$ for every bundle $z \mapsto e_{i_k}$. Hence, $\alpha|_x$ is an event trace (it is already conflict free). $\qquad\square$

Lemma 2. *Let $x, y \in \mathcal{C}$ such that $x \subseteq y$. For every event trace α such that $\overline{\alpha} = x$, there exists an event trace α' satisfying $\overline{\alpha'} = y$ and $\alpha'|_x = \alpha$.*

Proof. Let α, β be event traces such that $\overline{\alpha} = x$, $\overline{\beta} = y$ and $x \subseteq y$. Let β' be the concatenation of two sequences $\beta_1 \beta_2$, where events in β_1 are exactly those of x ordered with $\leq_\beta$ and β_2 is composed of events from $y \setminus x$ (set difference) ordered again with $\leq_\beta$. We now show that the concatenation $\alpha' = \alpha \beta_2$ is an event trace. That α' is conflict-free comes from the configuration y. To show the second property of an event trace, we need to show that every bundle pointing to an event e_2 in β_2 has to intersect $\overline{\alpha} \cup \overline{\beta_2}$ at an event occurring before e_2 with respect to the order $\leq_\beta$. That is clear because β is an event trace (notice that if $z \mapsto e_2$ holds, it is possible that the sole event in $z \cap \overline{\beta}$ belongs to $\overline{\alpha}$). Moreover, it is enough to show the property for events in β_2 only because α is already an event trace. Hence α' is an event trace and $\alpha'|_x = \alpha$. $\qquad\square$

With the help of these two lemmas, we now prove the envisaged characterisation of prefixing with configuration inclusion.

Proposition 8. *If $x, y \in \mathcal{C}$ and $x \subseteq y$, then $(x, \leq_x, \lambda_x) \trianglelefteq (y, \leq_y, \lambda_y)$.*

Proof. Let $x \subseteq y$. Let us first show that $\leq_x = \leq_y \cap (x \times x)$. Let $e, e' \in x$. We need to show that $e \leq_x e'$ iff $e \leq_y e'$. Assume $e \leq_x e'$, then Lemma 1 implies that $e \leq_y e'$ because every event trace for y restricts to an event trace for x. For the converse implication, let $e, e' \in x$ such that $e \leq_y e'$. Lemma 2 implies that every event trace for x can be obtained as a restriction of some event trace for y. Hence, $e \leq_x e'$. Therefore $\leq_x = \leq_y \cap (x \times x)$.

It remains to show that Property 1 holds. Let $e, e' \in y$, $e \leq_y e'$ and $e' \in x$. It is enough to show that $e \in x$ because, once that is established, we use $\leq_x = \leq_y \cap (x \times x)$ to deduce that $e \leq_x e'$. For a contradiction, assume that $e \notin x$. Then there exists an event trace $\beta' = \beta_1 \beta_2$ as specified in the proof of Lemma 2, that is, $\overline{\beta'} = y$, $\overline{\beta_1} = x$ and $e \in \overline{\beta_2}$. Thus, $e \not\leq_{\beta'} e'$ which contradicts the fact that $e \leq_y e'$. $\square$

B   Proof of Proposition 5

Proposition 9. *For every $(u, v) \in \downarrow \mathcal{L}^2$ such that $v \trianglelefteq u$, if $v \trianglelefteq \varphi(u)$ then $v_m^n \trianglelefteq \varphi(u_m^n)$ where $(u_m^n, v_m^n) = (\overline{\sigma}^m \circ \overline{\varphi})^n (u, v)$ and $m, n \in \mathbb{N}$.*

Proof. Let m be a fixed natural number and let us reason by induction on n.

- For $n = 0$, we have $(u_m^0, v_m^0) = (\overline{\sigma}^m \circ \overline{\varphi})^0 (u, v) = (u, v)$ and thus $v_m^0 = v \trianglelefteq \varphi(u) = u_m^0$.
- Let us assume $v_m^n \trianglelefteq \varphi(u_m^n)$. We have

$$(u_m^{n+1}, v_m^{n+1}) = (\overline{\sigma}^m \circ \overline{\varphi})^{n+1}(u, v) = (\overline{\sigma}^m \circ \overline{\varphi})(u_m^n, v_m^n) = \overline{\sigma}^m(u_m^n, \varphi(u_m^n)) \ .$$

Since $\overline{\sigma}$ only operates on the first component of the state, we have $v_m^{n+1} = \varphi(u_m^n)$ and

$$u_m^{n+1} = \underbrace{\sigma[\sigma[\ldots \sigma[u_m^n \leftarrow v_m^n] \ldots \leftarrow v_m^n] \leftarrow v_m^n]}_{m \text{ times}} \ . \tag{4}$$

The induction hypothesis implies $v_m^n \trianglelefteq u_m^n$, thus Eq. 4 is well defined and implies $u_m^n \trianglelefteq u_m^{n+1}$. It follows from the monotonicity of φ that $v_m^{n+1} = \varphi(u_m^n) \trianglelefteq \varphi(u_m^{n+1})$. $\square$

C   Proof of Proposition 7

Proposition 10. *If $f : \downarrow \mathcal{L} \to \downarrow \mathcal{L}$ is a partial function defined on a increasing sequence of lposets $\emptyset = u_0 \trianglelefteq u_1 \trianglelefteq \cdots$ and satisfies the two properties of a finisher, i.e. $u \trianglelefteq f(u)$ for all u and f is monotonic, then there exists a finisher φ (i.e. totally defined) such that $\varphi(u_i) = f(u_i)$ for every i.*

Proof. Let $\mathcal{E}$ be an event structure and f be a function satisfying the hypothesis of the proposition. We construct φ as follows

$$\varphi(u) = \begin{cases} f(u_i) & \text{if there is a maximal } i \text{ such that } u_i \trianglelefteq u \\ u & \text{if } u_i \trianglelefteq u \text{ for every } i \\ \emptyset & \text{otherwise} \end{cases}$$

Firstly, we show the prefixing property of finishers. Let $u \in \downarrow\mathcal{L}$:

- If there exists a maximal i such that $u_i \trianglelefteq u$ then $\varphi(u) = f(u_i) \trianglelefteq u_i \trianglelefteq u$.
- If $u_i \trianglelefteq u$ for all i, then $\varphi(u) = u \trianglelefteq u$.
- Otherwise, $\varphi(u) = \emptyset \trianglelefteq u$.

Secondly, we show that φ is monotonic. Let $u \trianglelefteq v$.

- If there exists a maximal i such that $u_i \trianglelefteq u$, then $\varphi(u) = f(u_i)$. There are three cases based on the value of $\varphi(v)$.
  - There exists a maximal j such that $u_j \trianglelefteq v$ and $\varphi(v) = f(u_j)$. Since $u \trianglelefteq v$, maximality of j implies that $u_i \trianglelefteq u_j$ and hence $\varphi(u) = f(u_i) \trianglelefteq f(u_j) = \varphi(v)$, by monotonicity of f.
  - For all j, $u_j \trianglelefteq v$ and therefore $\varphi(u) = f(u_i) \trianglelefteq u_i \trianglelefteq v = \varphi(v)$.
  - The case $\varphi(v) = \emptyset$ cannot happen, unless $f(u_i) = \emptyset$ because $u_i \trianglelefteq v$ for every i.
- If $u_i \trianglelefteq u$ for all i, then $u_i \trianglelefteq v$ for all i because $u \trianglelefteq v$. Hence $\varphi(u) = u \trianglelefteq v = \varphi(v)$.
- Otherwise, $\varphi(u) = \emptyset \trianglelefteq \varphi(v)$, whatever $\varphi(v)$ is. $\qquad\square$

D   Proof of Theorem 2

Theorem 3. *For every scheduler σ and finisher φ, there exists a (countable) family of finishers $\varphi_0, \varphi_1, \ldots$ such that the full resolution $(\sigma * \varphi)^*$ is the union of the family of resolutions $(\sigma\varphi_0)^*, (\sigma\varphi_1)^*, \ldots$.*

Proof. Let σ and φ be some scheduler and finisher on the event structure $\mathcal{E}$. The full resolution $(\sigma * \varphi)^*$ is depicted in Fig. 4.

Given a path π in the directed acyclic graph of Fig. 4, we generate a partial function f_π such that $f(u) = v$ iff $(u, v) \in \pi$. Therefore, f_π satisfies the first property of a finisher because each node of the tree is a state of $\mathcal{E}$ and it is monotonic because if $(u_i, v_i), (u_j, v_j) \in \pi$ such that $u_i \trianglelefteq u_j$, then there exist two indices k_i, k_j such that $f(u_i) = \varphi(u_{k_i})$ and $f(u_j) = \varphi(u_{k_j})$ and $u_{k_i} \trianglelefteq u_{k_j}$. Hence $f(u_i) \trianglelefteq f(u_j)$ and it extends to a finisher φ_π by Proposition 7. Since the directed acyclic graph can be recovered from the union of all paths, we deduce that

$$(\sigma * \varphi)^* = \cup_\pi (\sigma\varphi_\pi)^*$$

where π ranges over all paths in $(\sigma * \varphi)^*$ (which is of course countable). $\qquad\square$

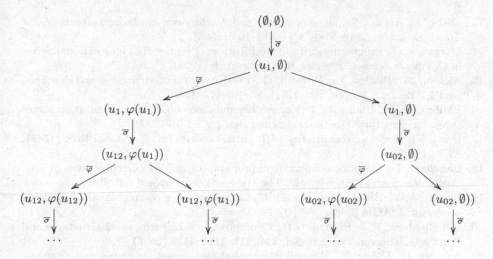

Fig. 4. Full resolution where unlabelled arrows are added for unchanging states.

References

1. Birkhoff, G.: Lattice theory. Number v. 25, pt. 2 in American Mathematical Society colloquium publications. American Mathematical Society (1940)
2. Cherief, F.: Back and forth bisimulations on prime event structures. In: Etiemble, D., Syre, J.-C. (eds.) PARLE 1992. LNCS, vol. 605, pp. 841–858. Springer, Heidelberg (1992). doi:10.1007/3-540-55599-4_128
3. Gischer, J.L.: The equational theory of pomsets. Theoret. Comput. Sci. **61**, 199–224 (1988)
4. Guttmann, W.: An algebraic approach to computations with progress. J. Logical Algebraic Methods Programm. **84**(3), 326–340 (2015)
5. Hoare, C.A.R.: Communicating sequential processes. Commun. ACM **21**(8), 666–677 (1978)
6. Hoare, C.A.R., Möller, B., Struth, G., Wehrman, I.: Concurrent Kleene algebra and its foundations. J. Logic Algebraic Programm. **80**(6), 266–296 (2011)
7. Katoen, J.-P.: Quantitative and qualitative extensions of event structures. Ph.D. thesis, University of Twente (1996)
8. Lamport, L.: Time, clocks, and the ordering of events in a distributed system. Commun. ACM **21**(7), 558–565 (1978)
9. Langerak, R.: Bundle event structures: a non-interleaving semantics for LOTOS. In: Formal Description Techniques for Distributed Systems and Communication Protocols, pp. 331–346 (1992)
10. McIver, A., Rabehaja, T., Struth, G.: An event structure model for probabilistic concurrent Kleene algebra. In: McMillan, K., Middeldorp, A., Voronkov, A. (eds.) LPAR 2013. LNCS, vol. 8312, pp. 653–667. Springer, Heidelberg (2013). doi:10.1007/978-3-642-45221-5_43
11. McIver, A., Rabehaja, T., Struth, G.: Probabilistic rely-guarantee calculus. In: Theoretical Computer Science (2016, In Press)
12. Milner, R.: A Calculus of Communicating Systems. Springer-Verlag New York Inc., Secaucus (1982)

13. Misra, J.: Axioms for memory access in asynchronous hardware systems. ACM Trans. Program. Lang. Syst. **8**(1), 142–153 (1986)
14. Morgan, C.C.: Programming from Specifications. Prentice Hall International Series in Computer Science. Prentice Hall, New York (1994)
15. Nielsen, M., Plotkin, G., Winskel, G.: Petri nets, event structures and domains, part I. Theoret. Comput. Sci. **13**(1), 85–108 (1981)
16. Phillips, I.C.C., Ulidowski, I.: Reverse bisimulations on stable configuration structures. In: Structural Operational Semantics, pp. 62–76 (2009)
17. Pratt, V.: Modeling concurrency with partial orders. Int. J. Parallel Prog. **15**(1), 33–71 (1986)
18. Rabehaja, T.: Algebraic verification of probabilistic and concurrent systems. Ph.D. thesis, Macquarie University and The University of Sheffield (2014)
19. Roscoe, A.W., Brookes, S.D., Hoare, C.A.R.: A theory of communicating sequential processes. J. ACM **3**, 560–599 (1984)
20. van Glabbeek, R.J., Plotkin, G.D.: Configuration structures, event structures and Petri nets. Theoret. Comput. Sci. **410**(41), 4111–4159 (2009)
21. Varacca, D., Völzer, H., Winskel, G.: Probabilistic event structures and domains. Theoret. Comput. Sci. **358**(2–3), 173–199 (2006)
22. Winskel, G.: Event structure semantics for CCS and related languages. In: Nielsen, M., Schmidt, E.M. (eds.) ICALP 1982. LNCS, vol. 140, pp. 561–576. Springer, Heidelberg (1982). doi:10.1007/BFb0012800
23. Winskel, G.: Event structures. In: Brauer, W., Reisig, W., Rozenberg, G. (eds.) ACPN 1986. LNCS, vol. 255, pp. 325–392. Springer, Heidelberg (1987). doi:10.1007/3-540-17906-2_31

On the Expressiveness of Symmetric Communication

Thomas Given-Wilson[✉] and Axel Legay

Inria, Rennes, France
{thomas.given-wilson,axel.legay}@inria.fr

Abstract. The expressiveness of communication primitives has been explored in a common framework based on the π-calculus by considering four features: *synchronism, arity, communication medium*, and *pattern-matching*. These all assume asymmetric communication between input and output primitives, however some calculi consider more *symmetric* approaches to communication such as fusion calculus and Concurrent Pattern Calculus. Symmetry can be considered either as supporting exchange of information between an action and co-action, or as unification of actions. By means of possibility/impossibility of encodings, this paper shows that the exchange approach is related to, or more expressive than, many previously considered languages. Meanwhile, the unification approach is more expressive than some, but mostly unrelated to, other languages.

1 Introduction

The expressiveness of process calculi based upon their communication primitives has been widely explored before [4,7,10,12,16,26]. In [12,16] this is detailed by examining combinations of four features, namely: *synchronism, arity, communication medium*, and *pattern-matching*. These features are able to represent many popular calculi including: monadic or polyadic π-calculus [23,24]; Linda [9]; asymmetric variations of Concurrent Pattern Calculus (CPC) [10,11,14]; and Psi calculi [1]. However, all these calculi exploit upon asymmetric input and output behaviour.

Symmetric behaviour has been considered before in process calculi. One example, fusion calculus [28] shifts away from explicit input and output of names to instead fuse them together in a symmetric equivalence relation. Another is CPC that shifted away from input and output primitives to a single primitive that can do both input or output (and equality tests) via the unification of patterns [14].

This paper abstracts away from specific calculi in the style of [12,16] to provide a general account of the expressiveness of *symmetric* communication primitives. Here symmetric communication does not require that input or output be associated to a particular action or co-action primitive, indeed all communication primitives can perform all possible input, output, or equality tests. This captures the spirit of both fusion calculus and CPC's interaction paradigms, while also generalising to something that can be applied to any calculus. However, there is some complexity when deciding how this should be represented in an abstract calculus since there are two reasonable choices. The first choice is to consider symmetry to support *exchange*, where an action and co-action interact and allow both input and output from either side. This exchange approach of action and co-action with both input and output on both sides aligns with the fusion

© Springer International Publishing AG 2016
A. Sampaio and F. Wang (Eds.): ICTAC 2016, LNCS 9965, pp. 139–157, 2016.
DOI: 10.1007/978-3-319-46750-4_9

calculus style of interaction. The second choice is to consider symmetry to support symmetric *unification*, where a single communication primitive is used for interaction. This approach of a symmetric unification via a single interaction primitive and allowing self recognition (as well as exchange) aligns with CPC style interaction.

The solution here is to consider both; leading to the symmetry feature having three possible instantiations. *Asymmetric* where there is explicit input (that can only contain input patterns) and output (that can only contain output terms), e.g.

$$n(\lambda x, \lambda y).P \mid \overline{n}\langle a, b \rangle.Q \quad \longmapsto \quad \{a/x, b/y\}P \mid Q \ .$$

Exchange where there are two explicit primitives action and co-action that can mix input patterns and output terms, e.g.

$$n(\lambda x, b).P \mid \overline{n}\langle a, \lambda y \rangle.Q \quad \longmapsto \quad \{a/x\}P \mid \{b/y\}Q \ .$$

Unification where this is a single communication primitive that contains a single class of patterns that unify with one-another, e.g.

$$n(\lambda x \bullet b \bullet c).P \mid n(a \bullet \lambda y \bullet c).Q \quad \longmapsto \quad \{a/x\}P \mid \{b/y\}Q \ .$$

By extending prior work with symmetry (and removing synchronism since all exchange and unification languages must be synchronous), the original twelve calculi of [12] are here expanded to thirty-six. This paper details the relations between the original twelve calculi and the twenty-four new calculi, yielding the following key results.

In general exchange languages are more expressive than their asymmetric counterparts. However, there are methods to encode exchange languages with bounded matching capabilities (i.e. a finite limit to the number of names that can be matched) into asymmetric languages. Thus indicating that pattern-matching is highly significant as a factor for determining encodings.

Within the exchange languages, expressiveness increases in a similar manner as the asymmetric languages. The exceptions occur when pattern-matching is intensional, since polyadic exchange languages cannot be encoded into monadic languages, but polyadic asymmetric languages can be encoded into monadic intensional languages.

No unification language can be encoded into an exchange or asymmetric language, this is due to a self recognising process S that can reduce with itself but not alone - something that cannot be defined in any asymmetric or exchange language.

Unification languages do not require name-matching to be able to encode name-matching languages, thus no-matching unification languages can encode name-matching asymmetric and exchange languages. An interesting result, since no asymmetric or exchange language without (at least) name-matching can encode name-matching.

Within the unification languages, relations between languages are identical to the asymmetric setting. This indicates that although unification is a different approach to interaction, the other features are largely unaffected by changing the interaction setting.

The structure of the paper is as follows. Section 2 introduces the considered calculi. Section 3 revises the encoding criteria used for comparing calculi. Section 4 provides a diagrammatic overview of the results. Section 5 explores new relations concerning asymmetric and exchange languages. Section 6 considers unification languages and their relations. Section 7 concludes, and discusses choices made here & in related work.

2   Calculi

This section defines the syntax, operational, and behavioural semantics of the calculi considered here. This relies heavily on the well-known notions developed for the π-calculus, the reference framework, and adapts them when necessary. With the exception of the symmetric constructs this is similar to prior definitions from [12].

Assume a countable set of names $\mathcal{N}$ (denoted a, b, c). *Name-matching patterns* (denoted m, n, o), and *symmetric patterns* (denoted p, q) are defined by:

$$m, n ::= \quad \lambda x \quad binding\ name \qquad p, q ::= \quad a \qquad name$$
$$\mid \ulcorner a \urcorner \quad name - match \qquad\qquad\qquad \mid m \qquad name - match patterns$$
$$\qquad\qquad\qquad\qquad\qquad\qquad\qquad \mid p \bullet q \quad compound.$$

Binding names (denoted $\lambda x, \lambda y, \lambda z$) are used to indicate input behaviour, name-matches $\ulcorner a \urcorner$ test for equality, and compounds combine two symmetric patterns into one (all as in [12,14]). The free names $fn(\cdot)$, binding names $bn(\cdot)$, and matched names $mn(\cdot)$ of name-matching and symmetric patterns are as expected, taking the union of sub-patterns for compound patterns. A symmetric pattern is linear iff all binding names within the pattern are pairwise distinct. The rest of this paper will only consider linear input patterns.

The symmetric patterns are chosen here to be very general and capture many concepts, however to clearly define the languages in this paper, define the following. The *terms* (denoted s, t) are the symmetric patterns that contain no binding names or name-matches. (These correspond to the terms of [12], the communicable patterns of CPC, and the output structures of Psi calculi.) The *intensional patterns* (denoted f, g) are the symmetric patterns that contain no names, i.e. they consist entirely of name-matching patterns and compounds. (These correspond to the intensional patterns of [12].)

The (parametric) syntax for the languages is:

$$P, Q, R ::= \mathbf{0} \mid ACT.P \mid COACT.P \mid (\nu n)P \mid P|Q \mid \mathbf{if}\ s = t\ \mathbf{then}\ P\ \mathbf{else}\ Q \mid *P \mid \sqrt{}.$$

The different languages are obtained by replacing the action ACT and co-action $COACT$ with the various definitions. The rest of the process forms as are usual: $\mathbf{0}$ denotes the null process; restriction $(\nu n)P$ restricts the visibility of n to P; and parallel composition $P|Q$ allows independent evolution of P and Q. The $\mathbf{if}\ s = t\ \mathbf{then}\ P\ \mathbf{else}\ Q$ represents conditional equivalence with $\mathbf{if}\ s = t\ \mathbf{then}\ P$ used when Q is $\mathbf{0}$ (like the name match of π-calculus, $\mathbf{if}\ s = t\ \mathbf{then}\ P\ \mathbf{else}\ Q$ blocks either P when $s \neq t$ or Q when $s = t$). The $*P$ represents replication of the process P. Finally, the $\sqrt{}$ is used to represent a success process or state, exploited for reasoning about encodings as in [10,18].

This paper considers the possible combinations of four features for communication: *arity* (monadic vs polyadic data), *communication medium* (message passing vs shared dataspaces), *pattern-matching* (simple binding vs name equality vs intensionality), and *symmetry* (asymmetric vs exchange vs unification). As a result there exist thirty-six languages denoted by $\mathcal{L}_{\alpha,\beta,\gamma,\delta}$ where:

$\alpha = M$ for monadic data P for polyadic data.
$\beta = D$ for dataspace-based communication, and C for channel-based communications.
$\gamma = NO$ for no matching capability, NM for name-matching, and I for intensionality.

$$\mathcal{L}_{-,-,NO,-} : IN ::= \lambda x \quad OUT ::= a \quad ALL ::= \lambda x \mid a \quad CH ::= a$$
$$\mathcal{L}_{-,-,NM,-} : IN ::= m \quad OUT ::= a \quad ALL ::= m \mid a \quad CH ::= a$$
$$\mathcal{L}_{-,-,I,-} : IN ::= f \quad OUT ::= t \quad ALL ::= p \quad CH ::= t$$
$$\mathcal{L}_{M,D,-,A} : ACT ::= (IN) \qquad\qquad COACT ::= \langle OUT \rangle$$
$$\mathcal{L}_{M,C,-,A} : ACT ::= CH(IN) \qquad\qquad COACT ::= \overline{CH}\langle OUT \rangle$$
$$\mathcal{L}_{P,D,-,A} : ACT ::= (\widetilde{IN}) \qquad\qquad COACT ::= \langle \widetilde{OUT} \rangle$$
$$\mathcal{L}_{P,C,-,A} : ACT ::= CH(\widetilde{IN}) \qquad\qquad COACT ::= \overline{CH}\langle \widetilde{OUT} \rangle$$
$$\mathcal{L}_{M,D,-,E} : ACT ::= (ALL) \qquad\qquad COACT ::= \langle ALL \rangle$$
$$\mathcal{L}_{M,C,-,E} : ACT ::= CH(ALL) \qquad\qquad COACT ::= \overline{CH}\langle ALL \rangle$$
$$\mathcal{L}_{P,D,-,E} : ACT ::= (\widetilde{ALL}) \qquad\qquad COACT ::= \langle \widetilde{ALL} \rangle$$
$$\mathcal{L}_{P,C,-,E} : ACT ::= CH(\widetilde{ALL}) \qquad\qquad COACT ::= \overline{CH}\langle \widetilde{ALL} \rangle$$
$$\mathcal{L}_{M,D,-,U} : ACT, COACT ::= (ALL)$$
$$\mathcal{L}_{M,C,-,U} : ACT, COACT ::= CH(ALL)$$
$$\mathcal{L}_{P,D,-,U} : ACT, COACT ::= (\widetilde{ALL})$$
$$\mathcal{L}_{P,C,-,U} : ACT, COACT ::= CH(\widetilde{ALL}) .$$

Fig. 1. Languages in this paper.

$\delta = A$ for asymmetric communication, E for exchange communication, and U for unification communication.

For simplicity a dash − is used when the instantiation of that feature is unimportant.

Thus the syntax of every language is obtained from the productions in Fig. 1. The first three lines define the components of communication primitives based upon the pattern-matching of the language; with *input patterns IN*, *output patterns OUT*, combined patterns *ALL*, and channel structures *CH*. The rest defines the languages by their action *ACT* and co-action *COACT* using the communication primitives. Here the denotation $\widetilde{\cdot}$ represents a sequence of the form $\cdot_1, \cdot_2, \ldots, \cdot_n$ and can be used for names, binding names, terms, and both kinds of patterns. As usual $(\nu x)P$ and binding names λx in any form (including *IN* and *ALL*) bind x in P. The corresponding notions of free and bound names of a process, denoted $\mathsf{fn}(P)$ and $\mathsf{bn}(P)$, are as usual. An action or co-action is linear if all binding names occur exactly once; this paper shall only consider linear actions and co-actions.

Observe that: monadic languages have a single *IN*, *OUT*, or *ALL* in their action and co-action, while polyadic languages have sequences. Dataspace-based languages are distinct from channel-based languages by not having a channel *CH* that is used for interaction. No-matching languages allow only binding names in *IN*, name-matching languages also allow name-matches, and intensional languages allow intensional patterns in *IN*. No-matching and name-matching languages only allow names in *OUT*, while intensional languages allow terms. Lastly, asymmetric languages only allow *IN* in actions and *OUT* in co-actions, while exchange and unification languages allow *ALL* in both (the latter by defining the co-action to be the action).

Note that α-conversion (denoted $=_\alpha$) is assumed in the usual manner. Finally, the structural equivalence relation $\equiv$ is defined as follows:

$$P \mid Q \equiv Q \mid P \qquad P \mid (Q \mid R) \equiv (P \mid Q) \mid R \qquad P \equiv P' \text{ if } P =_\alpha P' \qquad P \mid 0 \equiv P$$

$$(\nu a)0 \equiv 0 \qquad (\nu a)(\nu b)P \equiv (\nu b)(\nu a)P \qquad P \mid (\nu a)Q \equiv (\nu a)(P \mid Q) \quad \text{if } a \notin \mathsf{fn}(P) .$$

Most of the asymmetric languages correspond to the communication paradigm of popular process calculi, including (but not limited to): monadic or polyadic π-calculus; LINDA; asymmetric variations of CPC; and Psi calculi. For details on these and other languages see [12, 16]. With respect to symmetry: $\mathcal{L}_{P,C,NO,E}$ is closest in communication paradigm to the fusion calculus [28] although the scope of binding in communication is different. $\mathcal{L}_{M,D,I,U}$ corresponds to the communication paradigm of CPC; and $\mathcal{L}_{M,D,I,E}$, $\mathcal{L}_{M,C,I,E}$, and $\mathcal{L}_{M,C,I,U}$ to the communication paradigms of variants of CPC [10].

Remark 1. Most of the languages can be ordered; in particular $\mathcal{L}_{\alpha_1,\beta_1,\gamma_1,\delta_1}$ is a sub-language of $\mathcal{L}_{\alpha_2,\beta_2,\gamma_2,\delta_2}$ if it holds that $\alpha_1 \leq \alpha_2$ and $\beta_1 \leq \beta_2$ and $\gamma_1 \leq \gamma_2$ and $\delta_1 \leq \delta_2$, where $\leq$ is the least reflexive relation satisfying the following axioms:

$$M \leq P \qquad\qquad D \leq C \qquad\qquad NO \leq NM \leq I \qquad\qquad A \leq E.$$

This can be understood as a limited language variation being a special case of a more general language. Monadic communication is polyadic communication with all tuples of arity one. Dataspace-based communication is channel-based communication with all k-ary tuples communicating with channel name k. All name-matching communication is intensional communication without any compounds, and no-matching capability communication is both without any compounds and with only names or only binding names in patterns. Asymmetric communication is exchange with only input patterns in actions, and only output patterns in co-actions; and exchange languages are unification languages with restrictions upon the unification (this does not induce $\leq$, see Sect. 6).

The operational semantics of the languages is given here via reductions as in [12, 23]. An alternative style is via a *labelled transition system* (LTS) such as [16]. Here the reduction based style is chosen for simplicity. The LTS style can be used for intensional and symmetric languages [1, 10], and indeed captures many of the languages here [13].

Substitutions, denoted σ, ρ in non-pattern-matching and name-matching languages are mappings (with finite domain) from names to names. For intensional languages substitutions are mappings from names to terms. The application of a substitution σ to a pattern p is defined as follows:

$$\sigma x = \begin{cases} \sigma(x) & x \in \text{domain}(\sigma) \\ x & x \notin \text{domain}(\sigma) \end{cases} \qquad \sigma \ulcorner x \urcorner = \ulcorner (\sigma x) \urcorner \qquad \sigma(p \bullet q) = (\sigma p) \bullet (\sigma q).$$

Where substitution is as usual on names, and on the understanding that the name-match syntax can be applied to any term by defining: $\ulcorner (s \bullet t) \urcorner \overset{\text{def}}{=} \ulcorner s \urcorner \bullet \ulcorner t \urcorner$. Given a substitution σ and a process P, denote with σP the usual capture-avoiding application of σ to P. As usual capture can be avoided by exploiting α-equivalence [2].

Interaction between processes is handled by unification of patterns with other patterns. The core unification can be used for all languages as defined by the *unify* rule $\{p \| q\}$ of a single pattern p and a single pattern q to create two substitutions σ and ρ whose domains are the binding names of p and q, respectively. This is defined by:

$$\{a\|a\} = \{a\|\ulcorner a\urcorner\} = \{\ulcorner a\urcorner\|a\} = \{\ulcorner a\urcorner\|\ulcorner a\urcorner\} \overset{\text{def}}{=} (\{\},\{\})$$

$$\{\lambda x\|t\} \overset{\text{def}}{=} (\{t/x\},\{\}) \qquad \text{if } t \text{ is a term}$$

$$\{s\|\lambda x\} \overset{\text{def}}{=} (\{\},\{s/x\}) \qquad \text{if } s \text{ is a term}$$

$$\{p_1 \bullet p_2 \| q_1 \bullet q_2\} \overset{\text{def}}{=} (\sigma_1 \cup \sigma_2, \rho_1 \cup \rho_2) \qquad \text{if } \{p_i\|q_i\} = (\sigma_i,\rho_i) \text{ for } i \in \{1,2\}$$

$$\{p\|q\} \quad \text{undefined} \qquad \text{otherwise.}$$

Names and name-matches unify if they are for the same name. A binding name unifies with a term to produce a binding of the name to that term. Two compounds unify if their components unify; the resulting substitutions are the unions of those produced by unifying the components. Otherwise the unification is undefined (impossible). Note that the substitutions being combined have disjoint domain due to linearity of patterns, and this holds for the following two rules also.

The asymmetric and exchange languages exploit the *poly-match* rule $\text{MATCH}(\widetilde{p};\widetilde{q})$ that determines the matches of two sequences of patterns $\widetilde{p}$ and $\widetilde{q}$ to produce a pair of substitutions, as defined below:

$$\text{MATCH}(;) = (\emptyset,\emptyset) \qquad \frac{\{p_1\|q_1\} = (\sigma_1,\rho_1) \qquad \text{MATCH}(\widetilde{p};\widetilde{q}) = (\sigma_2,\rho_2)}{\text{MATCH}(p_1,\widetilde{p};q_1,\widetilde{q}) = (\sigma_1 \cup \sigma_2, \rho_1 \cup \rho_2)} \begin{array}{l} p_1 \text{ is a term} \\ q_1 \text{ is an intensional} \\ \text{pattern} \end{array}$$

$$\frac{\{p_1\|q_1\} = (\sigma_1,\rho_1) \qquad \text{MATCH}(\widetilde{p};\widetilde{q}) = (\sigma_2,\rho_2)}{\text{MATCH}(p_1,\widetilde{p};q_1,\widetilde{q}) = (\sigma_1 \cup \sigma_2, \rho_1 \cup \rho_2)} \begin{array}{l} p_1 \text{ is an intensional} \\ \text{pattern} \\ q_1 \text{ is a term.} \end{array}$$

The empty sequence matches with the empty sequence to produce empty substitutions. Otherwise when there are sequences $p_1,\widetilde{p}$ and $q_1,\widetilde{q}$ where p_1 is a term and q_1 is an intensional pattern (or vice versa) then they are unified $\{p_1 \| q_1\}$ and the remaining sequences use the poly-match rule. If both are defined and yield substitutions, the union of substitutions is yielded. Otherwise the poly-match is undefined, such as when; when a single unification fails, a term is aligned with a term, an intensional pattern with an intensional pattern, or when the sequences are of unequal arity.

The unification languages use the *poly-unify* rule $\text{UNIFY}(\widetilde{p};\widetilde{q})$ that is the same as the poly-match rule (without the side conditions) as shown below:

$$\text{UNIFY}(;) = (\emptyset,\emptyset) \qquad \frac{\{p_1\|q_1\} = (\sigma_1,\rho_1) \qquad \text{UNIFY}(\widetilde{p};\widetilde{q}) = (\sigma_2,\rho_2)}{\text{UNIFY}(p_1,\widetilde{p};q_1,\widetilde{q}) = (\sigma_1 \cup \sigma_2, \rho_1 \cup \rho_2)} \; .$$

Interaction is now defined by the following two axioms. The first

$$\overline{s}\langle\widetilde{p}\rangle.P \mid s(\widetilde{q}).Q \longmapsto (\sigma P) \mid (\rho Q) \qquad \text{MATCH}(\widetilde{p};\widetilde{q}) = (\sigma,\rho)$$

for asymmetric and exchange languages; and the second

$$s(\widetilde{p}).P \mid s(\widetilde{q}).Q \longmapsto (\sigma P) \mid (\rho Q) \qquad \text{UNIFY}(\widetilde{p};\widetilde{q}) = (\sigma,\rho)$$

for the unification languages. In both the s's are omitted for dataspace-based languages. Both axioms state that when the the symmetric patterns $\widetilde{p}$ and $\widetilde{q}$ poly-match or poly-unify, respectively, (and in the channel-based setting the input and output are along the same channel) to yield the substitutions σ and ρ, they reduce to σ applied to P in parallel with ρ applied to Q.

The reduction relation $\longmapsto$ also includes the following:

$$\frac{P \longmapsto P'}{P \mid Q \longmapsto P' \mid Q} \qquad \frac{P \longmapsto P'}{(va)P \longmapsto (va)P'} \qquad \frac{P \equiv Q \quad Q \longmapsto Q' \quad Q' \equiv P'}{P \longmapsto P'}$$

$$\frac{s = t \qquad P \mid Q \longmapsto S}{P \mid \text{if } s = t \text{ then } Q \text{ else } R \longmapsto S} \qquad \frac{s \neq t \qquad P \mid R \longmapsto S}{P \mid \text{if } s = t \text{ then } Q \text{ else } R \longmapsto S}$$

with $\Longmapsto$ denoting the reflexive, transitive closure of $\longmapsto$.

Lastly, for each language let $\cong$ denote a reduction-sensitive reference behavioural equivalence for that language, e.g. a barbed equivalence. That is, a behavioural equivalence $\cong$ such that whenever $P \cong P'$ and $P' \longmapsto$ imply $P \longmapsto$ as in Definition 5.3 of [18] (observe that his rules out weak bisimulations for example). For the asymmetric languages these are already known, either by their equivalent language in the literature or from [12, 13, 16]. For the non-asymmetric languages the results in [13] can be applied.

3   Encodings

This section recalls the definition of valid encodings for formally relating process calculi (details in [18]). The choice of valid encodings here is to align with prior works [12, 16, 18] and where possible reuse prior results. These valid encodings are those used, sometimes with mild adaptations, in [10, 14, 17, 18, 25] and have also inspired similar works [21, 22, 31]. However, there are alternative approaches to encoding criteria or comparing expressive power [3, 5, 7, 27, 30, 31]. Further discussion of the choices of encodings, and contrasting with other approaches can be found in [14, 17, 18, 29, 31].

An *encoding* of a language $\mathcal{L}_1$ into another language $\mathcal{L}_2$ is a pair $([\![\cdot]\!], \varphi_{[\![ ]\!]})$ where $[\![\cdot]\!]$ translates every $\mathcal{L}_1$-process into an $\mathcal{L}_2$-process and $\varphi_{[\![ ]\!]}$ maps every name (of the source language) into a tuple of k names (of the target language), for $k > 0$. In doing this, the translation may fix some names to play a precise rôle or may translate a single name into a tuple of names, this can be obtained by exploiting $\varphi_{[\![ ]\!]}$.

Now consider only encodings that satisfy the following properties. Let a $k-$ *ary context* $(\cdot_1; \ldots; \cdot_k)$ be a process with k holes. Denote with $\longmapsto^\omega$ an infinite sequence of reductions and let $P \Downarrow$ mean there exists P' such that $P \Longmapsto P'$ and $P' \equiv P'' \mid \sqrt{}$ for some P''. Moreover, let $\cong$ denote the reference behavioural equivalence. Finally, to simplify reading, let S range over processes of the source language (viz., $\mathcal{L}_1$) and T range over processes of the target language (viz., $\mathcal{L}_2$).

Definition 1 (Valid Encoding). *An encoding* $([\![\cdot]\!], \varphi_{[\![ ]\!]})$ *of* $\mathcal{L}_1$ *into* $\mathcal{L}_2$ *is valid if it satisfies the following five properties:*

1. Compositionality: *for every k-ary operator* op *of* $\mathcal{L}_1$ *and for every subset of names* N, *there exists a k-ary context* $C_{\text{op}}^N(\cdot_1; \ldots; \cdot_k)$ *of* $\mathcal{L}_2$ *such that, for all* $S_1, \ldots, S_k$ *with* $\text{fn}(S_1, \ldots, S_k) = N$, *it holds that* $[\![\text{op}(S_1, \ldots, S_k)]\!] = C_{\text{op}}^N([\![S_1]\!]; \ldots; [\![S_k]\!])$.
2. Name invariance: *for every* S *and substitution* σ, *it holds that* $[\![\sigma S]\!] = \sigma'[\![S]\!]$ *if* σ *is injective and* $[\![\sigma S]\!] \cong_2 \sigma'[\![S]\!]$ *otherwise where* σ' *is such that* $\varphi_{[\![ ]\!]}(\sigma(a)) = \sigma'(\varphi_{[\![ ]\!]}(a))$ *for every name a.*

3. Operational correspondence:
   – *for all S $\Longmapsto_1 S'$, it holds that $[\![S]\!] \Longmapsto_2 \cong_2 [\![S']\!]$;*
   – *for all $[\![S]\!] \Longmapsto_2 T$, there exists S' such that $S \Longmapsto_1 S'$ and $T \Longmapsto_2 \cong_2 [\![S']\!]$.*
4. Divergence reflection: *for every S such that $[\![S]\!] \longmapsto_2^\omega$, it holds that $S \longmapsto_1^\omega$.*
5. Success sensitiveness: *for every S, it holds that $S \Downarrow_1$ if and only if $[\![S]\!] \Downarrow_2$.*

Proposition 1. *Let $[\![\cdot]\!]$ be a valid encoding from $\mathcal{L}_1$ into $\mathcal{L}_2$; if there exist two $\mathcal{L}_1$ processes P of the form $p_1(p_2)P'$ and Q of the form either $q_1(q_2)Q'$ or $\overline{q_1}\langle q_2 \rangle Q'$ such that $P \mid Q \longmapsto$, then $[\![P \mid Q]\!] \longmapsto$.*

The following result exploits the *matching degree* of a language $\mathrm{M_D}(\cdot)$, defined as the least upper bound on the number of names that can be matched to yield reduction.

Proposition 2 (Theorem 5.9 from [18]). *If $\mathrm{M_D}(\mathcal{L}_1) > \mathrm{M_D}(\mathcal{L}_2)$ then there is no valid encoding of $\mathcal{L}_1$ into $\mathcal{L}_2$.*

Proposition 3 (Theorem 5.8 from [18]). *Assume there exists a $\mathcal{L}_1$-process S such that $S \longmapsto\!\!\!\!\!/\,_1$ and $S \Downarrow$ and $S \mid S \Downarrow$; moreover assume that every $\mathcal{L}_2$-process T that does not reduce is such that $T \mid T \longmapsto\!\!\!\!\!/\,_2$. Then there exists no valid encoding $[\![\cdot]\!]$ from $\mathcal{L}_1$ to $\mathcal{L}_2$.*

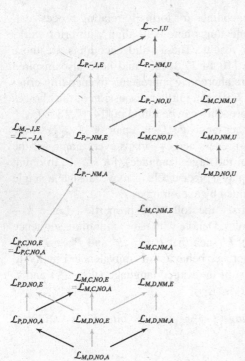

Fig. 2. Relations between all languages

The general way to prove the lack of a valid encoding is done as follows. By contradiction assuming there is a valid encoding $[\![\cdot]\!]$. Find a pair of processes P and Q that satisfy Proposition 1 such that $P \mid Q \longmapsto$ and $[\![P \mid Q]\!] \longmapsto$. From Q obtain some Q' such that $P \mid Q' \longmapsto\!\!\!\!\!/\,$ and $[\![P \mid Q']\!] \longmapsto$. Conclude by showing this in contradiction with some properties of the encoding or one of the propositions above.

The following result is a consequence of the choices of languages and encoding criteria, which corresponds to formalising Remark 1.

Proposition 4. *If a language $\mathcal{L}_1$ is a sublanguage of $\mathcal{L}_2$ then there exists a valid encoding $[\![\cdot]\!]$ from $\mathcal{L}_1$ into $\mathcal{L}_2$.*

Finally, the existence of encodings $[\![\cdot]\!]_1$ from $\mathcal{L}_1$ into $\mathcal{L}_2$ and $[\![\cdot]\!]_2$ from $\mathcal{L}_2$ into $\mathcal{L}_3$ does not ensure that $[\![[\![\cdot]\!]_1]\!]_2$ is a valid encoding from $\mathcal{L}_1$ into $\mathcal{L}_3$ [17]. However, this does hold when the encodings use $\equiv$ rather than $\cong$ in the target language, as is the case for all encodings presented in this work. This allows later assumption of composition of encodings here, although this is not true for all valid encodings in general.

4   Overview of Results

A diagram illustrating the results can be seen in Fig. 2. Arrows show increased expressive power and ='s show equivalence; black are from prior work, green from Sect. 5, and blue from Sect. 6. The lack of an arrow indicates no possible encoding in either direction (e.g. between $\mathcal{L}_{P,-,I,E}$ and $\mathcal{L}_{P,-,NM,U}$). Transitive relations are omitted (e.g. $\mathcal{L}_{P,-,NM,A}$ to $\mathcal{L}_{P,-,NO,U}$).

5   Asymmetry and Exchange

Exchange is a generalisation of asymmetric communication, i.e. $\mathcal{L}_{\alpha,\beta,\gamma,A}$ is trivially encoded by, $\mathcal{L}_{\alpha,\beta,\gamma,E}$ by Proposition 4. The rest of this section details other relations between asymmetric and exchange languages.

5.1   Exchange in Monadic Non-Intensional Languages

This section considers the simpler languages and demonstrates the proof techniques to show that exchange cannot be easily encoded into asymmetry.

For the monadic non-intensional languages changing from asymmetric to exchange communication alone is almost always an increase in expressive power. The following result is presented to demonstrate the proof technique for the most complex. Simpler variations can be used to show that there exists no encoding of: $\mathcal{L}_{M,D,NO,E}$ into $\mathcal{L}_{M,D,NO,A}$, or $\mathcal{L}_{M,D,NM,E}$ into $\mathcal{L}_{M,D,NM,A}$.

Theorem 1. *There exists no valid encoding of $\mathcal{L}_{M,C,NM,E}$ into $\mathcal{L}_{M,C,NM,A}$.*

The exception to the general $\mathcal{L}_{M,\beta,\gamma,E}$ is more expressive than $\mathcal{L}_{M,\beta,\gamma,A}$ when $\gamma \neq I$ is $\mathcal{L}_{M,C,NO,E}$ into $\mathcal{L}_{M,C,NO,A}$. This is detailed in Sect. 5.2.

Within the monadic non-intensional exchange languages the usual diamond of relations exists where adding channel-based communication or name-matching are both increases in expressive power. The separation results between $\mathcal{L}_{M,D,NM,E}$ and $\mathcal{L}_{M,C,NO,E}$ is the most interesting result, as the rest can be proved via matching degree.

Theorem 2. *The languages $\mathcal{L}_{M,D,NM,E}$ and $\mathcal{L}_{M,C,NO,E}$ are unrelated.*

5.2   Encoding Exchange into Asymmetry

This section considers where exchange languages can be encoded by asymmetric languages. Note that this does not ensure atomicity that motivates some languages [10].

An exchange language $\mathcal{L}_1$ can be encoded into an asymmetric language $\mathcal{L}_2$ if the matching degree of $\mathcal{L}_1$ is bounded, and: $\mathcal{L}_1$ and $\mathcal{L}_2$ are both channel-based no-matching languages; or $\mathcal{L}_2$ has a greater matching degree and is polyadic or channel-based.

In the first case, the key idea is to represent the channel name by a pair of names to indicate whether the input is on the action or co-action. Consider the following translation from $\mathcal{L}_{M,C,NO,E}$ into $\mathcal{L}_{M,C,NO,A}$:

$$[\![\,(\nu n)P\,]\!] \stackrel{\text{def}}{=} (\nu n_1)(\nu n_2)[\![\,P\,]\!]$$

$$[\![\,n(\lambda x).P\,]\!] \stackrel{\text{def}}{=} n_1(\lambda \text{rn}).\text{rn}(\lambda x_1).x(\lambda x_2).[\![\,P\,]\!]$$

$$[\![\,\overline{n}\langle\lambda x\rangle.P\,]\!] \stackrel{\text{def}}{=} n_2(\lambda \text{rn}).\text{rn}(\lambda x_1).x(\lambda x_2).[\![\,P\,]\!]$$

$$[\![\,n(a).P\,]\!] \stackrel{\text{def}}{=} (\nu \text{rn})\overline{n_2}\langle\text{rn}\rangle.\overline{\text{rn}}\langle a_1\rangle.\overline{\text{rn}}\langle a_2\rangle.[\![\,P\,]\!]$$

$$[\![\,\overline{n}\langle a\rangle.P\,]\!] \stackrel{\text{def}}{=} (\nu \text{rn})\overline{n_1}\langle\text{rn}\rangle.\overline{\text{rn}}\langle a_1\rangle.\overline{\text{rn}}\langle a_2\rangle.[\![\,P\,]\!]$$

$$[\![\,\textbf{if }s = t\textbf{ then }P\textbf{ else }Q\,]\!] \stackrel{\text{def}}{=} \textbf{if }s_1 = t_1\textbf{ then }[\![\,P\,]\!]\textbf{ else }[\![\,Q\,]\!]\,.$$

Here the names n_1 and n_2 represent two parts of the name n, and rn is a reserved name, these are all introduced by the renaming policy $\varphi_{[\![\,]\!]}$ [11, 18].

Theorem 3. *The encoding from $\mathcal{L}_{M,C,NO,E}$ into $\mathcal{L}_{M,C,NO,A}$ is valid.*

The above encoding illustrates how channel-based communication is sufficient when no name-matching or intensionality is included in the source language.

In the second case, the key idea is that a single name is sufficient to represent the shape of the encoded action or co-action, and so can ensure correct encoded interactions. Observe that in every case the reverse encoding is proved impossible easily via the matching degree and Proposition 2. The clearest illustration of this when the source language is monadic is the following encoding from $\mathcal{L}_{M,D,NO,E}$ into $\mathcal{L}_{M,C,NO,A}$. Consider the translation $[\![\,\cdot\,]\!]$ that is homeomorphic on all forms except for the action and co-action, and exploits two reserved names ia and ic that are translated as follows:

$$[\![\,(p).P\,]\!] \stackrel{\text{def}}{=} \begin{cases} \overline{\text{ic}}\langle a\rangle.[\![\,P\,]\!] & p = a \\ \text{ia}(\lambda x).[\![\,P\,]\!] & p = \lambda x \end{cases} \qquad\qquad [\![\,\langle p\rangle.P\,]\!] \stackrel{\text{def}}{=} \begin{cases} \overline{\text{ia}}\langle a\rangle.[\![\,P\,]\!] & p = a \\ \text{ic}(\lambda x).[\![\,P\,]\!] & p = \lambda x\,. \end{cases}$$

The channel name indicates the origin of the input, ia for action, and ic for co-action.

Theorem 4. *The encoding from $\mathcal{L}_{M,D,NO,E}$ into $\mathcal{L}_{M,C,NO,A}$ is valid.*

The existence of an encoding from $\mathcal{L}_{M,D,NM,E}$ into $\mathcal{L}_{M,C,NM,A}$ is achieved in the same manner by extending the initial translation to consider name matches $\ulcorner a\urcorner$ to also be inputs, e.g. i.e. $[\![\,(\ulcorner a\urcorner).P\,]\!] \stackrel{\text{def}}{=} \text{ia}(\ulcorner a\urcorner).[\![\,P\,]\!]$.

The existence of valid encodings from $\mathcal{L}_{M,C,NM,E}$ into $\mathcal{L}_{P,-,NM,A}$ can be shown with a similar technique, instead of the reserved names being used as a channel they are simply added as another part of the polyadic input or output (with name-matching on the input). For example, $[\![\,n(a).P\,]\!] \stackrel{\text{def}}{=} \overline{\text{ic}}\langle n, a\rangle$ and $[\![\,n(\lambda x).P\,]\!] \stackrel{\text{def}}{=} \text{ia}(\ulcorner n\urcorner, \lambda x)$.

A similar but more complex technique can be used to encode polyadic no-matching exchange languages into asymmetric languages. This is illustrated by the following encoding from $\mathcal{L}_{P,D,NO,E}$ into $\mathcal{L}_{P,C,NO,A}$. The encoding exploits a binary representation of the structure of an action or co-action. To this end define the *binary representation function* $\textsc{Bin}(\cdot)$ that converts a sequence of names and binding names into a bit-string and also the *complement* (or bitwise not) Not($\cdot$) of bit-strings (where ';' is concatenation):

$$\text{Bin}(a) = 0 \quad \text{Bin}(\lambda x) = 1 \quad \text{Bin}(n, \overline{n}) = \text{Bin}(n); \text{Bin}(\overline{n})$$
$$\text{Not}(0) = 1 \quad \text{Not}(1) = 0 \quad \text{Not}(X, \widetilde{X}) = \text{Not}(X); \text{Not}(\widetilde{X}) \ .$$

Given a sequence of binding names and names $\widetilde{p}$, the sequences of the binding names $\text{Bn}(\widetilde{p})$, and names $\text{Nm}(\widetilde{p})$ are defined by:

$$\text{Bn}(\lambda x, \widetilde{p}) = \lambda x, \text{Bn}(\widetilde{p}) \quad \text{Bn}(a, \widetilde{p}) = \text{Bn}(\widetilde{p})$$
$$\text{Nm}(\lambda x, \widetilde{p}) = \text{Nm}(\widetilde{p}) \quad \text{Nm}(a, \widetilde{p}) = a, \text{Nm}(\widetilde{p}) \ .$$

Now consider the translation $[\![\cdot]\!]$ that is homeomorphic on all forms except the action and co-action (and exploits a reserved name rn) that are translated as follows:

$$[\![(\widetilde{p}).P]\!] \stackrel{\text{def}}{=} a(\lambda \text{rn}, \text{Bn}(\widetilde{p})).\overline{\text{rn}}\langle \text{Nm}(\widetilde{p})\rangle.[\![P]\!] \qquad a = \text{Bin}(\widetilde{p})$$
$$[\![\langle\widetilde{p}\rangle.P]\!] \stackrel{\text{def}}{=} (\nu\text{rn})\overline{a}\langle \text{rn}, \text{Nm}(\widetilde{p})\rangle.\text{rn}(\text{Bn}(\widetilde{p})).[\![P]\!] \quad a = \text{Not}(\text{Bin}(\widetilde{p})) \ .$$

The idea is that the translated action and co-action match on the channel name that is the bit-string representation of their order of binding names and names. If they match the input performs all the action's bindings as well as an additional name (bound to) rn. The rôles are then reversed to complete the interaction.

Theorem 5. *The encoding from $\mathcal{L}_{P,D,NO,E}$ into $\mathcal{L}_{P,C,NO,A}$ is valid.*

The encoding from $\mathcal{L}_{P,C,NO,E}$ into $\mathcal{L}_{P,C,NO,A}$ exploits elements of the technique above. Define the function Val($\cdot$) that gives the numeric value of a binary string, e.g. Val(101) = 5 and Val(1010) = 10. Now the encoding from $\mathcal{L}_{P,C,NO,E}$ into $\mathcal{L}_{P,C,NO,A}$ can be constructed as follows exploiting a reserved name rn as usual:

$$[\![n(\widetilde{p}).P]\!] \stackrel{\text{def}}{=} n(\lambda \text{rn}, \text{Bn}(\widetilde{p}), \lambda z, \ldots, \lambda z_i).\overline{\text{rn}}\langle \text{Nm}(\widetilde{p})\rangle.[\![P]\!]$$
$$\text{where } i = \text{Val}(1; \text{Bin}(\widetilde{p})) - |\text{Bn}(\widetilde{p})|$$

$$[\![\overline{n}\langle\widetilde{p}\rangle.P]\!] \stackrel{\text{def}}{=} (\nu\text{rn})(\nu z_1)\ldots(\nu z_i)\overline{n}\langle \text{rn}, \text{Nm}(\widetilde{p}), z_1, \ldots, z_i\rangle.\text{rn}(\text{Bn}(\widetilde{p})).[\![P]\!]$$
$$\text{where } i = \text{Val}(1; \text{Not}(\text{Bin}(\widetilde{p}))) - |\text{Nm}(\widetilde{p})|$$

and translating all other processes homomorphically. Also $\widetilde{z}$ do not intersect one another, or any of the names in n and $\widetilde{p}$ and $fn(P)$.

The key idea is to map the binary representation of the structure of the action or co-action to the arity of the encoded action or co-action. To prevent conflicts between encodings, for example $n(a, \lambda x)$ and $n(\lambda x)$, the binary representation is pre-pended with 1. Thus, the arity of the action or co-action ensures correct interaction if the structure is correct, and the channel name is matched as usual.

Theorem 6. *The encoding from $\mathcal{L}_{P,C,NO,E}$ into $\mathcal{L}_{P,C,NO,A}$ is valid.*

Building on Theorem 5 and the equivalence between the languages $\mathcal{L}_{P,-,NM,A}$ [16] conclude that $\mathcal{L}_{P,-,NM,A}$ are able to encode all the: monadic non-intensional exchange languages; and the polyadic no-matching exchange languages.

5.3 Other Relations with Bounded Matching Degree

This section considers other relations between languages equally or less expressive than $\mathcal{L}_{P,-,NM,A}$, i.e. all the languages that can be encoded in $\mathcal{L}_{P,-,NM,A}$.

Within exchange languages, clearly $\mathcal{L}_{M,\beta,NO,E}$ is a sub-language of $\mathcal{L}_{P,\beta,NO,E}$ for any β and so can be validly encoded by Proposition 4. The following proves the separation results required to indicate an increase in expressiveness.

Theorem 7. *There exists no valid encoding of $\mathcal{L}_{P,D,NO,E}$ into $\mathcal{L}_{M,D,NO,E}$.*

Observe that this result can be used to show there exists no valid encoding of $\mathcal{L}_{P,C,NO,E}$ into $\mathcal{L}_{M,C,NO,E}$ by having all communication along a single channel name and preventing modification of this name by the encoding.

Regarding asymmetric languages, $\mathcal{L}_{P,D,NO,E}$ can validly encode $\mathcal{L}_{P,D,NO,A}$ by Proposition 4. The following proves an increase in expressiveness.

Theorem 8. *There exists no valid encoding of $\mathcal{L}_{P,D,NO,E}$ into $\mathcal{L}_{P,D,NO,A}$.*

5.4 Equivalent Languages with Unbounded Matching Degree

Once the matching degree is unbounded several languages become equivalent in expressiveness, this section formalises these results.

The intensional asymmetric languages all have equivalent expressiveness (by Theorem 6.5 of [12]) and to the monadic exchange languages. Consider the languages $\mathcal{L}_{M,D,I,E}$ and $\mathcal{L}_{M,C,I,E}$, there is a trivial valid encoding of $\mathcal{L}_{M,D,I,E}$ into $\mathcal{L}_{M,C,I,E}$ by Proposition 4. The following shows equivalence via the reverse encoding from $\mathcal{L}_{M,C,I,E}$ into $\mathcal{L}_{M,D,I,E}$. Take the encoding $[\![\cdot]\!]$ that is the homeomorphic on all processes except the action and co-action that are encoded as follows (exploiting reserved names as usual):

$$
\left.
\begin{aligned}
[\![\, p(q).P \,]\!] &\overset{\text{def}}{=} \langle \mathsf{ic} \bullet p \bullet q \rangle . [\![\, P \,]\!] \\
[\![\, \overline{p}\langle q \rangle.P \,]\!] &\overset{\text{def}}{=} \langle \mathsf{ia} \bullet p \bullet q \rangle . [\![\, P \,]\!]
\end{aligned}
\right\} \; q \text{ is a term}
$$

$$
\left.
\begin{aligned}
[\![\, p(q).P \,]\!] &\overset{\text{def}}{=} (\mathsf{ia} \bullet \ulcorner p \urcorner \bullet q) . [\![\, P \,]\!] \\
[\![\, \overline{p}\langle q \rangle.P \,]\!] &\overset{\text{def}}{=} (\mathsf{ic} \bullet \ulcorner p \urcorner \bullet q) . [\![\, P \,]\!]
\end{aligned}
\right\} \; q \text{ is an intensional pattern.}
$$

The translation compounds the channel pattern p with the term or intensional pattern q, converting to maintain being either a term or intensional pattern.

Theorem 9. *The encoding from $\mathcal{L}_{M,C,I,E}$ into $\mathcal{L}_{M,D,I,E}$ is valid.*

Now to complete the equivalences. Since all the languages $\mathcal{L}_{-,-,I,A}$ are equally expressive and since the languages $\mathcal{L}_{M,-,I,E}$ are equally expressive by Theorem 9 it suffices to consider examples from either group. The encodings from $\mathcal{L}_{-,-,I,A}$ into $\mathcal{L}_{M,-,I,E}$ follow by $\mathcal{L}_{M,D,I,A}$ being a sub-language of $\mathcal{L}_{M,D,I,E}$. In the other direction, there exists a valid encoding from $\mathcal{L}_{M,D,I,E}$ into $\mathcal{L}_{M,C,I,A}$, by a straightforward adaption of Theorem 4.

Considering polyadic non-intensional languages, $\mathcal{L}_{P,D,NM,E}$ can be encoded into $\mathcal{L}_{P,C,NM,E}$ by Proposition 4. For the converse, the standard approach [12,16] yields a valid encoding; one that is homeomorphic on all forms except the action $[\![\, a(\widetilde{p}).P \,]\!] \overset{\text{def}}{=}$

($\ulcorner a \urcorner, \overline{p}$).$[\![P]\!]$ and co-action $[\![\overline{a}\langle \overline{p}\rangle.P]\!] \overset{\text{def}}{=} \langle a, \overline{p}\rangle.[\![P]\!]$. Indeed this approach can be used for the polyadic intensional languages, showing the existence of a valid encoding from $\mathcal{L}_{P,C,I,E}$ into $\mathcal{L}_{P,D,I,E}$. Equivalence is completed by showing a valid encoding of $\mathcal{L}_{P,D,I,E}$ into $\mathcal{L}_{P,C,I,E}$ by Proposition 4.

5.5 Concluding Relations

This section concludes the relations between asymmetric and exchange languages by formalising those between languages with unbounded matching degree. In general this is showing separation results between different language groups.

Polyadic intensional exchange languages are more expressive than any other exchange or asymmetric languages. By the encodings in Sect. 5.4 in all languages considered here being dataspace-based or channel-based is immaterial to expressive power. The languages $\mathcal{L}_{P,-,NM,E}$ are sub-languages of $\mathcal{L}_{P,-,I,E}$ and so their expressiveness is included naturally, the reverse is from the following result.

Theorem 10. *There exists no valid encoding of $\mathcal{L}_{P,-,I,E}$ into $\mathcal{L}_{P,-,NM,E}$.*

Comparing within other intensional exchange languages, $\mathcal{L}_{P,-,I,E}$ can encode $\mathcal{L}_{M,-,I,E}$ by Proposition 4. The reverse separation result uses the technique of Theorem 11.

That the languages $\mathcal{L}_{M,-,I,E}$ are unrelated to $\mathcal{L}_{P,-,NM,E}$ follows from the separation results that show no valid encodings from $\mathcal{L}_{M,-,I,E}$ into $\mathcal{L}_{P,-,NM,E}$, and $\mathcal{L}_{P,-,NM,E}$ into $\mathcal{L}_{M,-,I,E}$ (proved using the techniques of Theorems 10 and 7, respectively). Note the groups of languages can be treated equivalently due to the encodings of Sect. 5.4.

Lastly, the languages $\mathcal{L}_{P,-,NM,A}$ can be encoded $\mathcal{L}_{P,-,NM,E}$ via Proposition 4. The reverse is prevented by the following result.

Theorem 11. *There exists no valid encoding from $\mathcal{L}_{P,-,NM,E}$ into $\mathcal{L}_{P,-,NM,A}$.*

6   Unification

This section considers the expressiveness of unification languages, and their relations to asymmetric and exchange languages.

6.1 Unification Cannot Be Simulated

The following result shows that no unification language can be encoded into an asymmetric or exchange language. Key is a *self recognising* process, defined to be $S = (a).\sqrt{}$ for the dataspace-based languages and $S = a(a).\sqrt{}$ for the channel-based languages, that has the behaviour $S \mid S \longmapsto \Downarrow$ but $S \longmapsto\!\!\!\!/$ and $S \not\Downarrow$. This can be exploited since no non-unification process can reduce in parallel with itself unless it reduces alone. The self recognising process can be used to yield the following result via Proposition 3.

Theorem 12. *There exists no valid encoding of a unification language $\mathcal{L}_{-,-,-,U}$ into any non-unification language $\mathcal{L}_{-,-,-,\delta}$ $\delta \neq U$.*

The above result can be used to prove a separation result from any unification language to a non-unification language, these results are omitted from the rest of the paper.

6.2   On Monadic Non-Intensional Unification Languages

All the languages $\mathcal{L}_{M,-,\gamma,E}$ where $\gamma \neq I$ are unrelated to any non-unification language. Similar to the languages $\mathcal{L}_{M,-,\gamma,A}$, these 4 form a diamond where expressiveness is increased by adding channel-based communication or pattern-matching.

The shift to unification leads to $\mathcal{L}_{M,D,NO,U}$ being unrelated to any other language $\mathcal{L}_{M,D,NO,-}$. The following result illustrates how to achieve such separation results and can be applied to other monadic non-intensional languages also.

Theorem 13. *There exists no valid encoding from $\mathcal{L}_{M,D,NO,\delta}$ where $\delta \neq U$ into $\mathcal{L}_{M,D,NO,U}$.*

The relations between the monadic non-intensional unification languages are as usual, although the usual proof techniques do not always hold. In particular, no-matching unification languages still have non-zero matching degree, so separation results that rely on matching degree alone no longer hold. $\mathcal{L}_{M,D,NO,U}$ can be validly encoded by $\mathcal{L}_{M,D,NM,U}$ via Proposition 4. The following proves the separation result.

Theorem 14. *There exists no valid encoding of $\mathcal{L}_{M,D,NM,U}$ into $\mathcal{L}_{M,D,NO,U}$.*

The above technique can be applied to prove that there exist no encodings from $\mathcal{L}_{M,C,NM,U}$ into $\mathcal{L}_{M,C,NO,U}$, or from $\mathcal{L}_{M,D,NM,U}$ into $\mathcal{L}_{M,C,NO,U}$ The rest of the separation results to prove that the relations are the same as in the asymmetric setting exploit the matching degree of the languages involved.

6.3   Equally Expressive Unification Languages

Once the matching degree is unbounded there is no difference in expressiveness between dataspace-based and channel-based communication for unification languages. Further, all the intensional unification languages have equal expressive power.

For the polyadic languages it is straightforward to represent channel-based communication by shifting the channel to the first position of a dataspace-based encoding. For both encodings from $\mathcal{L}_{P,C,NO,U}$ into $\mathcal{L}_{P,D,NO,U}$ and $\mathcal{L}_{P,C,NM,U}$ into $\mathcal{L}_{P,D,NM,U}$ are achieved by $[\![\, a(\widetilde{p}).P \,]\!] \stackrel{\text{def}}{=} (a, \widetilde{p}).[\![\, P \,]\!]$. The converse results are by Proposition 4.

This may at first appear unexpected since in the asymmetric and exchange languages $\mathcal{L}_{P,C,NO,\delta}$ ($\delta \neq U$) have matching degree 1 while $\mathcal{L}_{P,D,NO,\delta}$ ($\delta \neq U$) have matching degree 0. However, this does not hold for unification languages as due to the poly-unify rule their matching degree directly relates to their arity.

All the intensional unification languages are equally expressive. Clearly the languages $\mathcal{L}_{M,-,I,U}$ and $\mathcal{L}_{-,D,I,U}$ can be trivially validly encoded into the languages $\mathcal{L}_{P,-,I,U}$ and $\mathcal{L}_{-,C,I,U}$, respectively, by sub-language inclusion. An encoding from $\mathcal{L}_{P,-,I,U}$ into $\mathcal{L}_{M,-,I,U}$ can be easily achieved in the same manner as Theorem 5.4 of [12] by encoding the polyadic structure into a monadic intensional pattern. The key idea is that a sequence of patterns $\widetilde{p} = p_1, \ldots, p_i$ is encoded as a single pattern $(\text{rn} \bullet p_1) \bullet \ldots \bullet p_i$ where rn is a reserved name. For showing an encoding from $\mathcal{L}_{-,C,I,U}$ into $\mathcal{L}_{-,D,I,U}$ the same technique as used in Theorem 9 can be used.

6.4   Encodings into Polyadic Non-Intensional Languages

This section considers encodings into polyadic non-intensional unification languages. Despite being nominally no-matching it is still possible to encode polyadic name-matching into the languages $\mathcal{L}_{P,-,NO,U}$. Beyond this the usual increases in expressiveness hold for shifting from monadic to polyadic, and from no-matching to name-matching. The rest of this section details these relations.

Unification communication exploits pattern unification that allows equivalence of patterns. The key difference is that a single name can unify with itself unlike in the poly-match rule where $\text{MATCH}(a, a)$ is undefined. This breaks the directionality assumed in asymmetric and exchange primitives, and so invalidates many prior results.

The directionality of asymmetric or exchange languages can be maintained by an encoding when the target language is either polyadic or intensional. Define the *unprotect* function g that replaces all instances of $\ulcorner a \urcorner$ with a in a pattern. Consider the encoding $[\![\,\cdot\,]\!]$ from $\mathcal{L}_{P,D,NM,E}$ to $\mathcal{L}_{P,C,NO,U}$ that exploits the functions BIN and NOT of Sect. 5.2 and is homeomorphic on all forms except as defined below:

$$[\![\,(\widetilde{p}).P\,]\!] \overset{\text{def}}{=} a(\lambda\text{rn}, \widetilde{g(p)}).[\![\,P\,]\!] \qquad a = \text{BIN}(\widetilde{p})$$

$$[\![\,\langle\widetilde{p}\rangle.P\,]\!] \overset{\text{def}}{=} (\nu\text{rn})a(\text{rn}, \widetilde{g(p)}).[\![\,P\,]\!] \quad a = \text{NOT}(\text{BIN}(\widetilde{p}))$$

The binary encoding is used to ensure that inputs and outputs are properly aligned since otherwise two outputs may unify. The additional reserved name is to distinguish actions from co-actions in the translation. The unprotect function g converts name-matches into names since the former aren't defined in a no-matching language.

Theorem 15. *The encoding from $\mathcal{L}_{P,D,NM,E}$ into $\mathcal{L}_{P,C,NO,U}$ is valid.*

Observe that since $\mathcal{L}_{P,D,NM,E}$ generalises the languages $\mathcal{L}_{-,-,\gamma,\delta}$ where $\gamma \leq NM$ and $\delta \leq E$ this proof applies to all such languages.

Regarding other unification languages, clearly $\mathcal{L}_{P,-,NO,U}$ can be validly encoded by $\mathcal{L}_{P,-,NM,U}$, with shifts between dataspace-based and channel-based communication handled by the encodings of Sect. 6.3 and Proposition 4. The separation result required to indicate an increase in expressiveness from $\mathcal{L}_{P,-,NO,U}$ to $\mathcal{L}_{P,-,NM,U}$ can be proved using the same technique as Theorem 14. The relations between polyadic and monadic languages are as expected. $\mathcal{L}_{M,C,NO,U}$ can be encoded by $\mathcal{L}_{P,C,NO,U}$ via Proposition 4, and thus also $\mathcal{L}_{P,D,NO,U}$ by encoding from Sect. 6.3. The separation result that $\mathcal{L}_{M,C,NO,U}$ cannot encode $\mathcal{L}_{P,-,NO,U}$ is by Proposition 2. Similar results hold for the name-matching languages also. $\mathcal{L}_{M,C,NM,U}$ can be encoded by $\mathcal{L}_{P,C,NM,U}$ by Proposition 4 (and thus also $\mathcal{L}_{P,D,NM,U}$ by encoding from Sect. 6.3). The reverse separation that there exists no valid encoding of $\mathcal{L}_{P,-,NM,U}$ into $\mathcal{L}_{M,C,NM,U}$ is proven via Proposition 2.

6.5   Intensional Unification Languages

Since all the intensional unification languages are equivalent by exploiting encodings from Sect. 6.3, it remains to show their other relations.

The intensional unification languages can also encode directionality in a similar manner to Sect. 6.4 (Theorem 15). Consider the encoding $[\![\,\cdot\,]\!]$ from $\mathcal{L}_{P,D,I,E}$ to $\mathcal{L}_{M,C,I,U}$

that exploits the numerical encoding function Bɪɴ and Noᴛ of Sects. 5.2 and 6.4 and is the homeomorphic on all forms except the action and co-action:

$$[\![(p_1,\ldots,p_i).P]\!] \stackrel{\text{def}}{=} a(\lambda \mathsf{rn} \bullet (p_1 \bullet \ldots \bullet p_i)).[\![P]\!] \quad a = \mathrm{B\scriptstyle IN}(\widetilde{p})$$

$$[\![\langle p_1,\ldots,p_i \rangle.P]\!] \stackrel{\text{def}}{=} a(\mathsf{rn} \bullet (p_1 \bullet \ldots \bullet p_i)).[\![P]\!] \quad a = \mathrm{N\scriptstyle OT}(\mathrm{B\scriptstyle IN}(\widetilde{p}))$$

where rn is a reserved name as usual. The translations of actions and co-actions are as before except that compounding is used in place of polyadic sequencing.

Theorem 16. *The encoding from $\mathcal{L}_{P,D,I,E}$ into $\mathcal{L}_{M,C,I,U}$ is valid.*

$\mathcal{L}_{-,-,I,U}$ are more expressive than $\mathcal{L}_{P,-,I,E}$ since there exists an encoding from $\mathcal{L}_{P,D,I,E}$ into $\mathcal{L}_{M,C,I,U}$ by Theorem 16, conclude via encodings of Sects. 5.4 and 6.3.

Finally, within unification languages intensionality remains more expressive than non-intensionality. By encodings in Sect. 6.3 all languages considered here being channel-based or dataspace-based is immaterial. The languages $\mathcal{L}_{P,-,NM,U}$ can be encoded by $\mathcal{L}_{P,-,I,U}$ by Proposition 4, the final separation of $\mathcal{L}_{P,-,I,U}$ into $\mathcal{L}_{P,-,NM,U}$ can be proved using the techniques of Theorem 10 and completes the results.

7   Conclusions and Discussion

Symmetric communication primitives provide new and interesting perspectives on how languages and communication can occur. Considering exchange provides interesting insight into how much trading systems and atomic exchange actions can be captured within asymmetric languages by encoding. The ability to encode polyadic exchange without name matching into asymmetric languages indicates that it is the addition of (unbounded, or multiple) name-matching with exchange that really extends expressiveness. While exchange generally increases expressiveness over asymmetric languages, it is pattern-matching that provides the strongest expressiveness alone. The unification languages cannot be encoded into even exchange languages, the self recognising technique was used for CPC before [14] but is here generalised. The flexibility of unification allows for names to be matched even in a language that nominally does not have name-matching. This yields some interesting results where non-matching languages can encode name-matching languages by exploiting unification. However, name-matching still provides increased expressiveness within unification languages, and intensionality is the sole factor in determining the most expressive language.

Choices of Primitives. The choices of primitives here is to align with prior work and results [12, 16, 18]. However, there are other choices that would impact some results.

The patterns are chosen here to match those of CPC and it turns out that the (symmetric) patterns here are sufficient to represent most other approaches, such as Spi Calculus [14] and Psi Calculi terms [12]. More generally the core approach of compounding proves sufficient to represent many complex data structures and even (in practice) type information. This has been discussed and formalised in different settings [10, 19, 20] and in many works related to *pattern calculus*, *S F*-logic, and CPC.

For the process forms the most obvious alternative would be to consider a choice operator: $\alpha_1.P + \alpha_2.Q$ for some choice of α_i. Again the decision not to include this is to match with prior results [12, 16, 18]. The addition of such a choice operator could invalidate some findings, in particular Theorem 12. This provides illustration of which results would need to be reexamined with such a change, although it does not (a priori) indicate that the overall relative expressiveness would change. For example, previous simple results for the inability to encode CPC ($\mathcal{L}_{M,D,I,F}$) into π-calculus ($\mathcal{L}_{M,C,NO,A}$) have used this approach [14], however alternative proofs also exist such as Proposition 2 and Theorems 11, & 14 and in prior works [12, 16]. This lends weight to the rigour here that provides alternative approaches, and identifies which results rely on which primitives.

In this context there are many other possible choices of primitives for both the patterns and the processes. However, those here are sufficient to understand the core dynamics between the interaction features of languages. Also by using a common approach that is transitive for the encodings here, often more distant relations can be proved without relying on particular choices of primitives or proof techniques.

Related Work. This section provides a brief account of related works most close to the decisions and results here, since to cover all related works would take an entire paper.

There are already existing specific results for some symmetric process calculi that agree with the results here. CPC ($\mathcal{L}_{M,D,I,U}$) can homomorphically encode: π-calculus ($\mathcal{L}_{M,C,NO,A}$), Linda ($\mathcal{L}_{P,D,NM,A}$), and Spi Calculus (perhaps $\mathcal{L}_{M,C,I,A}$) while none of them can encode CPC [12, 14]. Similarly fusion calculus can encode π-calculi, although not the other way around [28]. Impossibility of encoding results for CPC and fusion calculus into many calculi can be derived from the results here. Fusion calculus and Psi calculi are unrelated to CPC in that neither can encode CPC, and CPC cannot encode either of them [10, 14]. However, these results rely upon the global effect of fusions in fusion calculus, and the inclusing of logic in Psi Calculi.

There are also related works on concurrent constraint languages (CCL) [6, 7]. The encoding criteria of de Boer and Palamidessi [7] have similarities to those here, but also some significant differences: they assume that parallel composition must be encoded homomorphically, include the choice operator, and exploit a different notion of computational correspondence. The homomorphic parallel composition holds for all encodings presented in this paper, but not for the separation results (making them stronger, although also making the proofs a little harder here). However, the inclusion of the choice operator and having different notions of computation correspondence in de Boer and Palamidessi's results yield a different setting to here, and so neither results directly subsume the other. Further, CCLs have a different communication paradigm, with interaction between a single process and a common store of constraints, which is quite different to the focus of this paper. However, such non-binary approaches to communications have been considered [8, 15]. In addition, the expressiveness of CCLs depends to some degree on the logic, which is again not considered as part of communication paradigms here (although since CCLs and Psi Calculi exploit logics, this may be an interesting path of future research). In [6] there is also unification of terms, however their approach is different in that unifying s and t by σ is achieved when $\sigma s = \sigma t$. It is possible to restructure the unification rule here to use a single substitution (although this is overly complex and requires reasoning over processes not just patterns), but the unification would still differ since there is no distinction for name-matches $\ulcorner a \urcorner$ in [6].

References

1. Bengtson, J., Johansson, M., Parrow, J., Victor, B.: Psi-calculi: a framework for mobile processes with nominal data and logic. Log. Methods Comput. Sci. **7**(1) (2011)
2. Bengtson, J., Parrow, J.: Formalising the pi-calculus using nominal logic. Log. Methods Comput. Sci. **5**(2) (2009)
3. Boudol, G.: Notes on algebraic calculi of processes. In: Apt, K.R. (ed.) Logics and Models of Concurrent Systems, pp. 261–303. Springer, New York (1985)
4. Busi, N., Gorrieri, R., Zavattaro, G.: On the expressiveness of Linda coordination primitives. Inf. Comput. **156**(1–2), 90–121 (2000)
5. Carbone, M., Maffeis, S.: On the expressive power of polyadic synchronisation in π-calculus. Nordic J. Comput. **10**(2), 70–98 (2003)
6. de Boer, F.S., Palamidessi, C.: Concurrent logic programming: asynchronism and language comparison. In: Proceedings of the 1990 North American Conference on Logic Programming, pp. 175–194. MIT Press, Cambridge (1990)
7. de Boer, F.S., Palamidessi, C.: Embedding as a tool for language comparison. Inf. Comput. **108**(1), 128–157 (1994)
8. Fournet, C., Gonthier, G.: The reflexive cham and the join-calculus. In: Proceedings of the 23rd ACM Symposium on Principles of Programming Languages, pp. 372–385. ACM Press (1996)
9. Gelernter, D.: Generative communication in Linda. ACM Trans. Program. Lang. Syst. **7**(1), 80–112 (1985)
10. Given-Wilson, T.: Concurrent Pattern Unification. Ph.D. thesis, University of Technology, Sydney, Australia (2012)
11. Given-Wilson, T.: An intensional concurrent faithful encoding of Turing machines. In: Proceedings of the ICE 2014, Berlin, Germany, 6 June 2014, pp. 21–37 (2014)
12. Given-Wilson, T.: On the expressiveness of intensional communication. In: Proceedings of EXPRESS/SOS, Rome, Italie, September 2014
13. Given-Wilson, T., Gorla, D.: Pattern matching and bisimulation. In: Nicola, R., Julien, C. (eds.) COORDINATION 2013. LNCS, vol. 7890, pp. 60–74. Springer, Heidelberg (2013). doi:10.1007/978-3-642-38493-6_5
14. Given-Wilson, T., Gorla, D., Jay, B.: A concurrent pattern calculus. Log. Methods Comput. Sci. **10**(3) (2014)
15. Given-Wilson, T., Legay, A.: On the expressiveness of joining. In: ICE 2015, Grenoble, France, June 2015
16. Gorla, D.: Comparing communication primitives via their relative expressive power. Inf. Comput. **206**(8), 931–952 (2008)
17. Gorla, D.: A taxonomy of process calculi for distribution and mobility. Distrib. Comput. **23**(4), 273–299 (2010)
18. Gorla, D.: Towards a unified approach to encodability and separation results for process calculi. Inf. Comput. **208**(9), 1031–1053 (2010)
19. Jay, B.: Pattern Calculus: Computing with Functions and Data Structures. Springer, Heidelberg (2009)
20. Jay, B., Given-Wilson, T.: A combinatory account of internal structure. J. Symbol. Logic **76**(3), 807–826 (2011)
21. Lanese, I., Pérez, J.A., Sangiorgi, D., Schmitt, A.: On the expressiveness of polyadic and synchronous communication in higher-order process calculi. In: Abramsky, S., Gavoille, C., Kirchner, C., Meyer auf der Heide, F., Spirakis, P.G. (eds.) ICALP 2010. LNCS, vol. 6199, pp. 442–453. Springer, Heidelberg (2010). doi:10.1007/978-3-642-14162-1_37

22. Lanese, I., Vaz, C., Ferreira, C.: On the expressive power of primitives for compensation handling. In: Gordon, A.D. (ed.) ESOP 2010. LNCS, vol. 6012, pp. 366–386. Springer, Heidelberg (2010). doi:10.1007/978-3-642-11957-6_20
23. Milner, R.: The polyadic π-calculus: a tutorial. In: Logic and Algebra of Specification, vol. 94. Series F. NATO ASI, 203–246. Springer, Heidelberg (1993)
24. Milner, R., Parrow, J., Walker, D.: A calculus of mobile processes I & II. Inf. Comput. **100**(1), 1–77 (1992)
25. Nielsen, L., Yoshida, N., Honda, K.: Multiparty symmetric sum types. In: Proceedings of EXPRESS, pp. 121–135 (2010)
26. Palamidessi, C.: Comparing the expressive power of the synchronous and asynchronous pi-calculi. Math. Struct. Comp. Sci. **13**(5), 685–719 (2003)
27. Parrow, J.: Expressiveness of process algebras. Electron. Not. Theoret. Comput. Sci. **209**, 173–186 (2008)
28. Parrow, J., Victor, B.: The fusion calculus: expressiveness and symmetry in mobile processes. In: Proceedings of 13th Annual IEEE Symposium on Logic in Computer Science, pp. 176–185, June 1998
29. Peters, K.: Translational expressiveness: comparing process calculi using encodings. Ph.D. thesis, Technische Universität Berlin, Fakultät IV, Germany (2012)
30. Shapiro, E.: Separating concurrent languages with categories of language embeddings. In: Proceedings of the Twenty-Third Annual ACM Symposium on Theory of Computing, STOC 1991, pp. 198–208. ACM, New York (1991)
31. van Glabbeek, R.J.: Musings on encodings and expressiveness. In: Proceedings of EXPRESS/SOS. EPTCS, vol. 89, pp. 81–98 (2012)

Towards MC/DC Coverage of Properties Specification Patterns

Ana C.V. de Melo[1]([⊠]), Corina S. Păsăreanu[2], and Simone Hanazumi[1]

[1] Department of Computer Science, University of São Paulo, São Paulo, Brazil
{acvm,hanazumi}@ime.usp.br
[2] Carnegie Mellon, NASA Ames Research Center - M/S 269-2,
Moffett Field, CA 94035, USA
corina.s.pasareanu@nasa.gov

Abstract. Model based testing is used to validate the actual system against its requirements described as formal specification, while formal verification proves that a requirement is not violated in the overall system. Verifying properties, in certain cases, becomes very expensive (or unpractical), mainly when the application of test techniques is enough for the users purposes. The *Modified Condition/Decision Coverage* (MC/DC), used in the avionics software industry, is recognised as a good technique to find out the possible mistakes on programs logics because it covers how each condition can affect the programs' decisions outcomes. It has also been adapted to provide the coverage of specifications in the requirements-based approach.

This paper proposes a technique to decompose properties (specifications), defined as regular expressions, into subexpressions representing test cases to cover the MD/DC for specifications (Unique First Word Recognition). Then, instead of proving an entire property, we can use a model checker to observe and select program executions that cover all the test cases given as the subexpressions. To support this approach, we give a syntactic characterisation of the properties decomposition, inductively defined over the syntax of regular expressions, and show how to use the technique to decompose Specification Patterns (SPS) and monitor their satisfiability using the Java PathFinder (JPF).

1 Introduction

Many systems today depend on the software industry, either because they are based on or embed in a kind of software. The variety of artefacts that depend on software to provide their services goes from an ordinary watch to complex medical devices, like pacemakers [12]. Assuring that a software behaves as "prescribed" is not a choice today, and this can only be granted by the application of techniques that guarantee the quality of the design and the final product, namely testing and formal verification.

Depending on the criticality of the software to the overall system, testing or formal verification are applied to guarantee its quality. The more rigorous are the techniques to assert the software quality, the more difficult to actually

A. Sampaio and F. Wang (Eds.): ICTAC 2016, LNCS 9965, pp. 158–175, 2016.
DOI: 10.1007/978-3-319-46750-4_10

apply them to the entire software and, in general, applying formal verification to software is more costly than testing. However, in many cases, testing a software is enough for the user purposes, including critical systems in the aviation industry. The *Modified Condition/Decision Coverage* (MC/DC) [1] test technique has been defined as a standard in the avionics software development [7]. It is an accurate software coverage criterion that can reveal many software bugs and has been applied at programming level.

Looking at the advantages of applying the MC/DC to programs, it has been adapted to specifications with the aim of covering the software requirements (Sect. 1.1), using a requirements-based testing approach. This paper focuses on providing a technique to cover formal, machine-checkable systems requirements undertaking the notion of MC/DC, and using a model checker to collect the test cases that satisfy the specification coverage criteria. Making this technique machine-checkable to collect the test cases requires a model-checker. We use JPF (Java Pathfinder) to this end because it can be used in practice to verify Java programs. For this purpose, the property specification must be converted into a minimal deterministic automaton (DFA), so that JPF will be able to read the automaton and traverse it while collecting the corresponding test cases. The conversion of a regular expression (representing a property) into a minimal DFA can be done by several algorithms in a semi-automated manner (using JFLAP tool, for instance), while converting LTL formula into DFA is not straightforward. Therefore, we chose to represent properties using regular languages despite it imposes the restriction of only describing program finite executions.

To provide a formal machine-checkable technique to cover MC/DC requirements-based testing, this paper presents three contributions: (i) define a coverage criterion on properties (specifications), defined as regular expressions, that decomposes a property into subexpressions to cover the MC/DC for specifications (Unique First Word Recognition); (ii) give a syntactic characterisation of the properties decomposition inductively defined over the syntax of regular expressions; and (iii) show how to use the technique to decompose properties described using Specification Patterns (SPS) and monitor their satisfiability using the Java PathFinder (JPF) model checker.

1.1 Related Work

The use of model checkers to generate test cases has been studied over the last two decades [3], including test cases generated from specifications. At programming level, model checkers are used to generate test cases by negating single properties (*trap properties*) and collecting the counter-examples to provide a test suite [4]. This idea is similarly used with program coverage criteria: the coverage criteria are described as a set of properties, the *trap properties* counterparts are created and submitted to a model checker that, in turn, generates a test suite. In this case, the test suite is created from the counter-examples of *trap properties* and the focus is to cover the behaviour of the System Under Test (SUT) undertaking a coverage criterion (the one defined as *trap properties*).

Apart from programs coverage criteria, model-based testing requires the notion of requirements-based coverage for specifications, described either in a logic or transition-based model. The focus is on the coverage of systems requirements, described as properties, instead of covering the SUT behaviour. Some experiments have been conducted to show the advantages on systems quality when metrics on requirements coverage [6,11,15] are considered together with the programs coverage metrics.

Since the model addressed and the focus on requirements differ from the ones for programs, some property-based coverage metrics were created. Looking at systems requirements described in Linear Temporal Logic (LTL), Tan et al. [14] defined metrics based on (non-) vacuity properties to cover LTL sub-formulae. He further defined state-based coverage metrics for specifications represented as Büchi automata [13], introducing the notion of vacuous states (similar to vacuity properties). Still using LTL as specification language, Whalen et al. [15] defined some coverage metrics based on the MC/DC criteria for programs. They are the Unique-First-Cause (UFC) coverage metrics, concerned with measuring whether a test suite is able to show that all atomic conditions (events) within the property affect the property outcome. Different from the MC/DC that is defined for propositional formulae, this new metric considers sequences of states to accommodate the LTL semantics and test cases are generated from sub-formulae calculated by LTL syntactic structure. Pecheur et al. [9] redefined a method for the syntactic characterisation of MC/DC test cases for LTL, and proved the soundness and completeness of their method under certain restrictions on the requirements formulation.

The present work is inspired by the previous works on the syntactic characterisation of LTL to generate test cases that cover the improved MC/DC criteria [9,15]. However, it differs from the previous works in several aspects. First, we define a syntactic characterisation method for regular expressions, based on the notion of *unique first word recognition*, instead of LTL. Second, we provide a more practical approach by focusing on defining and implementing the criterion for Dwyer's well-known Specification Patterns. These patterns are then used in the Java PathFinder (JPF) model checker to be instantiated by users. Then, instead of calculating the new sub-formulae for an arbitrary set of requirements, users can select and instantiate patterns to represent the system's requirements and submit them to the JPF to collect the test cases.

2   Background

2.1   Regular Expressions

A regular expression describes a set of strings (i.e. a regular language) and can be used to generate a finite-state automaton that represents the language or relation it describes.

Definition 1 (Regular Expressions). *A regular expression over a finite alphabet Σ is inductively described by*

Basis	Description
$\emptyset$	Denotes the empty event ({})
ε	Empty string ({ε})
e	Event e in alphabet Σ ({e})

Let E, E_1 and E_2 be regular expressions. The following operators can be applied to form new regular expressions that describe a set of strings containing:

Operator	Description	Additional	Description
$(E_1\|E_2)$	Choice of strings described by E_1 and E_2	$[-e]$	Any event in alphabet Σ, except e
$(E_1;E_2)$	Concatenation of strings described by E_1 and E_2	$[-e_1,...,e_n]$	Any event in alphabet Σ, except $e_1,...,e_n$
$(E)^*$	ε and the smallest superset described by E closed under string concatenation (;)	. $(E)?$	Any event in alphabet Σ ε and strings described by E

Note that additional operators given in Definition 1 are syntactic sugar. They can be defined over basic elements and operators. Besides that, certain properties on regular expressions are defined to conveniently rewrite expressions on the decomposition of test cases approach (they are presented in Appendix A). The precedence and associativity for the regular expressions are presented in Definition 2.

Definition 2 (Precedence and Associativity). Let E, E_1 and E_2 be regular expressions. The precedence and associativity of operators are defined as:

Operator	Precedence	Associativity
$(E)^*$	1	right to left
$(E_1;E_2)$	2	left to right
$(E_1\|E_2)$	3	left to right

Then, expression $(E_1 \mid E_2;(E_3)^*)$ has the same meaning of $(E_1 \mid (E_2;(E_3)^*))$ due to the precedence of operators, and $(E_1;E_2;E_3)$ is the same as $((E_1;E_2);E_3)$ due to the associativity of operator; (left to right).

2.2 Specification Pattern System (SPS)

Specification patterns [2] are formalism independent specification abstractions defined for finite-state verification. It helps practitioners in mapping descriptions of system behaviour into their formalism of choice, improving the transition of these formal methods to practice. To define a property using the SPS, one must define its scope and then, the corresponding pattern. Next, we give a short description of SPS. For a complete description of the system, please refer to [2,8].

Scopes. A scope is the extent of the program execution over which the pattern must hold, and it is determined by specifying a starting and an ending event for the pattern. The SPS has five types of scopes: (a) *global:* it covers the entire program computation; (b) *before R:* it starts at the beginning of the computation and ends with an occurrence of a state or event R; (c) *after L:* it starts from the occurrence of a state or event L until and ends with the program computation; (d) *between L and R:* it includes the intervals that start with the occurrence of state or event L and end with the occurrence of state or event R; and, (e) *after L - until R:* it includes the intervals that start from the occurrence of a state or event L and end with either the occurrence of a state or event R, or with the end of computation.

Patterns. A pattern is a specification abstraction that can be mapped to various formal representations, such as LTL [10] or regular expressions. The patterns are organised according to their semantics. Here, we give a brief description of the SPS patterns that are used in our work: (a) *absence:* a state or event P does not occur within a scope; (b) *universality:* a state or event P occurs throughout a scope; (c) *existence:* a state or event P must occur at least once within a scope; (d) *precedence:* a state or event P must always be preceded by a state or event T within a scope; and, (e) *response:* a state or event P must always be followed by a state or event T within a scope.

Example. Consider the example of a coffee machine, with the following program execution traces:

Trace 0: putCoin → getCoffee
Trace 1: putCoin → getCoffee → getMilk
Trace 2: putCoin → getCoffee → getSugar
Trace 3: putCoin → getMilk → getSugar

Suppose that we want to verify the property: *there is no occurrence of getMilk before getCoffee*. The regular expression that describes the absence of P in the before R scope is "$[-R]^*|[-P, R]^*; R; .^*$" [8]. It states that either the event R does not occur (and then there is no scope definition) or the following events occur in sequence: any event except P and R occurs 0 or more times, R occurs, and then any event of the alphabet occurs 0 or more times. Therefore, the regular expression specifies that no occurrence of event P can be observed before R. Returning to the property we want to specify, which says that no occurrence of *getMilk* should be observed before *getCoffee*, we can consider $P =$ getMilk and $R =$ getCoffee and then apply it to the SPS regular expression for the absence pattern in the before R scope. Thus, using SPS regular expressions, we have that the property can be specified as: "$[-getCoffee]^*|[-getMilk, getCoffee]^*; getCoffee; .^*$"

3   Unique First Positive Recognition

The MC/DC has been developed over programs conditions/decisions, and its coverage is concerned with showing how each condition can independently affect the decision's outcomes (for all decisions in a program). A test suite that satisfies the MC/DC requirements coverage makes all logical statements to be exercised in a program. To adapt this criterion to cover systems requirements defined as regular expressions, we must first develop the notion of how each event e contributes to the recognition of a language, in the same way that each condition can independently contribute to a decision in the MC/DC.

Conditions are treated here as events (e) observed in expressions (E - decision). In this way, we want to observe the occurrence of a given event e in an expression E or its absence, denoted as $[-e]$. Given a regular expression E for language L and a path π, an event e is the Unique First Recognition of an L's word, if in the first stage along π where an L word is recognised, it is recognised because of e, or its absence. For regular expressions, only a single event is observed at a time. This means that we do not deal with an "and" of events observation at any stage of computation. Then, the contribution of an event to provide a verdict of words recognised by a regular expression[1] must be defined. To this end, we first define the set of events that occur in regular expressions as last symbols in a word.

Definition 3 (Last Events in Expressions). *Let E be a regular expression that describes a language L over an alphabet Σ. LAST$(E) \subseteq \Sigma$ describes all events that occur in expression E as the last elements in words of language L.*

E	LAST(E)	E	LAST(E)
ε	$\{\}$	$(E_1 \mid E_2)$	LAST$(E_1) \cup$ LAST(E_2)
e	$\{e\}$	$(E_1; E_2)$	LAST$(E_2) \cup ($LAST$(E_1),\ if\ E_2 = (E_3)^*)$
$[-e]$	$\{e\}$	$(E)^*$	LAST(E)
$.$	Σ	$(E)?$	LAST(E)

Note that a special case is given for expressions built upon the sequence operator (;). If the last expression in the sequence is a star expression, it might be repeated *zero* or more times. Then, the previous expression must also be considered as a last one in the overall expression to account for the set of last events occurring in the recognised words. For instance, if $(E = E_1; (E_3)^*)$, E_1 must also be considered as last expression because E_3 might occur zero times and the last events in E_1 are last events also. The absence of an event is denoted as $[-e]$. Despite e can not appear as the last symbol in a word recognised by this expression, it is a syntactic element used to recognise/refuse a word defined by this expression. Then, $\{e\}$ is the set of symbols to be syntactically considered in expression $[-e]$.

For example, if $\Sigma = \{P, Q, R\}$ and $Ex_1 = ([-P]^*; P) \mid (Q; R)$, then LAST$(Ex_1) =$ LAST$(P) \cup$ LAST$(R) = \{P, R\}$. Q does not appear as a symbol in the expression that can recognise or refuse a word. On the other hand, for expression $Ex_2 = ([-P]^*; P) \mid (Q; R^*)$, LAST$(Ex_2) =$ LAST$(P) \cup ($LAST$(Q) \cup$ LAST$(R)) = \{P, Q, R\}$ because R^* is a star expression.

---

[1] Here, we use the relaxed term "recognised by a regular expression" meaning "recognised by a an automaton that recognises the language defined by a regular expression".

Based on **Definition** 3, the contribution of events to recognise a language's word is defined:

Definition 4 (Event Contribution - complete). *Given a regular expression E over alphabet Σ that defines a language L, an event $e \in \Sigma$ contributes to the complete recognition of an L's word if $e \in \text{LAST}(E)$.*

Now, the *unique first recognition of an L's word* definition looks at the independent contribution of each event in the alphabet to the words recognition outcomes (either positive or negative). For this definition, the notion of words prefixes and matching strings to regular expressions need to be defined.

Definition 5 (Matching Strings). *Let E, E_1 and E_2 be regular expressions defined over alphabet Σ, $e \in \Sigma$ and $u, u_1, u_2, ...u_n \in \Sigma^*$.*

E	$\text{MATCH}(u, E)$ if
ε	$u = \varepsilon$
e	$u = e$

E	$\text{MATCH}(u, E)$ if
$(E_1 \mid E_2)$	$\text{MATCH}(u, E_1) \vee \text{MATCH}(u, E_2)$
$(E_1; E_2)$	$u = u_1.u_2 \wedge \text{MATCH}(u_1, E_1) \wedge \text{MATCH}(u_2, E_2)$
$(E)^*$	$u = \varepsilon \vee \text{MATCH}(u, E) \vee$
	$(u = u_1.u_2....u_n \wedge \text{MATCH}(u_i, E) \wedge 1 \le i \le n)$

Definition 6 (Prefixes). *Let E be a regular expression defined over alphabet Σ, $e \in \Sigma$, $u, v, w \in \Sigma^*$ and $\text{MATCH}(u, E)$.*

$$\text{PREF}(u, E) = \{v \mid u = v.w\}$$
$$\text{PREF}_e(u, E) = \{v.e \mid u = v.e.w\}$$

Given a string (events path) u that matches a regular expression E, $\text{PREF}(u, E)$ and $\text{PREF}_e(u, E)$ calculate the set of u prefixes and the set of u prefixes that finish with an e symbol (event), respectively.

Definition 7 (Unique First Positive Recognition). *Given a regular expression E over alphabet Σ that defines a language L, an event $e \in \Sigma$ contributes to the Unique First Positive Recognition of an L's word $(UFPR(E, e))$ if*

– *by the time e is observed, it can complete a word recognition of L and none of its prefixes finished by e is a word in L:*

$$\exists u.e \in \Sigma^*.\text{MATCH}(u.e, E) \wedge$$
$$\nexists t \cdot t \in \text{PREF}_e(u, E)$$

Note that the notion of events contribution to the words recognition outcomes is related to the "instant" in which the verdict of recognition/nonrecognition can be delivered. Undertaking this notion, the event contribution to deliver a positive outcome is based on the complete contribution (**Definition 4**).

4 Regular Expressions Decomposed into Test Cases

To calculate the test cases necessary to cover UFPR (**Definition** 7) over a regular expression E, we provide a syntactic characterisation of events contribution on words recognition. The method decomposes a regular expression into subexpressions representing test cases for the set of events in the language alphabet: for each event, we syntactically calculate a set of test cases (given as subexpressions) to cover UFPR. It means that the requirements specifications are covered as these test cases are exercised in programs.

From a practical point-of-view, we are more interested in the test cases that can reveal the recognition of words (sequences of events) in a language. Here, we will consider the sets of subexpressions that positively contribute to recognise the allowed sequence of events, covering test cases based on the notion of UFPR.

Definition 8 (Positive Contribution). *Given an expression E over alphabet Σ, for each event $e \in \Sigma$, E^+ is the set of expressions defined over e that first affect E positively (it makes E to be recognised).*

First, the contribution of the basic elements are defined. In this definition, we focus on the contribution of event e to expressions. It is given for the positive and negative contributions to show the overall meaning of events contribution and the cases in which they are symmetric. To generate the observable test cases, however, we will focus on the positive contribution.

Definition 9 (Basis Contribution). *Let a regular expression E be defined over alphabet Σ, and $e, f \in \Sigma$. The contribution of event e to the basic expressions is defined as follows:*

$$(\varepsilon)_e^+ = \emptyset \tag{1}$$

$$(\varepsilon)_e^- = e \tag{2}$$

$$(.)_e^+ = e \tag{3}$$

$$(.)_e^- = \emptyset \tag{4}$$

$$(f)_e^+ = \begin{cases} e & \text{if } e = f \\ \emptyset & \text{if } e \neq f \end{cases} \tag{5}$$

$$(f)_e^- = \begin{cases} [-e] & \text{if } e = f \\ e & \text{if } e \neq f \end{cases} \tag{6}$$

$$([-f])_e^+ = \begin{cases} [-e] & \text{if } e = f \\ e & \text{if } e \neq f \end{cases} \tag{7}$$

$$([-f])_e^- = \begin{cases} e & \text{if } e = f \\ \emptyset & \text{if } e \neq f \end{cases} \tag{8}$$

The positive contribution of e to recognise the empty word (ε) is trivial; e can not deliver a positive outcome to this expression and then the $\emptyset$ is given as result. Conversely, e can provide a negative outcome to the empty word if it is recognised. To provide a negative contribution of the empty word, any event in alphabet Σ must be observed. Since we are dealing with unrecognised words in the negative contributions, any event observed is enough to provide this verdict, including e. The positive and negative contributions of e to the "any event in the alphabet expression" (denoted by expression .) are symmetric to the observation of ε. e gives a positive outcome for expression . if it is observed, while the negative outcome can not be delivered. Note that whenever the $\emptyset$ is given as a result, it means that e can not affect the recognition outcome.

For expressions made of an event f or its absence ($[-f]$), the positive and negative contributions of e depend on e coinciding with f. For expression, f, if event e equals to f, its positive contribution to the word recognition is e, while the negative observation of event e corresponds to observing any event in alphabet Σ other than e ($[-e]$). If event e differs from f, e can not affect the positive outcome but it can determine the negative outcome if it is observed. In the absence expression ($[-f]$), if e coincides with f, its absence gives a positive contribution while observing the expression e gives the negative outcome. If e and f differ, the provided expressions are e for the positive contribution and $\emptyset$ for the negative one.

The positive contribution takes into account events that can recognise an entire word, based on the concept of event complete contribution (**Definition** 4). Now, to define the test cases for the positive contribution, we use **Definition** 3 to syntactically characterise elements in an expression that actually complete the recognised words in the language.

Definition 10 (Operators Positive Contribution). *Let E, E_1 and E_2 be regular expressions defined over alphabet Σ. Then for each $e \in \Sigma$, the sets of test cases defined for the expression are:*

$$(E)_e^+ = \emptyset \qquad \qquad \text{if } e \notin \text{\scriptsize LAST}(E) \qquad (9)$$

$$(E_1 \mid E_2)_e^+ = (E_1)_e^+ \cup (E_2)_e^+ \qquad (10)$$

$$(E_1; E_2)_e^+ = \begin{cases} (E_1)_e^+; (E_2)_e^+ & \text{if } \varepsilon \in E_2^+ \\ E_1; (E_2)_e^+ & \text{otherwise} \end{cases} \qquad (11)$$

$$(E^*)_e^+ = \{\varepsilon\} \cup ((E)_e^+)^* \qquad (12)$$

If e does not occur to complete a word in E, no test case can be created to provide the independent positive contribution of e (it cannot be used to recognise an entire word). For the alternative operator, a test case can be produced from E_1 or E_2 as far as e occurs in them to complete a recognised word. Note that no guard on the occurrence of the event in these expressions has been defined: if e does not occur as the last element in one of these expressions, Eq. (9) is applied to produce the empty set.

The sequence operator will look for test cases that can complete a word using event e (or its absence). If E_2 is a star or ? expression[2], it might be used or not to complete a word (given by the guard if $\varepsilon \in E_2^+$). In these cases, E_1 can potentially generate the test cases, otherwise it is only used as a prefix to the test cases generated by E_2. These cases are depicted in Eq. (11). The test cases created for the star operator considers the prefix expression E.

5 Calculating SPS Subexpressions

To calculate the SPS subexpressions for each regular expression that represents the combination of SPS pattern (absence, universality, existence, precedence, response) and scope (global, before, after, between, after-until) under consideration, we have to analyse the positive contribution of each event in the regular expression. For instance, if we consider the pattern response in the scope between L and R, we have to calculate the positive contribution regarding each event in the regular expression: L and R, from the scope, P and T from the pattern.

To illustrate the decomposition of a regular expression that represents a combination of a SPS scope and pattern, we use the example of the absence pattern in the before R scope [8]:

$$([-R]^* \mid [-P, R]^*; R; .^*)$$

We analyse the expression regarding each event (P and R), following the rules for the calculation of the positive contribution presented previously. Considering that:

- $X_1 = [-R]^*$
- $X_2 = [-P, R]^*; R; .^*$

The expression we want to analyse has the form $(E_1 \mid E_2)$, thus, to calculate the subexpressions we follow the positive contribution rule $(E_1 \mid E_2)^+ = E_1^+ \cup E_2^+$. Therefore, for each event in the formula, we have to analyse X_1 and X_2, and combine the contribution of each expression accordingly.

Event P

- $\boxed{(X_1)^+ \text{ analysis:}}$
  $([-R]^*)^+$ $(E^*)^+ = \varepsilon \cup (E^+)^*$
    - ε
    - $([-R]^+)^*$ $([-f])_e^+ = e, \text{ if } e \neq f$
    $(P)^*$

---

[2] Operator ? does not appear in the definition because it can be defined using the | operator.

$-$ $\boxed{(X_2)^+ \text{ analysis:}}$

$([-P, R]^*; (R; .^*))^+$

$[-P, R]^*; (R; .^*)^+$

$[-P, R]^*; (R)^+; (.^*)^+$

$\emptyset$

$(E_1; E_2)^+ = E_1; E_2^+, \text{ if } \varepsilon \notin E_2^+$

$(E_1; E_2)^+ = E_1^+; E_2^+, \text{ if } \varepsilon \in E_2^+$

$(f)_e^+ = \emptyset, \text{ if } e \neq f$

$-$ $\boxed{\text{Positive contribution:}}$

1. ε
2. $(P)^*$

Event R

$-$ $\boxed{(X_1)^+ \text{ analysis:}}$

$([-R]^*)^+$

  ∘ ε

  ∘ $([-R]^+)^*$

  $[-R]^*$

$(E^*)^+ = \varepsilon \cup (E^+)^*$

$([-f])_e^+ = [-e], \text{ if } e = f$

$-$ $\boxed{(X_2)^+ \text{ analysis:}}$

$([-P, R]^*; (R; .^*))^+$

$[-P, R]^*; (R; .^*)^+$

$[-P, R]^*; (R)^+; (.^*)^+$

$[-P, R]^*; R; (.^*)^+$

  ∘ $[-P, R]^*; R; \varepsilon$

  ∘ $[-P, R]^*; R; (.^+)^*$

  $[-P, R]^*; R; (R)^*$

$(E_1; E_2)^+ = E_1; E_2^+, \text{ if } \varepsilon \notin E_2^+$

$(E_1; E_2)^+ = E_1^+; E_2^+, \text{ if } \varepsilon \in E_2^+$

$(f)_e^+ = e, \text{ if } e = f$

$(E^*)^+ = \varepsilon \cup (E^+)^*$

$(.)_e^+ = e$

$-$ $\boxed{\text{Positive contribution:}}$

1. ε
2. $[-R]^*$
3. $[-P, R]^*; R$
4. $[-P, R]^*; R; (R)^*$

Note that the positive contribution regarding event P in X_1 is the empty word ε and P^*, since X_1 covers all events except R. There is no positive contribution regarding event P in X_2. On the other hand, for event R, its positive contribution in X_1 is trivial, ε and $[-R]^*$. The positive contribution of event R in X_2 is represented by $[-P, R]^*; R; (R)^*$ (4) and $[-P, R]^*; R$ (3), which is a subexpression of the previous expression. Since we are interested in the positive contribution, we focus on the cases where the subexpressions are satisfied. Calling the set of satisfied traces $Sat(< expression\_id >)$, we have for event R, $Sat(3) \subseteq Sat(4)$. Therefore, in our work, when we reach this situation we only consider, in practice, the expression with the biggest satisfied traces set (in this case, we consider only regular expression 4 and ignore the expression 3).

6   SPS Subexpressions in JPF

To represent the subexpressions in JPF, we converted each one into a minimal deterministic automaton using the approach we describe in a previous work [5].

First, using the JFlap tool[3], we build the minimal deterministic finite automaton (DFA) with the accepting states. For instance, consider the automaton for P^*, which is the subexpression number 2 regarding the positive contribution of event P in the regular expression of the absence pattern in the before R scope. The minimal DFA for this subexpression is depicted in Fig. 1(a). The circle with a thick border represents the accepting state, and the initial state of the automaton is marked with a dashed arrow.

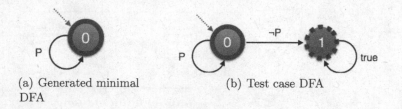

(a) Generated minimal   (b) Test case DFA
DFA

Fig. 1. Absence pattern in before R scope: positive contribution of event P: $(P)^*$

Since the subexpressions represent test cases, we need to know when we reach a rejecting state. Hence, we use the complement of the built automaton and complete the inputs to define the rejecting states. In this example, we completed the inputs for the automaton by adding the state 1 and the edge P, which was the missing input in the generated minimal DFA. Figure 1(b) presents the DFA for the test case of the example. The circle with a dashed border represents the rejecting states.

An important remark is about the empty word ε. Since it does not accept any real event (Java method), we do not build an automaton for it. Therefore, for the absence pattern in the before R scope, we consider, in practice, only the following subexpressions (Sect. 5):

- **Event P:** P^* (*subexpression 2*)
- **Event R:** $[-R]^*$ (*subexpression 2*), $[-P, R]^*$; R; $(R)^*$ (*subexpression 4*)

6.1   Verification with JPF

Recall the coffee machine example (Sect. 2.2), and the property: there is no occurrence of getMilk before getCoffee (i.e., the absence of $P = getMilk$ in the scope before $R = getCoffee$). As we can observe in the program traces, the property is not violated. This result is confirmed by JPF. Figure 2 presents the screenshot of the JPF verification results. On the right-hand side, we can see the property automaton, and the automaton edges that were covered by program execution traces (the program traces that covered an edge are identified by the label $Tr[< traceId_1, traceId_2, ...traceId_n >]$). On the left-hand side we have a partial view of the verification report, where we can see the property verdict that the property was satisfied.

---

[3] http://www.jflap.org.

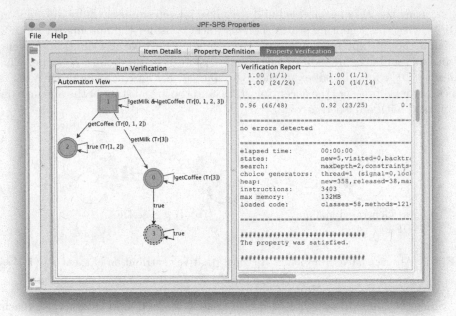

Fig. 2. JPF Verdict: the property is satisfied

The verification of the subexpressions (or test cases) occurs in the following manner: for each program trace execution, we traverse the corresponding set of subexpressions automata (in this example, we have three automata: one automaton regarding event P positive contribution (P^*), and two automata regarding event R positive contribution $([-R]^*, [-P, R]^*; R; (R)^*))$. When a subexpression automaton reach an accepting state, we store this verdict and remove the automaton from the initial set. Thus, we continue the verification process by traversing the remaining automata until the end of the process.

For instance, considering the test case 0 regarding event P: $(P)^*$. In this example, this test case refers to the occurrence of getMilk though the whole program execution. This test case fails in all program traces, because other events occur besides getMilk in all program traces, so we do not reach a success verdict (Fig. 3). In case of failure of a test case, the developer should analyze the program traces and whether this result was expected or not. If this result was not expected, the program should be fixed and the test case must be executed again. In this example, the failure was expected, so there is no need to fix anything in the program.

When considering test case 1 regarding event R: $[-R]^*$, which in our example means that no occurrence of $getCoffee$ should be observed, we reach a success verdict only in the last program execution trace (Trace 3). Figure 4 presents the verification result for test case 1, and the automaton edges coverage of a 100 %.

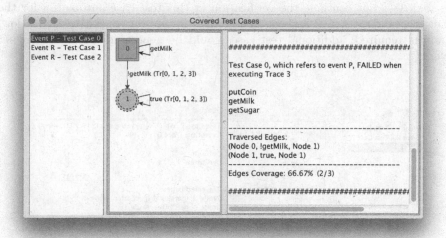

Fig. 3. Test case 0 regarding event P: $(P)^*$

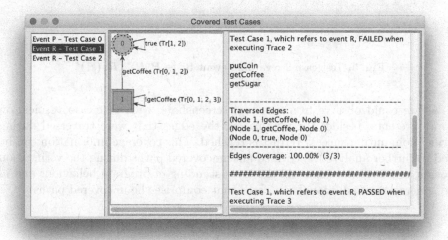

Fig. 4. Test case 1 regarding event R: $[-R]^*$

Finally, the test case 2 regarding event R, $[-P, R]^*$; R; $(R)^*$. In our example, it means that *getMilk* does not occur until the occurrence of *getCoffee*; and, after one occurrence of *getCoffee*, we can observe 0 or more occurrences of *getCoffee*. This test case reaches a success verdict at the first program execution trace (Trace 0). Figure 5 presents the results for test case 2, and the automaton edges coverage of 28.57 %. Since the first program trace satisfies this test case, the tool does not continue the verification of this test case automaton for the other traces (Traces 1, 2 and 3). Hence, the automaton edges coverage is low. If it was required, additional test cases could be generated to reach a higher coverage.

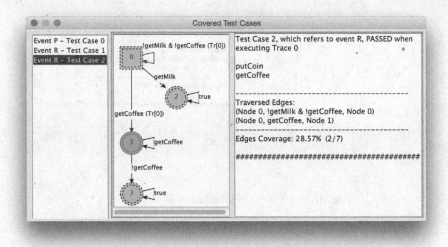

Fig. 5. Test case 2 regarding event R: $[-P, R]^*; R; (R)^*$

As we could observe from the tool screenshots, each test case verification report presents, besides the automaton, the edges that were traversed during verification, until a success verdict is reached. This coverage information can be used to further analyze the covered and uncovered paths during the verification process, and help developers in the understanding of program behaviour and in the generation of additional test cases that comprise the uncovered paths.

7   Concluding Remarks

In this paper, we described a method to calculate test cases to cover a property-based metrics for systems requirements based on the MC/DC, namely the Unique First Positive Recognition (UFPR). The system requirements are properties defined as regular expressions and the method is a syntactic characterisation of the properties decomposition, inductively defined over the syntax of regular expressions. Then, given a regular expression representing a system requirement, the subexpressions are calculated representing the test cases to cover the UFPR. If a test suite satisfies all these test cases, the system requirements are covered undertaking the UFPR.

A proof of concept for this method was developed using the Specification Patterns System (SPS). All those patterns were decomposed into their test cases (subexpressions) and to show the practical use of them, they were implemented in the Java PathFinder (JPF). For each pattern, the set of test cases are monitored by the (JPF) to collect the program test cases and create a test suite that covers the system requirements. From a practical point-of-view, users can choose and instantiate patterns to define the system requirements, submit them to the JPF and collect the test suite to cover the requirements. For this particular case (all properties in the SPS), the test cases are previously calculated and ready to be instantiated, no extra effort is needed to produce the test suite.

The contributions of this paper are threefold: the definition of the Unique First Positive Recognition for regular expressions; a method to syntacticly decompose systems requirements into test cases to cover the UFPR; and a tool to collect programs test suite that satisfy the UFPR of the Specification Patterns System.

Despite all contributions presented here, we want to improve this research in many aspects. The work presented here is limited to calculate the positive outcomes while the negative contribution still needs to be defined. The UFPR defines the positive contributions, and the method to calculate the test cases is restricted to the positive cases. The negative calculation is in progress. The positive decomposition calculation for the SPS properties was manually performed. We want to provide an automatic procedure to do this, together with a technique to reduce the number of test cases preserving the semantics. Considering that the SPS are the commonly used properties, having only them implemented in a tool does not impose a serious restriction in practice. However, if the calculation is automated, any property can be considered in practice. The SPS test cases are given as subexpressions and the automata counterparts are calculated independent from any given model checker. It means that this front-end can be used with model checkers other than JPF, and we plan to extend its usage to other model checkers.

Acknowledgments. This project has been funded by the State of São Paulo Research Foundation (FAPESP) - Processes: 2011/01928-1, 2012/23767-2, 2013/22317-6. We also would like to thank the NASA Amés Research Center and the Carnegie Mellon University - Silicon Valley, for providing a rich environment for the development of research activities.

A   Regular Expressions: Axioms and Definitions

Definition 11. *Let E and $E_1, \ldots, E_n$ be regular expressions defined over alphabet $\Sigma = \{e_1, ..., e_n\}$ and $1 \leq i \leq j \leq n$. Some properties and definitions are given:*

Identity and Annihilator:
$$\emptyset; E \equiv \emptyset$$
$$E; \emptyset \equiv \emptyset$$
$$E; \varepsilon \equiv E$$
$$\varepsilon; E \equiv E$$
$$E \mid \emptyset \equiv E$$

Associativity:
$$E_1; (E_2; E_3) \equiv (E_1; E_2); E_3$$
$$E_1 \mid (E_2 \mid E_3) \equiv (E_1 \mid E_2) \mid E_3$$

Distributivity:
$$E_1; (E_2 \mid E_3) \equiv (E_1; E_2) \mid (E_1; E_3)$$
$$(E_1 \mid E_2); E_3 \equiv (E_1; E_3) \mid (E_2; E_3)$$

Commutativity:
$$E_1 \mid E_2 \equiv E_2 \mid E_1$$

Idempotent:
$$E \mid E \equiv E$$

Closures:
$$(E^*)^* \equiv E^*$$
$$\emptyset^* \equiv \varepsilon$$
$$\varepsilon^* \equiv \varepsilon$$

Definitions:
$$. \triangleq e_1 \mid \ldots \mid e_n$$
$$[-e] \triangleq f_i \mid \ldots \mid f_j, \; ; \{f_i, \ldots f_j\} = (\Sigma - \{e\})$$
$$[-e_i, \ldots, e_j] \triangleq f_n \mid \ldots \mid f_m, \; ;$$
$$\{f_n, \ldots f_m\} = (\Sigma - \{e_i, \ldots, e_j\})$$
$$(E)? \triangleq \varepsilon \mid E$$

References

1. Ammann, P., Offutt, J.: Introduction to Software Testing, vol. 54. Cambridge University Press, Cambridge (2008)
2. Dwyer, M.B., Avrunin, G.S., Corbett, J.C.: Patterns in property specifications for finite-state verification. In: Proceedings of the 21st International Conference on Software Engineering, ICSE 1999, pp. 411–420. ACM, New York (1999)
3. Fraser, G., Wotawa, F., Ammann, P.E.: Testing with model checkers: a survey. Softw. Test. Verif. Reliab. **19**(3), 215–261 (2009)
4. Gargantini, A., Heitmeyer, C.: Using model checking to generate tests from requirements specifications. SIGSOFT Softw. Eng. Not. **24**(6), 146–162 (1999)
5. Hanazumi, S., de Melo, A.C.V., Păsăreanu, C.S.: From testing purposes to formal JPF properties. In: Java PathFinder Workshop. ACM (2014)
6. Hesari, S., Behjati, R., Yue, T.: Towards a systematic requirement-based test generation framework: industrial challenges and needs. In: Proceedings of the 2013 21st IEEE International Requirements Engineering Conference, RE 2013, pp. 261–266 (2013)

7. Holloway, C.: Towards understanding the DO-178C/ED-12C assurance case. In: 7th IET International Conference on System Safety, Incorporating the Cyber Security Conference 2012, p. 14. Institution of Engineering and Technology (2012)
8. S. Patterns, June 2015. http://patterns.projects.cis.ksu.edu/
9. Pecheur, C., Raimondi, F., Brat, G.: A formal analysis of requirements-based testing. In: Proceedings of the Eighteenth International Symposium on Software Testing and Analysis, ISSTA 2009, pp. 47–56 (2009)
10. Pnueli, A.: The temporal logic of programs. In: SFCS 1977: Proceedings of the 18th Annual Symposium on Foundations of Computer Science, pp. 46–57. IEEE Computer Society, Washington, DC (1977)
11. Rajan, A., Whalen, M., Staats, M., Heimdahl, M.P.E.: Requirements coverage as an adequacy measure for conformance testing. In: Liu, S., Maibaum, T., Araki, K. (eds.) ICFEM 2008. LNCS, vol. 5256, pp. 86–104. Springer, Heidelberg (2008). doi:10.1007/978-3-540-88194-0_8
12. Sametinger, J., Rozenblit, J., Lysecky, R., Ott, P.: Security challenges for medical devices. Commun. ACM **58**(4), 74–82 (2015)
13. Tan, L.: State coverage metrics for specification-based testing with Büchi automata. In: Gogolla, M., Wolff, B. (eds.) TAP 2011. LNCS, vol. 6706, pp. 171–186. Springer, Heidelberg (2011). doi:10.1007/978-3-642-21768-5_13
14. Tan, L., Sokolsky, O., Lee, I.: Specification-based testing with linear temporal logic. In: Proceedings of the 2004 IEEE International Conference on Information Reuse and Integration, IRI 2004 (2004)
15. Whalen, M.W., Rajan, A., Heimdahl, M.P., Miller, S.P.: Coverage metrics for requirements-based testing. In: International Symposium on Software Testing and Analysis, p. 25 (2006)

Calculi

Unification for λ-calculi Without Propagation Rules

Flávio L.C. de Moura[✉]

Departamento de Ciência da Computação, Universidade de Brasília, Brasília, Brazil
flaviomoura@unb.br

Abstract. We present a unification procedure for calculi with explicit substitutions (ES) without propagation rules. The novelty of this work is that the unification procedure was developed for the calculi with ES that belong to the paradigm known as "act at a distance", i.e. explicit substitutions are not propagated to the level of variables, as usual. The unification procedure is proved correct and complete, and enjoy a simple form of substitution, called *grafting*, instead of the standard capture avoiding variable substitution.

1 Introduction

Unification is an important operation extensively used in Computer Science whose goal is to find a substitution, when it exists, to identify terms in a certain theory. In particular, it is a central operation in automating higher-order reasoning. The correct framework for higher-order unification is the simply typed λ-calculus where this operation is known to be undecidable [21,23]. In addition, different unifiers of a given problem can be completely independent from each other, i.e. there is no unicity of solutions.

ES calculi internalize the substitution operation by extending the grammar of the λ-calculus and have been extensively studied during the last decades [1,19,22,26,26–28,30]. It has been mainly used as an intermediate formalism between the theory of the λ-calculus and its implementation. In this context many different approaches were proposed because it was surprisingly difficult to develop a formalism preserving important properties related to the simulation of the λ-calculus.

In this work we present a unification procedure for a family of extensions of the simply typed λ-calculus with explicit substitutions (ES) that act at a distance. This new approach to ES, which has motivations in the study of Proof-Nets [5,20], satisfy all expected properties for an ES calculus, such as preservation of strong normalization (PSN), confluence on open terms and simulation of one step β-reduction. This approach differs from the usual ES calculi because the explicit substitution is not percolated over terms, and the starting rule of the calculi can traverse ES. In this sense substitutions "act at a distance" or "do not have propagation rules". Several variants of calculi with ES at a distance

---

F.L.C. de Moura—Author partially supported by FAPDF.

A. Sampaio and F. Wang (Eds.): ICTAC 2016, LNCS 9965, pp. 179–195, 2016.
DOI: 10.1007/978-3-319-46750-4_11

have been proposed for different purposes, such as implicit computational complexity [8], the theory of abstract standardization [2,4], abstract machines [3], study of equational extensions of the λ-calculus [6], etc.

During the unification procedure, each term is transformed into its η-long normal form, which is a well known notion for λ-terms and some (meta)-confluent ES calculi. Nevertheless, the characterization of the η-long normal form for a calculus without propagation rules is not straightforward because explicit substitutions can appear everywhere in the term. In this sense, this work improves [12] as follows:

1. We define the notion of binding contexts that allows a precise characterization of the structure of the normal form and of the η-long normal form of a metaterm in calculi without propagation rules.
2. Renaming of bound variables is not taken for free. In fact, the cost of α-conversion needs to be explicitly considered in any implementation of a unification procedure for a calculus that works modulo α-conversion.

An extended version of this work is available at http://flaviomoura.mat.br.

2   Explicit Substitutions Without Propagation Rules

ES calculi without propagation rules include a family of calculi that allow a substitution to be performed even if it is not at the level of variables. The grammar for all calculi (with metaterms) in this family is defined as follows:

$$t, u ::= x \mid X_\Delta \mid t \ u \mid \lambda x.t \mid t[x/u] \tag{1}$$

where x is an ordinary variable, X_Δ is a metavariable with support set Δ (a set of ordinary variables), $t \ u$ is an application, $\lambda x.t$ is an abstraction and $t[x/u]$ is a term with an explicit substitution. Metavariables must come with a minimum amount of information in order to guarantee that some basic operations (like replacement of metavariables by metaterms) are sound in a typing context [27]. For the sake of clarity, we write X instead of $X_{\{\}}$. As we will see, this approach also simplifies the presentation of the unification rules, and permits to keep a clear separation between the substitutions generated by the unification procedure and the ones generated by the calculus itself during reduction. The separation between ordinary variables and metavariables allows an important simplification in the unification procedure because the substitution generated by the unification procedure is separated from the substitution generated by the calculus itself. Although the calculi considered here are not first-order in the sense that substitutions of variables need to take into account renamings of bound variables in order to avoid capture, the substitution generated by the unification procedure is first-order (*grafting*) [18] because metavariables are parameterized by a set of variables that is fundamental for the good behaviour of the procedure. As said before, the construction $t[x/u]$ denotes a term with an explicit substitution, and both $\lambda x.t$ and $t[x/u]$ bind x in t, i.e. t is the scope of x in both $\lambda x.t$ and $t[x/u]$, and in this case x is called a *bound* variable. For instance, both occurrences of x in the term $(x\ Y_{\{x\}})[x/u]$ are bounded by the explicit substitution. If a variable is not under the scope of an abstraction or an explicit substitution then it is called *free*. Therefore, we work with three sorts of variables: free variables, bound variables and metavariables. In addition, the set of free variables is divided into two kinds: the ones occurring in the support set of metavariables, and the real variables, which are defined as follows:

Definition 1. *The set of free variables occurring in the support sets of metavariables in a metaterm t, denoted by $\mathtt{fm}(t)$, and the set of free (real) variables occurring in the metaterm t, denoted by $\mathtt{fr}(t)$, are inductively defined as follows:*

- $\mathtt{fm}(X_\Delta) = \Delta$
- $\mathtt{fm}(x) = \{\}$
- $\mathtt{fm}(u\ v) = \mathtt{fm}(u) \cup \mathtt{fm}(v)$
- $\mathtt{fm}(\lambda x.u) = \mathtt{fm}(u)\backslash\{x\}$
- $\mathtt{fm}(u[y/v]) = (\mathtt{fm}(u)\backslash\{y\}) \cup \mathtt{fm}(v)$

- $\mathtt{fr}(X_\Delta) = \{\}$
- $\mathtt{fr}(x) = \{x\}$
- $\mathtt{fr}(u\ v) = \mathtt{fr}(u) \cup \mathtt{fr}(v)$
- $\mathtt{fr}(\lambda x.u) = \mathtt{fr}(u)\backslash\{x\}$
- $\mathtt{fr}(u[y/v]) = (\mathtt{fr}(u)\backslash\{y\}) \cup \mathtt{fr}(v)$

The set of free variables of a term t is defined as $\mathtt{fv}(t) = \mathtt{fm}(t) \cup \mathtt{fr}(t)$.

The number of occurrences of the (free) variable x in the term t is denoted by $|t|_x$, as defined above. So, for instance, $|(X_{\{x,y\}}\ x)[y/\lambda z.X_{\{x,z\}}]|_x = 3$.

The operational semantics of the calculi is given in terms of one-hole contexts, and two special classes of contexts called *substitution contexts* [2] and *binding contexts*:

$$
\begin{array}{lll}
\text{Contexts} & C ::= \langle\cdot\rangle \mid C\ t \mid t\ C \mid \lambda x.C \mid C[x/t] \mid t[x/C] \\
\text{Biding Contexts} & B ::= \langle\cdot\rangle \mid B[x/t] \mid \lambda x.B \\
\text{Substitution Contexts } L ::= \langle\cdot\rangle \mid L[x/t]
\end{array}
$$

We write $C\langle t\rangle$ for the term obtained by replacing the hole $\langle\cdot\rangle$ of C by t, thus *e.g.* $((\langle\cdot\rangle y)\langle x\rangle = (x\ y)$ and $(\lambda x.\langle\cdot\rangle)\langle x\rangle = \lambda x.x$. We write $C[\![u]\!]$ when the free variables of u are not captured by the context C, thus for example, $C[\![x]\!]$ denotes the term $(x\ y)$ if $C = \langle\cdot\rangle y$, and $\lambda y.x$, if $C = \lambda x.\langle\cdot\rangle$.

Action at a distance is characterized by the fact that ES are not percolated over terms, and variables can be substituted by terms even if the corresponding explicit substitution is not at the level of the variable itself. As usual for ES calculi, there is one rule to start the simulation of a β-reduction and a set of rules to complete the simulation of the β-step. The rule presented below is the starting rule for calculi acting at a distance, and its name comes from d*istance* B*eta*:

$$L\langle\lambda x.t\rangle\ u \mapsto_{\mathtt{dB}} L\langle t[x/u]\rangle \tag{2}$$

where L is a substitution context. Calculi with explicit substitutions that have (2) as starting rule are called *calculi without propagation rules*. It is interesting to note how the dB-rule generalizes β-reduction since dB-redexes include application whose left-hand side are not abstractions.

Binding contexts will be used to characterize normal forms. In fact, the first difficult in building a unification procedure for these calculi was to find an adequate notation for terms in normal form without forcing substitutions to be propagated over them. In [12], an ongoing work accepted for short presentation, the early stages of this work was presented assuming that ES do not occur outside the external abstractions of a normal form, which is possible since the operational semantics of the calculi is not changed. In Subsect. 2.2, we present a precise and elegant characterization of normal forms for calculi without propagation rules.

In the next subsection we shortly present some calculi without propagation rules. They differ in the way substitution is controlled as follows: the first calculus, called the *linear substitution calculus* ($\lambda_{\texttt{lsub}}$) [31], performs substitutions one at a time. The second calculus, called the *substitution calculus* ($\lambda_{\texttt{sub}}$) [6], performs the whole substitution in one step, i.e. it splits the (non-terminating) β-reduction of the λ-calculus into two (terminating) rules. Finally, the third calculus, known as the *structural λ-calculus* ($\lambda_{\texttt{str}}$) [5], duplicates ES by introducing new names for them until the variable of the substitution has just one occurrence in the term, which is then executed.

2.1   The Calculi with ES at a Distance

As explained before, all the calculi start the simulation of a β-reduction by rule (2), which introduces an explicit substitution. The execution and control of the explicit substitution can be done in different ways, which give rise to different calculi. The following tables in addition to rule (2) form the rules of the $\lambda_{\texttt{lsub}}$-, $\lambda_{\texttt{sub}}$- and $\lambda_{\texttt{str}}$-calculi:

lsub	sub
$C[\![x]\!][x/u] \mapsto_{\texttt{ls}} C[\![u]\!][x/u]$	$t[x/u] \mapsto_{\texttt{sub}} t\{x/u\}$, if $x \notin \texttt{fm}(t)$ or
$t[x/u] \quad\mapsto_{\texttt{w}} t, \qquad$ if $\lvert t \rvert_x = 0$	$x \in \texttt{fr}(t)$

str	
$t[x/u] \mapsto_{\texttt{c}} t_{\langle x \mid y \rangle}[x/u][y/u]$	if $\lvert t \rvert_x > 1$
$t[x/u] \mapsto_{\texttt{d}} t\{x/u\}$	if $\lvert t \rvert_x = 1$ and $x \notin \texttt{fm}(t)$
$t[x/u] \mapsto_{\texttt{w}} t$	if $\lvert t \rvert_x = 0$

In what follows, we write $\lambda_{\texttt{dB}}$-calculus to refer to any of these calculi. As usual, $\rightarrow$ denotes the contextual closure of $\mapsto$, $\twoheadrightarrow$ the reflexive-transitive closure of $\rightarrow$ and $=_{\lambda_{\texttt{dB}}}$ the equivalence relation generated by $\rightarrow$. In rule ls, $C[\![x]\!][x/u]$ denotes a term that contains at least one occurrence of the variable x and an explicit substitution that waits to substitute x by u. After a possible renaming to avoid capture of variables, the rule ls substitutes the variable x by u and leaves the explicit substitution $[x/u]$ at the very same place because it can still be used by other occurrences of x in the term. After the substitution of all occurrences of x in the term, the rule w can be applied to get rid of the explicit substitution. In the $\lambda_{\texttt{sub}}$-calculus, the sub-rule transforms in one step the explicit substitution into the implicit metasubstitution, as long as x does not occur in the support set of a metavariable, or it occurs as a real variable, in t. In the $\lambda_{\mathtt{str}}$-calculus, $t_{\langle x|y\rangle}$ denotes the non-deterministic replacement of i $(1 \leq i \leq |t|_x - 1)$ occurrences of x in t by a fresh variable y. A formal definition of the function $_{-\langle\_|\_\rangle}$ is given in what follows. Initially, we define an auxiliary relation that non-deterministically replaces exactly k occurrences of the variable x in t by the fresh variable y, denoted by $t^k_{\langle x|y\rangle}$.

Definition 2. *The non-deterministic replacement of $k \geq 0$ occurrences of the variable x by the fresh variable y in the metaterm t, denoted by $t^k_{\langle x|y\rangle}$, is inductively defined as follows:*

- $t^k_{\langle x|y\rangle} = t$, *if* $x \notin \mathtt{fv}(t)$ *or* $k = 0$, *otherwise*
- $x^k_{\langle x|y\rangle} = y$ • $X_{\Delta}{}^k_{\langle x|y\rangle} = X_{(\Delta\setminus\{x\})\cup\{y\}}$ • $\lambda z.u^k_{\langle x|y\rangle} = \lambda z.u^k_{\langle x|y\rangle}$
- $(u\ v)^k_{\langle x|y\rangle} = u^i_{\langle x|y\rangle}\ v^j_{\langle x|y\rangle}$, *where* $i + j = k$
- $u[z/v]^k_{\langle x|y\rangle} = u^i_{\langle x|y\rangle}[z/v^j_{\langle x|y\rangle}]$, *where* $i + j = k$.

According to this definition, the structural λ-calculus can be adapted as follows:

str	
$t[x/u] \mapsto_{\mathtt{c}} t^k_{\langle x|y\rangle}[x/u][y/u]$	if $|t|_x > 1$ and $0 < k < |t|_x$
$t[x/u] \mapsto_{\mathtt{d}} t\{x/u\}$	if $|t|_x = 1$ and $x \notin \mathtt{fm}(t)$
$t[x/u] \mapsto_{\mathtt{w}} t$	if $|t|_x = 0$

The metasubstitution, i.e. the substitution operation of the λ-calculus can be extended to the grammar of calculi with explicit substitutions as follows:

Definition 3. *Let t, u be metaterms. We inductively define the metasubstitution of x by u in t, written $t\{x/u\}$, as follows [27]:*

- $x\{x/u\} = u$
- $X_{\Delta}\{x/u\} = \begin{cases} X_{\Delta}[x/u], & \text{if } x \in \Delta \\ X_{\Delta}, & \text{if } x \notin \Delta \end{cases}$
- $(\lambda y.t_1)\{x/u\} = \lambda y.t_1\{x/u\}$, *if* $y \notin \mathtt{fv}(u)$
- $t_1[y/t_2]\{x/u\} = t_1\{x/u\}[y/t_2\{x/u\}]$ *if* $y \notin \mathtt{fv}(u)$

- $y\{x/u\} = y$
- $(t_1\ t_2)\{x/u\} = t_1\{x/u\}\ t_2\{x/u\}$

In the cases $(\lambda y.t_1)\{x/u\}$ and $t_1[y/t_2]\{x/u\}$ above, a renaming of bound variables before the propagation of the substitution can be necessary to avoid capture of variables.

Nevertheless, for the particular case of the substitution calculus presented above, this notion of metasubstitution leads to a non-confluent calculus. In fact, assuming that $x \notin \mathtt{fv}(v)$ and $y \notin \mathtt{fv}(u)$, one has the following non-joinable divergence:

$$(x\ y\ X_{\{x,y\}})[x/u][y/v]$$

sub $\qquad$ sub

$$(u\ y\ X_{\{x,y\}}[x/u])[y/v] \qquad (x\ v\ X_{\{x,y\}}[y/v])[x/u]$$

sub $\qquad$ sub

$$(u\ v\ X_{\{x,y\}}[y/v][x/u]) \qquad (u\ v\ X_{\{x,y\}}[x/u][y/v])$$

One way to close this diagram is to allow permutation of independent substitutions, which is usually formalized via an equation of the following form:

$$t[x/u][y/v] \equiv t[y/v][x/u], \text{ if } x \notin \mathtt{fv}(v) \text{ and } y \notin \mathtt{fv}(u) \qquad (3)$$

This approach is adopted in [33] in order to prove metaconfluence for the structural λ-calculus. In order to recover metaconfluence for the substitution calculus, we use a different approach based on a new notion of metasubstitution that do not change the position of an explicit substitution in a term for the case of metavariables. This new notion is given by the following definition, and more details can be found in [11].

Definition 4 (Metasubstitution). *Let t, u be metaterms. The **metasubstitution** $t\{x/u\}$ is defined as follows:*

$$t\{x/u\} = \begin{cases} t\{\!\{x/u\}\!\}[x/u], & \text{if } x \in \mathtt{fm}(t) \\ t\{\!\{x/u\}\!\}, & \text{if } x \notin \mathtt{fm}(t) \end{cases}$$

where $t\{\!\{x/u\}\!\}$ is inductively defined as follows:

$$
\begin{aligned}
t\{\!\{x/u\}\!\} &= t && \text{if } x \notin \mathtt{fr}(t), \text{ otherwise} \\
\bullet\ x\{\!\{x/u\}\!\} &= u \\
\bullet\ (\lambda y.v)\{\!\{x/u\}\!\} &= \lambda y.v\{\!\{x/u\}\!\} && \text{if } x \neq y\ \&\ y \notin \mathtt{fv}(u) \\
\bullet\ (t_1\ t_2)\{\!\{x/u\}\!\} &= t_1\{\!\{x/u\}\!\}\ t_2\{\!\{x/u\}\!\} \\
\bullet\ t_1[y/t_2]\{\!\{x/u\}\!\} &= t_1\{\!\{x/u\}\!\}[y/t_2\{\!\{x/u\}\!\}] && \text{if } x \neq y\ \&\ y \notin \mathtt{fv}(u)
\end{aligned}
$$

Using this new notion of metasubstitution, note that the previous divergence is no longer generated because the explicit substitution does not percolate over the term.

Despite the fact that calculi without propagation rules are not first order in the sense that they use a notion of substitution that renames bound variables in order to avoid capture, the substitution generated by the unification procedure is still first order. This is possible due to two facts:

1. Metavariables and (ordinary) variables belong to different classes, so that the substitution generated by the unification procedure are separated from the ones generated by the reduction relation.

2. The support set of metavariables contains the variables that can occur free in the term that will replace this metavariable, and in addition, metavariables parameterized by a set of variables allow us to express the functional dependency of the arguments w.r.t. the corresponding abstractions in a very straightforward and simple way.

We define *grafting* as a first order substitution whose domain is the set of metavariables:

Definition 5 (Grafting). *Let t and u be metaterms, and Δ a set of variables with $\mathtt{fv}(u) \subseteq \Delta$. We inductively define the grafting of u for X_Δ in t, denoted by $t[\![X_\Delta/u]\!]$, as follows:*

- $y[\![X_\Delta/u]\!] = y$
- $Y_{\Delta'}[\![X_\Delta/u]\!] = Y_{\Delta'}$, *if $X \neq Y$*
- $(\lambda x.t_1)[\![X_\Delta/u]\!] = \lambda x.(t_1[\![X_\Delta/u]\!])$
- $t_1[x/t_2][\![X_\Delta/u]\!] = (t_1[\![X_\Delta/u]\!])[x/t_2[\![X_\Delta/u]\!]]$

- $X_{\Delta'}[\![X_\Delta/u]\!] = \begin{cases} u & \text{if } \Delta \subseteq \Delta' \\ X_{\Delta'} & \text{otherwise.} \end{cases}$
- $(t_1\ t_2)[\![X_\Delta/u]\!] = (t_1[\![X_\Delta/u]\!])\ (t_2[\![X_\Delta/u]\!])$

In this way, the metavariable $X_{\{x,y,z\}}$ can be replaced by metaterms containing x, y and z as free variables, but no more than that. So, for instance,

$$(\lambda x.x\ X_{\{x,y\}})[\![X_{\{x\}}/x]\!] = \lambda x.x\ x.$$

We show next, grafting and reduction commute.

Lemma 1. *Let t, u be metaterms and X_Δ a metavariable with $\mathtt{fv}(u) \subseteq \Delta$. For $R \in \{\mathtt{dB}, \mathtt{sub}, \mathtt{lsub}, \mathtt{str}\}$, $t \to_R t'$ implies $t[\![X_\Delta/u]\!] \to_R t'[\![X_\Delta/u]\!]$*

Lemma 2. *Let t, u, v be metaterms and X_Δ a metavariable with $\mathtt{fv}(u) \subseteq \Delta$. Then $t\{x/v\}[\![X_\Delta/u]\!] = t[\![X_\Delta/u]\!]\{x/v[\![X_\Delta/u]\!]\}$.*

Corollary 1. *Grafting and reduction commute.*

2.2   Typing Rules

The right environment for unification is the simple type theory, in which simple types are defined by the following grammar:

$$\sigma ::= A \mid \sigma \to \sigma$$

where A range over a denumerable set of base types. Type variables will range over $\sigma, \tau, \gamma, \ldots$ with subscripts when necessary, and the constructor $\to$ of a functional type is right associative. Therefore, $\sigma \to \tau \to \gamma$ means $\sigma \to (\tau \to \gamma)$, and every typable metaterm t has a type of the form $\sigma_1 \to \sigma_2 \to \cdots \to \sigma_n \to A$ ($n \geq 0$), where A is the *target type* of t, written $\mathtt{tt}(t)$. If t is a metaterm of type σ, then we write $\mathtt{ty}(t) = \sigma$. A type assignment is a pair of the form $x : \sigma$, where x is a variable and σ is a type. The domain of a set of type assignments $\Gamma = \{x_1 : \sigma_1, x_2 : \sigma_2, \ldots, x_n : \sigma_n\}$, denoted by $Dom(\Gamma)$, is the set $\{x_1, x_2, \ldots, x_n\}$, whose variables are assumed to be pairwise distinct. We define *type contexts* as finite sets of type assignments. We assume that variables in a type context have at most one assignment, and the type contexts can be extended by adding new assignments, as long as this condition is preserved. If Σ is a type context then the restriction of Σ to the variables in the set S, denoted by $\Sigma|_S$, is defined as the set $\{x : \sigma \mid x \in S \text{ and } (x : \sigma) \in \Sigma\}$.

The fact that the term t has type σ in the type context Σ is expressed by the type judgment $\Sigma \vdash t : \sigma$. In addition, to each metavariable X_Δ we associate a unique type σ_X. We assume that for each type σ there is an infinite set of metavariables X_Δ such that $\sigma_X = \sigma$ and the typing rules are given by the following rules:

$$\frac{(x : \sigma) \in \Sigma}{\Sigma \vdash x : \sigma} \text{ (Var)} \qquad\qquad \frac{\Delta \subseteq Dom(\Sigma)}{\Sigma \vdash X_\Delta : \sigma_X} \text{ (MVar)}$$

$$\frac{\Sigma \vdash t : \sigma \to \tau \qquad \Sigma \vdash u : \sigma}{\Sigma \vdash t\, u : \tau} \text{ (App)}$$

$$\frac{\Sigma \cup \{x : \sigma\} \vdash t : \tau}{\Sigma \vdash \lambda x.t : \sigma \to \tau} \text{ (Abs)} \qquad \frac{\Sigma \cup \{x : \sigma\} \vdash t : \tau \qquad \Sigma \vdash u : \sigma}{\Sigma \vdash t[x/u] : \tau} \text{ (Clos)}$$

We present some properties of this typing system:

Lemma 3. *Let Σ be a type context, t a metaterm and σ a type. If $\Sigma' \supset \Sigma$ is another type context, then $\Sigma \vdash t : \sigma \implies \Sigma' \vdash t : \sigma$.*

Proof. By induction on $\Sigma \vdash t : \sigma$.

Lemma 4. *Let t be a metaterm such that $\Sigma \vdash t : \sigma$ for some type σ and type context Σ. Then $\Sigma|_{\mathtt{fv}(t)} \vdash t : \sigma$.*

Proof. By induction on $\Sigma \vdash t : \sigma$.

Lemma 5. *Let t be a metaterm such that $\Sigma \vdash t : \sigma$ for some type σ and type context Σ. If $t \to_{\mathtt{dB}} t'$ or $t \to_{\mathtt{w}} t'$ then $\Sigma \vdash t' : \sigma$.*

Lemma 6. *Let Σ and Σ' be type contexts such that its union is still a type context, and t, u be metaterms. If x is a fresh variable of type γ, $x \notin \mathtt{fm}(t)$, $\Sigma \cup \{x : \gamma\} \vdash t : \sigma$ and $\Sigma' \vdash u : \gamma$ then $\Sigma \cup \Sigma' \vdash t\{\!\{x/u\}\!\} : \sigma$*

Proof. By induction on t.

Lemma 7. *Let Σ and Σ' be type contexts such that its union is still a type context, and t, u be metaterms. If $x \in \mathtt{fm}(t)$, $\Sigma \cup \{x : \gamma\} \vdash t : \sigma$ and $\Sigma' \vdash u : \gamma$ then $\Sigma \cup \{x : \gamma\} \cup \Sigma' \vdash t\{\!\{x/u\}\!\} : \sigma$*

Proof. By induction on t.

Proposition 1. *Let Σ and Σ' be type contexts such that its union is still a type context. If t and u are metaterms such that $\Sigma \cup \{x : \gamma\} \vdash t : \sigma$ and $\Sigma' \vdash u : \gamma$ then $\Sigma \cup \Sigma' \vdash t\{x/u\} : \sigma$.*

Corollary 2. *Let Σ and Σ' be type contexts such that its union is still a type context. If C is a context, x a variable of type γ in the type context Σ, u is a metaterm such that $\Sigma' \vdash u : \gamma$ and $\Sigma \cup \{x : \gamma\} \vdash C[\![x]\!] : \sigma$ then $\Sigma \cup \Sigma' \vdash C[\![u]\!] : \sigma$.*

Theorem 1 (Subject reduction for the $\lambda_{\texttt{1sub}}$-calculus). *Let t be a metaterm such that $\Sigma \vdash t : \sigma$ for some type σ and type context Σ. If $t \to_{\lambda_{\texttt{1sub}}} t'$ then $\Sigma \vdash t' : \sigma$.*

Theorem 2 (Subject reduction for the $\lambda_{\texttt{sub}}$-calculus). *Let t be a metaterm such that $\Sigma \vdash t : \sigma$ for some type σ and type context Σ. If $t \to_{\lambda_{\texttt{sub}}} t'$ then $\Sigma \vdash t' : \sigma$.*

Lemma 8. *Let t be a metaterm. If $\Sigma \cup \{x : \gamma\} \vdash t : \sigma$ then $\Sigma \cup \{x : \gamma\} \cup \{y : \gamma\} \vdash t^k_{\langle x|y \rangle} : \sigma$, for all $0 \le k \le |t|_x$ and fresh variable y.*

Corollary 3. *If $|t|_x > 1$ and $\Sigma \cup \{x : \sigma_1\} \vdash t : \sigma$ then $\Sigma \cup \{x : \sigma_1\} \cup \{y : \sigma_1\} \vdash t_{\langle x|y \rangle} : \sigma$, where y is a fresh variable.*

Theorem 3 (Subject reduction for the $\lambda_{\texttt{str}}$-calculus). *Let t be a metaterm such that $\Sigma \vdash t : \sigma$ for some type σ and type context Σ. If $t \to_{\lambda_{\texttt{str}}} t'$ then $\Sigma \vdash t' : \sigma$.*

Our unification algorithm works with terms in *η-long normal form* (`lnf`) that is defined in what follows.

By definition, a term is in normal form if it has no reduct. The following lemma gives the general structure of a term in normal form.

Lemma 9 (Characterization of normal forms). *These terms can be characterized by the following structure:*

$$B\langle L_n \langle \cdots L_1 \langle L_0 \langle a \rangle\ t_1 \rangle \cdots t_n \rangle \rangle$$

where

- *a is either a variable or a metavariable;*
- *$L_0 = \langle \cdot \rangle$ if a is not a metavariable;*
- *the terms $t_1, t_2, \ldots, t_n$ are in normal form;*
- *every substitution of the form $[x/u]$ occurring in B or L_i $(0 \le i \le n)$ is such that x occurs in the support set of a metavariable that is in the scope of this substitution, and u is in normal form.*

Since the calculi considered here are metaconfluent, normal forms are unique, and we write $(t)\!\downarrow$ to denote the normal form of the term t.

Definition 6 (η-long normal form). *Let* $t = B\langle L_n \langle \cdots L_1 \langle L_0 \langle a \rangle\ t_1 \rangle \cdots t_n \rangle\rangle$ *be a term in normal form of type* $\sigma_1 \to \cdots \to \sigma_m \to A$ *with* A *atomic and* $n, m \geq 0$. *The* η*-long normal form* (lnf) *of* t, *written* $(t)\updownarrow$, *is given by*

$$\lambda x_1 \ldots x_m . B' \langle L'_n \langle \cdots L'_1 \langle L'_0 \langle a \rangle\ (t_1)\updownarrow \rangle \cdots (t_n)\updownarrow \rangle\rangle\ (x_1)\updownarrow \ldots (x_m)\updownarrow$$

where

- L'_i $(0 \leq i \leq n)$ *is obtained from* L_i *by taking each substitution of the form* $[x/u]$ *in* L_i *and replacing it by* $[x/(u)\updownarrow]$.
- B' *is obtained from* B *by preserving each abstractor of the form* λx, *and replacing each substitution of the form* $[x/u].in\ B$ *by* $[x/(u)\updownarrow]$.

Lemma 10. *The Definition 6 is well-defined.*

Definition 7 (Heading). *Let* $t = B\langle L_n \langle \cdots L_1 \langle L_0 \langle a \rangle\ t_1 \rangle \cdots t_n \rangle\rangle$ *be a term in* η*-long normal form. The heading of* t *is defined as* $(B\langle L_n \langle \cdots L_1 \langle L_0 \langle a \rangle\rangle \ldots \rangle\rangle)\downarrow$.

In this way, if $\lambda x.(\lambda y.a\ X_{\{z\}})[z/u]$ is a term in η-long normal form then it can be written as $B_2 \langle B_1 \langle B_0 \langle a\ X_{\{z\}}\rangle\rangle\rangle$ with $B_2 = \lambda x.\langle\cdot\rangle$, $B_1 = \langle\cdot\rangle[z/u]$ and $B_0 = \lambda y.\langle\cdot\rangle$. By definition, its heading is equal to $(\lambda x.(\lambda y.a)[z/u])\downarrow = \lambda x.\lambda y.a$.

Definition 8. *A metaterm in* lnf *is flexible if its head is a metavariable. Otherwise, it is rigid.*

3 Unification

In this section we present the unification procedure. It receives as argument a unification problem that is defined in what follows.

Definition 9. *A unification problem* P *is a finite set of pairs, also called disagreement pairs, of the form* $t =^? u$, *where* t *and* u *are terms in* lnf *of the same type. A disagreement pair* $t =^? u$ *is called trivial if* $t =_\alpha u$, *i.e. if* t *and* u *differ only by the name of bound variables. A grafting* σ *is a solution of the disagreement pair* $t =^? u$, *if* $t\sigma =_{\lambda_{dB}} u\sigma$. *A grafting* σ *is a solution of the unification problem* P *if it is a solution of each disagreement pair of* P. *A metavariable is unsolved w.r.t. the unification problem* P *if it appears in either a flexible-rigid or a rigid-rigid pair in* P, *otherwise, i.e. if it either appears only in flexible-flexible pairs or does not appear at all in* P, *it is said to be solved w.r.t. the unification problem* P. *A unification problem is pre-solved if it does not contain unsolved metavariables, i.e. if it contains only flexible-flexible pairs. A unification problem without solution will be denoted by* $\perp$. *The set of unifiers of a unification problem* P *is denoted by* $\mathcal{U}_{\lambda_{dB}}(P)$.

The unification rules are presented in Table 1. For a given unification problem P, one starts with the pair $P; \epsilon$, where ϵ is the empty grafting. The rules are applied non-deterministically to the current unification problem until no more rules apply, and then one reaches either $\perp$ (no solution is generated) or a

Table 1. Unification Rules

$\{t =^? t'\} \cup Q; \theta \overset{\text{Trivial}}{\Longrightarrow} Q; \theta$, if $t =_\alpha t'$
$\{t =^? t'\} \cup Q; \theta \overset{\text{Fail}}{\Longrightarrow} \bot$, if $t =^? t'$ is a rigid-rigid equation with headings that are not α-equivalent
$\{t =^? t'\} \cup Q; \theta \overset{\text{Dec}}{\Longrightarrow} (\bigcup_{i=1}^{n}\{B\langle L_n\langle \ldots L_i\langle t_i\rangle \ldots\rangle\rangle =^? B'\langle L'_n\langle \ldots L'_i\langle t'_i\rangle \ldots\rangle\rangle\}) \cup Q; \theta$ if $t =^? t'$ is a rigid-rigid equation, $t = B\langle L_n\langle \ldots L_2\langle L_1\langle a\ t_1\rangle\ t_2\rangle \ldots t_n\rangle\rangle$, $t' = B'\langle L'_n\langle \ldots L'_2\langle L'_1\langle b\ t'_1\rangle\ t'_2\rangle \ldots t'_n\rangle\rangle$ and $(B\langle a\rangle)\!\downarrow =_\alpha (B'\langle b\rangle)\!\downarrow$
$P; \theta \overset{\text{Exp}}{\Longrightarrow} (P\theta')\updownarrow; \theta\theta'$ if there is a flexible-rigid equation in P with the head of the flexible term equal to X_Δ with $\mathbf{ty}(X_\Delta) = \sigma_1 \to \cdots \to \sigma_n \to A$ and $n > 0$, and where $\theta' = [\![X_\Delta/\lambda x_1 \ldots x_n.Y_{\Delta \cup \{x_1,\ldots,x_n\}}]\!]$ and Y is a fresh metavariable
$P; \theta \overset{\text{Imit}}{\Longrightarrow} (P\theta')\updownarrow; \theta\theta'$ if there is a flexible-rigid equation with the head of   the flexible term equal to X_Δ (atomic metavariable of type A), and   the head of the rigid term equal to a free variable, say a,   of type $\sigma_1 \to \cdots \to \sigma_n \to A$ $(n \geq 0)$,   where $\theta' = [\![X_\Delta/a\ Y_{1\Delta} \ldots Y_{n\Delta}]\!]$ and $Y_1, \ldots, Y_n$ are fresh metavariables
$P; \theta \overset{\text{Select}}{\Longrightarrow} (P\theta')\updownarrow; \theta\theta'$ if there is a flexible-rigid equation with the flexible   term equal to $B\langle X_\Delta\rangle$, for a binding context B and $\mathbf{ty}(X_\Delta) = A$ (atomic),   and an ES $[y/u]$ in B with $\mathbf{ty}(y) = \sigma_1 \to \cdots \to \sigma_n \to A$ $(n \geq 0)$ and $\mathbf{fv}(u) \subseteq \Delta$ such that either u is flexible, or the heading of $B\langle u\rangle$ is α-equivalent to the heading of the rigid term of this equation,   where $\theta' = [\![X_\Delta/y\ Y_{1\Delta} \ldots Y_{n\Delta}]\!]$ and $Y_1, \ldots, Y_n$ are fresh metavariables

pre-solved form $P'; \sigma$. During the process to reach a pre-solved form, the procedure transforms the original problem by incrementally building partial graftings that will compose a solution to the original problem. In addition, each rule that can be applied the current unification problem gives rise to a new branch that potentially will generate a solution to the original problem.

The rules `Imit` and `Select` are generalizations of the imitation and projection rules of Huet's procedure for the simply typed λ-calculus [24]. A failure in the unification process will be characterized by unification problems containing rigid-rigid pairs with headings that are not α-equivalent.

The case of flexible-flexible pairs are usually not treated explicitly because they are always solvable and their solutions are not unique. In addition, flexible-flexible pairs are not relevant for most refutation methods [24], and hence our procedure consists in searching for pre-solved forms of the original problem.

Lemma 11. *The application of the unification rules to well-typed pairs gives rise only to well-typed pairs.*

Proof. By rule analysis.

3.1   Unification Procedure

In this section we detail the general behavior of the unification procedure, and we write $\Longrightarrow \equiv \xLongrightarrow{\text{Trivial}} \cup \xLongrightarrow{\text{Fail}} \cup \xLongrightarrow{\text{Dec}} \cup \xLongrightarrow{\text{Exp}} \cup \xLongrightarrow{\text{Imit}} \cup \xLongrightarrow{\text{Select}}$. Given a unification problem P, one builds a tree, called *unification tree* of P (cf. [13]), as follows:

1. The root of the tree is labeled with $P;\epsilon$
2. For each node, one non-deterministically selects a pair in P and applies a rule by adding a child node labeled with the new unification problem P'. In this case, we write $P;\sigma \xLongrightarrow{r} P\theta;\sigma\theta$.
3. The procedure stops when no more rule applies. A success branch is obtained if the procedure stops at a pre-solved form. Otherwise, the procedure generates a fail branch.

Example 1. Let P be the unification problem containing the sole pair:[1]

$$\{X_{\{u,v\}_{A\to A}} u_A =^? v_{A\to A} u_A\}$$

The unification tree is as follows where non-determinism corresponds to "or" branches:

$$\{X_{\{u,v\}} u =^? v\ u\};\epsilon$$

$$\text{Exp} \left\Vert [X_{\{u,v\}}/\lambda z.X_{1\{u,v,z\}}] \right.$$

$$\{X_{1\{u,v,z\}}[z/u] =^? v\ u\};[X_{\{u,v\}}/\lambda z.X_{1\{u,v,z\}}]$$

$$\text{Imit} \left\Vert [X_{1\{u,v,z\}}/v\ X_{2\{u,v,z\}}] \right.$$

$$\{(v\ X_{2\{u,v,z\}})[z/u] =^? v\ u\};[X_{\{u,v\}}/\lambda z.v\ X_{2\{u,v,z\}}]$$

$$\text{Dec} \Downarrow$$

$$\{X_{2\{u,v,z\}}[z/u] =^? u\};[X_{\{u,v\}}/\lambda z.v\ X_{2\{u,v,z\}}]$$

Imit $[X_{2\{u,v,z\}}/u]$ Select $\left\Vert [X_{2\{u,v,z\}}/z] \right.$

$$\{u =^? u\};[X_{\{u,v\}}/\lambda z.v\ u] \qquad\qquad \{u =^? u\};[X_{\{u,v\}}/\lambda z.v\ z]$$

Trivial $\Downarrow$ Trivial $\Downarrow$

$$\{\};[X_{\{u,v\}}/\lambda z.v\ u] \qquad\qquad\qquad \{\};[X_{\{u,v\}}/\lambda z.v\ z]$$

Example 2. Consider the unification problem

$$\{X_{\{u,v\}_{A\to A}}(v_{A\to A}\ u_A) =^? v_{A\to A}(X_{\{u,v\}_{A\to A}}\ u_A)\}$$

---

[1] We write the type information only in the initial terms of the examples for readability.

$$\{X_{\{u,v\}}(v\ u) =^? v(X_{\{u,v\}}\ u)\};\epsilon$$

$$\text{Exp} \bigg\Vert [\![X_{\{u,v\}}/\lambda w.X_{1\Delta}]\!]$$

$$\{X_{1\Delta}[w/(v\ u)] =^? v\ (X_{1\Delta}[w/u])\};[\![X_{\{u,v\}}/\lambda w.X_{1\Delta}]\!]$$

$$[\![X_{1\Delta}/v\ X_{2\Delta}]\!] \quad\quad [\![X_{1\Delta}/w]\!]\bigg\Vert \text{Select}$$

$$\text{Imit}$$

$$(*) \Longleftarrow \quad\quad \{v\ u =^? v\ u\};[\![X_{\{u,v\}}/\lambda w.w]\!]$$

$$\text{Trivial} \bigg\Vert$$

$$\{\};[\![X_{\{u,v\}}/\lambda w.w]\!]$$

where $\Delta = \{u,v,w\}$ and $(*)$ is given by

$$\{(v\ X_{2\Delta})[w/(v\ u)] =^? v((v\ X_{2\Delta})[w/u])\};[\![X_{\{u,v\}}/\lambda w.v\ X_{2\Delta}]\!]$$

which reduces to $\{X_{2\Delta}[w/(v\ u)] =^? v(X_{2\Delta}[w/u])\};[\![X_{\{u,v\}}/\lambda w.v\ X_{2\Delta}]\!]$ after an application of Dec. From this point, one can again apply both Imit and Select, and find out that this problem has infinitely many solutions: $[\![X_{\{u,v\}}/\lambda w.x^n\ w]\!]$ $(n \geq 0)$, where $x^0\ w = w$, and $x^{n+1}\ w = x(x^n\ w)$.

Example 3. Let

$$P\{\lambda y_A.X_{A\to(A\to A)\to A}\ y_A\ (\lambda w_A.w_A) =^? \lambda z_A.z_A\}$$

be a unification problem.

$$\{\lambda y.X\ y\ (\lambda w.w) =^? \lambda z.z\};\epsilon$$

$$\text{Exp} \bigg\Vert [\![X/\lambda z_1 z_2.X_{1\Delta}]\!]$$

$$\{\lambda y.X_{1\Delta}[z_2/\lambda w.w][z_1/y] =^? \lambda z.z\};[\![X/\lambda z_1 z_2.X_{1\Delta}]\!]$$

$$[\![X_{1\Delta}/z_2\ X_{2\Delta}]\!] \quad\quad [\![X_{1\Delta}/z_1]\!]\bigg\Vert \text{Select}$$

$$\text{Select}$$

$$(*) \Longleftarrow \quad\quad \{\lambda y.y =^? \lambda z.z\};[\![X/\lambda z_1 z_2.z_1]\!]$$

$$\text{Trivial} \bigg\Vert$$

$$\{\};[\![X/\lambda z_1 z_2.z_1]\!]$$

where $(*)$ is as follows:

$$\{\lambda y.X_{2\Delta}[z_2/\lambda w.w][z_1/y] =^? \lambda z.z\};[\![X/\lambda z_1 z_2.z_2\ X_{2\Delta}]\!]$$

$$[\![X_{2\Delta}/z_2\ X_{3\Delta}]\!] \quad\quad [\![X_{2\Delta}/z_1]\!]\bigg\Vert \text{Select}$$

$$\text{Select}$$

$$\vdots \Longleftarrow \quad\quad \{\lambda y.y =^? \lambda z.z\};[\![X/\lambda z_1 z_2.z_1]\!]$$

$$\text{Trivial} \bigg\Vert$$

$$\{\};[\![X/\lambda z_1 z_2.z_2\ z_1]\!]$$

where $\Delta = \{z_1, z_2\}$. This problem has infinite solutions. In fact, all Church numerals of the form $\lambda xy.y^k \, x \, (k \geq 0)$ are solutions.

Correctness and Completeness of the Unification Procedure. In this subsection, we state the correctness and completeness theorems of the presented unification procedure.

Lemma 12. *Let P be a unification problem. If $P; \theta \xrightarrow{R} P'; \theta'$ then $\mathcal{U}_{\lambda_{dB}}(P') = \mathcal{U}_{\lambda_{dB}}(P)$, where $R \in \{\texttt{Trivial}, \texttt{Fail}, \texttt{Dec}\}$.*

Lemma 13. *Let P be a solvable unification problem. If $P; \theta \xrightarrow{R} P'; \theta'$ then $\mathcal{U}_{\lambda_{dB}}(P') \subseteq \mathcal{U}_{\lambda_{dB}}(P)$, where $R \in \{\texttt{Exp}, \texttt{Imit}, \texttt{Select}\}$.*

Theorem 4. *Let P be a solvable unification problem. If $P; \theta \Longrightarrow^* P'; \theta'$ then $\mathcal{U}_{\lambda_{dB}}(P') \subseteq \mathcal{U}_{\lambda_{dB}}(P)$.*

Proof. By induction on the length of the derivation $P; \theta \Longrightarrow^* P'; \theta'$.

Theorem 5 (Correctness of the unification procedure). *Let P be a nontrivial unification problem, and θ a grafting. If $P; \theta \Longrightarrow^* P'; \theta'$ with P' in presolved form, then $\theta' \in \mathcal{U}_{\lambda_{dB}}(P)$.*

Theorem 6 (Completeness of the unification procedure). *Let P be a unification problem with solution σ. Then there exists a derivation of $P; \epsilon$ to a pre-solved form $P'; \theta$ such that $\sigma = \theta\gamma$, for some grafting γ.*

4   Conclusion and Future Work

Higher-order unification and matching are important operations extensively used in Mathematics and Computer Science that have already been studied in the context of ES [9,10,13,18] but the substitution operation in all the calculi considered so far acts by proximity. This means that the substitution is propagated over the terms, while in calculi that act at a distance, they are not. In spite of having the desired good properties, such as confluence, full composition and PSN, these calculi are useful to study the λ-calculus rather than to implement it [2,7]. In particular, the proof of confluence for the calculi with metaterms is simple and elegant.

In this work we presented a unification procedure for a family calculi of ES without propagation rules. Our procedure is not just an adaptation of Huet's [24] or Snyder's [34] one because it does not develop as many fail branches (which can be computationally expensive) as Huet's does, but it is still correct and complete. In addition, a specialized notion of η-long normal form was defined, and is central for the unification procedure. Unification in calculi that act at a distance is also interesting because grafting and reduction permute, which permits the presentation of a procedure that is simpler than the known unification procedures for the λ-calculus [15,24,34] because grafting is used instead of substitution, and also simpler than the traditional unification approach to ES [18] because no transformation, like precooking, is required.

As future work, we will study higher-order matching in these formalisms [14,17]. This would form an interesting framework to study problems like the decidability of higher-order matching which is still an open question after more than 30 years of investigation [16,17,25,29,32,35].

Acknowledgements. I want to thank the anonymous referees for comments and suggestions on this work.

References

1. Abadi, M., Cardelli, L., Curien, P.-L., Lévy, J.-J.: Explicit substitutions. J. Funct. Program. **1**(4), 375–416 (1991)
2. Accattoli, B.: An abstract factorization theorem for explicit substitutions. In: Tiwari [36], pp. 6–21
3. Accattoli, B., Barenbaum, P., Mazza, D.: Distilling abstract machines. In: Jeuring, J., Chakravarty, M.M.T. (eds.) Proceedings of the 19th ACM SIGPLAN International Conference on Functional Programming, Gothenburg, Sweden, 1–3 September 2014, pp. 363–376. ACM (2014)
4. Accattoli, B., Bonelli, E., Kesner, D., Lombardi, C.: A nonstandard standardization theorem. In: Jagannathan, S., Sewell, P. (eds.) POPL, pp. 659–670. ACM (2014)
5. Accattoli, B., Kesner, D.: The structural λ-calculus. In: Dawar, A., Veith, H. (eds.) CSL 2010. LNCS, vol. 6247, pp. 381–395. Springer, Heidelberg (2010). doi:10.1007/978-3-642-15205-4_30
6. Accattoli, B., Kesner, D.: The permutative λ-calculus. In: Bjørner, N., Voronkov, A. (eds.) LPAR 2012. LNCS, vol. 7180, pp. 23–36. Springer, Heidelberg (2012). doi:10.1007/978-3-642-28717-6_5
7. Accattoli, B., Kesner, D.: Preservation of strong normalisation modulo permutations for the structural lambda-calculus. Logical Methods Comput. Sci. **8**(1), 1–44 (2012)
8. Accattoli, B., Dal Lago, U.: On the invariance of the unitary cost model for head reduction. In: Tiwari [36], pp. 22–37
9. Ayala-Rincón, M., Kamareddine, F.: Unification via the λs_e-style of explicit substitution. Logical J. IGPL **9**(4), 489–523 (2001)
10. Ayala-Rincón, M., Kamareddine, F.: On applying the λs_e-style of unification for simply-typed higher order unification in the pure lambda calculus. Matemática Contemporânea **24**, 1–22 (2003)
11. de Moura, F.L.C., Kesner, D., Ayala-Rincón, M.: Metaconfluence of calculi with explicit substitutions at a distance. In: Raman, V., Suresh, S.P. (eds.) 34th International Conference on Foundation of Software Technology and Theoretical Computer Science (FSTTCS 2014). Leibniz International Proceedings in Informatics (LIPIcs), vol. 29, pp. 391–402. Schloss Dagstuhl-Leibniz-Zentrum fuer Informatik, Dagstuhl (2014)
12. de Moura, F.L.C.: Higher-order unification via explicit substitutions at a distance. In: LSFA 2014 (2014). Accepted for short presentation

13. de Moura, F.L.C., Ayala-Rincón, M., Kamareddine, F.: Higher-order unification: a structural relation between Huet's method and the one based on explicit substitutions. J. Appl. Logic **6**(1), 72–108 (2008)
14. de Moura, F.L.C., Kamareddine, F., Ayala-Rincón, M.: Second-order matching via explicit substitutions. In: Baader, F., Voronkov, A. (eds.) LPAR 2005. LNCS (LNAI), vol. 3452, pp. 433–448. Springer, Heidelberg (2005). doi:10.1007/978-3-540-32275-7_29
15. Dougherty, D.J.: Higher-order unification via combinators. TCS **114**(2), 273–298 (1993)
16. Dowek, G.: Third order matching is decidable. APAL **69**, 135–155 (1994)
17. Dowek, G.: Higher-order unification and matching. In: Robinson, A., Voronkov, A. (eds.) Handbook of Automated Reasoning, vol. 2, pp. 1009–1062. MIT press and Elsevier (2001). Chap. 16
18. Dowek, G., Hardin, T., Kirchner, C.: Higher order unification via explicit substitutions. Inf. Comput. **157**(1–2), 183–235 (2000)
19. Briaud, D., Lescanne, P., Rouyer-Degli, J.: λυ, a calculus of explicit substitutions which preserves strong normalization. JFP **6**(5), 699–722 (1996)
20. Girard, J.-Y.: Linear logic. Theor. Comput. Sci. **50**, 1–102 (1987)
21. Goldfarb, W.: The undecidability of the second-order unification problem. Theoret. Comput. Sci. **13**(2), 225–230 (1981)
22. Guillaume, B.: The λs_e-calculus does not preserve strong normalization. J. Func. Program. **10**(4), 321–325 (2000)
23. Huet, G.: The undecidability of unification in third order logic. Inf. Control **22**(3), 257–267 (1973)
24. Huet, G.: A unification algorithm for typed lambda-calculus. TCS **1**(1), 27–57 (1975)
25. Huet, G.: Résolution d'équations dans les langages d'ordre 1,2,..,ω. Ph.D. thesis, University Paris-7 (1976)
26. Kamareddine, F., Ríos, A.: Extending a λ-calculus with explicit substitution which preserves strong normalisation into a confluent calculus on open terms. J. Func. Program. **7**, 395–420 (1997)
27. Kesner, D.: A theory of explicit substitutions with safe and full composition. Logical Methods Comput. Sci. **5**(31), 1–29 (2009)
28. Lins, R.D.: A new formula for the execution of categorical combinators. In: Siekmann, J.H. (ed.) CADE 1986. LNCS, vol. 230, pp. 89–98. Springer, Heidelberg (1986). doi:10.1007/3-540-16780-3_82
29. Loader, R.: Higher order β matching is undecidable. Logic J. Interest Group Pure Appl. Logics **11**(1), 51–68 (2003)
30. Mellies, P.-A.: Typed λ-calculi with explicit substitutions may not terminate. In: Dezani-Ciancaglini, M., Plotkin, G. (eds.) TLCA 1995. LNCS, vol. 902, pp. 328–334. Springer, Heidelberg (1995). doi:10.1007/BFb0014062
31. Milner, R.: Local bigraphs and confluence: two conjectures: (extended abstract). ENTCS **175**(3), 65–73 (2007)
32. Padovani, V.: Decidability of fourth-order matching. Math. Struct. Comput. Sci. **10**(3), 361–372 (2000)
33. Renaud, F.: Metaconfluence of λj: dealing with non-deterministic replacements (2011). http://www.pps.univ-paris-diderot.fr/~renaud/lambdaj_mconf.pdf
34. Snyder, W., Gallier, J.H.: Higher-order unification revisited: complete sets of transformations. J. Symb. Comput. **8**(1/2), 101–140 (1989)

35. Stirling, C.: A game-theoretic approach to deciding higher-order matching. In: Bugliesi, M., Preneel, B., Sassone, V., Wegener, I. (eds.) ICALP 2006. LNCS, vol. 4052, pp. 348–359. Springer, Heidelberg (2006). doi:10.1007/11787006_30

36. Tiwari, A. (ed.) 23rd International Conference on Rewriting Techniques and Applications (RTA 2012). LIPIcs, Nagoya, Japan, 28 May 2012 – 2 June 2012, vol. 15. Schloss Dagstuhl - Leibniz-Zentrum fuer Informatik (2012)

Soundly Proving B Method Formulæ Using Typed Sequent Calculus

Pierre Halmagrand[(✉)]

Cnam/Inria/Ens Cachan, Paris, France
pierre.halmagrand@inria.fr

Abstract. The B Method is a formal method mainly used in the railway industry to specify and develop safety-critical software. To guarantee the consistency of a B project, one decisive challenge is to show correct a large amount of proof obligations, which are mathematical formulæ expressed in a classical set theory extended with a specific type system. To improve automated theorem proving in the B Method, we propose to use a first-order sequent calculus extended with a polymorphic type system, which is in particular the output proof-format of the tableau-based automated theorem prover Zenon. After stating some modifications of the B syntax and defining a sound elimination of comprehension sets, we propose a translation of B formulæ into a polymorphic first-order logic format. Then, we introduce the typed sequent calculus used by Zenon, and show that Zenon proofs can be translated to proofs of the initial B formulæ in the B proof system.

1 Introduction

Automated transport systems have spread in many cities during last decades, becoming a leading sector for the development of highly trusted software using formal methods. The B Method [1] is a formal method mainly used in the railway industry to specify and develop safety-critical software. It allows the development of correct-by-construction programs, thanks to a refinement process from an abstract specification to a deterministic implementation of the program. The soundness of the refinement steps depends on the validity of logical formulæ called proof obligations, expressed in a specific typed set theory. Common industrial projects using the B Method generate thousands of proof obligations, thereby relying on automated tools to discharge as many as possible proof obligations. A specific tool, called Atelier B [18], designed to implement the B Method and provided with a theorem prover, helps users verify the validity of proof obligations, automatically or interactively.

Improving the automated verification of proof obligations is a crucial task. The BWare research project [10] proposed to use external automated provers, like first-order Automated Theorem Provers (ATPs) and Satisfiability Modulo

---

This work is supported by the BWare project (ANR-12-INSE-0010) funded by the INS programme of the French National Research Agency (ANR).

A. Sampaio and F. Wang (Eds.): ICTAC 2016, LNCS 9965, pp. 196–213, 2016.
DOI: 10.1007/978-3-319-46750-4_12

Theory (SMT) solvers, by building a common platform to run these tools. This platform, based on the software verification tool Why3 [3], requires proof obligations to be encoded in its native language called WhyML. Mentré *et al.* [16] proposed a translator program called bpo2why to address this issue. This tool focuses on the translation of B proof obligations output by Atelier B into the WhyML language. Besides, the B set theory is defined directly in WhyML. Then, Why3 uses specific drivers to translate proof obligations and the theory from WhyML to the specific format of each automated tool.

The first-order ATP Zenon [6,9], based on the tableau method and recently extended to deal with polymorphic types, has been used to prove B proof obligations in the BWare project and obtained good experimental results, compared to the regular version of Zenon and other automated deduction tools as well [7]. One important feature of Zenon is to be a certifying prover [8], in the sense that it generates proof certificates, *i.e.* proof objects that can be verified by external proof checkers. It relies on an encoding of the output proof-format of Zenon, a typed sequent calculus called LLproof, into the proof checker Dedukti [5], a tool designed to be a universal backend to certify and share proofs coming from automated or interactive provers. These proof certificates allow us to be very confident about the soundness of the proofs produced by Zenon.

An issue about using Zenon to verify B proof obligations arises in the upstream translation chain, from B to Zenon input format. There is currently no formal guarantee that this chain is sound, and it would be a tremendous work to formalize the several steps represented by bpo2why and Why3. Instead, we decide to confirm the soundness of using Zenon to prove B proof obligations by formalizing a more general and direct translation from the B logic into a polymorphic first-order logic (PFOL for short), close to the Zenon input format. This translation is to be proven sound, in the sense that if Zenon finds a proof then it can be turned into a proof of the initial B formula. A solution is to show a logical equivalence between Zenon and B proof system.

The B set theory is provided with a specific type system, expressed using set constructs, resulting in a lack of separation in a B formula between typing and set reasoning. To help us embed B typing constraints into PFOL, we define a procedure to annotate B variables with their types, using the type-checking algorithm of the B Method. The interpretation of these types will then be given by the translation function into PFOL. Axioms and hypotheses are generalized by translating B types to (universally quantified) type variables in PFOL. In contrast, types coming from the formula to be proved are interpreted as type constants in PFOL. In addition, we define the reverse translation from PFOL to B, letting us to reword the initial B formula. Thanks to this reverse translation and the derivations of Zenon inference rules expressed using the B proof system, we can translate Zenon proofs into B proofs, guaranteeing the soundness of our translation.

The concerns about the confidence given to ATP in the case of B proof have been resolved using the alternative approach of a certified prover and relying on a deep embedding of the B logic into the interactive prover Coq, by

Jaeger *et al.* [14]. It has also been studied in the context of Event-B by Schmalz in [17]. The problem of type inference in the B Method was studied in other contexts, see for instance [4] for an embedding into PVS, and [13] for Coq.

This paper is organized as follows: in Sect. 2, we introduce the B Method syntax, proof system and type system, then we introduce the type annotation procedure; in Sect. 3, we present the polymorphic first-order logic and the typed sequent calculus used by Zenon; in Sect. 4, we give the translation used to encode B formulæ into PFOL; finally, in Sect. 5, we present the translation of proofs expressed in the sequent calculus of Zenon into B proofs.

2   The B Set Theory

In this section, we present the core logic and theory of the B Method. We first introduce the syntax, the proof system, the set theory and the typing rules of the B Method. Then, we introduce a procedure to annotate variables with their corresponding types. Finally, we present an elimination procedure of comprehension sets.

2.1   Syntax, Proof System and Set Theory

The presentation below follows faithfully the first two chapters of the B-Book [1] dealing with mathematical reasoning and set theory.

Syntax. The syntax of the B Method is made of four syntactic categories, for formulæ, expressions, variables and sets. A formula, or predicate, P is built from the logical connectives conjunction, implication and negation and the universal quantification. A formula may also be the result of a substitution in a formula, the equality between two expressions and the membership to a set. An expression E may be a variable, the result of a substitution in an expression, an ordered pair, an arbitrary element in a set or a set. A variable x is either an identifier or a list of variables. Finally, a set s is built using the elementary set constructs, *i.e.* the cartesian product, the powerset and the comprehension set, or may be the set BIG, a given infinite set.

$$P ::= P_1 \wedge P_2 \mid P_1 \Rightarrow P_2 \mid \neg P \mid \forall x \cdot P \mid [x := E]P \mid E_1 = E_2 \mid E \in s$$
$$E ::= x \mid [x := E_1]E_2 \mid E_1, E_2 \mid \mathsf{choice}(s) \mid s$$
$$x ::= identifier \mid x_1, x_2$$
$$s ::= s_1 \times s_2 \mid \mathbb{P}(s) \mid \{x \mid P\} \mid \mathsf{BIG}$$

Proof System. In Fig. 1, we present the proof system of the B Method. This is an adaptation of classical natural deduction for the B syntax. It should be noted that the B-Book proposes some rules to define the notion of non-freeness of a variable x in a formula P, denoted by $x \backslash P$. Since these rules are standard, we omit them here.

In the following, if x is a variable and Γ a set of formulæ, $x \backslash \Gamma$ means that $x \backslash H$ for each H of Γ; if Γ' is another set of formulæ, $\Gamma \sqsubset \Gamma'$ means that Γ is included in Γ'; and if P is a formula, $P \in \Gamma$ means that P occurs in Γ.

$$\frac{}{P \vdash_B P} \text{ BR1} \qquad \frac{\Gamma \vdash_B P \quad \Gamma \sqsubseteq \Gamma'}{\Gamma' \vdash_B P} \text{ BR2} \qquad \frac{P \in \Gamma}{\Gamma \vdash_B P} \text{ BR3}$$

$$\frac{\Gamma \vdash_B P \quad \Gamma, P \vdash_B Q}{\Gamma \vdash_B Q} \text{ BR4} \qquad \frac{\Gamma \vdash_B P \quad \Gamma \vdash_B P \Rightarrow Q}{\Gamma \vdash_B Q} \text{ MP} \qquad \frac{\Gamma \vdash_B P \quad \Gamma \vdash_B Q}{\Gamma \vdash_B P \wedge Q} \text{ R1}$$

$$\frac{\Gamma \vdash_B P \wedge Q}{\Gamma \vdash_B P} \text{ R2} \qquad \frac{\Gamma \vdash_B P \wedge Q}{\Gamma \vdash_B Q} \text{ R2}' \qquad \frac{\Gamma, P \vdash_B Q}{\Gamma \vdash_B P \Rightarrow Q} \text{ R3}$$

$$\frac{\Gamma \vdash_B P \Rightarrow Q}{\Gamma, P \vdash_B Q} \text{ R4} \qquad \frac{\Gamma, \neg Q \vdash_B P \quad \Gamma, \neg Q \vdash_B \neg P}{\Gamma \vdash_B Q} \text{ R5} \qquad \frac{\Gamma, Q \vdash_B P \quad \Gamma, Q \vdash_B \neg P}{\Gamma \vdash_B \neg Q} \text{ R6}$$

$$\frac{x \backslash \Gamma \quad \Gamma \vdash_B P}{\Gamma \vdash_B \forall x \cdot P} \text{ R7} \qquad \frac{\Gamma \vdash_B \forall x \cdot P}{\Gamma \vdash_B [x := E]P} \text{ R8} \qquad \frac{}{\Gamma \vdash_B E = E} \text{ R10}$$

$$\frac{\Gamma \vdash_B E = F \quad \Gamma \vdash_B [x := E]P}{\Gamma \vdash_B [x := F]P} \text{ R9}$$

Fig. 1. The Proof System of the B Method

Set Theory. As presented in the B-Book, the B Method set theory is a simplification of classical set theory. Some common axioms, like the foundation axiom, are not needed in this context (see Sect. 2.2), leading to a theory made only of six axioms. Actually, axioms presented below are axiom schemata that have to be instantiated with some proper expressions. The first column represents non-freeness proviso.

$$E, F \in s \times t \Leftrightarrow (E \in s \wedge F \in t) \qquad \text{SET1}$$
$$x \backslash (s, t) \qquad s \in \mathbb{P}(t) \Leftrightarrow \forall x \cdot (x \in s \Rightarrow x \in t) \qquad \text{SET2}$$
$$x \backslash s \qquad E \in \{x \mid x \in s \wedge P\} \Leftrightarrow (E \in s \wedge [x := E]P) \qquad \text{SET3}$$
$$x \backslash (s, t) \qquad \forall x \cdot (x \in s \Leftrightarrow x \in t) \Rightarrow s = t \qquad \text{SET4}$$
$$x \backslash s \qquad \exists x \cdot (x \in s) \Rightarrow \text{choice}(s) \in s \qquad \text{SET5}$$
$$\text{infinite}(\text{BIG}) \qquad \text{SET6}$$

Remark 1. The B-Book defines rewrite rules for secondary common constructs:

$$P \vee Q \to \neg P \Rightarrow Q \qquad P \Leftrightarrow Q \to (P \Rightarrow Q) \wedge (Q \Rightarrow P) \qquad \exists x \cdot P \to \neg \forall x \cdot \neg P$$
$$s \subseteq t \to s \in \mathbb{P}(t) \qquad s \subset t \to s \subseteq t \wedge s \neq t$$

2.2 Type System

The B Method set theory differs from other ones, like the Zermelo-Fraenkel set theory. The main difference consists in the addition of typing constraints to expressions, and the application of a type-checking procedure before proving. This avoids ill-formed formulæ such as $\exists x \cdot (x \in x)$, whose negation is provable in ZF, due to the foundation axiom, unlike for the B Method.

$$\frac{\Delta \vdash_{tc} \mathsf{ch}(P) \quad \Delta \vdash_{tc} \mathsf{ch}(Q)}{\Delta \vdash_{tc} \mathsf{ch}(P \wedge Q)} \ \mathrm{T1} \qquad \frac{\Delta \vdash_{tc} \mathsf{ch}(P) \quad \Delta \vdash_{tc} \mathsf{ch}(Q)}{\Delta \vdash_{tc} \mathsf{ch}(P \Rightarrow Q)} \ \mathrm{T2}$$

$$\frac{\Delta \vdash_{tc} \mathsf{ch}(P)}{\Delta \vdash_{tc} \mathsf{ch}(\neg P)} \ \mathrm{T3} \qquad \frac{x \backslash s \quad x \backslash \Delta \quad \Delta, x \in s \vdash_{tc} \mathsf{ch}(P)}{\Delta \vdash_{tc} \mathsf{ch}(\forall x \cdot (x \in s \Rightarrow P))} \ \mathrm{T4}$$

$$\frac{\Delta \vdash_{tc} \mathsf{ch}(\forall x \cdot (x \in s \Rightarrow \forall y \cdot (y \in t \Rightarrow P)))}{\Delta \vdash_{tc} \mathsf{ch}(\forall (x,y) \cdot (x,y \in s \times t \Rightarrow P))} \ \mathrm{T5} \qquad \frac{\Delta \vdash_{tc} \mathsf{ch}(\forall x \cdot (P \Rightarrow (Q \wedge R)))}{\Delta \vdash_{tc} \mathsf{ch}(\forall x \cdot ((P \wedge Q) \Rightarrow R))} \ \mathrm{T6}$$

$$\frac{\Delta \vdash_{tc} \mathsf{ty}(E) \equiv \mathsf{ty}(F)}{\Delta \vdash_{tc} \mathsf{ch}(E = F)} \ \mathrm{T7} \qquad \frac{\Delta \vdash_{tc} \mathsf{ty}(E) \equiv \mathsf{su}(s)}{\Delta \vdash_{tc} \mathsf{ch}(E \in s)} \ \mathrm{T8}$$

$$\frac{\Delta \vdash_{tc} \mathsf{su}(s) \equiv \mathsf{su}(t)}{\Delta \vdash_{tc} \mathsf{ch}(s \subseteq t)} \ \mathrm{T8'} \qquad \frac{x \in s \in \Delta \quad \Delta \vdash_{tc} \mathsf{su}(s) \equiv U}{\Delta \vdash_{tc} \mathsf{ty}(x) \equiv U} \ \mathrm{T9}$$

$$\frac{\Delta \vdash_{tc} \mathsf{ty}(E) \times \mathsf{ty}(F) \equiv U}{\Delta \vdash_{tc} \mathsf{ty}(E, F) \equiv U} \ \mathrm{T10} \qquad \frac{\Delta \vdash_{tc} \mathsf{su}(s) \equiv U}{\Delta \vdash_{tc} \mathsf{ty}(\mathsf{choice}(s)) \equiv U} \ \mathrm{T11}$$

$$\frac{\Delta \vdash_{tc} \mathbb{P}(\mathsf{su}(s)) \equiv U}{\Delta \vdash_{tc} \mathsf{ty}(s) \equiv U} \ \mathrm{T12} \qquad \frac{x \in s \in \Delta \quad \Delta \vdash_{tc} \mathsf{su}(s) \equiv \mathbb{P}(U)}{\Delta \vdash_{tc} \mathsf{su}(x) \equiv U} \ \mathrm{T13}$$

$$\frac{\Delta \vdash_{tc} \mathsf{su}(s) \times \mathsf{su}(t) \equiv U}{\Delta \vdash_{tc} \mathsf{su}(s \times t) \equiv U} \ \mathrm{T14} \qquad \frac{\Delta \vdash_{tc} \mathbb{P}(\mathsf{su}(s)) \equiv U}{\Delta \vdash_{tc} \mathsf{su}(\mathbb{P}(s)) \equiv U} \ \mathrm{T15}$$

$$\frac{\Delta \vdash_{tc} \mathsf{ch}(\forall x \cdot (x \in s \Rightarrow P)) \quad \Delta \vdash_{tc} \mathsf{su}(s) \equiv U}{\Delta \vdash_{tc} \mathsf{su}(\{x \mid x \in s \wedge P\}) \equiv U} \ \mathrm{T16} \qquad \frac{\mathsf{gi}(I) \in \Delta \quad \Delta \vdash_{tc} I \equiv U}{\Delta \vdash_{tc} \mathsf{su}(I) \equiv U} \ \mathrm{T17}$$

$$\frac{\Delta \vdash_{tc} \mathsf{su}(s) \equiv \mathbb{P}(U)}{\Delta \vdash_{tc} \mathsf{su}(\mathsf{choice}(s)) \equiv U} \ \mathrm{T18} \qquad \frac{\Delta \vdash_{tc} T \equiv U}{\Delta \vdash_{tc} \mathbb{P}(T) \equiv \mathbb{P}(U)} \ \mathrm{T19}$$

$$\frac{\Delta \vdash_{tc} T \equiv U \quad \Delta \vdash_{tc} V \equiv W}{\Delta \vdash_{tc} T \times V \equiv U \times W} \ \mathrm{T20} \qquad \frac{\mathsf{gi}(I) \in \Delta}{\Delta \vdash_{tc} I \equiv I} \ \mathrm{T21}$$

Fig. 2. The type system of the B method

The typing discipline proposed relies on the monotonicity of set inclusion. For instance, if we have an expression E and two sets s and t such that $E \in s$ and $s \subseteq t$, then $E \in t$. Going further with another set u such that $t \subseteq u$, we have then $E \in u$. The idea, as explained in the B-Book, is that, given a formula to be type checked, there exists an upper limit for such set containment. This upper limit is called the super-set of s and the type of E. Then, if u is the super-set of s, we obtain the typing information $E \in u$ and $s \in \mathbb{P}(u)$.

Type checking is performed by applying, in a backward way and following the numerical order, the inference rules presented in Fig. 2. Rules dealing with the right-hand side of a typing equivalence $\equiv$ are named with the same number primed, for T9 to T18. If this decision procedure terminates and does not fail, then the formula is said to be well-typed. This procedure uses two syntactic categories *Type* and *Type_Pred*:

$$Type \quad ::= \quad \mathsf{type}(E) \mid \mathsf{super}(s) \mid Type \times Type \mid \mathbb{P}(Type) \mid identifier$$
$$Type\_Pred \; ::= \; \mathsf{check}(P) \mid Type \equiv Type$$

In the following, we use ty, su and ch as abbreviations for the keywords type, super and check respectively. As a consequence, the type of an expression E may be either an identifier (see the notion of given set below), the powerset of a type or the cartesian product of two types; and for the particular case of sets, the type of a set is necessarily the powerset of a type.

A type-checking sequent like $\Delta \vdash_{tc} \mathsf{ch}(P)$ means that, within the environment Δ, the formula P is well-typed. The environment Δ is made of atomic formulæ of the form $x \in s$, where x is non-free in s. All free variables in P have to be associated with some atomic formula in Δ. The only exception is for variables in P representing some abstract given sets, introduced at a meta-level discourse like: "Given a set s ...". Such a given set s, which will be used to type other sets, is introduced in the environment Δ by the keyword given(s) (gi(s) for short), telling us that s is free in the formula to be type-checked, and has the specific property su$(s) = s$.

Example 1. Given two sets s and t, the formula:

$$\forall (a,b) \cdot (a,b \in \mathbb{P}(s \times t) \times \mathbb{P}(s \times t) \Rightarrow \{x \mid x \in a \wedge x \in b\} \subseteq s \times t)$$

will be used as a running example in this paper. We want to verify that this formula is well-typed, *i.e.* verify that the following sequent is satisfied:

$$\mathsf{gi}(s), \mathsf{gi}(t) \vdash_{tc}$$
$$\mathsf{ch}(\forall (a,b) \cdot (a,b \in \mathbb{P}(s \times t) \times \mathbb{P}(s \times t) \Rightarrow \{x \mid x \in a \wedge x \in b\} \subseteq s \times t))$$

By applying the rules of Fig. 2, we obtain the following typing derivation (due to the large size of the tree, we present only the names of rules, starting from the left with T5):

$$\mathsf{T5\text{-}T4\text{-}T4\text{-}T8'\text{-}T14'\text{-}T16} \begin{cases} \mathsf{T4\text{-}T8\text{-}T9\text{-}T13\text{-}T15\text{-}T19\text{-}T13'\text{-}T15'\text{-}T19\text{-}T14\text{-}T14'\text{-}T20} \begin{cases} \mathsf{T17\text{-}T17'\text{-}T21} \\ \mathsf{T17\text{-}T17'\text{-}T21} \end{cases} \\ \mathsf{T13\text{-}T15\text{-}T19\text{-}T14\text{-}T20} \begin{cases} \mathsf{T17\text{-}T17'\text{-}T21} \\ \mathsf{T17\text{-}T17'\text{-}T21} \end{cases} \end{cases}$$

2.3 Type Annotation

In the B syntax presented in Sect. 2.1, there are two constructs which introduce new bound variables: universal quantification $\forall x \cdot P$ and comprehension set $\{x \mid P\}$. It should be noted that the typing rules T4 and T16 dealing with these two syntactical constructs use the specific forms $\forall x \cdot x \in s \Rightarrow P$ and

$\{x \mid x \in s \wedge P\}$. Membership $x \in s$ is used to type the bound variable x. Unfortunately, typing information is hidden at a set theoretic level. There is no clear distinction between sets and types in the B Method.

For the translation function presented in Sect. 4.2, we want to distinguish the notion of types from the one of sets. We introduce a new syntactic category T for types:

$$T ::= identifier \mid T_1 \times T_2 \mid \mathbb{P}(T)$$

And we introduce the notation x^T meaning that the variable x has type T.

We now present a procedure to annotate variables with their type. Once the type-checking of a formula is done, the typing tree has environments Δ at each node, and in particular at leaves, following the syntax:

$$\Delta ::= \varnothing \mid \Delta, \mathsf{gi}(s) \mid \Delta, x \in s$$

In addition, Δ is augmented only by rule T4: if a formula $x \in s$ is added, then s has to be already associated in Δ (in particular because of rules T9 and T13), as a given set or in a formula like $s \in t$ for some already associated set t.

The annotation procedure transforms all the leaf environments Δ, *i.e.* the environments of the leaves, into annotated environments $\Delta^\star$, where all variables and given sets are annotated with their type, then uses these annotated environments to rebuild the typing tree of the (annotated) initial formula in a forward way. It should be noted that in a formula $x \in s$, the set s may be a composition of the two type constructors $\times$ and $\mathbb{P}$. We denote this kind of composition by a function symbol f with an arity n. Here is the syntax for $\Delta^\star$:

$$\Delta^\star ::= \varnothing \mid \Delta^\star, \mathsf{gi}(s^{\mathbb{P}(s)}) \mid \Delta^\star, x^{f(T_1,\dots,T_n)} \in f(s_1^{\mathbb{P}(T_1)}, \dots, s_n^{\mathbb{P}(T_n)})$$

We can now introduce the annotation procedure:

1. For all the leaf environments Δ:
   1.1. For all $\mathsf{gi}(s)$, we annotate s by its type $\mathbb{P}(s)$, and then substitute all occurrences of s in Δ by $s^{\mathbb{P}(s)}$;
   1.2. Following the introduction order in Δ, for all $x \in f(s_1^{\mathbb{P}(T_1)}, \dots, s_n^{\mathbb{P}(T_n)})$, we annotate x with its type $f(T_1, \dots, T_n)$, and we substitute all occurrences of x in Δ by $x^{f(T_1,\dots,T_n)}$;
2. Rebuild the (annotated) initial formula by applying the type-checking tree in a forward way, *i.e.* from the leaves to the root.

In the following, we denote by $P^\star$ the formula P where all variables are annotated. We extend this notation to sets of formulæ Γ, and expressions E.

Proposition 1. *The annotation is sound.*
We have, for a variable x, an expression E and a formula P:

1. *If x^T is associated in $\Delta^\star$, $\Delta^\star \vdash_{\mathsf{tc}} \mathsf{ty}(x^T) \equiv T$;*
2. *If $\Delta \vdash_{\mathsf{tc}} \mathsf{ty}(E) \equiv U$, then $\Delta^\star \vdash_{\mathsf{tc}} \mathsf{ty}(E^\star) \equiv U$;*
3. *If $\Delta \vdash_{\mathsf{tc}} \mathsf{ch}(P)$, then $\Delta^\star \vdash_{\mathsf{tc}} \mathsf{ch}(P^\star)$.*

The B proof system of Fig. 1 is neutral with respect to variable annotation, so it is always possible to apply the same proof derivation to an annotated formula. The provability of well-typed formulæ is then preserved: $\Gamma \vdash_B P$ if and only if $\Gamma^* \vdash_B P^*$.

Finally, we take the universal closure of all free variables corresponding to given sets. To lighten the presentation in examples, we annotate only the first occurrence of a variable.

Example 2. Going back to the running example, we obtained the following environment Δ for the leave of the upper branch:

$$\mathbf{gi}(s), \mathbf{gi}(t), a \in \mathbb{P}(s \times t), b \in \mathbb{P}(s \times t), x \in a$$

It leads to the annotated environment Δ^*:

$$\mathbf{gi}(s^{\mathbb{P}(s)}), \mathbf{gi}(t^{\mathbb{P}(t)}), a^{\mathbb{P}(s \times t)} \in \mathbb{P}(s \times t), b^{\mathbb{P}(s \times t)} \in \mathbb{P}(s \times t), x^{s \times t} \in a$$

Finally, we obtain the annotated formula:

$$\forall s^{\mathbb{P}(s)} \cdot (\forall t^{\mathbb{P}(t)} \cdot (\forall (a^{\mathbb{P}(s \times t)}, b^{\mathbb{P}(s \times t)}).$$
$$(a, b \in \mathbb{P}(s \times t) \times \mathbb{P}(s \times t) \Rightarrow \{x^{s \times t} \mid x \in a \wedge x \in b\} \subseteq s \times t)))$$

2.4 The Annotated Set Theory

Axioms SET5 and SET6 are introduced in the B Method set theory for theoretical reasons, like building natural numbers, and are never used in practice, in particular in proof obligations. So, we remove them from this work.

We now define the annotated version of the axioms presented in Sect. 2.1. In addition, we take the universal closure for all free variables.

$$\forall s^{\mathbb{P}(s)} \cdot (\forall t^{\mathbb{P}(t)} \cdot (\forall x^s \cdot (\forall y^t \cdot (x, y \in s \times t \Leftrightarrow (x \in s \wedge y \in t))))) \qquad \text{SET1}$$
$$\forall s^{\mathbb{P}(s)} \cdot (\forall t^{\mathbb{P}(s)} \cdot (s \in \mathbb{P}(t) \Leftrightarrow \forall x^s \cdot (x \in s \Rightarrow x \in t))) \qquad \text{SET2}$$
$$\forall s^{\mathbb{P}(s)} \cdot (\forall y^s \quad \cdot (y \in \{x^s \mid x \in s \wedge P\} \Leftrightarrow (y \in s \wedge [x := y]P))) \qquad \text{SET3}$$
$$\forall s^{\mathbb{P}(s)} \cdot (\forall t^{\mathbb{P}(s)} \cdot (\forall x^s \cdot (x \in s \Leftrightarrow x \in t) \Rightarrow s = t)) \qquad \text{SET4}$$

2.5 Skolemization of Comprehension Sets

We propose an elimination procedure of comprehension sets inside formulæ, based on the definition of new function symbols. The idea to skolemize comprehension sets is not new, see for instance [12]. In an expression, when meeting a set u of the shape: $u = \{x^T \mid P(x, s_1^{T_1}, \ldots, s_n^{T_n})\}$ we apply the following procedure:

1. Define a fresh function symbol $f^{\mathbb{P}(T)}$ of arity n and annotated by $\mathbb{P}(T)$;
2. Add to the B set theory, the axiom:

$$\forall s_1^{T_1} \cdot (\ldots \cdot (\forall s_n^{T_n} \cdot (\forall x^T \cdot (x \in f^{\mathbb{P}(T)}(s_1, \ldots, s_n) \Leftrightarrow P(x, s_1, \ldots, s_n)))))$$

3. Replace all the occurrences of u by $f^{\mathbb{P}(T)}(s_1, \ldots, s_n)$.

Remark 2. This skolemization procedure is sound (the new axiom is an instance of axiom SET3), but not complete (it is no more possible to define a set by comprehension during proof search).

Example 3. Applying skolemization to the running example leads to add the following axiom to the theory:

$$\forall a^{\mathbb{P}(s \times t)} \cdot (\forall b^{\mathbb{P}(s \times t)} \cdot (\forall x^{s \times t} \cdot (x \in f^{\mathbb{P}(s \times t)}(a, b) \Leftrightarrow x \in a \wedge x \in b)))$$

And we obtain the skolemized formula:

$$\forall s^{\mathbb{P}(s)} \cdot (\forall t^{\mathbb{P}(t)}.$$
$$(\forall (a^{\mathbb{P}(s \times t)}, b^{\mathbb{P}(s \times t)}) \cdot (a, b \in \mathbb{P}(s \times t) \times \mathbb{P}(s \times t) \Rightarrow f^{\mathbb{P}(s \times t)}(a, b) \subseteq s \times t)))$$

2.6 Updated Syntax and Proof System

To conclude this section, we present the new version of the B syntax, with annotated variables, function symbols and without comprehension sets, choice function and BIG. In addition, we suppose that expressions are normalized in the sense that substitutions are reduced, as it is for proof obligations, so we remove substitutions from the syntax. We also merge the two categories for expressions and sets in a single category called E. Finally, we introduce $\perp := P \wedge \neg P$ and $\top := \neg \perp$, where P is a fixed formula.

$$T ::= identifier \mid T_1 \times T_2 \mid \mathbb{P}(T)$$
$$P ::= \perp \mid \top \mid P_1 \wedge P_2 \mid P_1 \Rightarrow P_2 \mid \neg P \mid \forall x \cdot P \mid E_1 = E_2 \mid E_1 \in E_2$$
$$E ::= x \mid E_1, E_2 \mid E_1 \times E_2 \mid \mathbb{P}(E) \mid f^{\mathbb{P}(T)}(E_1, \ldots, E_n)$$
$$x ::= identifier \mid x^T \mid x_1^{T_1}, x_2^{T_2}$$

Finally, we enrich the B proof system of Fig. 1 with the two basic rules BR5 and BR6 dealing with $\perp$ and $\top$:

$$\frac{}{\Gamma, \perp \vdash_B Q} \text{BR5} := \frac{\dfrac{\dfrac{}{\Gamma, P \wedge \neg P, \neg Q \vdash_B P \wedge \neg P} \text{BR3}}{\Gamma, P \wedge \neg P, \neg Q \vdash_B P} \text{R2} \quad \dfrac{\dfrac{}{\Gamma, P \wedge \neg P, \neg Q \vdash_B P \wedge \neg P} \text{BR3}}{\Gamma, P \wedge \neg P, \neg Q \vdash_B \neg P} \text{R2'}}{\Gamma, P \wedge \neg P \vdash_B Q} \text{R5}$$

$$\frac{}{\Gamma \vdash_B \top} \text{BR6} := \frac{\dfrac{}{\Gamma, \perp \vdash_B Q} \text{BR5} \quad \dfrac{}{\Gamma, \perp \vdash_B \neg Q} \text{BR5}}{\Gamma \vdash_B \neg \perp} \text{R6}$$

3 LLproof: Typed Sequent Calculus of Zenon

3.1 Polymorphic First-Order Logic

We present in this section the polymorphic first-order logic, PFOL for short, used by the sequent calculus proof system LLproof. This presentation is highly inspired by [2].

A polymorphic signature is a triple $\Sigma = (\mathcal{K}, \mathcal{F}, \mathcal{P})$, where $\mathcal{K}$, $\mathcal{F}$ and $\mathcal{P}$ are countable sets of respectively type constructors k with their arity m, denoted $k :: m$, function symbols f and predicate symbols P with their type signature σ.

$$\sigma ::= f : \Pi \alpha_1 \ldots \alpha_m . \tau_1 \to \ldots \to \tau_n \to \tau \mid P : \Pi \alpha_1 \ldots \alpha_m . \tau_1 \to \ldots \to \tau_n \to o$$

where $\alpha_1 \ldots \alpha_m$ are the m first arguments of f or P and correspond to the type parameters; $\tau_1, \ldots, \tau_n$ are the following n arguments of f or P and correspond to the types of the term parameters; τ is the return type of f and o is the return pseudo-type of predicates P (but it is not a type of the language).

The syntax of PFOL is made of types, terms, formulæ and polymorphic formulæ. A type τ is either a type variable α or the application of a type constructor k. A term e is either a variable x or the application of a function symbol f to types and terms. A formula φ is inductively built from $\bot$, $\top$, conjunction, implication, negation, universal quantification over (term) variable, equality between terms and application of a predicate symbol. A polymorphic formula φ_α is a universal quantification over type variable. The typing rules of PFOL are standard

---

Closure and Quantifier-free Rules

$$\frac{}{\Gamma, \bot \vdash_{\mathsf{LL}} \bot}\ \bot \qquad \frac{}{\Gamma, \neg\top \vdash_{\mathsf{LL}} \bot}\ \neg\top \qquad \frac{}{\Gamma, t =_\tau u, u \neq_\tau t \vdash_{\mathsf{LL}} \bot}\ \mathrm{Sym}$$

$$\frac{}{\Gamma, P, \neg P \vdash_{\mathsf{LL}} \bot}\ \mathrm{Ax} \qquad \frac{}{\Gamma, t \neq_\tau t \vdash_{\mathsf{LL}} \bot}\ \neq \qquad \frac{\Gamma, P \vdash_{\mathsf{LL}} \bot \quad \Gamma, \neg P \vdash_{\mathsf{LL}} \bot}{\Gamma \vdash_{\mathsf{LL}} \bot}\ \mathrm{Cut}$$

$$\frac{\Gamma, \neg\neg P, P \vdash_{\mathsf{LL}} \bot}{\Gamma, \neg\neg P \vdash_{\mathsf{LL}} \bot}\ \neg\neg \qquad \frac{\Gamma, P \wedge Q, P, Q \vdash_{\mathsf{LL}} \bot}{\Gamma, P \wedge Q \vdash_{\mathsf{LL}} \bot}\ \wedge \qquad \frac{\Gamma, \neg(P \Rightarrow Q), P, \neg Q \vdash_{\mathsf{LL}} \bot}{\Gamma, \neg(P \Rightarrow Q) \vdash_{\mathsf{LL}} \bot}\ \neg\Rightarrow$$

$$\frac{\Gamma, P \Rightarrow Q, \neg P \vdash_{\mathsf{LL}} \bot \quad \Gamma, P \Rightarrow Q, Q \vdash_{\mathsf{LL}} \bot}{\Gamma, P \Rightarrow Q \vdash_{\mathsf{LL}} \bot}\ \Rightarrow$$

$$\frac{\Gamma, \neg(P \wedge Q), \neg P \vdash_{\mathsf{LL}} \bot \quad \Gamma, \neg(P \wedge Q), \neg Q \vdash_{\mathsf{LL}} \bot}{\Gamma, \neg(P \wedge Q) \vdash_{\mathsf{LL}} \bot}\ \neg\wedge$$

Quantifier Rules Over Variables

$$\frac{\Gamma, \neg\forall x : \tau.\ P(x), \neg P(c) \vdash_{\mathsf{LL}} \bot}{\Gamma, \neg\forall x : \tau.\ P(x) \vdash_{\mathsf{LL}} \bot}\ \neg\forall \qquad \text{where } c : \tau \text{ is a fresh constant}$$

$$\frac{\Gamma, \forall x : \tau.\ P(x), P(t) \vdash_{\mathsf{LL}} \bot}{\Gamma, \forall x : \tau.\ P(x) \vdash_{\mathsf{LL}} \bot}\ \forall \qquad \text{where } t : \tau \text{ is any closed term}$$

Quantifier Rules Over Type Variables

$$\frac{\Gamma, \forall\alpha.\ P(\alpha), P(\tau) \vdash_{\mathsf{LL}} \bot}{\Gamma, \forall\alpha.\ P(\alpha) \vdash_{\mathsf{LL}} \bot}\ \forall_{\text{type}} \qquad \text{where } \tau \text{ is any closed type}$$

Special Rule

$$\frac{\Gamma, P(t), t \neq_\tau u \vdash_{\mathsf{LL}} \bot \quad \Gamma, P(t), P(u) \vdash_{\mathsf{LL}} \bot}{\Gamma, P(t) \vdash_{\mathsf{LL}} \bot}\ \mathrm{Subst}$$

Fig. 3. The typed sequent calculus LLproof

and can be found in [2]. In the following, we may omit the m first type arguments for function and predicate symbols when they are clear from the context.

$$\tau \;::=\; \alpha \mid k(\tau_1, \ldots, \tau_m)$$
$$e \;::=\; x \mid f(\tau_1, \ldots, \tau_m; e_1, \ldots, e_n)$$
$$\varphi \;::=\; \bot \mid \top \mid \varphi_1 \wedge \varphi_2 \mid \varphi_1 \Rightarrow \varphi_2 \mid \neg\varphi \mid \forall x : \tau.\varphi \mid e_1 =_\tau e_2$$
$$\mid\; P(\tau_1, \ldots, \tau_m; e_1, \ldots, e_n)$$
$$\varphi_\alpha \;::=\; \forall \alpha.\varphi_\alpha \mid \forall \alpha.\varphi$$

3.2   The Typed Sequent Calculus Proof System **LLproof**

In Fig. 3, we present the typed sequent calculus **LLproof** used by the automated theorem prover **Zenon** to output proofs. This sequent calculus is close to a tableau method proof system; we are looking for a contradiction, given the negation of the goal as an hypothesis. All formulæ are on the left hand side of the sequent, and the negation of the goal has to be unsatisfiable. In addition, the contraction rule is always applied, leading to a growing context Γ.

This presentation differs with the one in [8], which also introduces the proof system **LLproof** and its embedding into the proof-checker **Dedukti**. We remove the rules for equivalence and existential quantification, because these constructs are defined using other ones in the B Method (see Sect. 2.1). Moreover, we replace all rules from the category Special Rules by the new one Subst, since the Subst rule is easier to translate and can be used to define other Special rules [8].

The rules $\forall$ and $\neg\forall$ dealing with quantification over variables both get a side condition about the type of the chosen instance.

Rule $\forall_{\mathsf{type}}$ is applied to instantiate the type variables in axioms with the closed types coming from the translation of the proof obligation to be proved.

4   Translation of **B** Formulæ into PFOL

4.1   Type Signatures of Primitive Constructs

We start by defining a general skeleton for the type signatures of the B basic constructs. We introduce two type constructors **Set** and **Pair** corresponding respectively to the B type constructors $\mathbb{P}$ and $\times$. Then, we can define the function symbols $(\text{-}, \text{-})$ for ordered pair, $\mathbb{P}(\text{-})$ for powerset and $\text{-} \times \text{-}$ for product set. Finally, we define two predicate symbols for membership and equality. For easier reading, we use an infix notation with type arguments subscripted. For instance, $\text{-} \in_\alpha \text{-}$ corresponds to $\in (\alpha, \text{-}, \text{-})$.

$$\mathcal{T}_{\mathsf{ske}} := \begin{cases} \mathsf{Set}(\text{-}) :: 1, \; \mathsf{Pair}(\text{-},\text{-}) :: 2 \\ (\text{-},\text{-})_{\alpha_1,\alpha_2} & : \Pi\alpha_1\alpha_2.\; \alpha_1 \to \alpha_2 \to \mathsf{Pair}(\alpha_1, \alpha_2) \\ \mathbb{P}_\alpha(\text{-}) & : \Pi\alpha.\; \mathsf{Set}(\alpha) \to \mathsf{Set}(\mathsf{Set}(\alpha)) \\ \text{-} \times_{\alpha_1,\alpha_2} \text{-} : \Pi\alpha_1\alpha_2.\; \mathsf{Set}(\alpha_1) \to \mathsf{Set}(\alpha_2) \to \mathsf{Set}(\mathsf{Pair}(\alpha_1, \alpha_2)) \\ \text{-} \in_\alpha \text{-} & : \Pi\alpha.\; \alpha \to \mathsf{Set}(\alpha) \to o \\ \text{-} =_\alpha \text{-} & : \Pi\alpha.\; \alpha \to \alpha \to o \end{cases}$$

4.2   Translating Formulæ from B to PFOL

We present in Fig. 4 the translation function of B formulæ into PFOL formulæ. This translation, denoted $\langle P \rangle$ for some B formula P, is made of the three translations $\langle T \rangle_t$ for types, $\langle P \rangle_f$ for formulæ and $\langle E \rangle_e$ for expressions, and a function $\theta(E)$ that returns the PFOL type of a B expression E.

One important point in this embedding is the interpretation given to B type identifiers coming from the type annotation procedure (see Sect. 2.3). We interpret B type identifiers coming from axioms and hypotheses as type variables (and take the universal closure with respect to them), and B type identifiers of the formula to prove (also called goal) as new constants, *i.e.* nullary type constructors. This allows us to get polymorphic axioms in PFOL and a monomorphic/many-sorted goal. To achieve this, we add to all B formulæ to translate a flag ax for axioms and hypotheses and gl for the goal.

Before presenting the three translation functions, we have to define a function called $Sig(f(\ldots))$, where f is a B function symbol coming from the skolemization of comprehension sets (see Sect. 2.5), that returns the type signature of f. Let $FV(e)$ be the set of free variables of an expression e.

$$Sig(f^{\mathbb{P}(T)}(E_1, \ldots, E_n)) = \underset{\alpha \in FV_1^n(\theta(E_i))}{\Pi} \alpha. \; \theta(E_1) \to \ldots \to \theta(E_n) \to \theta(\mathbb{P}(T))$$

During the translation procedure, we carry a target PFOL theory $\mathcal{T}$ composed by the skeleton $\mathcal{T}_{ske}$ defined in Sect. 4.1, previously translated formulæ, new type constructors and new type signatures. Also, for each formula to be translated, we carry a PFOL local context Δ of bound variables and their type, and a set Ω of pairs of B type identifiers and their corresponding PFOL types, *i.e.* type variables for axioms and type constants for goals.

Example 4. Continuing with the running example, we first translate axioms SET1, SET2 and SET4, then the axiom coming from the skolemization, and finally the goal. To lighten the presentation, we omit the subscripted type arguments of function and predicate symbols of $\mathcal{T}_{ske}$ and we factorize the symbol $\forall$. The three set theory axioms become:

$\forall \alpha_1, \alpha_2. \; \forall s : \mathsf{Set}(\alpha_1), t : \mathsf{Set}(\alpha_2), x : \alpha_1, y : \alpha_2. \; (x, y) \in s \times t \Leftrightarrow (x \in s \wedge y \in t)$
$\forall \alpha. \; \forall s : \mathsf{Set}(\alpha), t : \mathsf{Set}(\alpha). \; s \in \mathbb{P}(t) \Leftrightarrow \forall x : \alpha. \; x \in s \Rightarrow x \in t$
$\forall \alpha. \; \forall s : \mathsf{Set}(\alpha), t : \mathsf{Set}(\alpha). \; (\forall x : \alpha. \; x \in s \Leftrightarrow x \in t) \Rightarrow s = t$

The remainder of the theory, *i.e.* the signature of f, the axiom defining f and the declaration of the two type constants coming from the translation of the goal, is:

$$\begin{cases} k_1 :: 0, \; k_2 :: 0 \\ f : \Pi \alpha_1 \alpha_2. \; \mathsf{Set}(\mathsf{Pair}(\alpha_1, \alpha_2)) \to \mathsf{Set}(\mathsf{Pair}(\alpha_1, \alpha_2)) \to \mathsf{Set}(\mathsf{Pair}(\alpha_1, \alpha_2)) \\ \forall \alpha_1, \alpha_2. \; \forall a : \mathsf{Set}(\mathsf{Pair}(\alpha_1, \alpha_2)), b : \mathsf{Set}(\mathsf{Pair}(\alpha_1, \alpha_2)), x : \mathsf{Pair}(\alpha_1, \alpha_2). \\ \quad x \in f(a, b) \Leftrightarrow (x \in a \wedge x \in b) \end{cases}$$

$$\theta(E) = \text{match } E \text{ with}$$
$$\begin{array}{ll}
| \ x^T & \to \Delta(x) \\
| \ E_1, E_2 & \to \text{Pair}(\theta(E_1), \theta(E_2)) \\
| \ E_1 \times E_2 & \to \text{Set}(\text{Pair}(\theta(E_1), \theta(E_2))) \\
| \ \mathbb{P}(E) & \to \text{Set}(\theta(E)) \\
| \ f^{\mathbb{P}(T)}(\ldots) & \to \text{Set}(\langle T \rangle_t)
\end{array}$$

$$\langle T \rangle_t = \text{match } T \text{ with}$$
$$\begin{array}{ll}
| \ id \text{ when } flag = ax \to & \begin{cases} \text{if } id \in \Omega \text{ then return } \Omega(id) \\ \text{else } \Omega := \Omega, (id, \alpha_{id}) \text{ return } \alpha_{id} \end{cases} \\
| \ id \text{ when } flag = gl \to & \begin{cases} \text{if } id \in \Omega \text{ then return } \Omega(id) \\ \text{else } \mathcal{T} := \mathcal{T}, k_{id} :: 0 \ ; \ \Omega := \Omega, (id, k_{id}) \text{ return } k_{id} \end{cases} \\
| \ T_1 \times T_2 & \to \text{Pair}(\langle T_1 \rangle_t, \langle T_2 \rangle_t) \\
| \ \mathbb{P}(T) & \to \text{Set}(\langle T \rangle_t)
\end{array}$$

$$\langle P \rangle_f = \text{match } P \text{ with}$$
$$\begin{array}{ll}
| \ \bot \ | \ \top & \to \bot \ | \ \top \\
| \ P_1 \wedge P_2 & \to \langle P_1 \rangle_f \wedge \langle P_2 \rangle_f \\
| \ P_1 \Rightarrow P_2 & \to \langle P_1 \rangle_f \Rightarrow \langle P_2 \rangle_f \\
| \ \neg P & \to \neg \langle P \rangle_f \\
| \ \forall x^T \cdot P & \to \forall x : \langle T \rangle_t \cdot \langle P \rangle_f \text{ and } \Delta := \Delta, x : \langle T \rangle_t \\
| \ \forall (x_1^{T_1}, x_2^{T_2}) \cdot P & \to \begin{cases} \forall x_1 : \langle T_1 \rangle_t \cdot \forall x_2 : \langle T_2 \rangle_t \cdot \langle P \rangle_f \\ \text{and } \Delta := \Delta, x_1 : \langle T_1 \rangle_t, x_2 : \langle T_2 \rangle_t \end{cases} \\
| \ E_1 = E_2 & \to \langle E_1 \rangle_e =_{\theta(E_1)} \langle E_2 \rangle_e \\
| \ E_1 \in E_2 & \to \langle E_1 \rangle_e \in_{\theta(E_1)} \langle E_2 \rangle_e
\end{array}$$

$$\langle E \rangle_e = \text{match } E \text{ with}$$
$$\begin{array}{ll}
| \ x^T & \to x \\
| \ E_1, E_2 & \to (\langle E_1 \rangle_e, \langle E_2 \rangle_e)_{\theta(E_1), \theta(E_2)} \\
| \ E_1 \times E_2 & \to \langle E_1 \rangle_e \times_{\tau_1, \tau_2} \langle E_2 \rangle_e \text{ where } \begin{cases} \theta(E_1) = \text{Set}(\tau_1) \\ \theta(E_2) = \text{Set}(\tau_2) \end{cases} \\
| \ \mathbb{P}(E) & \to \mathbb{P}_\tau(\langle E \rangle_e) \text{ where } \theta(E) = \text{Set}(\tau) \\
| \ f^{\mathbb{P}(T)}(E_1, \ldots, E_n) \to \\
\quad \text{if } f : \Pi\alpha_1 \ldots \alpha_m. \ \tau_1 \to \ldots \to \tau_n \to \tau \notin \mathcal{T} \\
\quad \text{then } \mathcal{T} := \mathcal{T}, f : Sig(f^{\mathbb{P}(T)}(E_1, \ldots, E_n)) \\
\quad \text{return } f(\tau_1', \ldots, \tau_m'; \langle E_1 \rangle_e, \ldots, \langle E_n \rangle_e) \text{ where } \begin{cases} \theta(E_1) = \tau_1(\tau_1', \ldots, \tau_m') \\ \ldots \\ \theta(E_n) = \tau_n(\tau_1', \ldots, \tau_m') \end{cases}
\end{array}$$

Fig. 4. Translation from B to PFOL

Finally, the translation of the goal (we unfold the $\subseteq$ definition, see Sect. 2.1) is:

$$\forall s : \text{Set}(k_1), t : \text{Set}(k_2), a : \text{Set}(\text{Pair}(k_1, k_2)), b : \text{Set}(\text{Pair}(k_1, k_2)).$$
$$(a, b) \in \mathbb{P}(s \times t) \times \mathbb{P}(s \times t) \Rightarrow f(a, b) \in \mathbb{P}(s \times t)$$

5   Translating **LLproof** Proofs into **B** Proofs

In Fig. 5, we present the reverse translation, denoted $\langle \varphi \rangle^{-1}$, to translate monomorphic PFOL formulæ into B formulæ. This reverse translation is simpler than the one presented in Sect. 4.2 because we do not need to translate types, annotations for bound variables and function symbols not being necessary anymore.

$$
\begin{aligned}
\langle \varphi \rangle_f^{-1} &= \text{match } \varphi \text{ with} \\
&\mid \bot \mid \top & &\rightarrow \bot \mid \top \\
&\mid \varphi_1 \wedge \varphi_2 & &\rightarrow \langle \varphi_1 \rangle_f^{-1} \wedge \langle \varphi_2 \rangle_f^{-1} \\
&\mid \varphi_1 \Rightarrow \varphi_2 & &\rightarrow \langle \varphi_1 \rangle_f^{-1} \Rightarrow \langle \varphi_2 \rangle_f^{-1} \\
&\mid \neg \varphi & &\rightarrow \neg \langle \varphi \rangle_f^{-1} \\
&\mid \forall x : \tau.\ \varphi & &\rightarrow \forall x \cdot \langle \varphi \rangle_f^{-1} \\
&\mid e_1 =_\tau e_2 & &\rightarrow \langle e_1 \rangle_e^{-1} = \langle e_2 \rangle_e^{-1} \\
&\mid e_1 \in_\tau e_2 & &\rightarrow \langle e_1 \rangle_e^{-1} \in \langle e_2 \rangle_e^{-1} \\
\\
\langle e \rangle_e^{-1} &= \text{match } E \text{ with} \\
&\mid x & &\rightarrow x \\
&\mid (e_1, e_2)_{\tau_1, \tau_2} & &\rightarrow \langle e_1 \rangle_e^{-1}, \langle e_2 \rangle_e^{-1} \\
&\mid e_1 \times_{\tau_1, \tau_2} e_2 & &\rightarrow \langle e_1 \rangle_e^{-1} \times \langle e_2 \rangle_e^{-1} \\
&\mid \mathbb{P}_\tau(e) & &\rightarrow \mathbb{P}(\langle e \rangle_e^{-1}) \\
&\mid f(\tau_1', \ldots, \tau_m'; e_1, \ldots, e_n) & &\rightarrow f(\langle e_1 \rangle_e^{-1}, \ldots, \langle e_n \rangle_e^{-1})
\end{aligned}
$$

Fig. 5. Translation from PFOL to B

Theorem 1. *For a set of B formulæ Γ and a B goal P, if there exists a LLproof proof of the sequent $\langle \Gamma \rangle, \langle \neg P \rangle \vdash_{LL} \bot$, then there exists a set Γ' of monomorphic instances of $\langle \Gamma \rangle$, and a B proof of the sequent $\langle \Gamma' \rangle^{-1}, \neg P \vdash_B \bot$.*

Proof. We present a sketch of the proof.

1. We show that if P is a B goal, then we have $\langle \langle P \rangle \rangle^{-1} \Leftrightarrow P$.
2. Given a proof Π of the sequent $\langle \Gamma \rangle, \langle \neg P \rangle \vdash_{LL} \bot$, there exists a proof Π_{Kleene} of the sequent, starting with all applications of $\forall_{type}$ rules on polymorphic formulæ, thanks to the permutation of inference rules in sequent calculus [15].
3. We take the subproof Π_{mono} of Π_{Kleene}, where we removed all the $\forall_{type}$ nodes and the remaining polymorphic formulæ.
4. The set Γ' of monomorphic instances of $\langle \Gamma \rangle$ is made of the root node formulæ of Π_{mono}, except $\langle \neg P \rangle$.
5. We extend the reverse translation to LLproof sequents, $\langle P_1, \ldots, P_n \vdash_{LL} Q \rangle^{-1} \rightarrow \langle P_1 \rangle^{-1}, \ldots, \langle P_n \rangle^{-1} \vdash_B \langle Q \rangle^{-1}$, and to LLproof proof nodes in Figs. 6 and 7.
6. $\langle \Pi_{mono} \rangle^{-1}$ is a B proof of the sequent $\langle \Gamma' \rangle^{-1}, \neg P \vdash_B \bot$.

Axiom

$$\dfrac{\overline{\langle P \vdash P \rangle^{-1}} \; \text{BR3} \quad \overline{\langle \neg P \vdash \neg P \rangle^{-1}} \; \text{BR3}}{\langle P, \neg P \vdash \bot \rangle^{-1}} \; \text{R5}$$

$\neq$

$$\dfrac{\overline{\langle \vdash t =_\tau t \rangle^{-1}} \; \text{R10} \quad \overline{\langle \neg(t =_\tau t) \vdash \neg(t =_\tau t) \rangle^{-1}} \; \text{BR3}}{\langle \neg(t =_\tau t) \vdash \bot \rangle^{-1}} \; \text{R5}$$

Sym

$$\dfrac{\overline{\langle t =_\tau u \vdash t =_\tau u \rangle^{-1}} \; \text{BR3} \quad \dfrac{\overline{\langle \vdash t =_\tau t \rangle^{-1}} \; \text{R10} \quad \overline{\langle \neg(t =_\tau t) \vdash \neg(t =_\tau t) \rangle^{-1}} \; \text{BR3}}{\langle \neg(t =_\tau t) \vdash \bot \rangle^{-1}} \; \text{R5}}{\langle t =_\tau u, \neg(u =_\tau t) \vdash \bot \rangle^{-1}} \; \text{R9}$$

$\neg\neg$

$$\dfrac{\dfrac{\overline{\langle \neg P \vdash \neg P \rangle^{-1}} \; \text{BR3} \quad \overline{\langle \neg\neg P \vdash \neg\neg P \rangle^{-1}} \; \text{BR3}}{\langle \neg\neg P \vdash P \rangle^{-1}} \; \text{R5} \quad \overline{\langle \neg\neg P, P \vdash \bot \rangle^{-1}} \; \text{BR4}}{\langle \neg\neg P \vdash \bot \rangle^{-1}}$$

$\wedge$

$$\dfrac{\dfrac{\overline{\langle P \wedge Q \vdash P \wedge Q \rangle^{-1}} \; \text{BR3}}{\langle P \wedge Q \vdash P \rangle^{-1}} \; \text{R2} \quad \dfrac{\dfrac{\overline{\langle P \wedge Q \vdash P \wedge Q \rangle^{-1}} \; \text{BR3}}{\langle P \wedge Q \vdash Q \rangle^{-1}} \; \text{R2'} \quad \overline{\langle P \wedge Q, P, Q \vdash \bot \rangle^{-1}} \; \text{BR4}}{\langle P \wedge Q, P \vdash \bot \rangle^{-1}} \; \text{BR4}}{\langle P \wedge Q \vdash \bot \rangle^{-1}}$$

$\Rightarrow$

$$\dfrac{\dfrac{\langle P \Rightarrow Q, \neg P \vdash \bot \rangle^{-1} \quad \overline{\langle \vdash \neg \bot \rangle^{-1}} \; \text{BR6}}{\langle P \Rightarrow Q \vdash P \rangle^{-1}} \; \text{R5} \quad \dfrac{\overline{\langle P \Rightarrow Q \vdash P \Rightarrow Q \rangle^{-1}} \; \text{BR3}}{\langle P \Rightarrow Q \vdash Q \rangle^{-1}} \; \text{MP} \quad \overline{\langle P \Rightarrow Q, Q \vdash \bot \rangle^{-1}} \; \text{BR4}}{\langle P \Rightarrow Q \vdash \bot \rangle^{-1}}$$

Fig. 6. Translations of LLproof Rules into B Proof System (part 1)

We give in Figs. 6 and 7 the translations for each LLproof proof node. Each node can be translated to a B derivation where all PFOL sequents are translated into B sequents, leading to a B proof tree. To lighten the presentation, we omit to indicate the context Γ and some useless formulæ (removable by applying BR2) on the left-hand side of sequents, and we use $\vdash$ for $\vdash_{LL}$. For instance, the translation of the LLproof Axiom rule should be:

$$\dfrac{\overline{\langle \Gamma, P, \neg P, \neg\bot \vdash_{LL} P \rangle^{-1}} \; \text{BR3} \quad \overline{\langle \Gamma, P, \neg P, \neg\bot \vdash_{LL} \neg P \rangle^{-1}} \; \text{BR3}}{\langle \Gamma, P, \neg P \vdash_{LL} \bot \rangle^{-1}} \; \text{R5}$$

Example 5. The proof of the running example is too big to be presented here. Instead, we present the proof translation for the following B formula, given s:

$$\forall x \cdot (x \in s \Rightarrow x \in s)$$

$\neg\wedge$

$$\dfrac{\dfrac{\langle\neg(P\wedge Q),\neg P\vdash\bot\rangle^{-1}\quad\overline{\langle\vdash\neg\bot\rangle^{-1}}\,\mathrm{BR6}}{\langle\neg(P\wedge Q)\vdash P\rangle^{-1}}\,\mathrm{R5}\quad\dfrac{\langle\neg(P\wedge Q),\neg Q\vdash\bot\rangle^{-1}\quad\overline{\langle\vdash\neg\bot\rangle^{-1}}\,\mathrm{BR6}}{\langle\neg(P\wedge Q)\vdash Q\rangle^{-1}}\,\mathrm{R5}}{\dfrac{\langle\neg(P\wedge Q)\vdash P\wedge Q\rangle^{-1}}{\langle\neg(P\wedge Q)\vdash\bot\rangle^{-1}}\,\mathrm{R5}}\mathrm{R1}\qquad\varPi$$

where $\varPi := \dfrac{}{\langle\neg(P\wedge Q)\vdash\neg(P\wedge Q)\rangle^{-1}}\,\mathrm{BR3}$

$\neg\Rightarrow$

$$\dfrac{\dfrac{\langle\neg(P\Rightarrow Q),P,\neg Q\vdash\bot\rangle^{-1}\quad\overline{\langle\vdash\neg\bot\rangle^{-1}}\,\mathrm{BR6}}{\langle\neg(P\Rightarrow Q),P\vdash Q\rangle^{-1}}\,\mathrm{R5}}{\langle\neg(P\Rightarrow Q)\vdash P\Rightarrow Q\rangle^{-1}}\,\mathrm{R3}\qquad\dfrac{\dfrac{}{\langle\neg(P\Rightarrow Q)\vdash\neg(P\Rightarrow Q)\rangle^{-1}}\,\mathrm{BR3}}{\langle\neg(P\Rightarrow Q)\vdash\bot\rangle^{-1}}\,\mathrm{R5}$$

$\neg\forall$

$$\dfrac{\dfrac{\langle\neg\forall x:\tau.P(x),\neg P(c)\vdash\bot\rangle^{-1}\quad\overline{\langle\vdash\neg\bot\rangle^{-1}}\,\mathrm{BR6}}{\langle\neg\forall x:\tau.P(x)\vdash P(c)\rangle^{-1}}\,\mathrm{R5}}{\langle\neg\forall x:\tau.P(x)\vdash\forall x:\tau.P(x)\rangle^{-1}}\,\mathrm{R7}\qquad\dfrac{\dfrac{}{\langle\neg\forall x:\tau.P(x)\vdash\neg\forall x:\tau.P(x)\rangle^{-1}}\,\mathrm{BR3}}{\langle\neg\forall x:\tau.P(x)\vdash\bot\rangle^{-1}}\,\mathrm{R5}$$

$\forall$

$$\dfrac{\dfrac{\dfrac{}{\langle\forall x:\tau.\ P(x)\vdash\forall x:\tau.\ P(x)\rangle^{-1}}\,\mathrm{BR3}}{\langle\forall x:\tau.\ P(x)\vdash P(t)\rangle^{-1}}\,\mathrm{R8}\qquad\langle\forall x:\tau.\ P(x),P(t)\vdash\bot\rangle^{-1}}{\langle\forall x:\tau.\ P(x)\vdash\bot\rangle^{-1}}\,\mathrm{BR4}$$

Subst

$$\dfrac{\dfrac{\dfrac{\langle P(t),\neg(t=_\tau u)\vdash\bot\rangle^{-1}\quad\overline{\langle\vdash\neg\bot\rangle^{-1}}\,\mathrm{BR6}}{\langle P(t)\vdash t=_\tau u\rangle^{-1}}\,\mathrm{R5}\qquad\dfrac{}{\langle P(t)\vdash P(t)\rangle^{-1}}\,\mathrm{BR3}}{\langle P(t)\vdash P(u)\rangle^{-1}}\,\mathrm{R9}\qquad\langle P(t),P(u)\vdash\bot\rangle^{-1}}{\langle P(t)\vdash\bot\rangle^{-1}}\,\mathrm{BR4}$$

Fig. 7. Translation of LLproof Rules into B Proof System (part 2)

The latter leads to the PFOL formula, where k is a constant:

$$\forall s:\mathsf{Set}(k).\ \forall x:k.\ x\in s\Rightarrow x\in s$$

The LLproof proof is:

$$\dfrac{\dfrac{\dfrac{\dfrac{c_x\in_k c_s,\ c_x\notin_k c_s\vdash_{\mathsf{LL}}\bot}{\neg(c_x\in_k c_s\Rightarrow c_x\in_k c_s)\vdash_{\mathsf{LL}}\bot}\,\mathrm{Ax}}{\neg\forall x:k.\ x\in_k c_s\Rightarrow x\in_k c_s\vdash_{\mathsf{LL}}\bot}\,{\neg\Rightarrow}}{\neg\forall s:\mathsf{Set}(k).\ \forall x:k.\ x\in_k s\Rightarrow x\in_k s\vdash_{\mathsf{LL}}\bot}\,{\neg\forall}}{}\,{\neg\forall}$$

We obtain the B proof (we removed the universal quantification over the given set s, the first R5 node in the translation of $\neg\forall$, some useless formulæ on the left-hand side of sequents and used $\vdash$ for $\vdash_B$, c for c_x and s for c_s):

$$\cfrac{\cfrac{\cfrac{\overline{c \in s \vdash c \in s}^{\text{BR3}} \quad \overline{c \notin s \vdash c \notin s}^{\text{BR3}}}{c \in s, c \notin s \vdash \bot}^{\text{R5}} \quad \overline{\vdash \neg\bot}^{\text{BR6}}}{\cfrac{c \in s \vdash c \in s}{\vdash c \in s \Rightarrow c \in s}^{\text{R3}} \quad \cfrac{\cfrac{\neg(c \in s \Rightarrow c \in s) \vdash \neg(c \in s \Rightarrow c \in s)}{\neg(c \in s \Rightarrow c \in s) \vdash \bot}^{\text{BR3}} \quad \overline{\vdash \neg\bot}^{\text{BR6}}}{\vdash c \in s \Rightarrow c \in s}^{\text{R5}}}{\vdash \forall x \cdot (x \in s \Rightarrow x \in s)}^{\text{R7}}$$

6   Conclusion

Automated theorem provers are in general made of thousands lines of code, using elaborate decision procedures and specific heuristics. The confidence in such tools may therefore be questioned. The correctness of Zenon proofs is already guaranteed by the checking of proof certificates by an external proof checker. But to prove B proof obligations, Zenon relies on two external tools, bpo2why and Why3, to translate proof obligations into its input format, which raises the question whether the proof found still corresponds to a proof of the original statement.

In this paper, we have formalized a different and direct translation from the B Method to a polymorphic first-order logic. The main purpose of this work is not to replace bpo2why, but to validate the use of Zenon to prove B proof obligations. One of the most challenging part of this translation deals with the encoding of the B notion of types. Our solution to make the axioms polymorphic allows us to benefit from the flexibility of polymorphism. Furthermore, we showed that this translation is sound and gave a procedure to translate Zenon proofs in the B proof system.

As future work, we want to prove the soundness and completeness of the deduction modulo theory [11] extension of the proof system LLproof with regard to those of LLproof, in particular in the case of the B Method.

References

1. Abrial, J.R.: The B-Book: Assigning Programs to Meanings. Cambridge University Press, Cambridge (1996)
2. Blanchette, J.C., Böhme, S., Popescu, A., Smallbone, N.: Encoding monomorphic and polymorphic types. In: Piterman, N., Smolka, S.A. (eds.) TACAS 2013. LNCS, vol. 7795, pp. 493–507. Springer, Heidelberg (2013). doi:10.1007/978-3-642-36742-7_34
3. Bobot, F., Filliâtre, J.C., Marché, C., Paskevich, A.: Why3: shepherd your herd of provers. In: International Workshop on Intermediate Verification Languages (Boogie) (2011)

4. Bodeveix, J.-P., Filali, M.: Type synthesis in B and the translation of B to PVS. In: Bert, D., Bowen, J.P., Henson, M.C., Robinson, K. (eds.) ZB 2002. LNCS, vol. 2272, pp. 350–369. Springer, Heidelberg (2002). doi:10.1007/3-540-45648-1_18

5. Boespflug, M., Carbonneaux, Q., Hermant, O.: The λΠ-calculus modulo as a universal proof language. In: Proof Exchange for Theorem Proving (PxTP) (2012)

6. Bonichon, R., Delahaye, D., Doligez, D.: Zenon: an extensible automated theorem prover producing checkable proofs. In: Dershowitz, N., Voronkov, A. (eds.) LPAR 2007. LNCS (LNAI), vol. 4790, pp. 151–165. Springer, Heidelberg (2007). doi:10.1007/978-3-540-75560-9_13

7. Bury, G., Delahaye, D., Doligez, D., Halmagrand, P., Hermant, O.: Automated deduction in the B set theory using typed proof search and deduction modulo. In: LPAR 20 : 20th International Conference on Logic for Programming, Artificial Intelligence and Reasoning, Suva, Fiji (2015)

8. Cauderlier, R., Halmagrand, P.: Checking Zenon modulo proofs in Dedukti. In: Fourth Workshop on Proof eXchange for Theorem Proving (PxTP), Berlin, Germany (2015)

9. Delahaye, D., Doligez, D., Gilbert, F., Halmagrand, P., Hermant, O.: Zenon modulo: when achilles outruns the tortoise using deduction modulo. In: McMillan, K., Middeldorp, A., Voronkov, A. (eds.) LPAR 2013. LNCS, vol. 8312, pp. 274–290. Springer, Heidelberg (2013). doi:10.1007/978-3-642-45221-5_20

10. Delahaye, D., Dubois, C., Marché, C., Mentré, D.: The Bware project: building a proof platform for the automated verification of B proof obligations. In: Ameur, Y.A., Schewe, K.-S. (eds.) Abstract State Machines, Alloy, B, VDM, and Z (ABZ). LNCS, vol. 8477, pp. 290–293. Springer, Heidelberg (2014)

11. Dowek, G., Hardin, T., Kirchner, C.: Theorem proving Modulo. J. Autom. Reasoning (JAR) **31**, 33–72 (2003)

12. Dowek, G., Miquel, A.: Cut elimination for zermelo set theory. Archive for Mathematical Logic. Springer, Heidelberg (2007, submitted)

13. Jacquel, M., Berkani, K., Delahaye, D., Dubois, C.: Verifying B proof rules using deep embedding and automated theorem proving. Softw. Eng. Formal Methods **7041**, 253–268 (2011)

14. Jaeger, É., Dubois, C.: Why would you trust B? In: Dershowitz, N., Voronkov, A. (eds.) LPAR 2007. LNCS (LNAI), vol. 4790, pp. 288–302. Springer, Heidelberg (2007). doi:10.1007/978-3-540-75560-9_22

15. Kleene, S.C.: Permutability of inferences in Gentzens calculi LK and LJ. In: Bulletin Of The American Mathematical Society, vol. 57, pp. 485–485. Amer Mathematical Soc, Providence (1951)

16. Mentré, D., Marché, C., Filliâtre, J.-C., Asuka, M.: Discharging proof obligations from Atelier B using multiple automated provers. In: Derrick, J., Fitzgerald, J., Gnesi, S., Khurshid, S., Leuschel, M., Reeves, S., Riccobene, E. (eds.) ABZ 2012. LNCS, vol. 7316, pp. 238–251. Springer, Heidelberg (2012). doi:10.1007/978-3-642-30885-7_17

17. Schmalz, M.: Formalizing the logic of event-B. Ph.D. thesis, Diss., Eidgenössische Technische Hochschule ETH Zürich, Nr. 20516, 2012 (2012)

18. ClearSy: Atelier B 4.1 (2013). http://www.atelierb.eu/

Deriving Inverse Operators for Modal Logic

Michell Guzmán[1]([✉]), Salim Perchy[1], Camilo Rueda[3], and Frank D. Valencia[2,3]

[1] Inria-LIX, École Polytechnique de Paris, Palaiseau, France
michell.guzman@inria.fr
[2] CNRS-LIX, École Polytechnique de Paris, Palaiseau, France
[3] Pontificia Universidad Javeriana de Cali, Cali, Colombia

Abstract. Spatial constraint systems are algebraic structures from concurrent constraint programming to specify spatial and epistemic behavior in multi-agent systems. We shall use spatial constraint systems to give an abstract characterization of the notion of normality in modal logic and to derive right inverse/reverse operators for modal languages. In particular, we shall identify the weakest condition for the existence of right inverses and show that the abstract notion of normality corresponds to the preservation of finite suprema. We shall apply our results to existing modal languages such as the weakest normal modal logic, Hennessy-Milner logic, and linear-time temporal logic. We shall discuss our results in the context of modal concepts such as bisimilarity and inconsistency invariance.

Keywords: Modal logic · Inverse operators · Constraint systems · Modal algebra · Bisimulation

1 Introduction

Constraint systems (cs's) provide the basic domains and operations for the semantic foundations of several declarative models and process calculi from *concurrent constraint programming (ccp)* [3,8,9,11,15,18,23,25]. In these calculi, processes can be thought of as both concurrent computational entities and logic specifications (e.g., process composition can be seen as parallel execution and conjunction). All ccp process calculi are parametric in a cs that specifies partial information upon which programs (processes) may act.

A cs is often represented as a complete algebraic lattice $(Con, \sqsubseteq)$. The elements of Con, the *constraints*, represent partial information and we shall think of them as being *assertions*. The intended meaning of $c \sqsubseteq d$ is that d specifies at least as much information as c (i.e., d entails c). The join operation $\sqcup$, the

This work has been partially supported by the ANR project 12IS02001 PACE, the Colciencias project 125171250031 CLASSIC, and Labex DigiCosme (project ANR-11-LABEX-0045-DIGICOSME) operated by ANR as part of the program "Investissement d'Avenir" Idex Paris-Saclay (ANR-11-IDEX-0003-02).

A. Sampaio and F. Wang (Eds.): ICTAC 2016, LNCS 9965, pp. 214–232, 2016.
DOI: 10.1007/978-3-319-46750-4_13

bottom *true* and the top *false* of the lattice $(Con, \sqsubseteq)$ correspond to conjunction, the empty information and the join of all information, respectively. The ccp operations and their logical counterparts typically have a corresponding elementary construct or operation on the elements of the constraint system. In particular, parallel composition and conjunction correspond to the *join* operation, and existential quantification and local variables correspond to a cylindrification operation on the set of constraints [25].

Similarly, the notion of computational space and the epistemic notion of belief in the sccp process calculi [15] correspond to a family of functions $[\cdot]_i : Con \rightarrow Con$ on the elements of the constraint system Con that preserve finite suprema. These functions are called *space functions*. A cs equipped with space functions is called a *spatial constraint system* (scs). From a computational point of view the assertion (constraint) $[c]_i$ specifies that c resides within the space of agent i. From an epistemic point of view, the assertion $[c]_i$ specifies that agent i considers c to be true (i.e. that in the world of agent i the assertion c is true). Both intuitions convey the idea of c being local to agent i.

The Extrusion Problem. Given a space function $[\cdot]_i$, the *extrusion problem* consists in finding/constructing a *right inverse* of $[\cdot]_i$, called *extrusion function*, satisfying some basic requirements (e.g., preservation of finite suprema). By right inverse of $[\cdot]_i$ we mean a function $\uparrow_i : Con \rightarrow Con$ such that $[\uparrow_i c]_i = c$. From a computational point of view, the intended meaning of $[\uparrow_i c]_i = c$ is that within a space context $[\cdot]_i$, $\uparrow_i c$ extrudes c from agent i's space. From an epistemic point of view, we can use $[\uparrow_i c]_i$ to express *utterances* by agent i, i.e., to specify that agent i wishes to say c to the outside world. One can then think of extrusion/utterance as the *right inverse* of space/belief.

Modal logics [21] extend classical logic to include operators expressing modalities. Depending on the intended meaning of the modalities, a particular modal logic can be used to reason about space, knowledge, belief or time, among others. Some modal logics have been extended with *inverse modalities* to specify, for example, past tense assertions in temporal logic [24], utterances in epistemic logic [13], and backward moves in modal logic for concurrency [19], among others. Although the notion of spatial constraint system is intended to give an algebraic account of spatial and epistemic assertions, we shall show that it is sufficiently robust to give an algebraic account of more general modal assertions.

Contributions. We shall study the extrusion problem for a meaningful family of scs's that can be used as semantic structures for modal logics. These scs's are called *Kripke spatial constraint systems* because its elements are *Kripke structures*. We shall show that the extrusion functions of Kripke scs's, i.e. the right inverses of the space functions, correspond to right inverse modalities in modal logic. We shall derive a complete characterization for the existence of right inverses of space functions: The weakest restriction on the elements of Kripke scs's that guarantees the existence of right inverses. We shall also give an algebraic characterization of the modal logic notion of normality as maps that preserve finite suprema. We then give a complete characterization and derivations of extrusion functions that are normal (and thus they correspond to normal inverse modalities). Finally, we use the above-mentioned contributions to the problem of whether a given modal language can be extended with right inverse operators. We discuss the implications of our results for specific modal languages and modal concepts such the minimal modal logic K_n [10], Hennessy-Milner logic [14], a modal logic of linear-time [20], and bisimulation.

2   Background: Spatial Constraint Systems

In this section we recall the notion of basic constraint system [3] and the more recent notion of spatial constraint system [15]. We presuppose basic knowledge of order theory and modal logic [1,2,10,21].

The concurrent constraint programming model of computation [25] is parametric in a *constraint system* (cs) specifying the structure and interdependencies of the partial information that computational agents can ask of and post in a *shared store*. This information is represented as *assertions* traditionally referred to as *constraints*.

Constraint systems can be formalized as *complete algebraic lattices* [3][1]. The elements of the lattice, the *constraints*, represent (partial) information. A constraint c can be viewed as an *assertion* (or a *proposition*). The lattice order $\sqsubseteq$ is meant to capture entailment of information: $c \sqsubseteq d$, alternatively written $d \sqsupseteq c$, means that the assertion d represents as much information as c. Thus we may think of $c \sqsubseteq d$ as saying that d *entails* c or that c can be *derived* from d. The *least upper bound (lub)* operator $\sqcup$ represents join of information; $c \sqcup d$, the least element in the underlying lattice above c and d. Thus $c \sqcup d$ can be seen as an assertion stating that both c and d hold. The top element represents the lub of all, possibly inconsistent, information, hence it is referred to as *false*. The bottom element *true* represents the empty information.

Definition 1 (Constraint Systems [3]). *A constraint system (cs)* **C** *is a complete algebraic lattice (Con, $\sqsubseteq$). The elements of Con are called* constraints. *The symbols $\sqcup$, true and false will be used to denote the least upper bound (lub) operation, the bottom, and the top element of* **C**, *respectively.*

We shall use the following notions and notations from order theory.

Notation 1 (Lattices). *Let* **C** *be a partially ordered set (poset) (Con, $\sqsubseteq$). We shall use $\bigsqcup S$ to denote the least upper bound (lub) (or supremum or join) of the elements in S, and $\bigsqcap S$ is the greatest lower bound (glb) (infimum or meet) of the elements in S. We say that* **C** *is a complete lattice iff each subset of Con has a supremum and an infimum in Con. A non-empty set $S \subseteq Con$ is* directed *iff every finite subset of S has an upper bound in S. Also $c \in Con$ is* compact *iff for any directed subset D of Con, $c \sqsubseteq \bigsqcup D$ implies $c \sqsubseteq d$ for*

---

[1] An alternative syntactic characterization of cs, akin to Scott information systems, is given in [25].

some $d \in D$. A complete lattice **C** *is said to be* algebraic *iff for each $c \in Con$, the set of compact elements below it forms a directed set and the lub of this directed set is c. A* self-map *on Con is a function $f : Con \rightarrow Con$. Let $(Con, \sqsubseteq)$ be a complete lattice. The self-map f on Con preserves the supremum of a set $S \subseteq Con$ iff $f(\bigsqcup S) = \bigsqcup \{f(c) \mid c \in S\}$. The preservation of the infimum of a set is defined analogously. We say f preserves finite/infinite suprema iff it preserves the supremum of arbitrary finite/infinite sets. Preservation of finite/infinite infima is defined similarly.*

Spatial Constraint Systems. The authors of [15] extended the notion of cs to account for distributed and multi-agent scenarios where agents have their own space for local information and for performing their computations.

Intuitively, each agent i has a *space* function $[\cdot]_i$ from constraints to constraints. Recall that constraints can be viewed as assertions. We can then think of $[c]_i$ as an assertion stating that c is a piece of information residing *within a space attributed to agent i*. An alternative *epistemic logic* interpretation of $[c]_i$ is an assertion stating that agent i *believes* c or that c holds within the space of agent i (but it may not hold elsewhere). Both interpretations convey the idea that c is local to agent i. Similarly, $[[c]_j]_i$ is a hierarchical spatial specification stating that c holds within the local space the agent i attributes to agent j. Nesting of spaces can be of any depth. We can think of a constraint of the form $[c]_i \sqcup [d]_j$ as an assertion specifying that c and d hold within two *parallel/neighboring* spaces that belong to agents i and j, respectively. From a computational/ concurrency point of view, we think of $\sqcup$ as parallel composition. As mentioned before, from a logic point of view the join of information $\sqcup$ corresponds to conjunction.

Definition 2 (Spatial Constraint System [15]). *An n-agent spatial constraint system (n-scs)* **C** *is a cs $(Con, \sqsubseteq)$ equipped with n self-maps $[\cdot]_1, \ldots, [\cdot]_n$ over its set of constraints Con such that: (S.1) $[true]_i = true$, and (S.2) $[c \sqcup d]_i = [c]_i \sqcup [d]_i$ for each $c, d \in Con$.*

Axiom S.1 requires space functions to be strict maps (i.e. bottom preserving). Intuitively, it states that having an empty local space amounts to nothing. Axiom S.2 states that the information in a given space can be distributed. Notice that requiring S.1 and S.2 is equivalent to requiring that each $[\cdot]_i$ preserves *finite suprema*. Also S.2 implies that each $[\cdot]_i$ is monotonic: I.e., if $c \sqsupseteq d$ then $[c]_i \sqsupseteq [d]_i$.

Extrusion and utterance. We can also equip each agent i with an *extrusion* function $\uparrow_i : Con \rightarrow Con$. Intuitively, within a space context $[\cdot]_i$, the assertion $\uparrow_i c$ specifies that c must be posted outside of (or extruded from) agent i's space. This is captured by requiring the *extrusion* axiom $[\uparrow_i c]_i = c$. In other words, we view *extrusion/utterance* as the right inverse of *space/belief* (and thus space/belief as the left inverse of extrusion/utterance).

Definition 3 (Extrusion). *Given an n-scs $(Con, \sqsubseteq, [\cdot]_1, \ldots, [\cdot]_n)$, we say that $\uparrow_i$ is extrusion function for the space $[\cdot]_i$ iff $\uparrow_i$ is a right inverse of $[\cdot]_i$, i.e., iff $[\uparrow_i c]_i = c$.*

From the above definitions it follows that $[c \sqcup \uparrow_i d]_i = [c]_i \sqcup d$. From a spatial point of view, agent i *extrudes* d from its local space. From an epistemic view this can be seen as an agent i that believes c and *utters* d to the outside world. If d is inconsistent with c, i.e., $c \sqcup d = false$, we can see the utterance as an intentional *lie* by agent i: The agent i utters an assertion inconsistent with their own beliefs.

The Extrusion/Right Inverse Problem. A legitimate question is: Given space $[\cdot]_i$ can we derive an extrusion function $\uparrow_i$ for it? From set theory we know that there is an extrusion function (i.e., a right inverse) $\uparrow_i$ for $[\cdot]_i$ iff $[\cdot]_i$ is *surjective*. Recall that the *pre-image* of $y \in Y$ under $f : X \to Y$ is the set $f^{-1}(y) = \{x \in X \mid y = f(x)\}$. Thus the extrusion $\uparrow_i$ can be defined as a function, called *choice* function, that maps each element c to some element from the pre-image of c under $[\cdot]_i$.

The existence of the above-mentioned choice function assumes the Axiom of Choice. The next proposition from [13] gives some constructive extrusion functions. It also identifies a distinctive property of space functions for which a right inverse exists.

Proposition 1. *Let $[\cdot]_i$ be a space function of scs. Then*

1. *If $[false]_i \neq false$ then $[\cdot]_i$ does not have any right inverse.*
2. *If $[\cdot]_i$ is surjective and preserves arbitrary suprema then $\uparrow_i : c \mapsto \bigsqcup [c]_i^{-1}$ is a right inverse of $[\cdot]_i$ and preserve arbitrary infima.*
3. *If $[\cdot]_i$ is surjective and preserves arbitrary infima then $\uparrow_i : c \mapsto \bigsqcap [c]_i^{-1}$ is a right inverse of $[\cdot]_i$ and preserve arbitrary suprema.*

We have presented spatial constraint systems as algebraic structures for spatial and epistemic behaviour as that was their intended meaning. Nevertheless, we shall see that they can also provide an algebraic structure to reason about Kripke models with applications to modal logics.

In Sect. 4 we shall study the existence, constructions and properties of right inverses for a meaningful family of scs's; the Kripke scs's. The importance of such a study is the connections we shall establish between right inverses and reverse modalities which are present in temporal, epistemic and other modal logics. Property (1) in Proposition 1 can be used as a test for the non-existence of a right-inverse. The space functions of Kripke scs's preserve arbitrary suprema, thus Property (2) will be useful. They do not preserve in general arbitrary (or even finite) infima so we will not apply Property (3).

It is worth to point out that the derived extrusion $\uparrow_i$ in Property (3), preserves arbitrary suprema, this implies $\uparrow_i$ is *normal* in a sense we shall make precise next. Normal self-maps give an abstract characterization of normal modal operators, a fundamental concept in modal logic. We will be therefore interested in deriving normal inverses.

3   Constraint Frames and Normal Self Maps

Spatial constraint systems are algebraic structures for spatial and mobile behavior. By building upon ideas from Geometric Logic and Heyting Algebras [26] we can also make them suitable as semantics structures for modal logic. In this section we give an algebraic characterization of the concept of normal modality as those maps that preserve finite suprema.

We can define a general form of implication by adapting the corresponding notion from Heyting Algebras to constraint systems. Intuitively, a *Heyting implication* $c \to d$ in our setting corresponds to the *weakest constraint* one needs to join c with to derive d: i.e., the greatest lower bound $\prod \{e \mid e \sqcup c \sqsupseteq d\}$. Similarly, the negation of a constraint c, written $\sim c$, can be seen as the *weakest constraint* *inconsistent* with c, i.e., the greatest lower bound $\prod \{e \mid e \sqcup c \sqsupseteq false\} = c \to false$.

Definition 4 (Constraint Frames). *A constraint system* $(Con, \sqsubseteq)$ *is said to be a* constraint frame *iff its joins distribute over arbitrary meets: More precisely,* $c \sqcup \prod S = \prod \{c \sqcup e \mid e \in S\}$ *for every* $c \in Con$ *and* $S \subseteq Con$. *Given a constraint frame* $(Con, \sqsubseteq)$ *and* $c, d \in Con$, *define Heyting implication* $c \to d$ *as* $\prod \{e \in Con \mid c \sqcup e \sqsupseteq d\}$ *and Heyting negation* $\sim c$ *as* $c \to false$.

The following basic properties of Heyting implication are immediate consequences of the above definitions.

Proposition 2. *Let* $(Con, \sqsubseteq)$ *be a constraint frame. For every* $c, d, e \in Con$ *we have: (1)* $c \sqcup (c \to d) = c \sqcup d$, *(2)* $c \sqsupseteq (d \to e)$ *iff* $c \sqcup d \sqsupseteq e$, *and (3)* $c \to d = true$ *iff* $c \sqsupseteq d$.

In modal logics one is often interested in *normal modal* operators. The formulae of a modal logic are those of propositional logic extended with modal operators. Roughly speaking, a modal logic operator m is normal iff (1) the formula $m(\phi)$ is a theorem (i.e., true in all models for the underlying modal language) whenever the formula ϕ is a theorem, and (2) the implication formula $m(\phi \Rightarrow \psi) \Rightarrow (m(\phi) \Rightarrow m(\psi))$ is a theorem. Since constraints can be viewed as logic assertions, we can think of modal operators as self-maps on constraints. Thus, using Heyting implication, we can express the normality condition in constraint frames as follows.

Definition 5 (Normal Maps). *Let* $(Con, \sqsubseteq)$ *be a constraint frame. A self-map* m *on Con is said to be* normal *if (1)* $m(true) = true$ *and (2)* $m(c \to d) \to (m(c) \to m(d)) = true$ *for each* $c, d \in Con$.

We now prove that the normality requirement is equivalent to the requirement of preserving finite suprema. The next theorem basically states that Condition (2) in Definition 5 is equivalent to the seemingly simpler condition: $m(c \sqcup d) = m(c) \sqcup m(d)$.

Theorem 1 (Normality & Finite Suprema). *Let* **C** *be a constraint frame* $(Con, \sqsubseteq)$ *and let* f *be a self-map on* Con. *Then* f *is normal if and only if* f *preserves finite suprema.*

Proof. It suffices to show that for any bottom preserving self-map f, $\forall c, d \in Con : f(c \rightarrow d) \rightarrow (f(c) \rightarrow f(d)) = true$ iff $\forall c, d \in Con : f(c \sqcup d) = f(c) \sqcup f(d)$. (Both conditions require f to be bottom preserving, i.e., $f(true) = true$, and preservation of non-empty finite suprema is equivalent to the preservation of binary suprema.) Here we show the *only-if* direction (the other direction is simpler).

Assume that $\forall c, d \in Con : f(c \rightarrow d) \rightarrow (f(c) \rightarrow f(d)) = true$. Take two arbitrary $c, d \in Con$. We first prove $f(c \sqcup d) \sqsupseteq f(c) \sqcup (d)$. From the assumption and Proposition 2(3) we obtain

$$f((c \sqcup d) \rightarrow d) \sqsupseteq f(c \sqcup d) \rightarrow f(d). \tag{1}$$

From Proposition 2(3) $(c \sqcup d) \rightarrow d = true$. Since $f(true) = true$ we have $f((c \sqcup d) \rightarrow d) = true$. We must then have, from Eq. 1, $f(c \sqcup d) \rightarrow f(d) = true$ as well. Using Proposition 2(3) we obtain $f(c \sqcup d) \sqsupseteq f(d)$. In a similar fashion, by exchanging c and d in Eq. 1, we can obtain $f(d \sqcup c) \sqsupseteq f(c)$. We can then conclude $f(c \sqcup d) \sqsupseteq f(c) \sqcup f(d)$ as wanted.

We now prove $f(c) \sqcup f(d) \sqsupseteq f(c \sqcup d)$. From the assumption and Proposition 2(3) we have

$$f(c \rightarrow (d \rightarrow c \sqcup d)) \sqsupseteq f(c) \rightarrow f(d \rightarrow c \sqcup d). \tag{2}$$

Using Proposition 2 one can verify that $c \rightarrow (d \rightarrow c \sqcup d) = true$. Since $f(true) = true$ then $f(c \rightarrow (d \rightarrow c \sqcup d)) = true$. From Eq. 2, we must then have $f(c) \rightarrow f(d \rightarrow c \sqcup d) = true$ and by using Proposition 2(3) we conclude $f(c) \sqsupseteq f(d \rightarrow c \sqcup d)$. From the assumption and Proposition 2(3) $f(d \rightarrow c \sqcup d) \sqsupseteq f(d) \rightarrow f(c \sqcup d)$. We then have $f(c) \sqsupseteq f(d \rightarrow c \sqcup d) \sqsupseteq f(d) \rightarrow f(c \sqcup d)$. Thus $f(c) \sqsupseteq f(d) \rightarrow f(c \sqcup d)$ and then using Proposition 2(2) we obtain $f(c) \sqcup f(d) \sqsupseteq f(c \sqcup d)$ as wanted. $\qquad\square$

By applying the above theorem, we can conclude that space functions from constraint frames are indeed normal self-maps, since they preserve finite suprema.

4   Extrusion Problem for Kripke Constraint Systems

This is the main and more technical part of the paper. Here we will study the extrusion/right inverse problem for a meaningful family of spatial constraint systems (scs's); the Kripke scs. In particular we shall derive and give a *complete* characterization of normal extrusion functions as well as identify the *weakest* condition on the elements of the Kripke scs's under which extrusion functions may exist. To illustrate the importance of this study it is convenient to give some intuition first.

Kripke structures (KS) [16] are a fundamental mathematical tool in logic and computer science. They can be seen as transition systems and they are used to give semantics to modal logics. A KS M provides a relational structure with a set of states and one or more accessibility relations $\xrightarrow{i}_M$ between them: $s \xrightarrow{i}_M t$ can be seen as a transition, labelled with i, from s to t in M. Broadly speaking, the Kripke semantics interprets each modal formula ϕ as a certain set $[\![\phi]\!]$ of pairs (M, s), called pointed KS's, where s is a state of the KS M. In modal logics with one or more modal (box) operators $\Box_i$, the formula $\Box_i \phi$ is interpreted as $[\![\Box_i \phi]\!] = \{(M, s) \mid \forall t : s \xrightarrow{i}_M t, (M, t) \in [\![\phi]\!]\}$.

Analogously, in a Kripke scs each constraint c is equated to a set of pairs (M, s) of pointed KS. Furthermore, we have $[c]_i = \{(M, s) \mid \forall t : s \xrightarrow{i}_M t, (M, t) \in c\}$. This means that formulae can be interpreted as constraints and in particular $\Box_i$ can be interpreted by $[\cdot]_i$ as $[\![\Box_i \phi]\!] = [\,[\![\phi]\!]\,]_i$.

Inverse modalities $\Box_i^{-1}$, also known as reverse modalities, are used in many modal logics. In tense logics they represent past operators [22] and in epistemic logic they represent utterances [13]. The basic property of a (right) inverse modality is given by the axiom $\Box_i(\Box_i^{-1} \phi) \Leftrightarrow \phi$. In fact, given a modal logic one may wish to see if it can be extended with reverse modalities (e.g., is there a reverse modality for the always operator of temporal logic?).

Notice that if we have an extrusion function $\uparrow_i$ for $[\cdot]_i$ we can provide the semantics for inverse modalities $\Box_i^{-1}$ by letting $[\![\Box_i^{-1} \phi]\!] = \uparrow_i(\,[\![\phi]\!]\,)$. We then have $[\![\Box_i(\Box_i^{-1} \phi)]\!] = [\![\phi]\!]$ thus validating the axiom $\Box_i(\Box_i^{-1} \phi) \Leftrightarrow \phi$. This illustrates the relevance of deriving extrusion functions and establishing the weakest conditions under which they exist. Furthermore, the algebraic structure of Kripke scs may help us stating derived properties of the reverse modality such as that of being normal (Definition 5).

4.1   KS and Kripke SCS

We begin by recalling some notions and notations related to Kripke models.

Definition 6 (Kripke Structures). *An n-agent Kripke Structure (KS) M over a set of atomic propositions Φ is a tuple $(S, \pi, \mathcal{R}_1, \ldots, \mathcal{R}_n)$ where S is a nonempty set of states, $\pi : S \to (\Phi \to \{0, 1\})$ is an interpretation associating with each state a truth assignment to the primitive propositions in Φ, and $\mathcal{R}_i$ is a binary relation on S. A pointed KS is a pair (M, s) where M is a KS and s is a state of M.*

We shall use the following notation in the rest of the paper.

Notation 2. *Each $\mathcal{R}_i$ is referred to as the accessibility relation for agent i. We shall use $\xrightarrow{i}_M$ to refer to the accessibility relation of agent i in M. We write $s \xrightarrow{i}_M t$ to denote $(s, t) \in \mathcal{R}_i$. We use $\triangleright_i(M, s) = \{(M, t) \mid s \xrightarrow{i}_M t\}$ to denote the pointed KS reachable from the pointed KS (M, s). The interpretation function π tells us what primitive propositions are true at a given state: p holds at state s iff $\pi(s)(p) = 1$. We shall use S_M and π_M to denote the set of states and interpretation function of M.*

We now define the Kripke scs wrt a set $\mathcal{S}_n(\Phi)$ of pointed KS.

Definition 7 (Kripke Spatial Constraint Systems [15]). *Let $\mathcal{S}_n(\Phi)$ be a non-empty set of n-agent Kripke structures over a set of primitive propositions Φ and let Δ be the set of all pointed Kripke structures (M, s) such that $M \in \mathcal{S}_n(\Phi)$. We define the Kripke n-scs for $\mathcal{S}_n(\Phi)$ as $\mathbf{K}(\mathcal{S}_n(\Phi)) = (Con, \sqsubseteq, [\cdot]_1, \ldots, [\cdot]_n)$ where $Con = \mathcal{P}(\Delta)$, $\sqsubseteq\ =\ \supseteq$, and*

$$[c]_i \overset{\text{def}}{=} \{(M, s) \in \Delta \mid \rhd_i(M, s) \subseteq c\}. \tag{3}$$

The structure $\mathbf{K}(\mathcal{S}_n(\Phi)) = (Con, \sqsubseteq, [\cdot]_1, \ldots, [\cdot]_n)$ is a complete algebraic lattice given by a powerset ordered by reversed inclusion $\supseteq$. The join $\sqcup$ is set intersection, the meet $\sqcap$ is set union, the top element *false* is the empty set $\emptyset$, and bottom *true* is the set Δ of all pointed Kripke structures (M, s) with $M \in \mathcal{S}_n(\Phi)$. Notice that $\mathbf{K}(\mathcal{S}_n(\Phi))$ is a frame since meets are unions and joins are intersections so the distributive requirement is satisfied. Furthermore, each $[\cdot]_i$ preserves arbitrary suprema (intersection) and thus, from Theorem 1 it is a normal self-map.

Proposition 3. *Let $\mathbf{K}(\mathcal{S}_n(\Phi)) = (Con, \sqsubseteq, [\cdot]_1, \ldots, [\cdot]_n)$ as in Definition 7. Then (1) $\mathbf{K}(\mathcal{S}_n(\Phi))$ is a spatial constraint frame and (2) each $[\cdot]_i$ preserves arbitrary suprema.*

4.2 Existence of Right Inverses

We shall now address the question of whether a given Kripke constraint system can be extended with extrusion functions. We shall identify a sufficient and necessary condition on accessibility relations for the existence of an extrusion function $\uparrow_i$ given the space $[\cdot]_i$. We shall also give explicit right inverse constructions.

Notation 3. *For notational convenience, we take the set Φ of primitive propositions and n to be fixed from now on and omit them from the notation. E.g., we write $\mathcal{M}$ instead of $\mathcal{M}_n(\Phi)$.*

The following notions play a key role in our complete characterization, in terms of classes of KS, of the existence of right inverses for Kripke space functions.

Definition 8 (Determinacy and Unique-Determinacy). *Let S and $\mathcal{R}$ be the set of states and an accessibility relation of a KS M, respectively. Given $s, t \in S$, we say that s determines t wrt $\mathcal{R}$ if $(s, t) \in \mathcal{R}$. We say that s uniquely determines t wrt $\mathcal{R}$ if s is the only state in S that determines t wrt $\mathcal{R}$. A state $s \in S$ is said to be determinant wrt $\mathcal{R}$ if it uniquely determines some state in S wrt $\mathcal{R}$. Furthermore, $\mathcal{R}$ is determinant-complete if every state in S is determinant wrt $\mathcal{R}$.*

Example 1. Figure 1 illustrates some typical determinant-complete accessibility relations for agent i. Notice that any determinant-complete relation $\overset{i}{\longrightarrow}_M$ is

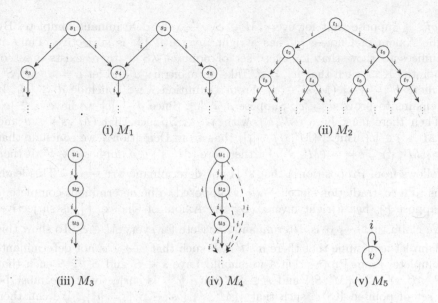

Fig. 1. Accessibility relations for an agent i. In each sub-figure we omit the corresponding KS M_k from the edges and draw $s \xrightarrow{i} t$ whenever $s \xrightarrow{i}_{M_k} t$.

necessarily *serial* (or left-total): i.e., For every $s \in S_M$, there exists $t \in S_M$ such that $s \xrightarrow{i}_M t$. Tree-like accessibility relations where all paths are infinite are determinant-complete (Fig. 1(ii) and (iii)). Also some non-tree like structures such as Fig. 1(i) and (v). Figure 1(iv) shows a non determinate-complete accessibility relation by taking the transitive closure of Fig. 1(iii).

We need to introduce some notation.

Notation 4. *Recall that $\rhd_i(M, s) = \{(M, t) \mid s \xrightarrow{i}_M t\}$ where $\xrightarrow{i}_M$ denotes the accessibility relation of agent i in the KS M. We extend this definition to sets of states as follows $\rhd_i(M, S) = \bigcup_{s \in S} \rhd_i(M, s)$. Furthermore, we shall write $s \xrightarrow{i}_M t$ to mean that s uniquely determines t wrt $\xrightarrow{i}_M$.*

The following proposition gives an alternative definition of determinant states.

Proposition 4. *Let $s \in S_M$. The state s is determinant wrt $\xrightarrow{i}_M$ if and only if for every $S' \subseteq S_M$: If $\rhd_i(M, s) \subseteq \rhd_i(M, S')$ then $s \in S'$.*

The following theorem provides a complete characterization, in terms of classes of KS, of the existence of right inverses for space functions.

Theorem 2 (Completeness). *Let $[\cdot]_i$ be a spatial function of a Kripke scs $\mathbf{K}(\mathcal{S})$. Then $[\cdot]_i$ has a right inverse if and only if for every $M \in \mathcal{S}$ the accessibility relation $\xrightarrow{i}_M$ is determinant-complete.*

Proof. – Suppose that for every $M \in \mathcal{S}$, $\xrightarrow{i}_M$ is determinant-complete. By the Axiom of Choice, $[\cdot]_i$ has a right inverse if $[\cdot]_i$ is surjective. Thus, it suffices to show that for every set of pointed KS d, there exists a set of pointed KS c such that $[c]_i = d$. Take an arbitrary d and let $c = \rhd_i(M', S')$ where $S' = \{s \mid (M, s) \in d\}$. From Definition 7 we conclude $d \subseteq [c]_i$. It remains to prove $d \supseteq [c]$. Suppose $d \not\supseteq [c]$. Since $d \subseteq [c]$ we have $d \subset [c]$. Then there must be a (M', s'), with $M' \in \mathcal{S}$, such that $(M', s') \notin d$ and $(M', s') \in [c]$. But if $(M', s') \in [c]_i$ then from Definition 7 we conclude that $\rhd_i(M', s') \subseteq c = \rhd_i(M', S')$. Furthermore $(M', s') \notin d$ implies $s' \notin S'$. It then follows from Proposition 4 that s' is not determinant wrt $\xrightarrow{i}_{M'}$. This leads us to a contradiction since $\xrightarrow{i}_{M'}$ is supposed to be determinant-complete.

– Suppose $[\cdot]_i$ has a right inverse. By the Axiom of Choice, $[\cdot]_i$ is surjective. We claim that $\xrightarrow{i}_M$ is determinant-complete for every $M \in \mathcal{S}$. To show this claim let us assume that there is $M' \in \mathcal{S}$ such that $\xrightarrow{i}_M$ is not determinant-complete. From Proposition 4 we should have $s \in S$ and $S' \subseteq S$ such that $\rhd_i(M', s) \subseteq \rhd_i(M', S')$ and $s \notin S'$. Since $[c']_i$ is surjective there must be a set of pointed KS c' such that $\{(M', s') \mid s' \in S'\} = [c']_i$. We can then verify, using Definition 7, that $\rhd_i(M, S') \subseteq c'$. Since $\rhd_i(M', s) \subseteq \rhd_i(M', S')$ then $\rhd_i(M', s) \subseteq c'$. It then follows from Definition 7 that $(M', s) \in [c']_i$. But $[c']_i = \{(M', s') \mid s' \in S'\}$ then $s \in S'$, a contradiction. $\square$

Henceforth we use $\mathcal{M}^D$ to denote the class of KS's whose accessibility relations are determinant-complete. It follows from Theorem 2 that $\mathcal{S} = \mathcal{M}^D$ is the largest class for which space functions of a Kripke scs $\mathbf{K}(\mathcal{S})$ have right inverses.

4.3   Right Inverse Constructions

Let $\mathbf{K}(\mathcal{S}) = (Con, \sqsubseteq, [\cdot]_1, \ldots, [\cdot]_n)$ be the Kripke scs. The Axiom of Choice and Theorem 2 tell us that each $[\cdot]_i$ has a right inverse (extrusion function) if and only if $\mathcal{S} \subseteq \mathcal{M}^D$. We are interested, however, in explicit constructions of the right inverses.

Remark 1. Recall that any Kripke scs $\mathbf{K}(\mathcal{S}) = (Con, \sqsubseteq, [\cdot]_1, \ldots, [\cdot]_n)$ is ordered by reversed inclusion (i.e., $c \sqsubseteq d$ iff $d \subseteq c$). Thus, for example, saying that some f is the least function wrt $\subseteq$ satisfying certain conditions is equivalent to saying that f is the greatest function wrt $\sqsubseteq$ satisfying the same conditions. As usual given two self-maps f and g over Con we define $f \sqsubseteq g$ iff $f(c) \sqsubseteq g(c)$ for every $c \in Con$.

Since any Kripke scs space function preserve arbitrary suprema (Proposition 3), we can apply Proposition 1.2 to obtain the following canonical greatest right-inverse construction. Recall that the pre-image of c under $[\cdot]_i$ is given by $[c]_i^{-1} = \{d \mid c = [d]_i\}$.

Definition 9 (Max Right Inverse). *Let* $\mathbf{K}(\mathcal{S}) = (Con, \sqsubseteq, [\cdot]_1, \ldots, [\cdot]_n)$ *be a Kripke scs over* $\mathcal{S} \subseteq \mathcal{M}^D$. *We define* $\uparrow_i^M$ *as the following self-map on* Con:
$$\uparrow_i^M : c \mapsto \bigsqcup [c]_i^{-1}.$$

It follows from Proposition 1.2 that $\uparrow_i^M$ is a right inverse for $[\cdot]_i$, and furthermore, from its definition it is clear that $\uparrow_i^M$ is the greatest right inverse of $[\cdot]_i$ wrt $\sqsubseteq$.

Nevertheless, as stated in the following proposition, $\uparrow_i^M$ is not necessarily *normal* in the sense of Definition 5. To state this more precisely, let us first extend the terminology in Definition 8.

Definition 10 (Indeterminacy and Multiple Determinacy). *Let* S *and* $\mathcal{R}$ *be the set of states and an accessibility relation of a KS* M, *respectively. Given* $t \in S$, *we say that* t *is* determined *wrt* $\mathcal{R}$ *if there is* $s \in S$ *such that* s *determines* t *wrt* $\mathcal{R}$, *else we say that* t *is* indetermined *(or* initial*) wrt* $\mathcal{R}$. *Similarly, we say that* t *is* multiply, *or* ambiguously, determined *if it is determined by at least two different states in* S *wrt* $\mathcal{R}$.

The following statement and Theorem 1 lead us to conclude that the presence of indetermined/initial states or multiple-determined states causes $\uparrow_i^M$ not to be normal.

Proposition 5. *Let* $\mathbf{K}(\mathcal{S}) = (Con, \sqsubseteq, [\cdot]_1, \ldots, [\cdot]_n)$ *and* $\uparrow_i^M$ *as in Definition 9. Let* $\mathbf{nd}(\mathcal{S}) = \{(M, t) \mid M \in \mathcal{S}$ *&* t *is indetermined wrt* $\xrightarrow{i}_M\}$ *and* $\mathbf{md}(\mathcal{S}) = \{(M, t) \mid M \in \mathcal{S}$ *&* t *is multiply determined wrt* $\xrightarrow{i}_M\}$:

– *If* $\mathbf{nd}(\mathcal{S}) \neq \emptyset$ *then* $\uparrow_i^M(true) \neq true$.
– *If* $\mathbf{md}(\mathcal{S}) \neq \emptyset$ *then* $\uparrow_i^M(c \sqcup d) \neq \uparrow_i^M(c) \sqcup \uparrow_i^M(d)$ *for some* $c, d \in Con$.

In what follows we shall identify right inverse constructions that are normal. The notion of indeterminacy and multiply determinacy we just introduced in Definition 10 will play a central role.

4.4   Normal Right Inverses

The following central lemma provides distinctive properties of any normal right inverse.

Lemma 1. *Let* $\mathbf{K}(\mathcal{S}) = (Con, \sqsubseteq, [\cdot]_1, \ldots, [\cdot]_n)$ *be the Kripke scs over* $\mathcal{S} \subseteq \mathcal{M}^D$. *Suppose that* f *is a normal right-inverse of* $[\cdot]_i$. *Then for every* $M \in \mathcal{S}$, $c \in Con$:

1. $\rhd_i(M, s) \subseteq f(c)$ *if* $(M, s) \in c$,
2. $\{(M, t)\} \subseteq f(c)$ *if* t *is multiply determined wrt* $\xrightarrow{i}_M$, *and*
3. $true \subseteq f(true)$.

The above property tell us what sets should necessarily be included in every $f(c)$ if f is to be both normal and a right inverse of $[\cdot]_i$. It turns out that it is sufficient to include exactly those sets to obtain a normal right inverse of $[\cdot]_i$. In other words the above lemma gives us a *complete* set of conditions for normal right inverses. In fact, the least self-map f wrt $\sqsubseteq$, i.e., the greatest one wrt the lattice order $\sqsubseteq$, satisfying Conditions 1, 2 and 3 in Lemma 1 is indeed a normal right-inverse. We call such a function the *max normal right inverse* $\uparrow_i^{\mathtt{MN}}$ and is given below.

Definition 11 (Max Normal-Right Inverse). *Let* $\mathbf{K}(\mathcal{S}) = (Con, \sqsubseteq, [\cdot]_1, \ldots, [\cdot]_n)$ *be a Kripke scs over* $\mathcal{S} \subseteq \mathcal{M}^D$. *We define the* max normal right inverse *for agent* i, $\uparrow_i^{\mathtt{MN}}$ *as the following self-map on* Con:

$$\uparrow_i^{\mathtt{MN}}(c) \stackrel{\text{def}}{=} \begin{cases} true & if\ c = true \\ \{(M,t) \mid t \text{ is determined wrt } \xrightarrow{i}_M\ \&\ \forall s : s \xrightarrow{i}_M t, (M,s) \in c\} \end{cases} \tag{4}$$

(Recall that $s \xrightarrow{i}_M t$ *means that* s *uniquely determines* t *wrt* $\xrightarrow{i}_M$.)*

We now state that $\uparrow_i^{\mathtt{MN}}(c)$ is the greatest normal right inverse of $[\cdot]_i$ wrt $\sqsubseteq$.

Theorem 3. *Let* $\mathbf{K}(\mathcal{S}) = (Con, \sqsubseteq, [\cdot]_1, \ldots, [\cdot]_n)$ *and* $\uparrow_i^{\mathtt{MN}}$ *as in Definition 11.*

- *The self-map* $\uparrow_i^{\mathtt{MN}}$ *is a normal right inverse of* $[\cdot]_i$,
- *For every normal right-inverse* f *of* $[\cdot]_i$, *we have* $f \sqsubseteq \uparrow_i^{\mathtt{MN}}$.

Notice that $\uparrow_i^{\mathtt{MN}}(c)$ excludes undetermined states if $c \neq true$. It turns out that we can add them and obtain a more succinct normal right inverse:

Definition 12 (Normal Inverse). *Let* $\mathbf{K}(\mathcal{S}) = (Con, \sqsubseteq, [\cdot]_1, \ldots, [\cdot]_n)$ *be a Kripke scs over* $\mathcal{S} \subseteq \mathcal{M}^D$. *Define* $\uparrow_i^{\mathtt{N}} : Con \to Con$ *as* $\uparrow_i^{\mathtt{N}}(c) \stackrel{\text{def}}{=} \{(M,t) \mid \forall s : s \xrightarrow{i}_M t, (M,s) \in c\}$.

Clearly $\uparrow_i^{\mathtt{N}}(c)$ includes every (M,t) such that t is indetermined wrt $\xrightarrow{i}_M$.

Theorem 4. *Let* $\mathbf{K}(\mathcal{S}) = (Con, \sqsubseteq, [\cdot]_1, \ldots, [\cdot]_n)$ *and* $\uparrow_i^{\mathtt{N}}$ *as in Definition 12. The self-map* $\uparrow_i^{\mathtt{N}}$ *is a normal right inverse of* $[\cdot]_i$.

We conclude this section with the order of the right-inverses we identified.

Corollary 1. *Let* $\mathbf{K}(\mathcal{S}) = (Con, \sqsubseteq, [\cdot]_1, \ldots, [\cdot]_n)$ *be a Kripke scs over* $\mathcal{S} \subseteq \mathcal{M}^D$. *Let* $\uparrow_i^{\mathtt{M}}, \uparrow_i^{\mathtt{MN}}$, *and* $\uparrow_i^{\mathtt{N}}$ *as in Definitions 9, 11 and 12, respectively. Then* $\uparrow_i^{\mathtt{N}} \sqsubseteq \uparrow_i^{\mathtt{MN}} \sqsubseteq \uparrow_i^{\mathtt{M}}$.

5   Applications

In this section we will apply and briefly discuss the results obtained in the previous section in the context of modal logic. First we recall the notion of modal language.

Definition 13 (Modal Language). *Let Φ be a set of primitive propositions. The modal language $\mathcal{L}_n(\Phi)$ is given by the following grammar: $\phi, \psi, \ldots := p \mid \phi \wedge \psi \mid \neg\phi \mid \square_i\phi$ where $p \in \Phi$ and $i \in \{1, \ldots, n\}$. We shall use the abbreviations $\phi \vee \psi$ for $\neg(\neg\phi \wedge \neg\psi)$, $\phi \Rightarrow \psi$ for $\neg\phi \vee \psi$, $\phi \Leftrightarrow \psi$ for $(\phi \Rightarrow \psi) \wedge (\psi \Rightarrow \phi)$, the constant false ff for $p \wedge \neg p$, and the constant tt for $\neg$ff.*

We say that a pointed KS (M, s) *satisfies* ϕ iff $(M, s) \models \phi$ where $\models$ is defined inductively as follows: $(M, s) \models p$ iff $\pi_M(s)(p) = 1$, $(M, s) \models \phi \wedge \psi$ iff $(M, s) \models \phi$ and $(M, s) \models \psi$, $(M, s) \models \neg\phi$ iff $(M, s) \not\models \phi$, and $(M, s) \models \square_i\phi$ iff $(M, t) \models \phi$ for every t such that $s \xrightarrow{i}_M t$. This notion of satisfiability is invariant under a standard equivalence on Kripke structures: *Bisimilarity*, itself a central equivalence in concurrency theory [14].

Definition 14 (Bisimilarity). *Let $\mathcal{B}$ be a symmetric relation on pointed KS's. The relation is said to be a* bisimulation *iff for every $((M, s), (N, t)) \in \mathcal{B}$: (1) $\pi_M(s) = \pi_N(t)$ and (2) if $s \xrightarrow{i}_M s'$ then there exists t' s.t. $t \xrightarrow{i}_N t'$ and $((M, s'), (N, t')) \in \mathcal{B}$. We say that (M, s) and (N, t) are bisimilar, written $(M, s) \sim (N, t)$ if there exists a bisimulation $\mathcal{B}$ such that $((M, s), (N, t)) \in \mathcal{B}$.*

The well-known result of *bisimilarity-invariance* for modal satisfiability implies that (M, s) and (M, t) satisfy the same formulae in $\mathcal{L}_n(\Phi)$ whenever $(M, s) \sim (N, t)$ [14].

Modal logics are typically interpreted over different classes of KS's obtained by imposing conditions on their accessibility relations. Let $\mathcal{S}_n(\Phi)$ be a non-empty set of n-agent Kripke structures over a set of primitive propositions Φ. A modal formula ϕ is said to be *valid* in $\mathcal{S}_n(\Phi)$ iff $(M, s) \models \phi$ for each (M, s) such that $M \in \mathcal{S}_n(\Phi)$.

We can interpret modal formulae as constraints in a given Kripke scs $\mathbf{C} = \mathbf{K}(\mathcal{S}_n(\Phi))$ as follows.

Definition 15 (Kripke Constraint Interpretation). *Let $\mathbf{C}$ be a Kripke scs $\mathbf{K}(\mathcal{S}_n(\Phi))$. Given a modal formula ϕ in the modal language $\mathcal{L}_n(\Phi)$, its interpretation in the Kripke scs $\mathbf{C}$ is the constraint $\mathbf{C}[\![\phi]\!]$ inductively defined as follows:*

$$\mathbf{C}[\![p]\!] = \{(M, s) \mid \pi_M(s)(p) = 1\}$$
$$\mathbf{C}[\![\phi \wedge \psi]\!] = \mathbf{C}[\![\phi]\!] \sqcup \mathbf{C}[\![\psi]\!]$$
$$\mathbf{C}[\![\neg\phi]\!] = \sim \mathbf{C}[\![\phi]\!]$$
$$\mathbf{C}[\![\square_i\phi]\!] = [\, \mathbf{C}[\![\phi]\!] \,]_i$$

Remark 2. One can verify that for any Kripke scs $\mathbf{K}(\mathcal{S}_n(\Phi))$, the Heyting negation $\sim c$ (Definition 4) is $\Delta \setminus c$ where Δ is the set of all pointed Kripke structures (M, s) such that $M \in \mathcal{S}_n(\Phi)$ (i.e., boolean negation). Similarly, Heyting implication $c \to d$ is equivalent to $(\sim c) \cup d$ (i.e., boolean implication).

It is easy to verify that the constraint $\mathbf{C}[\![\phi]\!]$ includes those pointed KS (M, s), where $M \in \mathcal{S}_n(\Phi)$, such that $(M, s) \models \phi$. Thus, ϕ is valid in $\mathcal{S}_n(\Phi)$ if and only if $\mathbf{C}[\![\phi]\!] = true$.

Notice that from Proposition 3 and Theorem 1, each space function $[\cdot]_i$ of $\mathbf{K}(\mathcal{S}_n(\Phi))$ is a normal self-map. From Definitions 5 and 15 we can derive the following standard property stating that $\Box_i$ is a normal modal operator: (*Necessitation*) If ϕ is valid in $\mathcal{S}_n(\Phi)$ then $\Box_i\phi$ is valid in $\mathcal{S}_n(\Phi)$, and (*Distribution*) $\Box_i(\phi \Rightarrow \psi) \Rightarrow (\Box_i\phi \Rightarrow \Box_i\psi)$ is valid in $\mathcal{S}_n(\Phi)$.

Right-Inverse Modalities. Reverse modalities, also known as inverse modalities, arise naturally in many modal logics. For example in temporal logics they are past operators [20], in modal logics for concurrency they represent backward moves [19], in epistemic logic they correspond to utterances [13].

To illustrate our results in the previous sections, let us fix a modal language $\mathcal{L}_n(\Phi)$ (whose formulae are) interpreted in an arbitrary Kripke scs $\mathbf{C} = \mathbf{K}(\mathcal{S}_n(\Phi))$. Suppose we wish to extend it with modalities $\Box_i^{-1}$, called reverse modalities also interpreted over the same set of KS's $\mathcal{S}_n(\Phi)$ and satisfying some minimal requirement. The new language is given by the following grammar.

Definition 16 (Modal Language with Reverse Modalities). *Let Φ be a set of primitive propositions. The modal language $\mathcal{L}_n^{+r}(\Phi)$ is given by the following grammar: $\phi, \psi, \ldots := p \mid \phi \wedge \psi \mid \neg\phi \mid \Box_i\phi \mid \Box_i^{-1}\phi$ where $p \in \Phi$ and $i \in \{1, \ldots, n\}$.*

The minimal semantic requirement for each $\Box_i^{-1}$ is that, regardless of the interpretation we give to $\Box_i^{-1}\phi$, we should have:

$$\Box_i\Box_i^{-1}\phi \Leftrightarrow \phi \quad \text{valid in } \mathcal{S}_n(\Phi). \tag{5}$$

We then say that $\Box_i^{-1}$ is a *right-inverse* modality for $\Box_i$ (by analogy to the notion of right-inverse of a function).

Since $\mathbf{C}[\![\Box_i\phi]\!] = [\mathbf{C}[\![\phi]\!]]_i$, we can use the results in the previous sections to derive semantic interpretations for $\Box_i^{-1}\phi$ by using a right inverse $\uparrow_i$ for the space function $[\cdot]_i$ in Definition 15. Assuming that such a right inverse exists, we can then interpret the reverse modality in $\mathbf{C}$ as

$$\mathbf{C}[\![\Box_i^{-1}\phi]\!] = \uparrow_i(\mathbf{C}[\![\phi]\!]). \tag{6}$$

Since each $\uparrow_i$ is a right inverse of $[\cdot]_i$, it is easy to verify that the interpretation satisfies the requirement in Eq. 5. Furthermore, from Theorem 2 we can conclude that for each $M \in \mathcal{S}_n(\Phi)$, $\xrightarrow{i}_M$ must necessarily be determinant-complete.

Normal Inverse Modalities. We can choose $\uparrow_i$ in Eq. 6 from the set $\{\uparrow_i^{\mathrm{N}}, \uparrow_i^{\mathrm{MN}}, \uparrow_i^{\mathrm{M}}\}$ of right-inverse constructions in Sect. 4.3. Assuming that $\uparrow_i$ is a normal self-map (e.g., either $\uparrow_i^{\mathrm{N}}$ or $\uparrow_i^{\mathrm{MN}}$), we can show from Definition 5 and Eq. 6 that $\Box_i^{-1}$ is itself a normal modal operator in the following sense: (1) If ϕ is valid in $\mathcal{S}_n(\Phi)$ then $\Box_i^{-1}\phi$ is valid in $\mathcal{S}_n(\Phi)$, and (2) $\Box_i^{-1}(\phi \Rightarrow \psi) \Rightarrow (\Box_i^{-1}\phi \Rightarrow \Box_i^{-1}\psi)$ is valid in $\mathcal{S}_n(\Phi)$.

Inconsistency Invariance. Since we assumed a right inverse for $[\cdot]_i$, from Proposition 1(1) we should have

$$\neg\Box_i\mathit{ff} \text{ valid in } \mathcal{S}_n(\Phi) \tag{7}$$

(recall that $f\!f$ is the constant false). Indeed using the fact that $[\cdot]_i$ is a normal self-map with an inverse $\uparrow_i$ and Theorem 1, we can verify the following:

$$\mathbf{C}[\![\Box_i f\!f]\!] = \mathbf{C}[\![\Box_i(f\!f \wedge \Box_i^{-1} f\!f)]\!] = \mathbf{C}[\![\Box_i f\!f \wedge \Box_i \Box_i^{-1} f\!f]\!] = \mathbf{C}[\![\Box_i f\!f \wedge f\!f]\!] = \mathbf{C}[\![f\!f]\!]$$

This implies $\Box_i f\!f \Leftrightarrow f\!f$ is valid in $\mathcal{S}_n(\Phi)$ and this means that $\neg\Box_i f\!f$ is valid in $\mathcal{S}_n(\Phi)$.

Modal systems such K_n or Hennessy-Milner logic [14] where $\neg\Box_i f\!f$ is not an axiom cannot be extended with a reverse modality satisfying Eq. 5 (without restricting their models). The issue is that the axiom $\neg\Box_i f\!f$, typically needed in epistemic, doxastic and temporal logics, would require the accessibility relations of agent i to be serial (recall that determinant-complete relations are necessarily serial). In fact $\Box_i f\!f$ is used in HM logic to express deadlocks wrt to i; $(M, s) \models \Box_i f\!f$ iff there is no s' such that $s \xrightarrow{i}_M s'$. Clearly there cannot be state deadlocks wrt i if $\xrightarrow{i}_M$ is required to be serial for each M.

Bisimilarity Invariance. Recall that bisimilarity invariance says that bisimilar pointed KS's satisfy the same formulae in $\mathcal{L}_n(\Phi)$. The addition of a reverse modality $\Box_i^{-1}$ may violate this invariance: Bisimilar pointed KS's may not longer satisfy the same formulae in $\mathcal{L}_n^{+r}(\Phi)$. This can be viewed as saying that the addition of inverse modalities increases the distinguishing power of the original modal language. We prove this next.

Let us suppose that the chosen right inverse $\uparrow_i$ in Eq. 6 is any normal self-map whatsoever. It follows from Lemma 1(2) and Eq. 6 that if t is *multiply-determined* wrt $\xrightarrow{i}_M$ then $(M, t) \models \Box_i^{-1} f\!f$. We can use Lemma 1(1) and Eq. 6 to show that if t is *uniquely determined* wrt $\xrightarrow{i}_M$ then $(M, t) \not\models \Box_i^{-1} f\!f$.

Now take v and s_4 as in Fig. 1. Suppose that $\pi_{M_5}(v) = \pi_{M_1}(s_i)$ for every s_i in the states of M_1. Clearly $(M_1, s_4) \sim (M_5, v)$. Since s_4 is multiply determined and v is uniquely determined, we conclude that $(M_1, s_4) \models \Box_i^{-1} f\!f$ but $(M_1, v) \not\models \Box_i^{-1} f\!f$. Thus $\Box_i^{-1} f\!f$ can tell uniquely determined states from multiply determined ones but bisimilarity cannot.

Temporal Operators. We conclude this section with a brief discussion on some right-inverse linear-time modalities. Let us suppose that $n = 2$ in our modal language $\mathcal{L}_n(\Phi)$ under consideration (thus interpreted in Kripke scs $\mathbf{C} = \mathbf{K}(\mathcal{S}_2(\Phi))$). Assume further that the intended meaning of the two modalities $\Box_1$ and $\Box_2$ are the *next* operator ($\bigcirc$) and the *henceforth/always* operator ($\Box$), respectively, in a *linear-time* temporal logic. To obtain the intended meaning we take $\mathcal{S}_2(\Phi)$ to be the largest set such that: If $M \in \mathcal{S}_2(\Phi)$, M is a 2-agent KS where $\xrightarrow{1}_M$ is isomorphic to the successor relation on the natural numbers and $\xrightarrow{2}_M$ is the reflexive and transitive closure of $\xrightarrow{1}_M$. The relation $\xrightarrow{1}_M$ is intended to capture the linear flow of time. Intuitively, $s \xrightarrow{1}_M t$ means t is the only next state for s. Similarly, $s \xrightarrow{2}_M t$ for $s \neq t$ is intended to capture the fact that t is one of the infinitely many future states for s.

Let us first consider the next operator $\Box_1 = \bigcirc$. Notice that $\xrightarrow{1}_M$ is determinant-complete. If we apply Eq. 6 with $\uparrow_1 = \uparrow_1^M$, i.e., the greatest right inverse of $[\cdot]_1$, we obtain $\square_1^{-1} = \ominus$, a past modality known in the literature as *strong* previous operator [20]. The operator $\ominus$ is given by $(M,t) \models \ominus \phi$ iff there exists s such that $s \xrightarrow{M}_1 t$ and $(M,s) \models \phi$. If we take $\uparrow_i$ to be the normal right inverse $\uparrow_i^{N}$, we obtain $\square_1^{-1} = \widetilde{\ominus}$ the past modality known as *weak* previous operator [20]. The operator $\widetilde{\ominus}$ is given by $(M,t) \models \widetilde{\ominus} \phi$ iff for every s if $s \xrightarrow{M}_1 t$ then $(M,s) \models \phi$. Notice that the only difference between the two operators is the following: If s is an indetermined/initial state wrt $\xrightarrow{1}_M$ then $(M,s) \not\models \ominus \phi$ and $(M,s) \models \widetilde{\ominus} \phi$ for any ϕ.

Let us now consider the always operator $\square_2 = \square$. Notice that $\xrightarrow{2}_M$ is not determinant-complete: Take any sequence $s_0 \xrightarrow{1}_M s_1 \xrightarrow{1}_M \dots$ The state s_1 is not determinant because for every s_j such that $s_1 \xrightarrow{2}_M s_j$ we have $s_0 \xrightarrow{2}_M s_j$. Theorem 2 says that there is no right-inverse $\uparrow_2$ of $[\cdot]_i$ that can give us a $\square_2^{-1}$ satisfying Eq. 5.

By analogy to the above-mentioned past operators, one may think that the past operator *it-has-always-been* $\boxminus$ [24] may provide a reverse modality for $\square$ in the sense of Eq. 5. The operator is given by $(M,t) \models \boxminus\phi$ iff $(M,s) \models \phi$ for every s such that $s \xrightarrow{2}_M t$. Clearly $\square\boxminus\phi \Rightarrow \phi$ is valid in $\mathcal{S}_2(\Phi)$ but $\phi \Rightarrow \square\boxminus \phi$ is not.

6   Concluding Remarks and Related Work

We studied the existence and derivation of right inverses (extrusion) of space functions for the Kripke spatial constraint systems. We showed that being *determinant-complete* is the weakest condition on KS's that guarantees the existence of such right inverses. We identified the greatest normal right inverse of any given space function. We applied these results to modal logic by using space functions and their right inverses as the semantic counterparts of box modalities and their right inverse modalities. We discussed our results in the context of modal concepts such as bisimilarity invariance, inconsistency invariance and temporal modalities.

Most of the related work was discussed in the previous sections. In previous work [13] the authors derived an inverse modality but only for the specific case of a logic of belief. The work was neither concerned with giving a *complete* characterization of the existence of right inverse nor deriving *normal* inverses. The constraint systems in this paper can be seen as modal extension of geometric logic [26]. Modal logics have also been studied from an algebraic perspective by using modal extensions of boolean and Heyting algebras in [2,4,17]. These works, however, do not address issues related to inverse modalities. Inverse modalities have been used in temporal, epistemic and logic for concurrency. In [24] the authors discuss inverse temporal and epistemic modalities from a proof theory perspective. The works [5,12,19] use modal logic with reverse modalities for specifying true concurrency and [6,7] use backward modalities for characterizing branching bisimulation. None of these works is concerned with an algebraic approach or with deriving inverse modalities for modal languages.

References

1. Abramsky, S., Jung, A.: Domain theory. In: Abramsky, S., et al. (ed.) Handbook of Logic in Computer Science, vol. 3, pp. 1–168. Oxford University Press, Oxford (1994)
2. Blackburn, P., De Rijke, M., Venema, Y.: Modal Logic, 1st edn. Cambridge University Press, Cambridge (2002)
3. Boer, F.S., Di Pierro, A., Palamidessi, C.: Nondeterminism and infinite computations in constraint programming. Theor. Comput. Sci. **151**, 37–78 (1995)
4. Chagrov, A., Zakharyaschev, M.: Modal Logic. Oxford Logic Guides, vol. 35 (1997)
5. De Nicola, R., Ferrari, G.L.: Observational logics and concurrency models. In: Nori, K.V., Veni Madhavan, C.E. (eds.) FSTTCS 1990. LNCS, vol. 472, pp. 301–315. Springer, Heidelberg (1990). doi:10.1007/3-540-53487-3_53
6. De Nicola, R., Montanari, U., Vaandrager, F.: Back and forth bisimulations. In: Baeten, J.C.M., Klop, J.W. (eds.) CONCUR 1990. LNCS, vol. 458, pp. 152–165. Springer, Heidelberg (1990). doi:10.1007/BFb0039058
7. De Nicola, R., Vaandrager, F.: Three logics for branching bisimulation. J. ACM (JACM) **42**(2), 458–487 (1995)
8. Díaz, J.F., Rueda, C., Valencia, F.D.: Pi+- calculus: a calculus for concurrent processes with constraints. CLEI Electron. J. **1**(2), 2 (1998)
9. Fages, F., Ruet, P., Soliman, S.: Linear concurrent constraint programming: operational and phase semantics. Inf. Comput. **165**, 14–41 (2001)
10. Fagin, R., Halpern, J.Y., Moses, Y., Vardi, M.Y.: Reasoning About Knowledge, 4th edn. MIT Press, Cambridge (1995)
11. Falaschi, M., Olarte, C., Palamidessi, C., Valencia, F.: Declarative diagnosis of temporal concurrent constraint programs. In: Dahl, V., Niemelä, I. (eds.) ICLP 2007. LNCS, vol. 4670, pp. 271–285. Springer, Heidelberg (2007). doi:10.1007/978-3-540-74610-2_19
12. Goltz, U., Kuiper, R., Penczek, W.: Propositional temporal logics and equivalences. In: Cleaveland, W.R. (ed.) CONCUR 1992. LNCS, vol. 630, pp. 222–236. Springer, Heidelberg (1992). doi:10.1007/BFb0084794
13. Haar, S., Perchy, S., Rueda, C., Valencia, F.D.: An algebraic view of space/belief and extrusion/utterance for concurrency/epistemic logic. In: PPDP 2015, pp. 161–172. ACM (2015)
14. Hennessy, M., Milner, R.: Algebraic laws for nondeterminism and concurrency. J. ACM (JACM) **32**(1), 137–161 (1985)
15. Knight, S., Palamidessi, C., Panangaden, P., Valencia, F.D.: Spatial and epistemic modalities in constraint-based process calculi. In: Koutny, M., Ulidowski, I. (eds.) CONCUR 2012. LNCS, vol. 7454, pp. 317–332. Springer, Heidelberg (2012). doi:10.1007/978-3-642-32940-1_23
16. Kripke, S.A.: Semantical considerations on modal logic. Acta Philos. Fennica **16**, 83–94 (1963)
17. Macnab, D.: Modal operators on heyting algebras. Algebra Univers. **12**(1), 5–29 (1981)
18. Nielsen, M., Palamidessi, C., Valencia, F.D.: Temporal concurrent constraint programming: denotation, logic and applications. Nord. J. Comput. **9**(1), 145–188 (2002)
19. Phillips, I., Ulidowski, I.: A logic with reverse modalities for history-preserving bisimulations. In: EXPRESS 2011. EPTCS, vol. 64, pp. 104–118 (2011)

20. Pnueli, A., Manna, Z.: The Temporal Logic of Reactive and Concurrent Systems. Springer, New York (1992)
21. Popkorn, S.: First Steps in Modal Logic, 1st edn. Cambridge University Press, Cambridge (1994)
22. Prior, A.N.: Past, Present and Future, vol. 154. Oxford University Press, Oxford (1967)
23. Réty, J.: Distributed concurrent constraint programming. Fundam. Inf. **34**, 323–346 (1998)
24. Ryan, M., Schobbens, P.-Y.: Counterfactuals and updates as inverse modalities. J. Logic Lang. Inf. **6**, 123–146 (1997)
25. Saraswat, V.A., Rinard, M., Panangaden, P.: Semantic foundations of concurrent constraint programming. In: POPL 1991, pp. 333–352. ACM (1991)
26. Vickers, S.: Topology via Logic, 1st edn. Cambridge University Press, Cambridge (1996)

Specifications

Specifying Properties of Dynamic Architectures Using Configuration Traces

Diego Marmsoler[✉] and Mario Gleirscher

Technische Universität München, Munich, Germany
{diego.marmsoler,mario.gleirscher}@tum.de

Abstract. The architecture of a system describes the system's overall organization into components and connections between those components. With the emergence of mobile computing, dynamic architectures became increasingly important. In such architectures, components may appear or disappear, and connections may change over time.

Despite the growing importance of dynamic architectures, the specification of properties for those architectures remains a challenge. To address this problem, we introduce the notion of *configuration traces* to model properties of dynamic architectures. Then, we investigate these properties to identify different types thereof. We show *completeness* and *consistency* of these types, i.e., we show that (almost) every property can be separated into these types and that a property of one type does not impact properties of other types.

Configuration traces can be used to specify general properties of dynamic architectures and the separation into different types provides a systematic way for their specification. To evaluate our approach we apply it to the specification and verification of the Blackboard pattern in Isabelle/HOL.

1 Introduction

A systems architecture provides a set of components and connections between their ports. With the emergence of mobile computing, dynamic architectures became more and more important [5, 10, 20]. In such architectures, components can appear and disappear and connections can change, both over time.

Despite the increasing importance of dynamic architectures some questions regarding their specification still remain:

- How can *properties* of dynamic architectures be specified in general?
- How can those properties be separated into different *types*?

A property of dynamic architectures characterizes executions of such architectures. Consider, for example, the following property for a publisher-subscriber [8] system: *Whenever a component p of type `Publisher` provides a message for which a `Subscriber` component s was subscribed, s is connected to p.* Another example describes a property of a Blackboard architecture [8]: *Whenever a component of type `BlackBoard` provides a message containing a problem to be solved,*

© Springer International Publishing AG 2016
A. Sampaio and F. Wang (Eds.): ICTAC 2016, LNCS 9965, pp. 235–254, 2016.
DOI: 10.1007/978-3-319-46750-4_14

a component of type `KnowledgeSource`, *able to solve this problem, is eventually activated.* Usually, such properties can be separated into different types, such as: (i) *Behavior properties* characterizing the behavior of certain components. (ii) *Activation properties* characterizing the activation/deactivation of components. (iii) *Connection properties* characterizing the dynamic connection between components.

To answer the above questions, we first introduce a formal model of dynamic architectures. Thereby we model an architecture as a set of configuration traces which, in turn, is a sequence over architecture configurations. An architecture configuration is modeled as a set of components, valuations of the component ports with messages, and connections between these ports.

In a second step, we characterize *behavior*, *activation*, and *connection* properties. We show the distinct nature of those types of properties and investigate their expressive power. Thereby we characterize the notion of *separable architecture property* and show that each of them can be *uniquely* described through the intersection of a corresponding behavior, activation, and connection property.

We evaluate our approach by specifying (and analyzing) the Blackboard pattern for dynamic architectures using the Isabelle/HOL [22] interactive theorem prover. Therefore, we first specify behavior, activation, and connection properties for Blackboard architectures. Then, we specify the pattern's guarantee as an architecture property. Finally, we verify the pattern by proving its guarantee from the original properties using Isabelle's structured proof language Isar [28].

The remainder of the article is organized as follows: Sect. 2 reviews existing work in this area. Section 3 introduces the Blackboard pattern as a running example. Section 4 introduces our model for dynamic architectures. Section 5 describes and investigates different types of properties for those architectures. Section 6 presents an approach to systematically specify properties and applies it to specify the Blackboard pattern. In Sect. 7 we provide a critical discussion of possible weaknesses of the approach. Finally, Sect. 8 summarizes our results and discusses potential implications and future work.

2   Background and Related Work

Related work can be found in three different areas: 1. Architecture Description Languages, 2. Modeling of Architectural Styles, and 3. Modeling of Constraints for Dynamic Architectures. In the following we briefly discuss each of them.

2.1   Architecture Description Languages

Over the last three decades, a number of so-called Architecture Description Languages (ADLs) emerged to support the formal specification of architectures. Some of them also support the specification of dynamic aspects such as Rapide [17], Darwin [18], Dynamic Wright [2,3], Π-ADL [23], xADL [11], and ACME [13].

While ADLs support the formal specification of architectures, they were developed with the aim to specify individual architecture solutions, rather than properties for architectures which require more abstract specification techniques. Nevertheless, these works provide the conceptual foundation for our work since many of the abstractions used in our model are based on the concepts introduced by ADLs.

2.2 Modeling Architectural Styles

Architectural styles focus on the specification of architectural constraints, rather than specific architectures.

One of the first approaches to formalize architectural styles is discussed by Abowd et al. [1]. There, the authors apply a denotational semantics approach to software architectures by using the specification language Z [26]. Other examples used to specify architectural styles include the Chemical Abstract Machine [15] or Wright [3] which allow for the specification of architectural constraints for static architectures. Two further ideas come from Moriconi et al. [21] and Penix et al. [24]. Both apply algebraic specification to software architectures. Finally, Bernardo et al. [4] use process algebras to specify architectural types which can be seen as a form of architectural styles.

While these approaches focus on the specification of architectural constraints rather than architectures, they do usually not allow for the specification of dynamic architectural constraints which is the focus of this work. Nevertheless, these works provide many important conceptual insights into the specification of architectural constraints on which we build.

2.3 Specification of Constraints for Dynamic Architectures

Work in this area is most closely related to our work.

The approach of Le Métayer [16] applies graph theory to specify architectural evolution. The author proposes the use of graph grammars to specify architectural evolution. A similar approach comes from Hirsch and Montanari [14] who employ hypergraphs as a formal model to represent styles and their reconfigurations. While we also apply a graph-based approach to model architectural properties, the major difference lies in the specification of behavior. While the discussed approaches focus on structural aspects, we aim at a combination of structural and behavioral aspects.

Another, closely related approach is the one of Wermlinger et al. [29]. The authors combine behavior and structure to model dynamic reconfigurations. One major difference to our work concerns the underlying model of interaction. While the authors use an action synchronization communication model, our model is based on time-synchronous communication. Both communication models have their advantages and drawbacks. Thus, by providing a time-synchronous alternative, we actually complement their work.

Recently, categorical approaches to dynamic architecture reconfiguration appeared such as the work of Castro et al. [9] or Fiadeiro and Lopes [12].

While these approaches provide fundamental insights into the specification of dynamic architecture properties, their model remains implicitly in the categorical constructions. Thus, we complement their work by providing an explicit model of dynamic architecture properties.

Finally, we do not know of any existing work investigating different types of properties of dynamic architectures. However, as stated in the introduction, this is an important aspect to systematically specify properties of dynamic architectures. In this work we provide a formal investigation of properties which is another contribution to current literature.

3   Running Example: Specifying Blackboard Architectures

In this paper, we use the Blackboard architecture design pattern as a running example to show our approach to the specification and verification of dynamic architectures. This pattern was described, for example, by Shaw and Garlan [25], Buschmann et al. [8], and Taylor et al. [27].

Blackboards work with *problems* and *solutions* for them. Hence, we denote by PROB the set of all problems and by SOL the set of all solutions. Complex problems consist of *subproblems* which can be complex themselves. To solve a problem, its subproblems have to be solved first. Therefore, we assume the existence of a *subproblem relation* $\prec\ \subseteq$ PROB $\times$ PROB. For complex problems, this relation may not be known in advance. Indeed, one of the benefits of a Blackboard architecture is that a problem can be solved also without knowing this relation in advance. However, the subproblem relation has to be well-founded (*wf*) for a problem to be solvable. In particular, we do not allow cycles in the transitive closure of $\prec$.

While there may be different approaches to solve a problem (i.e. several ways to split a problem into subproblems), we assume that the final solution for a problem is unique. Thus, we assume the existence of a function solve: PROB $\rightarrow$ SOL which assigns the *correct* solution to each problem. Note, however, that this function is not known in advance and it is one of the reasons of using this pattern to calculate this function.

4   A Model of Dynamic Architectures

In the following we introduce our model of dynamic architectures. It is based on Broy's FOCUS theory [6] and an adaptation of its dynamic extension [7]. A property is modeled as a set of *configuration traces* which are sequences of *architecture configurations* that, in turn, consist of a set of *active components*, valuation of their ports with type-conform messages, and *connections* between their ports. The model serves the specification of properties for dynamic architectures as shown by the running example.

4.1   Foundations

This section introduces basic concepts of our model such as ports which can be valuated by messages.

Convention 1. *In the following, we denote by $X \dashrightarrow Y$, the set of partial functions from a set X to a set Y. For a partial function $f \colon X \dashrightarrow Y$, we denote by:*

- $\operatorname{dom}(f)$ *the domain of f,*
- $\operatorname{ran}(f)$ *the range of f, and by*
- $f \mid_{X'}$ *the restriction of f to the set $X' \subseteq X$. If $X = \mathbb{N}$ and $x \in \mathbb{N}$ we denote by $f \downarrow_x \stackrel{def}{=} f \mid_{\{n \in \mathbb{N} \mid n \leqslant x\}}$ the restriction of f to the first x numbers.*

If $\operatorname{dom}(f) = X$, f is called total *and denoted by $f \colon X \to Y$.*

Messages and ports. In our model, components communicate by exchanging *messages* over *ports*. Thereby, ports are typed by a set of messages which can go through the corresponding port. Thus, we assume the existence of the following sets:

- set M containing all messages,
- sets $\mathsf{P_i}$ and $\mathsf{P_o}$ containing all input and output ports, respectively, and set $\mathsf{P} = \mathsf{P_i} \cup \mathsf{P_o}$ containing all ports. We require a port to be either input or output, but not both:

$$\mathsf{P_i} \cap \mathsf{P_o} = \varnothing. \tag{1}$$

Moreover, we assume the existence of a type function which assigns a set of messages to each port:

$$(T_p)_{p \in \mathsf{P}}, \text{ with } T_p \subseteq \mathsf{M} \text{ for each } p \in \mathsf{P}. \tag{2}$$

Valuation. In our model, components communicate by sending and receiving messages through ports. This is achieved through the notion of *port valuation*. Roughly speaking, a valuation for a set of ports is an assignment of messages to each port. Note that in our model, ports can be valuated by a set of messages meaning that a component can send/receive no message, a single message, or multiple messages at each point in time.

For ports $P \subseteq \mathsf{P}$, we denote by $\overline{P}$ the set of all possible *port-valuations*, formally,

$$\overline{P} \stackrel{def}{=} \{\mu \colon P \to \wp(\mathsf{M}) \mid \forall p \in P \colon \mu(p) \subseteq T_p\}. \tag{3}$$

Moreover, we denote by $[p_1, p_2, \ldots \mapsto \{m_1\}, \{m_2\}, \ldots]$ the valuation of ports $p_1, p_2, \ldots$ with sets $\{m_1\}, \{m_2\}, \ldots$, respectively. For singleton sets we shall sometimes omit the set parentheses and simply write $[p_1, p_2, \ldots \mapsto m_1, m_2, \ldots]$.

4.2   Components and Interfaces

This section introduces the basic notions of component and interface.

Components. In our model, the basic unit of computation is a *component*. A component is identified by a component identifier which is why we postulate the existence of the set of all component identifiers C.

Component port valuation. In our model, the same port can be reused by different components. Thus, to uniquely identify a *component port*, we need to combine it with the corresponding component. Therefore, we generalize the notion of port valuation introduced in Eq. (3) to component ports $P \subseteq C \times P$ as follows:

$$\overline{P} \overset{\text{def}}{=} \{\mu \colon P \to \wp(\mathsf{M}) \mid \forall (c, p) \in P \colon \mu((c, p)) \subseteq T_p\}.$$

Interfaces. A component communicates with its environment through an interface by sending and receiving messages over ports.

Definition 2. *An* interface *is a pair* (P_i, P_o) *with:*

- *input ports* $P_i \subseteq \mathsf{P_i}$, *and*
- *output ports* $P_o \subseteq \mathsf{P_o}$.

The set of all interfaces is denoted by $\mathcal{I}$.

Similar to components, interfaces have an identifier which is why we postulate the existence of the set of all interface identifiers I.

Interface port valuation. As for components, the same port can be used by different interfaces. Thus, to uniquely identify an *interface port*, we need to combine it with the corresponding interface identifier. Therefore, we can generalize the notion of valuation introduced in Eq. (3) to interface ports $\mathsf{I} \times \mathsf{P}$ as done for component port valuations.

4.3   Interface Specifications

An interface specification declares a set of component and interface identifiers. Moreover, it associates an interface identifier with each component identifier and an interface with each interface identifier.

Definition 3. *An* interface specification *is a 4-tuple* (C, I, t^c, t^i) *consisting of:*

- *a set of component identifiers* $C \subseteq \mathsf{C}$,
- *a set of interface identifiers* $I \subseteq \mathsf{I}$,
- *a mapping* $t^c \colon C \to I$, *assigning an interface identifier to each component,*
- *a mapping* $t^i \colon I \to \mathcal{I}$, *which assigns an interface to each interface identifier.*

The set of all interface specifications is denoted by $\mathcal{S}_I$.

Convention 4. *For an n-tuple* $Z = (z_1, \ldots, z_n)$, *we denote by* $[z]^i = z_i$ *with* $1 \leqslant i \leqslant n$ *the projection to the i-th component of* Z.

Convention 5. *For interface specification* $S_i = (C, I, t^c, t^i) \in \mathcal{S}_I$ *we denote by:*

- $\mathsf{in}(I', S_i) \overset{\text{def}}{=} \bigcup_{i \in I'}(\{i\} \times [t^i(i)]^1)$ *the set of input ports,*
- $\mathsf{out}(I', S_i) \overset{\text{def}}{=} \bigcup_{i \in I'}(\{i\} \times [t^i(i)]^2)$ *the set of output ports,*
- $\mathsf{port}(I', S_i) \overset{\text{def}}{=} \mathsf{in}(I', S_i) \cup \mathsf{out}(I', S_i)$ *the set of all ports,*

for a set of interface identifiers $I' \subseteq I$, *respectively.*

The same notation can be used to denote the ports for a set of component identifiers $C' \subseteq C$ by substituting $t^i(i)$ with $t^i(t^c(c))$ for each $c \in C'$ in the above definitions.

4.4   Architecture Configurations and Configuration Traces

Architectures are modeled as sets of configuration traces which are sequences over architecture configurations.

Architecture Configurations. In our model, an architecture configuration connects ports of active components. It consists of a set of active components and a so-called connection relation connecting the component ports.

Definition 6. *An* architecture configuration *over interface specification $S_i = (C, I, t^c, t^i) \in \mathcal{S}_I$ is a triple (C', N, μ), consisting of:*

- *a set of active components $C' \subseteq C$,*
- *a connection $N : \underline{\text{in}(C', S_i)} \dashrightarrow \wp(\text{out}(C', S_i))$,*
- *a valuation $\mu \in \text{port}(C', S_i)$.*

We require connected ports to be consistent in their valuation, i.e. if a component provides messages at its output port, these messages are transferred to the corresponding connected input ports:

$$\forall p_i \in \text{dom}\,(N) : \mu(p_i) = \bigcup_{p_o \in N(p_i)} \mu(p_o). \tag{4}$$

The set of all possible architecture configurations for interface specification $S_i \in \mathcal{S}_I$ is denoted by $\mathcal{K}(S_i)$.

Note that connection N is modeled as a set-valued, partial function from component input ports to component output ports, meaning that:

- input/output ports can be connected to several output/input ports, respectively, and
- not every input/output port needs to be connected to an output/input port, respectively.

Convention 7. *In the following we use $c :: I$ to denote that component variable c requires the corresponding component to have the assigned interface I. Moreover, port names are used to denote the corresponding port valuation.*

Example 1. Assuming $p_1, p_2, p_3, (p_1, s_1), (p_2, s_2) \in \mathsf{M}$, $ks_1, ks_2, bb \in \mathsf{C}$, $i_p, i_s \in \mathsf{P}_i$, and $o_p, o_s \in \mathsf{P}_o$. Figure 1 shows an architecture configuration (C', N, μ) for interface specification S_{BB} (as defined in Sect. 4.5 with $C = \{ks_1, ks_2, bb\}$), with:

- active components $C' = \{ks_1, bb\}$;
- connection N, with $N((bb, o_p)) = \{(ks_1, i_p)\}$, $N((bb, o_s)) = \{(ks_1, i_s)\}$, $N((ks_1, o_p)) = \{(bb, i_p)\}$, $N((ks_1, o_s)) = \{(bb, i_s)\}$; and
- valuation
  $\mu = [(ks_1, i_p), (ks_1, o_p), (bb, o_s), \cdots \mapsto \{p_1, p_2, p_3\}, \{(p_2, \{p_4\})\}, \{(p_1, s_1)\}, \cdots]$.

Convention 8. *For an architecture configuration $k = (C', N, \mu) \in \mathcal{K}(S_i)$ over interface specification $S_i = (C, I, t^c, t^i) \in \mathcal{S}_I$ we denote by*

$$\text{in}_c^o(S_i, k) \;\overset{def}{=}\; \text{in}(C', S_i) \setminus \text{dom}\,(N), \tag{5}$$

the set of open input configuration ports.

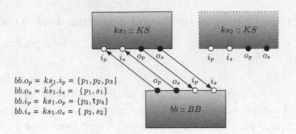

Fig. 1. Architecture configuration

Equivalences Between Architecture Configurations. Architecture configurations can be related according to several aspects. In the following we introduce several notions of architecture configuration equivalence.

Definition 9. *Two architecture configurations* $k = (C', N, \mu)$, $k' = (C'', N', \mu')$ *over interface specification* $S_i \in \mathcal{S}_I$, *with* $k, k' \in \mathcal{K}(S_i)$, *are behavior equivalent, written* $k \approx^b k'$, *iff*

$$\forall p \in \mathsf{port}(C' \cap C'', S_i) \colon \mu(p) = \mu'(p). \tag{6}$$

Definition 10. *Two architecture configurations* $k = (C', N, \mu)$, $k' = (C'', N', \mu')$ *over interface specification* $S_i \in \mathcal{S}_I$, *with* $k, k' \in \mathcal{K}(S_i)$, *are activation equivalent, written* $k \approx^a k'$, *iff*

$$C' = C''. \tag{7}$$

Definition 11. *Two architecture configurations* $k = (C', N, \mu)$, $k' = (C'', N', \mu')$ *over interface specification* $S_i \in \mathcal{S}_I$, *with* $k, k' \in \mathcal{K}(S_i)$, *are connection equivalent, written* $k \approx^n k'$, *iff*

$$\forall p \in \mathsf{in}(C' \cap C'', S_i) \colon N(p) = N'(p). \tag{8}$$

These relations suffice to determine architecture configuration equivalence.

Property 1. Two ACs $k, k' \in \mathcal{K}(S_i)$ are the same iff they are behavior equivalent, connection equivalent and activation equivalent:

$$k = k' \iff k \approx^b k' \wedge k \approx^n k' \wedge k \approx^a k'.$$

However, not every relation is indeed an equivalence relation.

Property 2. Activation equivalence is an equivalence relation. Behavior and connection equivalence are reflexive, symmetric, but not transitive.

Example 2 (Why behavior and connection equivalence are not necessarily transitive). Consider three architecture configurations $k' = (C'', N', \mu')$, $k = (C', N, \mu)$, $k'' = (C''', N'', \mu'') \in \mathcal{K}(S_i)$, such that $C' \subseteq C''$ and $C' \subseteq C'''$ but there exists a $c \in C'' \cap C'''$ which is not in C' and $\mu'(c, p) \neq \mu''(c, p)$ for some port p. Furthermore, assume $k' \approx^b k$ and $k \approx^b k''$. Since $\mu'(c, p) \neq \mu''(c, p)$, we have $k' \not\approx^b k''$.

A similar example can be given for connection equivalence.

Configuration traces. A configuration trace consists of a series of configuration snapshots of an architecture during system execution. Thus, a configuration trace is modeled as a sequence of architecture configurations at a certain point in time.

Definition 12. *A configuration trace (CT) over interface specification $S_i \in \mathcal{S}_I$ is a mapping $\mathbb{N} \to \mathcal{K}(S_i)$. The set of all CTs for S_i is denoted by $\mathcal{K}^t(S_i)$.*

Example 3. Figure 2 shows a configuration trace $t \in \mathcal{K}^t(S_i)$ with corresponding configurations $t(0) = k_0$, $t(1) = k_1$, and $t(2) = k_2$. Configuration k_0, for example, is shown in Example 1.

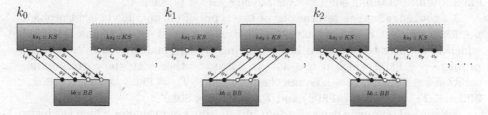

Fig. 2. Configuration trace (port valuations not shown, see Fig. 1 for an example)

Note that an architecture property is modeled as a *set of configuration traces*, rather than just one single trace. This is due to the fact that component behavior, as well as the appearance and disappearance of components, and the reconfiguration of the architecture is usually non-deterministic and dependent on the current input provided to an architecture.

Moreover, note that our notion of architecture is highly dynamic in the following sense:

– *components* may appear and disappear over time and
– *architecture configurations* may change over time.

4.5   Running Example: Blackboard Interface Specification

A Blackboard architecture consists of a `BlackBoard` component and several `KnowledgeSource` components. Figure 3 shows an interface specification $S_{BB} = (C, I, t^c, t^i) \in \mathcal{S}_I$ of the pattern.

BlackBoard interface. A `BlackBoard` (BB) is used to capture the current state on the way to a solution of the original problem. Its state consists of all currently open subproblems and solutions for subproblems.

A `BlackBoard` expects two types of input: 1. via i_p: a problem $p \in$ PROB which a `KnowledgeSource` is able to solve, together with a set of subproblems $P \subseteq$ PROB the `KnowledgeSource` requires to be solved before solving the original problem P, 2. via i_s: a problem $p \in$ PROB solved by a `KnowledgeSource`, together with the corresponding solution $s \in$ SOL.

A BlackBoard returns two types of output: 1. via o_p: a set $P \subseteq \text{PROB}$ which contains all the problems to be solved, 2. via o_s: a set of pairs $PS \subseteq \text{PROB} \times \text{SOL}$. Thus, we require the port types: $T_{i_p} = \text{PROB} \times \wp(\text{PROB})$, and $T_{i_s} = \text{PROB} \times \text{SOL}$, $T_{o_p} = \text{PROP}$ and $T_{o_s} = \text{PROB} \times \text{SOL}$.

KnowledgeSource interface. A KnowledgeSource (KS) is a domain expert able to solve problems in that domain. It may lack expertise of other domains. Moreover, it can recognize problems which it is able to solve and subproblems which have to be solved first by other KnowledgeSources.

A KnowledgeSource expects two types of input: 1. via i_p: a set $P \subseteq \text{PROB}$ which contains all the problems to be solved, 2. via i_s: a set of pairs $PS \subseteq \text{PROB} \times \text{SOL}$ containing solutions for already solved problems.

A KnowledgeSource returns one of two types of output: 1. via o_p: a problem $p \in \text{PROB}$ which it is able to solve together with a set of subproblems $P \subseteq \text{PROB}$ which it requires to be solved before solving the original problem, 2. via o_s: a problem $p \in \text{PROB}$ which it was able to solve together with the corresponding solution $s \in \text{SOL}$. Thus, we require the port types: $T_{i_p} = \text{PROB}$ and $T_{i_s} = \text{PROB} \times \text{SOL}$ and $T_{o_p} = \text{PROB} \times \wp(\text{PROB})$ and $T_{o_s} = \text{PROB} \times \text{SOL}$.

A KnowledgeSource can solve only certain types of problems which is why we assume the existence of a mapping $prob \colon C \to \text{PROB}$ to associate a set of problems with each KnowledgeSource. Then we require for each KnowledgeSource that it only solves problems given by this mapping:

$$\forall k \in \mathcal{K}(S_{BB}), (c,p) \in \text{out}(S_i, k) \colon t^c(c) = KS \implies \left[\left[k\right]^3(p)\right]^1 \in prob(c). \quad (9)$$

While we assume only one BlackBoard component $bb \in C$, the number of KnowledgeSource components is not restricted.

Fig. 3. Interface specification for Blackboards.

5   Specifying Properties of Dynamic Architectures

Properties of dynamic architectures can be specified as sets of configuration traces over an interface specification. In the following, we investigate the nature of such properties and introduce the notion of *behavior*, *activation*, and *connection* properties as special kinds of *architecture properties* to our model. Moreover, we introduce the notion of *separable architecture property* and show that such a property can always be represented as the intersection of corresponding behavior, activation, and connection properties. Then, we show that the intersection of such properties is guaranteed to be non-empty, given that the properties themselves are non-empty.

This way, we get a step-wise method for the specification of properties for dynamic architectures by concentrating on the three different property-types as shown below by our running example.

5.1   Architecture Properties

We first introduce a basic notion of architecture property which serves as a foundation for all classes of architecture properties discussed below. An architecture property is a set of configuration traces which does not constrain valuation of open input ports. Thus, an architecture property is defined as a set of configuration traces fulfilling a special closure property.

Definition 13. *An architecture property (AP) is a set of configuration traces P, such that input port valuations are not restricted:*

$$\forall t \in P, n \in \mathbb{N}, \mu \in \overline{\text{in}_c^o(S_i, t(n))} \ \exists t' \in P \colon t' \downarrow_{n-1} = t \downarrow_{n-1} \wedge$$

$$\forall p \in \text{in}_c^o(S_i, t(n)) \colon [t'(n)]^3 (p) = \mu(p). \tag{10}$$

5.2   Behavior Properties

A behavior property is an architecture property which does not constrain connections and activations. Thus, a behavior property is defined as a set of configuration traces fulfilling a special closure property.

Definition 14. *A behavior property (BP) for an interface specification $S_i = (C, I, t^c, t^i) \in \mathcal{S}_I$, is an AP $B \subseteq \mathcal{K}^t(S_i)$, such that connections and activations are not restricted:*

$$\forall t \in B, n \in \mathbb{N}, k \in \{k \in \mathcal{K}(S_i) \mid k \approx^b t(n)\}$$

$$\exists t' \in B \colon t' \downarrow_{n-1} = t \downarrow_{n-1} \wedge t'(n) \approx^a k \wedge t'(n) \approx^n k. \tag{11}$$

Example 4. Figure 4 shows how an architecture property B can violate Definition 14: Assume that B allows a configuration trace t with $t(0)$ and denies some k with $k \approx^b t(0)$ at $n = 0$, i.e. $\nexists t' \in B \colon t'(0) \approx^a k \vee t'(n) \approx^n k$. Hence, B constrains activation and, thus, contains unnecessary parts of an activation property.

Running Example: Behavior Property Specification. We provide behavior properties for both, BlackBoard and KnowledgeSource components. Thereby we use a temporal-logic notation (based on [19]) to specify sets of configuration traces. Variables denote component identifiers, problems and solutions. Port names are used to denote port valuations and $c :: I$ is used to denote that

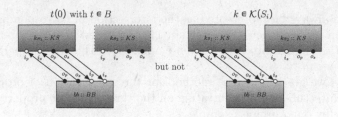

Fig. 4. Example of an ill-formed behavior property.

component identifier c has interface I.

BlackBoard behavior. A BlackBoard provides the *current state* towards solving the original problem. If a KnowledgeSource requires subproblems to be solved, the BlackBoard redirects those problems to other KnowledgeSources. Moreover, the BlackBoard provides available solutions to all KnowledgeSources.

We view a BlackBoard as a set of configuration traces $\mathcal{K}^t(S_{BB})$ specified by three behavior properties:

– if a solution to a subproblem is received on its input, then it is eventually provided at its output:

$$\Box\left((p,s) \in (bb, i_s) \implies \Diamond\left((p,s) \in (bb, o_s)\right)\right), \tag{12}$$

– if solving a problem requires a set of subproblems to be solved first, those problems are eventually provided at its output:

$$\Box\left((p,P) \in (bb, i_p) \implies (\forall p' \in P \colon \Diamond\left(p' \in (bb, o_p)\right))\right), \tag{13}$$

– a problem is provided as long as it is not solved:

$$\Box\left(p \in (bb, o_p) \implies ((p \in (bb, o_p)) \; \mathcal{W} \; ((p, \mathtt{solve}(p)) \in (bb, i_s)))\right). \tag{14}$$

KnowledgeSource behavior. A KnowledgeSource receives open problems via i_p and solutions for other problems via i_s. It might contribute to the solution of the original problem by solving subproblems. Hence, it performs one of two possible actions: 1. If it has solutions for all the required subproblems, it solves the problem and publishes the solution via o_s, 2. If it requires solutions to subproblems, it notifies the BlackBoard about its ability to solve the problem and about these subproblems via o_p.

We view a KnowledgeSource as a set of configuration traces $\mathcal{K}^t(S_{BB})$ specified by the following behavior properties:

– if a KnowledgeSource gets correct solutions for all the required subproblems, then it solves the problem eventually:

$$\Box \forall ks \colon\colon KS, (p, P) \in (ks, o_p) \colon$$
$$\left((\forall p \in P \colon \Diamond\left((p, \mathtt{solve}(p)) \in (ks, i_s)\right)) \implies \Diamond(p, \mathtt{solve}(p)) \in (ks, o_s)\right), \tag{15}$$

– in order to solve a problem, a `KnowledgeSource` requires solutions only for smaller problems:

$$\Box \forall ks :: KS: \ ((p, P) \in (ks, o_p) \Longrightarrow \forall p' \in P : p' \prec p), \tag{16}$$

– if a `KnowledgeSource` is able to solve a problem it will eventually communicate this:

$$\Box \forall ks :: KS: \ p \in prob(ks) \wedge p \in (ks, i_p) \Longrightarrow \Diamond (\exists P \subseteq \mathtt{PROB}: \ (p, P) \in (ks, o_p)). \tag{17}$$

Note that Eqs. (12)-(17) constrain only the behavior of components. They do neither restrict activation nor connections. Thus, the resulting architecture property is indeed an example of a behavior property as defined in Definition 14.

5.3   Activation Properties

An architecture property is an activation property if it does neither restrict behavior nor connection. Thus, activation properties are again defined by means of a special closure property.

Definition 15. *An* activation property *(AP) for interface specification $S_i \in \mathcal{S}_I$, is an AP $A \subseteq \mathcal{K}^t(S_i)$, such that connections and behavior are not restricted:*

$$\forall t \in A, n \in \mathbb{N}, k \in \{k \in \mathcal{K}(S_i) \mid k \approx^a t(n)\}$$
$$\exists t' \in A: t' \downarrow_{n-1} = t \downarrow_{n-1} \wedge t'(n) \approx^n k \wedge t'(n) \approx^b k. \tag{18}$$

Running Example: Activation Property Specification. Activation properties of the Blackboard pattern are described in a *configuration diagram* (Fig. 5): The double solid frame for an interface (e.g. *BB*) denotes the condition that components have to be active from the beginning on whereas interfaces with a single frame (e.g. *KS*) allow components to be de-/activated over time.

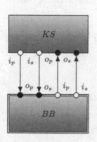

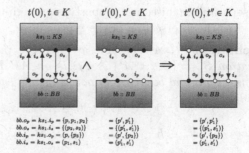

Fig. 5. Configuration diagram of Blackboards for activation and connection

Fig. 6. Architecture violating Eq. (21)

Moreover, we require that whenever a knowledge source offers to solve some problem, it is always activated when solutions to the required subproblems are provided[1]:

$$\Box \forall c :: KS, (p, P) \in (k, o_p):$$

$$(\forall q \in P : \Diamond(q, solve(q)) \in (bb, o_s)) \implies \Diamond(q, solve(q)) \in (bb, o_s) \wedge \|c\|. \quad (19)$$

Note that the activation constraints induced by the diagram in Fig. 5 as well as Eq. (19) constrain only the activation of components. They do neither restrict connections nor behavior which is why the resulting architecture property is indeed an example of an activation property as defined in Definition 15.

5.4 Connection Properties

A connection property is not allowed to restrict neither behavior nor activation. Again this is described by a special closure property.

Definition 16. *A connection property (CP) for interface specification $S_i \in \mathcal{S}_I$, is an AP $N \subseteq \mathcal{K}^t(S_i)$, such that activations and behavior are not restricted:*

$$\forall t \in N, n \in \mathbb{N}, k \in \{k \in \mathcal{K}(S_i) \mid k \approx^n t(n)\}$$

$$\exists t' \in N : t' \downarrow_{n-1} = t \downarrow_{n-1} \wedge t'(n) \approx^a k \wedge t'(n) \approx^b k. \quad (20)$$

Running Example: Connection Property Specification. Connection properties are also specified graphically in the configuration diagram in Fig. 5. The solid arcs denote a constraint requiring that the ports of a `KnowledgeSource` component are connected with the corresponding ports of a `BlackBoard` component as depicted, whenever both components are active.

Note that the connection constraints induced by the diagram in Fig. 5 constrain only the connection of components. They do neither restrict activation nor behavior. Thus, the resulting architecture property is indeed an example of a connection property as defined in Definition 16.

5.5 Separable Architecture Properties

A separable architecture property is an architecture property which can be specified as the intersection of the types above.

Definition 17. *A separable architecture property (SAP) for interface specification $S_i = (C, I, t^c, t^i) \in \mathcal{S}_I$, is an AP $K \subseteq \mathcal{K}^t(S_i)$, such that activation, connection, and behavior do not influence each other:*

$$\forall t \in \mathcal{K}^t(S_i), n \in \mathbb{N} : \Big((\exists t_b \in K : t_b \downarrow_{n-1} = t \downarrow_{n-1} \wedge t_b(n) \approx^b t(n)) \wedge$$

$$(\exists t_n \in K : t_n \downarrow_{n-1} = t \downarrow_{n-1} \wedge t_n(n) \approx^n t(n)) \wedge$$

$$(\exists t_a \in K : t_a \downarrow_{n-1} = t \downarrow_{n-1} \wedge t_a(n) \approx^a t(n)) \Big)$$

$$\implies \exists t' \in K : t' \downarrow_n = t \downarrow_n . \quad (21)$$

---

[1] We use $\|c\|$ to denote that component c is active at the corresponding time.

Example 5 (Architecture violating Eq. (21)). Figure 6 shows an example of an architecture property K which violates the condition required by Eq. (21): $t''(0)$ is connection and activation equivalent with $t(0)$, and behavior equivalent with $t'(0)$. Hence, architectural property K has to permit t''.

Running Example: Blackboard Guarantee. In the following, we specify a guarantee of blackboard architectures as a separable architecture property over the interface specification S_{BB}.

Theorem 1. *Assuming that knowledge sources are active when required:*

$$\Box\Big(p \in (bb, o_p) \implies \Diamond\Big(\exists ks :: KS : p \in prob(ks) \land$$

$$(\forall p' \in P : (\Diamond((p', s) \in (bb, o_s) \implies \|ks\|)))\Big)\Big), \tag{22}$$

a Blackboard architecture guarantees to solve the original problem:

$$\Box\Big(p \in (bb, i_p) \implies \Diamond\Big((p, solve(p)) \in (bb, o_s))\Big)\Big). \tag{23}$$

Proof (Sketch. A detailed proof is given in Isabelle/HOL.). The proof is by well-founded induction over the problem relation $\prec$: We are sure that for each problem eventually a `KnowledgeSource` exists which is capable to solve the problem, Eq. (22). The required subproblems are provided to the `BlackBoard` by the connection constraint of Fig. 5. The `BlackBoard` will provide these subproblems eventually on its output o_p, Eq. (13). Since the subproblems provided to the `BlackBoard` are strictly less, Eq. (16), they will be solved and provided by the `BlackBoard` by induction over the steps 1 to 4. A `KnowledgeSource` will eventually be activated for each solution, Eq. (22), and connected to the `BlackBoard` (Fig. 5). This `KnowledgeSource` eventually has all solutions to its subproblems and will then solve the original problem by Eq. (15). The solution is received eventually by the `BlackBoard` due to Fig. 5. Finally, this solution is provided by the `BlackBoard` due to Eq. (12).

5.6   Completeness

In the following we discuss an important property of the proposed methodology which ensures that each separable architectural property can be described as the intersection of a corresponding behavior, connection, and activation property.

Theorem 2. *Each SAP $K \subseteq \mathcal{K}^t(S_i)$ for interface specification $S_i \in \mathcal{S}_I$ can be uniquely described through the intersection of a BP $B \subseteq \mathcal{K}^t(S_i)$, CP $N \subseteq \mathcal{K}^t(S_i)$, and AP $A \subseteq \mathcal{K}^t(S_i)$:*

$$B \cap N \cap A = K. \tag{24}$$

Proof (Sketch). Given an AP, construct the corresponding BP, AP, and CP. Then show equality of the original property and the intersection.

5.7   Consistency

Another important property of the proposed methodology regards the consistency of the different properties. It ensures that the methodology does indeed not introduce any inconsistencies. Formally, we show that the intersection of behavior, activation, and connection properties is always non-empty if the corresponding properties are non-empty.

Theorem 3. *For each BP* $B \subseteq \mathcal{K}^t(S_i)$, *CP* $N \subseteq \mathcal{K}^t(S_i)$, *and AP* $A \subseteq \mathcal{K}^t(S_i)$, *such that the properties are non-empty:*

$$B, N, A \neq \varnothing, \tag{25}$$

the intersection is non-empty: $B \cap N \cap A \neq \varnothing$.

Proof (Sketch). Show

$$\forall n \in \mathbb{N} \, \exists t \in \mathcal{K}^t(S_i), t_b \in B, t_n \in N, t_a \in A : t \downarrow_n = t_b \downarrow_n = t_a \downarrow_n = t_n \downarrow_n$$

by induction over n to have $\exists t \in B \cap N \cap A$.

6   Specifying Properties of Dynamic Architectures

In this section, we describe an approach to the specification of separable properties of dynamic architectures based on the theory discussed so far.

Properties can be specified directly by a set of configuration traces. Moreover, Fig. 7 depicts an overview of the proposed approach to separate the specification into the different types.

In a first step one has to specify an interface. Based on the interface specification one can then define behavior properties, connection properties, and activation properties. The intersection of the corresponding configuration traces represent the specified architectures.

Specifying interfaces. To specify interfaces first one has to specify a set of ports and corresponding types of messages. This can be achieved by traditional specification techniques such as algebraic specifications [30]. Interfaces can then be specified by grouping a set of ports.

Specifying behavior properties. Based on an interface specification, one can specify behavior properties. This can be achieved e.g. by specifying execution traces over the ports of an interface.

Specifying activation properties. Finally, activation properties may be specified by traces over a set of components. Such traces specify the set of active components at each point in time.

Specifying connection properties. Connection properties have to be specified as special kind of configuration traces.

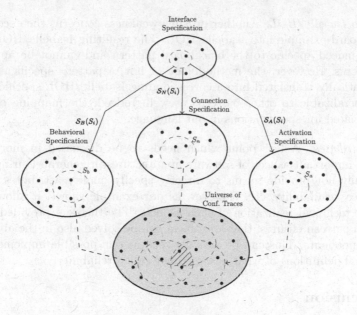

Fig. 7. Specifying Architectural Styles

Running Example: Blackboard Verification. For the verification of the blackboard architecture pattern, we transferred behavior, activation, and connection properties (see Sects. 5.2, 5.3, and 5.4) as well as the pattern's guarantee (see Sect. 5.5) into Isabelle/HOL [22]. There, we proved that each implementation complying with the three individual properties fulfills the architecture property describing the guarantee underlying any blackboard architecture[2]

7   Discussion

In the following, we briefly discuss our approach and possible limitations. Thereby, we critically examine some of its potential weaknesses in more detail.

Dynamic interfaces. One possible weakness concerns the nature of our underlying model. Definition 12 does not allow components to change their interface over time. This could be seen as a restriction of the model, however, it was a deliberate decision since for now, we did not yet find the need for components to change their interfaces. Indeed, it remains an open question whether dynamic interfaces are useful, at all. However, if the need for them arises, it should be noted, that the underlying model can be adapted to allow for dynamic interfaces as well.

---

[2] The script can be downloaded at http://www.marmsoler.com/pattern/Blackboard. thy

Mapping to Isabelle/HOL. Another possible weakness concerns the encoding of the blackboard example into Isabelle/HOL. The resulting Isabelle/HOL specification is indeed specific to the blackboard pattern and cannot be applied to other patterns. However, the methodology of how a pattern specification can be systematically translated into a corresponding Isabelle/HOL specification is indeed generalizable to other patterns as well. Indeed, the mapping could be fully automated for specifications in our language.

Quality attributes. A last point which needs to be discussed in more detail regards an important aspect of software architectures in general. Our approach does actually not provide means to directly specify quality attributes such as performance, availability, etc. However, as our example shows, it allows us to specify the technical realization of such aspects. The theorem provided for the blackboard pattern ensures, that a problem can be solved also in the absence of certain components. This can be actually seen as one possible implementation (or technical definition) of what is sometimes called reliability.

8   Conclusion

In this article, we provide a formal notion of properties for dynamic architectures and investigate different types of properties. The major results can be summarized as follows:

- We provide a *novel model for dynamic architectures* and a formal notion of *properties* for these kind of architectures (Sect. 4). Thereby we introduce the notion of architecture configuration and configuration traces (Definition 6). Then we model an architecture property as a set of configuration traces (Definition 12) for which open input port valuations are not restricted (Definition 13).
- We provide a characterization of *behavior properties*, *activation properties*, and *connection properties* for dynamic architectures (Sect. 5). Each property-type is defined as an architecture property fulfilling a special closure property: A behavior property is not allowed to restrict activations or connections (Definition 14). An activation property, on the other hand, is neither allowed to restrict connections nor behavior (Definition 15). Finally, a connection property is not allowed to restrict activation or behavior (Definition 16).
- We provide a characterization of *separable architecture properties*, architecture properties which can be separated into behavior, activation, and connection parts (Definition 17). We show that each separable architecture property can indeed be separated into behavior, activation, and connection properties (Theorem 2). We show that the intersection of behavior, activation, and connection properties always yields a non-empty architecture property (Theorem 3).

We evaluated our results by deriving a systematic way to specify properties for dynamic architectures and apply it to the specification of blackboard architectures (Sect. 6):

- We specified the constraints imposed by the pattern as behavior, activation, and connection properties.
- We formulated the pattern's guarantee as an architecture property.
- We verified the correctness of the pattern by proving its guarantee from the pattern's constraints in Isabelle/HOL.

We imagine the following implications of our results: (i) The proposed model can be used to specify properties for dynamic architectures. (ii) The results on the different types of properties provide a systematic way to specify separable architecture properties for those kinds of architectures by focusing on different aspects of a dynamic architecture.

We perceive future work in three major areas:

- Based on the insights provided by our results, we aim to build specialized specification and analysis techniques for the three identified property-types: activation, connection, and behavior properties. Especially the specification of behavior properties remains an open issue since they are usually specified locally to a component instead of over the whole architecture. Thus, we are currently investigating how such local specifications can be transformed to specifications over dynamic architectures where the specified component can be activated/deactivated over time.
- Another direction of work concerns the transformation of specifications to (interactive) theorem provers to support the verification of specifications. Currently we are working on a systematic way to transform specifications in the presented formalism to corresponding Isabelle/HOL specifications.
- Finally, the approach should be applied to specify and investigate patterns for dynamic architectures to further evaluate our approach and (maybe even more important) to provide detailed insights into the nature of existing patterns as well as to discover new patterns for dynamic architectures.

Acknowledgments. We would like to thank Jonas Eckhardt, Vasileios Koutsoumpas, and the reviewers of ICTAC 2016 for their comments and helpful suggestions.

References

1. Abowd, G.D., Allen, R., Garlan, D.: Formalizing style to understand descriptions of software architecture. ACM TOSEM **4**, 319–364 (1995)
2. Allen, R., Douence, R., Garlan, D.: Specifying and analyzing dynamic software architectures. In: Astesiano, E. (ed.) FASE 1998. LNCS, vol. 1382, pp. 21–37. Springer, Heidelberg (1998). doi:10.1007/BFb0053581
3. Allen, R.J.: A formal approach to software architecture. Technical report, DTIC Document (1997)
4. Bernardo, M., Ciancarini, P., Donatiello, L.: On the formalization of architectural types with process algebras. ACM SIGSOFT SEN **25**, 140–148 (2000)
5. Bradbury, J.S., Cordy, J.R., Dingel, J., Wermelinger, M.: A survey of self-management in dynamic software architecture specifications. In: WOSS (2004)
6. Broy, M.: A logical basis for component-oriented software and systems engineering. Comput. J. **53**(10), 1758–1782 (2010)

7. Broy, M.: A model of dynamic systems. In: Bensalem, S., Lakhneck, Y., Legay, A. (eds.) ETAPS 2014. LNCS, vol. 8415, pp. 39–53. Springer, Heidelberg (2014). doi:10.1007/978-3-642-54848-2_3

8. Buschmann, F., Meunier, R., Rohnert, H., Sommerlad, P., Stal, M.: A system of patterns: Pattern-oriented software architecture (1996)

9. Castro, P.F., Aguirre, N.M., López Pombo, C.G., Maibaum, T.S.E.: Towards managing dynamic reconfiguration of software systems in a categorical setting. In: Cavalcanti, A., Deharbe; D., Gaudel, M.-C., Woodcock, J. (eds.) ICTAC 2010. LNCS, vol. 6255, pp. 306–321. Springer, Heidelberg (2010). doi:10.1007/978-3-642-14808-8_21

10. Clements, P.C.: A survey of architecture description languages. In: IWSSD (1996)

11. Dashofy, E.M., Van der Hoek, A., Taylor, R.N.: A highly-extensible, xml-based architecture description language. In: WICSA, IEEE (2001)

12. Fiadeiro, J.L., Lopes, A.: A model for dynamic reconfiguration in service-oriented architectures. Softw. Syst. Model. **12**(2), 349–367 (2013)

13. Garlan, D.: Formal modeling and analysis of software architecture: components, connectors, and events. In: Bernardo, M., Inverardi, P. (eds.) SFM 2003. LNCS, vol. 2804, pp. 1–24. Springer, Heidelberg (2003). doi:10.1007/978-3-540-39800-4_1

14. Hirsch, D., Montanari, U.: Two graph-based techniques for software architecture reconfiguration. Electron. Notes Theor. Comput. Sci. **51**, 177–190 (2002)

15. Inverardi, P., Wolf, A.L.: Formal specification and analysis of software architectures using the chemical abstract machine model. IEEE TSE **21**, 373–386 (1995)

16. Le Métayer, D.: Describing software architecture styles using graph grammars. IEEE TSE **24**, 521–533 (1998)

17. Luckham, D.C., Kenney, J.J., Augustin, L.M., Vera, J., Bryan, D., Mann, W.: Specification and analysis of system architecture using Rapide. IEEE TSE **21**, 336–355 (1995)

18. Magee, J., Kramer, J.: Dynamic structure in software architectures. ACM SIG-SOFT SEN **21**, 3–14 (1996)

19. Manna, Z., Pnueli, A.: The Temporal Logic of Reactive and Concurrent Systems: Specification. Springer, New york (2012)

20. Medvidovic, N.: ADLs and dynamic architecture changes. In: ISAW (1996)

21. Moriconi, M., Qian, X., Riemenschneider, R.A.: Correct architecture refinement. IEEE TSE **21**, 356–372 (1995)

22. Nipkow, T., Paulson, L.C., Wenzel, M.: Isabelle/HOL: A Proof Assistant for Higher-Order Logic. Springer Science & Business Media, Heidelberg (2002)

23. Oquendo, F.: π-ADL: an architecture description language based on the higher-order typed π-calculus for specifying dynamic and mobile software architectures. ACM SIGSOFT SEN **29**, 1–14 (2004)

24. Penix, J., Alexander, P., Havelund, K.: Declarative specification of software architectures. In: ASE (1997)

25. Shaw, M., Garlan, D.: Software Architecture: Perspectives on an Emerging Discipline, vol. 1. Prentice Hall Englewood Cliffs, Upper Saddle River (1996)

26. Spivey, J.M., Abrial, J.: The Z notation (1992)

27. Taylor, R.N., Medvidovic, N., Dashofy, E.M.: Software Architecture: Foundations, Theory, and Practice. Wiley Publishing, Hoboken (2009)

28. Wenzel, M.: Isabelle/Isar: a generic framework for human-readable proof documents. From Insight to Proof: Festschrift in Honour of Andrzej Trybulec **10**, 277–298 (2007)

29. Wermelinger, M., Lopes, A., Fiadeiro, J.L.: A graph based architectural (re) configuration language. ACM SIGSOFT SEN **26**(5), 21–32 (2001)

30. Wirsing, M.: Algebraic Specification. MIT Press, Cambridge (1991)

Behavioural Models for FMI Co-simulations

Ana Cavalcanti[✉], Jim Woodcock, and Nuno Amálio

University of York, York, UK
ana.cavalcanti@york.ac.uk

Abstract. Simulation is a favoured technique for analysis of cyber-physical systems. With their increase in complexity, co-simulation, which involves the coordinated use of heterogeneous models and tools, has become widespread. An industry standard, FMI, has been developed to support orchestration; we provide the first behavioural semantics of FMI. We use the state-rich process algebra, *Circus*, to present our modelling approach, and indicate how models can be automatically generated from a description of the individual simulations and their dependencies. We illustrate the work using three algorithms for orchestration. A stateless version of the models can be verified using model checking via translation to CSP. With that, we can prove important properties of these algorithms, like termination and determinism, for example. We also show that the example provided in the FMI standard is not a valid algorithm.

Keywords: Verification · Modelling · *Circus* · CSP

1 Introduction

The Functional Mock-up Interface (FMI) [12] is an industry standard for co-simulation: collaborative simulation of separately developed models. It has been applied across a variety of domains, including automotive, energy, aerospace, and real-time systems integration; dozens of tools support the standard.

An FMI co-simulation [4] is organised around black-box slave FMUs (Functional Mockup Units): effectively, wrappings of models that are interconnected through their inputs and outputs. FMUs are passive entities whose simulation is triggered and orchestrated by a master algorithm. A simulation is divided into steps that serve as synchronisation and data exchange points; between these steps, the FMUs are simulated independently. The master algorithm communicates with the FMUs via a number of functions that compose the FMI API.

Here, we present the first behavioural formal semantics for FMI-based co-simulations. We use *Circus* [21], a state-rich process algebra that combines Z [26] for data modelling and CSP [23] for behavioural specification. We characterise formally master algorithms and FMUs that make appropriate use of the FMI API. These abstract models of a co-simulation can be automatically generated from the number of FMUs, their inputs and outputs and dependencies.

The general models can be used to verify specific master algorithms and the adequacy of simulation models for FMUs. We have verified a classic algorithm

© Springer International Publishing AG 2016
A. Sampaio and F. Wang (Eds.): ICTAC 2016, LNCS 9965, pp. 255–273, 2016.
DOI: 10.1007/978-3-319-46750-4_15

from the FMI standard for Simulink [19], and a more robust algorithm that caters for FMU failures [4]. This revealed that the example in the standard implicitly assumes that FMUs do not raise fatal errors; it is not a valid algorithm.

Circus models, with abstracted state, can be translated to CSP and verified using the FDR3 model checker [16]. We prove important properties discussed in the FMI literature, like termination and determinism using the FDR3 model checker. Richer models can be verified using a *Circus* theorem prover [14]. Given a choice of master algorithm and formal models of the FMUs, our work can also be used to prove properties of an overall system described by the separate simulations. *Circus* can currently cater only for discrete-time models. On the other hand, a continuous time extension of *Circus* that can be used to give semantics to continuous-systems simulations [13] is under development.

Broman [4] has presented the most influential formalisation of FMI to date: a state-based model of the three main API functions that set and get FMU variables and trigger a simulation step with two master algorithms and a proof of core properties. Our model of a co-simulation also has its interface defined by the interactions corresponding to the simulation steps and the exchange of data associated with them. Our behavioural model covers a large portion of the FMI API, defining valid patterns for its usage and error treatment.

Sections 2 and 3 describe FMI for co-simulation and *Circus*. Section 4 describes the *Circus* semantics of FMI. The specification and verification of master algorithms and co-simulations is discussed in Sect. 5. Section 6 presents our conclusions.

2   FMI

Modelling and simulating cyber-physical systems (CPSs) [10] involves different engineering fields: a global system with components tackled by domain engineers using specialised tools. Co-simulation [18] involves tool interoperability for modelling and simulating heterogeneous components. FMI avoids the need for tool-specific integration, by exchanging dynamic models, co-simulating heterogeneous models, and protecting intellectual property. We deal with co-simulation, but we can also reason about simulations with model exchange.

A master algorithm orchestrates a collection of FMUs that may be stand-alone, containing runnable code, or be coupled, in which case it contains a wrapper to a simulation tool. Like FMI, our model is agnostic to the particular realisation of an FMU, and does not cover any communication infrastructure that may be in place to support distributed co-simulation. We assume that communication between the master algorithm and the various FMUs is reliable.

When the co-simulation is started, the models of the FMUs are solved independently between two discrete communication points defined by a step. For that, the master algorithm reads the outputs of the FMUs, sets their inputs, and then waits for all FMUs to simulate up to the defined communication point, before advancing the simulation time. Master algorithms differ in their approach to handling the definition of the step sizes and any simulation errors.

Although the FMI standard does not specify any particular master algorithms, or the technology for development of FMUs, it specifies an API that can be used to orchestrate the various simulations. Restrictions on the use of the API functions specify, indirectly and informally, how a master algorithm can be defined and how an FMU may respond. Our model captures a significant subset of the FMI API, and defines formally validity for algorithms and FMUs.

3 Circus

The main construct of *Circus* is a process, used to specify a system and its components. Processes communicate with each other via channels. Communications are instantaneous and synchronous events. A process can have a state, defined using a Z schema, and a behaviour, defined using an action.

To illustrate *Circus*, Fig. 1 presents the model of a *Timer* from a valid master algorithm. *Timer* takes as parameters the current time ct, the step size hc, and the end time tN of the simulation. Although it is possible to set up experiments without an end time, we restrict ourselves to experiments that are time bounded.

channel : $setT$: $TIME$; $updateSS$: $NZTIME$; $step$: $TIME \times NZTIME$; end
process $Timer \mathrel{\widehat{=}} ct, hc, tN$: $TIME$ • **begin**
state $State == [currentTime, stepSize$: $TIME]$
$Step =$
$\quad setT?t : t \leq tN \longrightarrow currentTime := t;\ Step$
$\quad \square\ updateSS?ss \longrightarrow stepSize := ss;\ Step$
$\quad \square\ step!currentTime!stepSize \longrightarrow currentTime := currentTime + stepSize;\ Step$
$\quad \square\ currentTime = tN\ \&\ end \longrightarrow \textbf{Stop}$
• $currentTime, stepSize := ct, hc;\ Step$
end

Fig. 1. *Circus* specification of a *Timer* process

Timer's state contains two components: *currentTime* and *stepSize*. Its behaviour is defined by the action at the end. After initialising *currentTime* and *stepSize* using ct and hc, it calls the local action *Step*. It takes inputs on channels *setT* and *updateSS* to update the current time and step size. The channel declarations define the type of the values that can be communicated through them: $TIME$ is the set of natural numbers, and $NZTIME$ excludes 0. Step sizes cannot be 0. It uses a channel *step* to output the current time and step size. After a communication on *step*, the current time is advanced to the next simulation step; at the end of the experiment ($currentTime = tN$), it synchronises on *end*.

The action *Step* offers communications on the above channels in external choice ($\square$). The time t input through *setT* cannot exceed the end time tN of the simulation. The offer of synchronisation on *end* is guarded by $currentTime = tN$ and only becomes available if this condition holds. After the event *end*, the timer deadlocks: behaves like the action **Stop**.

Processes can also be defined by combination of other processes. For example, the specification of the process *TimedInteractions* below combines three processes *Timer*, *endSimulation* and *Interaction*.

$$TimedInteractions \cong t0, tN : TIME \bullet$$
$$\left(\begin{array}{l} (\ Timer(t0, 1, tN)\ \Delta\ endSimulation) \\ \quad [\![\{\ step, end, setT, updateSS, endsimulation\ \}]\!] \\ Interaction \end{array} \right) \setminus \{\!|\ step, end, setT, updateSS\ |\!\}$$

TimedInteractions has two parameters: a start and an end time $t0$ and tN. It uses *Timer* defined above with arguments $t0$, 1, and tN. *Timer* can be interrupted (Δ) by the process *endSimulation*. It, however, runs in parallel ($[\![\ ]\!]$) with the process *Interaction*. They synchronise on communications on *step*, *end*, *setT*, *updateSS*, and *endsimulation*, but otherwise proceed independently. The process that results from the parallelism hides ($\setminus$) communications on *step*, *end*, *setT*, and *updateSS*, which are used just internally by *Timer* and *Interaction*.

A complete account of Circus can be found in [8]. We explain any extra notation not explained here as needed.

4   A Model of FMI

The FMI API consists of functions used by the master algorithm to orchestrate the FMUs. In our model, these functions are defined as channels whose types correspond to the input and output types of the functions; see Table 1.

We use the given type $FMI2COMP$ to represent an instance of an FMU. In FMI, these are pointers to an FMU-specific structure that contains the information needed to simulate it. Here, we use identifiers for such components.

Valid variable names and values are represented by the sets VAR and VAL. We do not model the FMI type system, which includes reals, integers, booleans,

Table 1. Channels that model FMI API functions

`fmi2Get`	$FMI2COMP.VAR.VAL.FMI2ST$
`fmi2Set`	$FMI2COMP.VAR.VAL.FMI2STF$
`fmi2DoStep`	$FMI2COMP.TIME.NZTIME.FMI2STF$
`fmi2Instantiate`	$FMI2COMP.Bool$
`fmi2SetUpExperiment`	$FMI2COMP.TIME.Bool.TIME.FMI2ST$
`fmi2EnterInitializationMode`	$FMI2COMP.FMI2ST$
`fmi2ExitInitializationMode`	$FMI2COMP.FMI2ST$
`fmi2GetBooleanStatusfmi2Terminated`	$FMI2COMP.Bool.FMI2ST$
`fmi2GetMaxStepSize`	$FMI2COMP.TIME.FMI2ST$
`fmi2Terminate`	$FMI2COMP.FMI2ST$
`fmi2FreeInstance`	$FMI2COMP.FMI2ST$
`fmi2GetFMUState`	$FMI2COMP.FMUSTATE.FMI2ST$
`fmi2SetFMUState`	$FMI2COMP.FMUSTATE.FMI2ST$

characters, strings, and bytes; however, it is not difficult to cater for this type system. Extensions to the type system are expected in future versions of FMI.

The type $FMI2ST$ contains flags of the FMI type fmi2Status that are returned by the API functions. We include fmi2OK, fmi2Error, and fmi2Fatal, which indicate, respectively, that all is well, the FMU encountered an error, and the computations are irreparable for all FMUs. The extra flag fmi2Discard is also included in the superset $FMI2STF$; it can only be returned by fmi2Set and fmi2DoStep. fmi2Set indicates that a status cannot be returned, and in the case of fmi2DoStep that a smaller step size is required or the requested information cannot be returned. We do not include fmi2Warning, used for logging, and fmi2Pending, used for asynchronous simulation steps.

$FM\ddot{U}STATE$ contains values that represent an internal state of an FMU. It comprises all values (of parameters, inputs, buffers, and so on) needed to continue a simulation. It can be recorded by a master algorithm to support rollback.

The signature of the channels impose restrictions on the use of the API. It is not possible to call fmi2DoStep with a non-positive step size. Given a particular configuration of FMUs, we can define the types of the fmi2Get and fmi2Set channels so that setting or getting a variable that is not in the given FMU is undefined. Without this fine tuning, such attempts lead to deadlocks in our model: a check for deadlock freedom ensures the absence of such problems. The API actually includes specialised fmi2Get and fmi2Set functions for each data type available. As already said, we do not cater for the FMI type system.

The function fmi2Instantiate returns a pointer to a component, and null if the instantiation fails. Since we do not model pointers, we use a boolean to cater for the possibility of failure. The function fmi2GetMaxStepSize is not part of the standard; we use it to implement the rollback algorithm in [4].

The overall structure of our models of a co-simulation is shown in Fig. 2. The visible channels are fmi2Get, fmi2Set, and fmi2DoStep. So, we can use our model to verify properties of co-simulations that can be described in terms of these interactions, and involving variables from any of the FMUs involved.

The other channels enforce the expected control flow of a master algorithm. They are used for communication between the process $MAlgorithm$ that models a master algorithm and each process $FMUInterface(i)$ that models the FMU identified by i. We call $FMIWrapper$ the collection of FMU interfaces: they execute independently in parallel, that is, in interleaving.

The control channel $endsimulation$ is used to shutdown the simulation. Since an FMU may fail, its termination may not be carried out gracefully (with fmi2Terminate and fmi2FreeInstance). So, $endsimulation$ is used to indicate the end of the experiment in all cases and shutdown the model processes.

In what follows, we describe our specifications of $MAlgorithm$ (Sect. 4.1) and $FMUInterface$ (Sect. 4.2), which provide a correctness criterion for these components. In Sect. 4.3, we describe how to construct models of specific FMUs. Applications of our models are described in Sect. 5.

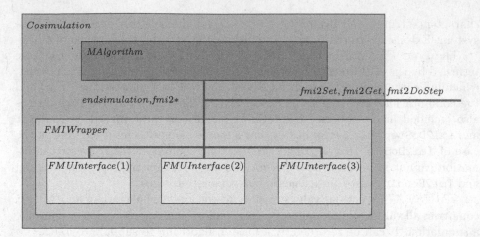

Fig. 2. Structure of a co-simulation model

4.1 Master Algorithms

A master algorithm is a monolithic program that defines the connections between the FMUs and the time of the simulation steps, and handles any errors raised by an FMU. In our model, we consider each of these aspects of a master algorithm separately. The overall structure of the *MAlgorithm* process is described in Fig. 3. It provides a general characterisation of the valid history of interactions of a master algorithm. It does not commit to specific policies to define step sizes and error handling in case an API function returns `fmi2Discard`. The treatment of `fmi2Error` and `fmi2Fatal` is restricted by the standard.

MAlgorithm has three main components described next. *TimedInteractions* specifies the co-simulation steps and orchestration of the FMUs. *FMUStatesManager* controls access to the internal state of the FMUs. *ErrorHandler* monitors the occurrence of an *fmi2Error* or *fmi2Fatal* from the API functions.

TimedInteractions has two components. *Timer* is presented in Sect. 3. It uses *step* and *end* to drive the *Interaction* process, which defines the orchestration of the FMUs. This is the core process that restricts the order in which API functions can be used. *Timer* also exposes channels *setT* and *updateSS* to allow *Interaction* to define algorithms will rollback or a variable step size. The timer can be terminated by the signal *endsimulation* raised by *Interaction*.

Interaction is the sequential composition of *Instantiation*, *InstantiationMode*, *InitializationMode*, and *slaveInitialized*, which correspond to states that define the stages of a co-simulation [12, p.103]. The definitions of these processes depend on the configuration of the FMUs. Given such a configuration, they can be automatically generated as indicated below. A configuration is characterised by a sequence of FMU identifiers ($FMUs : \text{seq}FMI2COMP$), and sequences that define the parameters and their values ($parameters : \text{seq}(FMI2COMP \times VAR \times VAL)$), inputs and their initial values ($inputs : \text{seq}(FMI2COMP \times VAR \times VAL)$), outputs ($outputs : \text{seq}(FMI2COMP \times VAR)$), and an input/output

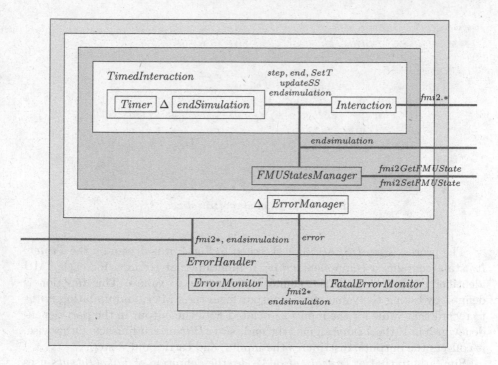

Fig. 3. Structure of a model of a master algorithm

port dependency graph [4] *pdg*. Some of this information is also needed to generate automatically a sketch of the models of the FMUs (see Sect. 4.3).

The port dependency graph *pdg* is a relation between outputs and inputs defined by a pair of type *FMI2COMP* × *VAR*. The graph establishes how the inputs of each of the FMUs depend on the outputs of the others. It must be acyclic, and this can be automatically checked using the CSP model checker. Using the port dependency graph, once we retrieve the outputs, via the `fmi2Get` function, we know how to provide the inputs, via the `fmi2Set` function.

Instantiation, defined below, instantiates the FMUs. It is an iterated sequential composition (;) of actions *fmi2Instantiate.i?sc* → **Skip**, where *i* comes from *FMUs* and **Skip** is the action that terminates immediately.

InstantiationMode and *InitializationMode* allow the setting up of parameters and initial values of inputs before calling the API function that signals the start of the next phase. We show below *InitializationMode*. For an element *inp* of *inputs*, we use projection functions *FMU*, *name* and *val* to get its components.

(; *inp* : *inputs* • *fmi2Set*!(*FMU inp*)!(*name inp*)!(*val inp*)?*st* ⟶ **Skip**);
(; *i* : *FMUs* • *fmi2ExitInitializationMode*!*i*?*st* ⟶ **Skip**)

We can easily generalise the model to allow an interleaving of the events involved. The value of such a generalisation, however, is unclear (and it harms the possibility of automated verification via model checking).

process $slaveInitialized \; \hat{=}$
state $State == [rinps : FMI2COMP \nrightarrow (VAR \nrightarrow VAL)]$
$\ldots$

$TakeOutputs \; \hat{=}$
$; \; out : outputs \bullet fmi2Get.(FMU \; out).(name \; out)?v \longrightarrow$
$\quad ; \; inp : pdg(out) \bullet$
$\qquad rinps := rinps \oplus \{ (FMU \; inp) \mapsto ((rinps \, (FMU \; inp)) \oplus \{(name \; inp) \mapsto v\}) \}$
$Main \; \hat{=} \; end \longrightarrow \mathbf{Skip}$
$\qquad \square \; step?t?hc \longrightarrow TakeOutputs; \; DistributeInputs; \; Step$
$\bullet \; Main$
end

Fig. 4. Sketch of $slaveInitialized$

The process $slaveInitialized$ is sketched in Fig. 4; it is driven by the $Timer$. Its state contains a component $rinps$: a function that records, for each FMU identifier a function from the names of its inputs to values. This function is defined by taking the value of each output from the FMUs, and updating $rinps$ to record that value for the inputs associated with the output in the port dependency graph. If the $Timer$ signals the end, $slaveInitialized$ finishes. Otherwise, it collects the outputs, distributes the inputs, and carries out a step.

Similarly to that of $InitializationMode$, the definition of $TakeOutputs$ uses an iterated sequence, now over $outputs$: the sequence of pairs that identify an FMU and an output name. Once the value v of an output out is obtained, it is assigned to each input inp in the sequence $pdf(out)$ associated with out in the port dependency graph pdg. We use $\oplus$ to denote function overriding.

$DistributeInputs$ uses inp to set the inputs of the FMUs using $fmi2Set$. $Step$ proceeds with the calls to $fmi2DoStep$ and if all goes well, recurses back to the $Main$ action of $slaveInitialized$. Their definitions are omitted for brevity.

$FMUStatesManager$ controls the use of the functions fmi2GetFMUState and fmi2SetFMUState for each of the FMUs. It is an interleaving of instances of the process $FMUStateManager(i)$ in Fig. 5 for each of the FMUs. Once an FMU is instantiated, then it is possible to retrieve its state. After that, both gets and sets are allowed. The actual values of the state are defined in the FMUs, but recorded in the master algorithm via $fmi2GetFMUState$ for later use with $fmi2SetFMUState$ as defined in $FMUStateManager(i)$.

For complex internal states, model checking can become infeasible (although we have managed it for simple examples). To carry out verifications that are independent of the values of the internal state of the FMUs, we need to adjust only this component. Some examples, explored in the next section, are properties of algorithms that do not support retrieval and resetting of the FMU states, determinism and termination of algorithms, and so on.

The $ErrorHandler$ process contains two components: monitors for $fmi2Error$ and $fmi2Fatal$. If any of the API functions returns an error, they signal that to the $ErrorManager$ via a channel $error$. Upon an error, the $ErrorManager$

process $FMUStatesManager \;\widehat{=}\; i : FMI2COMP \bullet$ **begin**

$AllowAGet \;\widehat{=}\; fmi2GetFMUState.i?s?st \longrightarrow AllowsGetsAndSets(s)$

$AllowsGetsAndSets \;\widehat{=}\; s : FMUSTATE \bullet$
  $fmi2GetFMUState.i?t?st \longrightarrow AllowsGetsAndSets(t)$
  $\Box\; fmi2SetFMUState.i!s?st \longrightarrow AllowsGetsAndSets(s)$

$\bullet\; fmi2Instantiate.i?b \longrightarrow AllowAGet$

end

Fig. 5. Model of $FMUStateManager$

interrupts the main flow of execution. In the case of an $fmi2Fatal$ error, the simulation is stopped via $endsimulation$. In the case of an $fmi2Error$, a call to $fmi2FreeInstance$ is allowed, before the simulation is ended.

4.2   FMU Interfaces

The model of a valid FMU is simpler. It captures the control flow of an FMU, specifying, at each stage, the API functions to which it can respond. Unsurprisingly, it has some of the restrictions of a master algorithm, but it is much more lax, in that it captures just the expected capabilities of an FMU.

At first, the only API function that is available is `fmi2Instantiate`. The simple action below specifies this behaviour.

$Instantiation =$
$$fmi2Instantiate.i?b \longrightarrow \begin{pmatrix} b\; \& \; status := fmi2OK;\; Instantiated \\ \Box \\ \neg\; b\; \& \; status := fmi2Fatal;\; RUN(FMUAPI(i)) \end{pmatrix}$$

A state component $status$ records the result of the last call to an API function. In this case, it is updated based on the boolean b returned by $fmi2Instantiate$. If the instantiation is successful, the behaviour is described by $Instantiated$, sketched below; otherwise, it is unrestricted: specified by $RUN(FMUAPI(i))$, which allows the occurrence of any API functions, in any order.

$Instantiated =\; status = fmi2Fatal\; \&\; RUN(FMUAPI(i))$
  $\Box\; status \notin \{fmi2Error, fmi2Fatal\}\&$
$$\begin{pmatrix} fmi2Get.i?n?v?st \longrightarrow status := st;\; Instantiated \\ \Box\; fmi2DoStep.i?t?hc?st \longrightarrow status := st;\; Instantiated \\ \Box \cdots \end{pmatrix}$$
  $\Box\; st \neq fmi2Fatal\; \&\; fmi2FreeInstance!i?st \longrightarrow \cdots$

Again, if there is a fatal error, the behaviour is unrestricted. If there is no error, all functions except `fmi2Instantiate` are available. Finally, if there is a non-fatal error, only `fmi2FreeInstance` is possible.

While a pattern of calls is defined by a master algorithm, so that, for example, all outputs are obtained before the inputs are distributed, the FMU is passive and does not impose such a policy on its use. So, the various actions enforce only the restrictions in the standard [12, p.105].

Although it is possible to specify a more restricted behaviour for FMUs, such a specification rules out robust FMU implementations that handle calls to the API functions that do not necessarily follow the strict pattern of a co-simulation. Next, we describe how to generate FMU models that follow a more restricted pattern that is adequate for use with valid master algorithms.

4.3 Specific FMU Models

In the previous section, we have presented a general model for an FMU. The particular model of an FMU depends, of course, on its functionality, and must conform to (trace refine) our general model. This can be proved via model checking for stateless models of FMUs that do not offer the facility to retrieve and set its internal state. In this case, the models do not offer the choices of communications $fmi2GetFMUState.i?st$ and $fmi2SetFMUState.i?st$. The availability of such facilities is defined by capability flags of the FMU.

We can, however, generate a sketch of the model of an FMU using information about its structure: lists of parameters p_i, inputs inp_i, and outputs out_i. This information is used to construct a master algorithm (see Sect. 4.1). Figure 6

process $FMUSketch \mathrel{\widehat{=}} i : FMI2COMP \bullet$ **begin**

state $State = [currentTime, endTime : TIME;\ cp_i, cinp_i, cev_i, cout_i]$

$Instantiation = fmi2Instantiate.i!true \longrightarrow$ **Skip**

$InstantiationMode =$
$\quad fmi2Set.i.p_i?v!fmi2OK \longrightarrow cp_i := v;\ InstantiationMode$
$\quad \square\ fmi2SetUpExperiment.i?t0!true?tN!fmi2OK \longrightarrow$
$\quad\quad currentTime, endTime := t0, tN;$
$\quad\quad fmi2EnterInitializationMode.i!fmi2OK \longrightarrow$ **Skip**

$InitializationMode =$
$\quad fmi2Set.i.inp_i?v!fmi2OK \longrightarrow cinp_i := v;\ InitializationMode$
$\quad \square\ fmi2ExitInitializationMode.i!fmi2OK \longrightarrow UpdateState$

$slaveInitialized =$
$\quad fmi2Get.i.out_i!cout_i!fmi2OK \longrightarrow slaveInitialized$
$\quad \square\ fmi2Set.i.inp_i?v.fmi2OK \longrightarrow cinp_i := v;\ slaveInitialized$
$\quad \square\ fmi2DoStep.i?t?ss!fmi2OK \longrightarrow (UpdateState;\ slaveInitialized)$

$\bullet\ Instantiation;\ InstantiationMode;\ InitializationMode;$
$\quad (slaveInitialized\ \Delta$
$\quad\quad fmi2Terminate.i!fmi2OK \longrightarrow fmi2FreeInstance.i!fmi2OK \longrightarrow$ **Stop**)

end

Fig. 6. Sketch of a model for a specific FMU

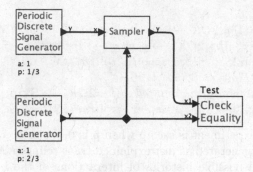

Fig. 7. Test case for sampling of discrete event signals [5]

shows the sketch of a *Circus* process with the FMU behaviour. Its state includes components cp_i, $cinp_i$, and $cout_i$, besides the current and end simulation time.

Its structure is similar to that of the *Interaction* process used to model a master algorithm. In all cases, the interactions flag success ($fmi2OK$). If an FMU makes assumptions about its inputs, the possibility of error can be modelled. For example, *Instantiation* indicates success, but to explore the possibility of failure, we can define it as $fmi2Instantiate.i?b \rightarrow$ **Skip**. The action *UpdateState* is left unspecified. It is this action that specifies the functionality of the FMU. It can be automatically generated if there is a more complete model of the FMU. For example, [7] shows the case if a discrete-time Simulink model is available.

If the FMU supports retrieval and update of its state, we need to add the following choices to *InstantiationMode*, *InitializationMode*, and *slaveInitialized*.

$\cdots$

$\square\, fmi2GetFMUState.i!\,\theta\, State!fmi2OK \longrightarrow \cdots$
$\square\, fmi2SetFMUState.i?s?st \longrightarrow \theta\, State := s;\ \cdots$

Via $fmi2GetFMUState$, it outputs the whole state record, that is, $\theta State$, and via $fmi2SetFMUState$, we can update it.

If the state, either via setting of parameters and input or via an update, may become invalid, we can flag $fmi2Fatal$ and deadlock. For example, we consider the test case shown in Fig. 7 taken from [5]. It has been designed to show that components with discrete timed behaviour coordinate their representation of time. There are three main components: two periodic discrete signal generators, both generating the same signal, one with period one time unit and the other two time units; and a discrete sampler. The test criterion is that the output of the Sampler should equal the output of the second periodic discrete signal generator at all superdense times. There is an implicit constraint that the period p should not be 0; therefore, we specify its *InstantiationMode* action as follows.

$InstantiationMode =$
  $fmi2Set.i.a?v!fmi2OK \longrightarrow a := v \longrightarrow InstantiationMode$.
  $\Box\, fmi2Set.i.p?v!fmi2OK \longrightarrow p := v \longrightarrow InstantiationMode$
  $\Box\, p \neq 0\ \&\ fmi2SetUpExperiment.i?t0!true?tN!fmi2OK \longrightarrow$
    $currentTime, endTime := t0, tN;$
    $fmi2EnterInitializationMode.i!fmi2OK \longrightarrow$ **Skip**
  $\Box\, p = 0\ \&\ fmi2SetUpExperiment.i?t0!true?tN!fmi2Fatal \longrightarrow$ **Stop**

In this case, if the experiment is set up when p is 0, we have a fatal error.

An FMU model generated as just explained trace refines $FMUInterface(i)$. This means that all possible histories of interactions of the FMU are possible for $FMUInterface(i)$ and, therefore, valid according to that criterion. We have proofs of refinement for all FMUs in Fig. 7 and for a data-flow network.

5   Evaluation: Verification Applications

In this section, we show how we can use our formal semantics for FMI to verify master algorithms and to study system properties via their co-simulations. For automation, our semantics can be translated from *Circus* to CSPM (the input language for the model checker FDR3), using a strategy similar to that of [20], so that it can be both model checked in FDR3 and executed in ProBe (FDR's process behaviour explorer), for suitably chosen model parameters.

5.1   Master Algorithms

As well as giving a correctness criterion for a master algorithm, the model presented in Sect. 4 gives an indication of how to construct models for particular algorithms. We consider here three examples.

Classic Brute-Force. The simplest algorithm uses a fixed step size, has no access to the state of the FMUs, and queries them for termination if `fmi2Discard` is flagged. To model this algorithm, we define a process *ClassicMAlgorithm* with the same structure shown in Fig. 3, but more specific components.

ClassicMAlgorithm uses a simple timer that does not use *setT* or *updateSS*. For the *FMUStatesManager*, we use a simple process that just terminates immediately. Finally, for *Interaction*, we use the parallel composition of *Interaction* itself with a process *DiscardMonitor*, whose main action is *Monitor* defined below, followed by an action *Terminated* that shuts down the FMUs.

$Monitor \,\hat{=}$
  $fmi2DoStep?i?t?hc?st : st \neq fmi2Discard \longrightarrow Monitor$
  $\Box\, fmi2DoStep?i?t?hc.fmi2Discard \longrightarrow$
    $\left(\begin{array}{l} fmi2GetBooleanStatusfmi2Terminated.i.true?st \longrightarrow ToDiscard \\ \Box\, fmi2GetBooleanStatusfmi2Terminated.i.false?st \longrightarrow Monitor \end{array} \right)$
  $\Box\, stepAnalysed \longrightarrow Monitor\ \Box\ step?t?hc \longrightarrow Monitor$
  $\Box\, end \longrightarrow$ **Skip**

Monitor ignores all flags *st* returned by *fmi2DoStep* except *fmi2Discard*. If this flag is returned, it queries the FMU using *fmi2GetBooleanStatusfmi2 Terminated*. If the FMU requests termination, *Monitor* behaves like *ToDiscard* whose simple definition we omit. In *ToDiscard*, when completion of the step is indicated via either a *stepAnalysed* or a *step?t?hc* event, the co-simulation is terminated. The signal *stepAnalysed* is not part of the *Interaction* interface, but is used to indicate that `fmi2DoStep` has been carried out for all FMUs, and we are now in a position to decide how to continue with the co-simulation.

Since *ClassicMAlgorithm* has the same structure as *MAlgorithm*, we can prove refinement by considering each of the components in isolation. While proof of refinement by model checking for the whole model is not feasible, it is feasible for the individual components. In the sequel, we use the same approach to analyse more complex algorithms. It is also feasible to prove that *ClassicMAlgorithm* terminates, but otherwise does not deadlock, and is deterministic.

The example in the FMI standard is a classic algorithm with a fixed step and handling of `fmi2Discard`, but does not include error management. So, its specification does not include the *ErrorHander* and the *ErrorManager*. Model checking can show that this is not a valid algorithm. A simple counterexample shows that it continues and calls `fmi2Instantiate` a second time even after the first call returns an `fmi2Fatal` flag. This is explicitly ruled out in the standard.

Simulink. This is a widely used tool for simulation based on control law diagrams [19]. A popular solver uses a variable-step policy based on change rate of the state. To model this algorithm, we use a process *SimulinkMAlgorithm*, which is similar to *ClassicMAlgorithm*, but has another monitor *VaryStep*, specified in Fig. 8. It is composed in parallel with *Interaction* to define a process *VariableStepInteraction* used in *SimulinkMAlgorithm*.

VaryStep takes as parameters a *threshold* for change and the initial value of the step size *initialSS*. Taking a simple approach, we define a state that records the old (*oldOuts*) and new (*newOuts*) values of the outputs, besides the current step size *currentSS*. After the state is initialised (using the action *Init*) to record undefined (ϵ) old values for the outputs, no new values (empty function $\emptyset$), and the initial step size, the monitor steps by recording the new output values (*Monitor*) and then changing the step size (*Adjust*). Adjustment is based just on a comparison between the old and new values defined by an (omitted) function *delta*. If the *threshold* is reached, a new step size is defined by another function *newstep* and informed to the *Timer*.

We have established that *SimulinkMAlgorithm* is valid, that is, it refines *MAlgorithm*, by proving that the new *VariableStepInteraction* refines *Interaction*. We have also proved termination, deadlock freedom, and determinism.

Rollback. In the same way as illustrated by *VaryStep* in Fig. 8, we can model a sophisticated algorithm suggested in [4]. We define a *Rollback* monitor that has the same structure as *VaryStep*. Its *Monitor* (a) saves the state using

process $VaryStep \cong threshold : VAL;\ initialSS : NZTIME \bullet$ **begin**

state
$\quad State = [oldOuts, newOuts : (FMI2COMP \times VAR) \rightarrow VAL;\ currentSS : NZTIME]$

$\underline{\quad Init \quad\rule{8cm}{0.4pt}}$
$State'$
$\mathrm{dom}\ oldOuts' = \mathrm{ran}\ outputs \wedge \mathrm{ran}\ oldOuts = \epsilon \wedge newOuts' = \varnothing$
$currentSS' = initialSS$

$Monitor \cong;\ out : outputs \bullet$
$\quad fmi2Get.(FMU\ out).(name\ out)?nv?st \longrightarrow newOuts := newOuts \oplus \{out \mapsto nv\}$

$Adjust \cong$ **if** $delta(oldOuts, newOuts) \geq threshold \longrightarrow$
$\qquad\qquad currentSS := newstep(delta(oldOuts, newOuts), currentSS);$
$\qquad\qquad updateSS!currentSS \longrightarrow$ **Skip**
$\qquad [\!] \ delta(oldOuts, newOuts) > threshold \longrightarrow$ **Skip**
$\qquad$ **fi**

$Step = Monitor;\ Adjust;\ Step$

$\bullet\ Init;\ (Step \vartriangle endSimulation)$

end

Fig. 8. Model of $VaryStep$

$fmi2GetFMUState$ before each step of co-simulation, and (b) queries the maximum step size that each FMU is prepared to take. This uses an extra FMI API function `fmi2GetMaxStepSize`. In $Adjust$, if any of the maximum values returned is lower than that originally proposed, the states of the FMUs are reset using $fmi2SetFMUState$, and the time as well as the step size are adjusted (using $setT$ and $updateSS$). We have again proved validity, termination, and determinism.

In [4], determinism is also based on the FMU states, which are visible via `fmi2Get` and `fmi2Set`. On the other hand, that work considers determinism with respect to the order of retrieval and update of variables and execution of the FMUs. In our models, this order is fixed. To establish determinism in that sense, we need to consider a highly parallel model with all valid execution orders respecting the port dependency graph. This is the approach in [7], where verification uses theorem proving. The approach taken here is more amenable to model checking and sufficient to verify sequential implementations of simulations.

As explained in the previous section, the definition of $Interaction$ is determined by structural information about the FMUs configuration. Using that information, and a choice of master algorithm (fixed or variable step, treatment of `fmi2Discard`, and so on), we can obtain a model. For the FMUs, in the previous section, we have explained how to derive (sketches of) models.

5.2 Co-simulations

Our semantics is also useful for analysis using FDR of the FMU compositions in co-simulations for deadlock, livelock, and determinism. We have done this verification, for instance, for the discrete event signal example in Fig. 7.

The semantics can also be used to validate the results of co-simulation runs. For example, Fig. 9 describes a short scenario involving two co-simulation steps. We specify it using CSP-M, rather than *Circus*, and write the traces refinement ([T=) assertion we use for verification. The assertion says that this scenario is a possible trace of the model: it is a correct co-simulation run. (We may check this by noting that the final two operations set the same inputs for FMU 4 (Check Equality)—the FMU that checks equality in the simulation model.) To facilitate model checking, we use numbers for the names of the variables. With this approach, we validate our model against an actual co-simulation.

```
DSynchronousEventsSpec =
  -- Set parameters
  fmi2Set.1.1.1.fmi2OK -> fmi2Set.1.2.1.fmi2OK ->
  fmi2Set.2.1.1.fmi2OK -> fmi2Set.2.2.2.fmi2OK ->
  -- Set initial values of inputs
  fmi2Set.3.1.1.fmi2OK -> fmi2Set.3.2.1.fmi2OK ->
  fmi2Set.4.1.1.fmi2OK -> fmi2Set.4.2.1.fmi2OK ->
  -- Steps
  fmi2Get.1.1.1.fmi2OK -> fmi2Get.2.1.1.fmi2OK -> fmi2Get.3.3.1.fmi2OK ->
  fmi2Set.3.1.1.fmi2OK -> fmi2Set.3.2.1.fmi2OK ->
  fmi2Set.4.2.1.fmi2OK -> fmi2Set.4.1.1.fmi2OK ->
  fmi2DoStep.1.0.2.fmi2OK -> fmi2DoStep.2.0.2.fmi2OK ->
  fmi2DoStep.3.0.2.fmi2OK -> fmi2DoStep.4.0.2.fmi2OK ->
  fmi2Get.1.1.1.fmi2OK -> fmi2Get.2.1.1.fmi2OK -> fmi2Get.3.3.1.fmi2OK ->
  fmi2Set.3.1.1.fmi2OK -> fmi2Set.3.2.1.fmi2OK ->
  fmi2Set.4.2.1.fmi2OK -> fmi2Set.4.1.1.fmi2OK ->
  fmi2DoStep.1.2.2.fmi2OK -> fmi2DoStep.2.2.2.fmi2OK ->
  fmi2DoStep.3.2.2.fmi2OK -> fmi2DoStep.4.2.2.fmi2OK -> SKIP

assert Cosimulation(0,2) [T= SynchronousEventsSpec
```

Fig. 9. Scenarios for Fig. 7: sampling of discrete event signals

Moreover, we can go further and check behavioural correctness too. The specification of an FMI composition C is an assertion over traces of events corresponding to the FMI API, principally doStep, get, and set. A similar technique is used for specification of processes in CSPm based on traces of events [17], and in CCS, using temporal logic over actions [3].

An alternative is to use a more abstract composition of FMUs A as a specification. A can be used as an oracle in testing the simulation: do a step of C and then compare it with a step of A. A and C can be used even more directly in our model by carrying out a refinement check in FDR3.

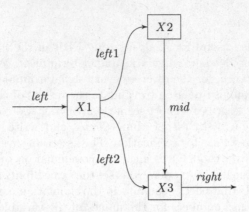

Fig. 10. A data-flow example

Consider a dataflow process taken from [17, p.124] and depicted in Fig. 10 that computes the weighted sums of consecutive pairs of inputs. So, if the input is $x_0, x_1, x_2, x_3, \ldots$, then the output is $(a * x_0 + b * x_1)$, $(a * x_1 + b * x_2), (a * x_2 + b * x_3), \ldots$, for weights a and b. The network has two external channels, $left$ and $right$, and three internal channels. $X2$ multiplies an input on channel $left1$ by a and passes the result to $X3$ on mid. $X3$ multiplies an input on the $left2$ channel by b and adds the result to the corresponding value from the mid channel. $X1$ duplicates its inputs and passes them to the other two processes (since all values except the first and last are used twice), where the multiplications can be performed in parallel. A little care needs to be taken to get the order of communications on the $left1$ and $left2$ channels right, otherwise a deadlock soon ensues.

The CSP specification of this network remembers the previous input.

$$DFProc(a, b) = left?x \longrightarrow P(x)$$
$$P(x) = left?y \longrightarrow right!(a * x + b * y) \longrightarrow P(y)$$

The key part of the main FMU in this specification is shown in Fig. 11.

Once the slave FMU has been initialised, the master algorithm can instruct it to perform a simulation step (`fmi2DoStep`). The FMU fetches the state item, gets the next input, fetches the parameters a and b, performs the necessary computation, and stores it as the current output.

We have been able to encode both the specification and implementation of the data flow network, with small values for $maxint$, and check behavioural refinement. We have identified the problem alluded to above, in getting the communications on $left1$ and $left2$ in the wrong order; issues to do with determinism concerning hidden state in our model; and termination issues to do with the end of the experiment and closing down resources. We have also been able to demonstrate in a small way the consistency of the semantics model.

```
DFSPECFMUProc(i) =
  let
    slaveInitialized(hc) =
      ...
      []
      fmi2DoStep.i?t?ss!fmi2OK -> (UpdateState; slaveInitialized(ss))
    UpdateState =
      get.i.1?x:INPUTVALP ->
        getinput.i.1?y:INPUTVALP ->
          getparam.i.1?a:PARAMVAL -> getparam.i.2?b:PARAMVAL ->
            setoutput.i.1!(a*x+b*y) -> SKIP
  within
    Instantiation; InstantiationMode(eps,eps);
    InitializationMode; slaveInitialized(0)
```

Fig. 11. Data flow specification

The transformation from *Circus* to CSPM corresponding to the FMI API requires the identification of barrier synchronisations that correspond to the doStep commands. An appropriate strategy is outlined in [6].

6 Conclusions

We have provided a comprehensive model of the FMI API, characterising formally valid master algorithms and FMUs. We can use our models to prove validity of master algorithms and FMU models. For stateless models, model checking is feasible, and we can use that to establish properties of interest of algorithms and FMU models. For state-rich models, we need theorem proving.

Given information about the network of FMUs and a choice of master algorithm, it is possible to construct a model of their co-simulation automatically for reasoning about the whole system. This is indicated by how our models are defined in terms of information about parameters, inputs, and so on, for each FMU, and about the FMU connections. A detailed account of the generation process and its mechanisation are, however, left as future work.

We have discussed a few example master algorithms. This includes a sophisticated rollback algorithm presented in [4] using a proposed extension of the FMI. It uses API functions to get and set the state of an FMU. In [4], this algorithm uses a doStep function that returns an alternative step size, in case the input step size is not possible. Here, instead, we use an extra function that can get the alternative step size. This means that our standard algorithms respect the existing signature of the fmi2DoStep function. As part of our future work, we plan to model one additional master algorithm proposed in [4].

There has been very practical work on new master algorithms, generation of FMUs and simulations, and hybrid models [2,9,11,22]. Tripakis [25] shows how components with different underlying models (state machines, synchronous data flow, and so on) can be encoded as FMUs. Savicks [24] presents a framework for co-simulation of Event-B and continuous models based on FMI, using a fixed-step master algorithm and a characterisation of simulation components as a class specialised by Event-B models or FMUs. This work has no semantics for the FMI API, but supplements reasoning in Event-B with simulation of FMUs.

Pre-dating FMI, the work in [15] presents models of co-simulations using timed automata, with validation and verification carried out using UPPAAL, and support for code generation. It concentrates on the combination of one continuous and one discrete component using a particular orchestration approach. The work in [5] discusses the difficulties for treatment of hybrid models in FMI.

There are several ways in which our models can be enriched: definition of the type system, consideration of asynchronous FMUs, sophisticated error handling policies that allow resetting of the FMU states, and increased coverage of the API. FMI includes capability flags that define the services supported by FMUs, like asynchronous steps, and retrieval and update of state, for example. We need a family of models to consider all combinations of values of the capability flags. We have explained here how a typical combination can be modelled.

Our long-term goal is to use our semantics to reason about the overall system composed of the various simulation models. In particular, we are interested in hybrid models, involving FMUs defined by languages for discrete and for continuous modelling. To cater for models involving continuous FMUs, we plan to use a *Circus* extension [13]. Using current support for *Circus* in Isabelle [14], we may also be able to explore code generation from the models. We envisage fully automated support for generation and verification of models and programs.

Acknowledgements. The work is funded by the EU INTO-CPS project (Horizon 2020, 664047). Ana Cavalcanti and Jim Woodcock are also funded by the EPSRC grant EP/M025756/1. Anonymous referees have made insightful suggestions. No new primary data were created during this study.

References

1. Abrial, J.R.: Modeling in Event-B-System and Software Engineering. Cambridge University Press, Cambridge (2010)
2. Bastian, J., Clauß, C., Wolf, S., Schneider, P.: Master for co-simulation using FMI. In: Modelica Conference (2011)
3. Bradfield, J., Stirling, C.: Verifying temporal properties of processes. In: Baeten, J.C.M., Klop, J.W. (eds.) CONCUR 1990. LNCS, vol. 458, pp. 115–125. Springer, Heidelberg (1990). doi:10.1007/BFb0039055
4. Broman, D., et al.: Determinate composition of FMUs for co-simulation. In: ACM SIGBED International Conference on Embedded Software. IEEE (2013)
5. Broman, D., et al.: Requirements for hybrid cosimulation standards. In: 18th International Conference on Hybrid Systems: Computation and Control, pp. 179–188. ACM (2015)
6. Butterfield, A., Sherif, A., Woodcock, J.: Slotted-circus. In: Davies, J., Gibbons, J. (eds.) IFM 2007. LNCS, vol. 4591, pp. 75–97. Springer, Heidelberg (2007). doi:10.1007/978-3-540-73210-5_5

7. Cavalcanti, A.L.C., Clayton, P., O'Halloran, C.: From control law diagrams to ada via Circus. Formal Aspects Comput. **23**(4), 465–512 (2011)
8. Cavalcanti, A.L.C., Sampaio, A.C.A., Woodcock, J.C.P.: A refinement strategy for Circus. Formal Aspects Comput. **15**(2–3), 146–181 (2003)
9. Denil, J., et al.: Explicit semantic adaptation of hybrid formalisms for FMI co-simulation. In: Spring Simulation Multi-Conference (2015)
10. Derler, P., Lee, E.A., Vincentelli, A.S.: Modeling cyber-physical systems. Proc. IEEE **100**(1), 13–28 (2012)
11. Feldman, Y.A., Greenberg, L., Palachi, E.: Simulating Rhapsody SysML blocks in hybrid models with FMI. In: Modelica Conference (2014)
12. FMI development group: Functional mock-up interface for model exchange and co-simulation, 2.0. (2014). https://www.fmi-standard.org
13. Foster, S., et al.: Towards a UTP semantics for Modelica. In: Unifying Theories of Programming. LNCS. Springer (2016)
14. Foster, S., Zeyda, F., Woodcock, J.: Isabelle/UTP: a mechanised theory engineering framework. In: Naumann, D. (ed.) UTP 2014. LNCS, vol. 8963, pp. 21–41. Springer, Heidelberg (2015). doi:10.1007/978-3-319-14806-9_2
15. Gheorghe, L., et al.: A formalization of global simulation models for continuous/discrete systems. In: Summer Computer Simulation Conference, pp. 559–566. Society for Computer Simulation International (2007)
16. Gibson-Robinson, T., et al.: FDR3–a modern refinement checker for CSP. In: Tools and Algorithms for the Construction and Analysis of Systems, pp. 187–201 (2014)
17. Hoare, C.A.R.: Communicating Sequential Processes. Prentice-Hall, Upper Saddle River (1985)
18. Kübler, R., Schiehlen, W.: Two methods of simulator coupling. Math. Comput. Modell. Dynamical Syst. **6**(2), 93–113 (2000)
19. The MathWorks Inc: Simulink. www.mathworks.com/products/simulink
20. Oliveira, M., Cavalcanti, A.: From Circus to JCSP. In: Davies, J., Schulte, W., Barnett, M. (eds.) ICFEM 2004. LNCS, vol. 3308, pp. 320–340. Springer, Heidelberg (2004). doi:10.1007/978-3-540-30482-1_29
21. Oliveira, M.V.M., Cavalcanti, A.L.C., Woodcock, J.C.P.: A UTP semantics for Circus. Formal Aspects Comput. **21**(1–2), 3–32 (2009)
22. Pohlmann, U., et al.: Generating functional mockup units from software specifications. In: Modelica Conference (2012)
23. Roscoe, A.W.: Understanding concurrent systems. In: Texts in Computer Science. Springer (2011)
24. Savicks, V., et al.: Co-simulating event-b and continuous models via FMI. In: Summer Simulation Multiconference, pp. 37:1–37:8. Society for Computer Simulation International (2014)
25. Tripakis, S.: Bridging the semantic gap between heterogeneous modeling formalisms and FMI. In: International Conference on Embedded Computer Systems: Architectures, Modeling, and Simulation, pp. 60–69. IEEE (2015)
26. Woodcock, J.C.P., Davies, J.: Using Z-Specification, Refinement, and Proof. Prentice-Hall, New York (1996)

An Abstract Model for Proving Safety of Autonomous Urban Traffic

Martin Hilscher and Maike Schwammberger[✉]

Department of Computing Science, University of Oldenburg, Oldenburg, Germany
{hilscher,schwammberger}@informatik.uni-oldenburg.de

Abstract. The specification of *Multi-lane Spatial Logic* (MLSL) was introduced in [1, 2] for proving safety (collision freedom) on multi-lane motorways and country roads. We now consider an extension of MLSL to deal with urban traffic scenarios, thereby focusing on crossing manoeuvres at intersections. To this end, we modify the existing abstract model by introducing a generic topology of urban traffic networks. We then show that even at intersections we can use purely spatial reasoning, detached from the underlying car dynamics, to prove safety of controllers modelled as extended timed automata.

Keywords: Multi-dimensional spatial logic · Urban traffic · Bended view · Virtual lanes · Autonomous cars · Collision freedom · Timed Automata

1 Introduction

Traffic safety is a relevant topic as driving assistance systems and probably soon fully autonomously driving cars are increasingly capturing the market. In this context, safety means collision freedom and thus reasoning about car dynamics and spatial properties. An approach to separate the car dynamics from the spatial considerations and thereby simplify reasoning, was introduced in [1] with the Multi-lane Spatial Logic (MLSL) for expressing spatial properties on multi-lane motorways. This logic and its dedicated abstract model was extended with length measurement in [2] for country roads with oncoming traffic.

MLSL is inspired by Moszkowski's interval temporal logic [3], Zhou, Hoare and Ravn's Duration Calculus [4] and Schäfer's Shape Calculus [5]. MLSL extends interval temporal logic by a second dimension, whilst considering continuous (positions on lanes) and discrete components (the number of a lane). With these two-dimensional features we can, for instance, express that a car is occupying a certain space on a lane.

Aside from highway traffic and country roads, safety in urban traffic scenarios is of high importance. There, drivers' mistakes lead to large amounts of human

---

This research was partially supported by the German Research Council (DFG) in the Transregional Collaborative Research Center SFB/TR 14 AVACS.

© Springer International Publishing AG 2016
A. Sampaio and F. Wang (Eds.): ICTAC 2016, LNCS 9965, pp. 274–292, 2016.
DOI: 10.1007/978-3-319-46750-4_16

casualties. Because of that, we now extend the approaches of [1,2] to urban traffic manoeuvres.

Whereas in [1,2] the abstract model consisted of adjacent lanes with infinite length and without any overlaps, we now consider intersecting lanes and the safety of crossing manoeuvres, like turning left or right, on such lanes. To deal with these new conditions, the extension of MLSL by length measurement given in [2] is quite convenient as we, e.g., need to talk about a car's distance to a crossing to ensure it is not just passing it without any consideration. Furthermore, our intersecting lanes are no longer infinite, as we consider lanes to be finite road segments between two intersections.

Related work. Loos and Platzer present an approach for intersections of single lanes with one car on each lane in [6]. They use traffic lights as a control mechanism, where a car is not permitted to enter an intersection when the light is red. The authors verify the safety of their hybrid systems with the tool KeYmaera.

In [7], Werling et al. examine safety of their algorithm for handling moving traffic in urban scenarios, which was deployed on the DARPA Urban Challenge 2007 finalist AnnyWAY and uses a hierarchical finite state machine. This approach has strong assumptions, e.g. a constant vehicular speed of all cars.

A number of solutions have been proposed to improve vehicular safety by using intelligent transportation systems. An example for this is given by Colombo and Del Vecchio in [8], where a supervisor for collision avoidance at intersections is synthesised. This supervisor is based on a hybrid algorithm and acts as a scheduler for the cars.

The key contribution of our paper is the adaption of the existing abstract model and the logic MLSL from [2] to a more powerful model with intersections for urban traffic scenarios. For this, in Sect. 2 we present the new concept of *bended views* to cope with the concept of turning left resp. right at intersections. In Sect. 3, we introduce semantics of an extension of timed automata [9], the *automotive-controlling timed automata* (ACTA). With these, we construct a controller for crossing manoeuvres and adapt the lane change controller from [2] for our purposes. Finally, we prove safety of this crossing controller in Sect. 4. A conclusion, more work directly related to our approach and ideas for future work is given in Sect. 5.

2 Abstract Model

Our abstract model for urban traffic focuses on modelling traffic situations at intersections and contains a set of *crossing segments* $\mathbb{CS} = c_0, c_1, \ldots$ and a set of *lanes* $\mathbb{L} = 0, 1, \ldots$ connecting different crossings, with typical elements l, m, n. Adjacent lanes are bundled to *road segments* $\mathbb{RS} = \{0,1\}, \{2,3\}, \ldots$ such that $\mathbb{RS}$ is a subset of the power set of $\mathbb{L}$. Typical elements of $\mathbb{RS}$ are r_0, r_1, r_2. Adjacent crossing segments form an intersection. Every car has a unique *car identifier* from the set $\mathbb{I} = A, B, \ldots$ and a real value for its position *pos* on a lane.

In the following, we will use car E as the *car under consideration* and we introduce the special constant *ego* with valuation $\nu(ego) = E$ to refer to this car. When we are talking about an arbitrary car, we will use the identifier C.

To simplify reasoning, only local parts of traffic are considered as every car has its own local *view*. If no crossing is within some given *horizon*, the *standard view* $V(E)$ of car E only contains a bounded extension of all adjacent lanes.

If an intersection is within the horizon of car E, its standard view covers a bounded part of the road segment it is driving on, the intersection itself, and a bounded extension of the road segment it is about to drive on after passing the crossing. We refer to this as a *bended view* as the car E may turn left or right at the crossing (cf. Fig. 1). We straighten this bended view to *virtual lanes*, to allow for purely spatial reasoning around the corner with our later introduced logic *Urban Multi-lane Spatial Logic* (UMLSL).

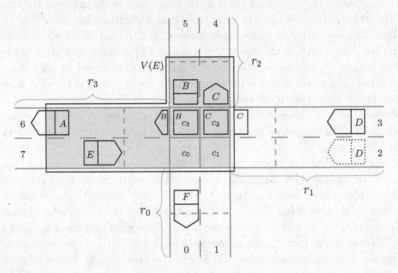

Fig. 1. In its view $V(E)$, indicated by the gray shading, car E sees car A driving on lane 6 and cars B and C which are currently both turning right at the intersection. $V(E)$ does not cover the road segments r_0 and r_1 an thus not the positions of car F and D.

UMLSL extends the logic introduced in [1] by atoms to formalise traffic situations on crossing segments and has a continuous (real positions on lanes) and a discrete dimension (number of lanes and crossing segments). In the following we will talk about car E as the *owner* of view $V(E)$.

Example. In a part of view $V(E)$ (cf. Fig. 1) the formula $\phi \equiv re(ego) \frown free \frown cs \wedge free$ holds. Here, $re(ego)$ is the space car E *reserves* on lane 7, the atom *free* represents the free space in front of car E, and $cs \wedge free$ stands for the unoccupied space on crossing segment c_0. The *horizontal chop operator* $\frown$ is similar to its equivalent in interval temporal logic and is used to deal with adjacent segments.   □

While a reservation re(ego) is the space car E is actually occupying, a *claim* cl(ego) is akin to setting the direction indicator (cf. dotted part of car D in Fig. 1, showing the desire of car D to change its lane). Thus, a claim represents the space a car plans to drive on in the future. Reserved and claimed spaces have the extension of the *safety envelopes* of the cars, which include a car's physical size and its braking distance. For now, we will use a concept of *perfect knowledge* which means thus assume that every car perceives the safety envelopes of all other cars in its view.

We distinguish between the movement of cars on lanes and on crossings. We allow for two-way traffic on continuous lanes of finite length, assuming every lane has one direction, but cars may temporarily drive in the opposite direction to perform an overtaking manoeuvre. As a car's direction will change while turning on an intersection, we cannot assign one specific direction to a crossing segment. Therefore, we consider crossing segments as discrete elements which are either fully occupied by a car or empty.

When a car is about to drive onto a discrete crossing segment and time elapses, the car's safety envelope will stretch to the whole crossing segment, while disappearing continuously on the lane it drove on. An example for this behaviour are cars B and C in Fig. 1, where B leaves lane 5 and enters lane 6 continuously, while it occupies the whole discrete crossing segment c_3.

2.1 Topology

For this paper we restrict the abstract model to road segments with two lanes, one in each direction. Intersections are four connected crossing segments with four road segments meeting at a crossing, comparable to our example in Fig. 1.

We describe the connections between lanes and crossing segments by a graph $\mathcal{N}$, whose nodes are elements from $\mathbb{L}$ and $\mathbb{CS}$. Additionally, as we are dealing with traffic that is presumably evolving over time, we need to capture the (finite and real) length of lanes and crossing segments in our graph. We group adjacent crossing segments to strongly connected components $\mathcal{I}_{cs}$ (intersections) and neighbouring lanes to components $\mathcal{I}_{rs}$ (road segments). Later, we will use the information given by $\mathcal{N}$ to determine the parts of lanes and crossing segments the safety envelope of an arbitrary car C occupies.

Definition 1 (Urban Road Network). *An* urban road network *is defined by a graph* $\mathcal{N} = (V, E_u, E_d, \omega, \mathcal{I}_{cs}, \mathcal{I}_{rs})$, *where*

- V *is a finite set of nodes, with* $V = \mathbb{L} \cup \mathbb{CS}$,
- $E_u \subseteq \mathbb{L} \times \mathbb{L}$ *is the set of undirected edges,*
- $E_d \subseteq (V \times V) \setminus (\mathbb{L} \times \mathbb{L})$ *is the set of directed edges,*
- $\omega : V \to \mathbb{R}^+$ *is a mapping assigning a weight to each node in V, describing the length of the related segment,*
- $\mathcal{I}_{cs} \subseteq \mathcal{P}(\mathbb{CS})$ *is a set of strongly connected components id_{cs}, where elements $v_1, v_2 \in \mathbb{CS}$ are part of the same component iff there exists a finite sequence $\pi_{v_{1,2}}$ of directed edges between v_1 and v_2 and for all elements π_i of $\pi_{v_{1,2}}$:*

$\pi_i \in \mathbb{CS}$. For an arbitrary crossing segment $cs' \in V$ the function $I_{cs} : \mathbb{CS} \to \mathcal{I}_{cs}$ with $cs' \mapsto id_{cs}$ returns the strongly connected component cs' is part of.

- $\mathcal{I}_{rs} \subseteq \mathcal{P}(\mathbb{L})$ is a set of strongly connected components id_{rs}, where $v_1, v_2 \in \mathbb{L}$ are in the same component iff there is a finite sequence $\pi_{v_{1,2}}$ of undirected edges between v_1 and v_2 and for all elements π_i of $\pi_{v_{1,2}}$: $\pi_i \in \mathbb{L}$. For an arbitrary lane $l \in \mathbb{L}$ the function $I_{rs} : \mathbb{L} \to \mathcal{I}_{rs}$ with valuation $l \mapsto id_{rs}$ returns the strongly connected component l is part of.

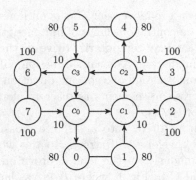

Fig. 2. Urban road network for Fig. 1.

Note that self-loops are excluded. Two elements $v_1, v_2 \in \mathbb{L}$ with $(v_1, v_2) \in E_u$ moreover constitute two parallel lanes, where the undirected edge allows for lane change manoeuvres between these two lanes. Every car C has a path $pth(C)$ which is infinite in both directions and comprises its travelling route.

Example. Considering the traffic situation depicted in Fig. 1 and the corresponding road network in Fig. 2, $pth(E) = \langle \ldots, 7, c_0, c_1, c_2, 4, \ldots \rangle$ is a suitable path for the car under consideration E, where it plans on turning left.   □

2.2 Traffic Snapshot

In this paragraph we present a model $\mathcal{TS}$ of a *traffic snapshot*, which captures the traffic on an urban road network at a given point in time. We recall definitions from [1], extending them where needed and assign an infinite path $pth(C) \in V^{\mathbb{Z}}$ to every car C.

The node car C is currently driving on we call $pth(C)_{curr(C)}$, where $curr(C)$ is the index of this node in $pth(C)$. The real-valued position $pos(C)$ defines the position of car C on the current node $pth(C)_{curr(C)}$. The node in $pth(C)$ that is reached after some time t elapses, we call $pth(C)_{next(C)}$. Note that $curr(C) = next(C)$, iff C did not move far enough to leave its current node.

To formalise two-way traffic on road segments, we assume that every lane has one direction and cars normally drive on a lane in the direction of increasing real values. Cars may only temporarily drive in the opposite direction of a lane to perform an overtaking manoeuvre. We differentiate between claims and reservations on continuous lanes (*clm*, *res*) and on discrete crossing segments (*cclm*, *cres*).

Definition 2 (Traffic Snapshot). *A* traffic snapshot $\mathcal{TS}$ *is a structure*
$\mathcal{TS} = (res, clm, cres, cclm, pth, curr, pos)$ *with*

- $res : \mathbb{I} \to \mathcal{P}(\mathbb{L})$ *such that* $res(C)$ *is the set of lanes that car* C *reserves,*
- $clm : \mathbb{I} \to \mathcal{P}(\mathbb{L})$ *such that* $clm(C)$ *is the set of lanes car* C *claims,*
- $cres : \mathbb{I} \to \mathcal{P}(\mathbb{CS})$ *such that* $cres(C)$ *is the set of crossing segments car* C *reserves,*
- $cclm : \mathbb{I} \to \mathcal{P}(\mathbb{CS})$ *such that* $cclm(C)$ *is the set of crossing segments car* C *claims,*
- $pth : \mathbb{I} \to V^{\mathbb{Z}}$ *such that* $pth(C)$ *is the path car* C *pursues,*
- $curr : \mathbb{I} \to \mathbb{Z}$ *such that* $curr(C)$ *is the index of the current element of* $pth(C)$ *the car* C *is driving on and*
- $pos : \mathbb{I} \to \mathbb{R}$ *such that* $pos(C)$ *is the position of the rear of car* C *on its lane.*

Let $\mathbb{TS}$ *denote the set of all traffic snapshots.*

When time elapses, every car C with a positive speed will change its current position according to the path $pth(C)$. The function $nextCrossing(C)$ is then used to identify all crossing segments of the next intersection for C along $pth(C)$.

$$nextCrossing(C) = \begin{cases} \emptyset & \text{if } pth(C)_{next(C)} \notin \mathbb{CS} \\ \langle \pi_i, \dots, \pi_{i+n} \rangle & \text{if } pth(C) = \langle \dots, \pi_{i-1}, \pi_i, \dots, \pi_{i+n}, \dots \rangle \\ & \wedge i \leq next(C) \leq i + n \\ & \wedge \forall j \in \{i, \dots, i+n\} : \pi_j \in \mathbb{CS} \\ & \wedge \pi_{i-1} \in \mathbb{RS} \end{cases}$$

We allow for the following transitions between traffic snapshots to model the behaviour of cars in urban traffic and use the overriding notation $\oplus$ of the mathematical notation language Z for function updates [10].

Definition 3 (Transitions). *The following transitions describe the changes that may occur at a traffic snapshot* $\mathcal{TS} = (res, clm, cres, cclm, pth, curr, pos)$.

$$\mathcal{TS} \overset{t}{\to} \mathcal{TS}' \quad \Leftrightarrow \quad \mathcal{TS}' = (res', clm, cres', cclm, pth, curr', pos')$$
$$\wedge \forall C \in \mathbb{I} : 0 \leq pos'(C) \leq \omega(pth(C)_{next(C)}) \quad (1)$$
$$\wedge \, cres'(C) = cres \oplus \{C \mapsto \{nextCrossing(C)\}\}$$
$$\wedge \, next(C) \neq curr(C) \to (|res(C)| = 1 \wedge |clm(C)| = 0)$$

$$\mathcal{TS} \xrightarrow{c(C,n)} \mathcal{TS}' \quad \Leftrightarrow \quad \mathcal{TS}' = (res, clm', cres, cclm, pth, curr, pos)$$
$$\wedge |clm(C)| = |cclm(C)| = |cres(C)| = 0$$
$$\wedge |res(C)| = 1 \quad (2)$$
$$\wedge \{n+1, n-1\} \cap res(C) \neq \emptyset$$
$$\wedge \, clm' = clm \oplus \{C \mapsto \{n\}\}$$

$$\mathcal{TS} \xrightarrow{wd \ c(C)} \mathcal{TS}' \quad \Leftrightarrow \quad \mathcal{TS}' = (res, clm', cres, cclm, pth, curr, pos)$$
$$\wedge \, clm' = clm \oplus \{C \mapsto \emptyset\} \quad (3)$$

$$TS \xrightarrow{r(C)} TS' \iff TS' = (res', clm', cres, cclm, pth, curr, pos)$$
$$\land clm' = clm \oplus \{C \mapsto \emptyset\}$$
$$\land |cclm(C)| = |cres(C)| = 0 \tag{4}$$
$$\land res' = res \oplus \{C \mapsto res(C) \cup clm(C)\}$$

$$TS \xrightarrow{\text{wd } r(C,n)} TS' \iff TS' = (res', clm, cres, cclm, pth, curr, pos)$$
$$\land res' = res \oplus \{C \mapsto \{n\}\} \tag{5}$$
$$\land n \in res(C) \land |res(C)| = 2$$

$$TS \xrightarrow{cc(C)} TS' \iff TS' = (res, clm, cres, cclm', pth, curr, pos)$$
$$\land |res(C)| = 1$$
$$\land |clm(C)| = |cclm(C)| = |cres(C)| = 0 \tag{6}$$
$$\land cclm' = cclm \oplus \{C \mapsto \{nextCrossing(C)\}\}$$

$$TS \xrightarrow{\text{wd } cc(C)} TS' \iff TS' = (res, clm, cres, cclm', pth, curr, pos)$$
$$\land cclm' = cclm \oplus \{C \mapsto \emptyset\} \tag{7}$$

$$TS \xrightarrow{rc(C)} TS' \iff TS' = (res, clm, cres', cclm', pth, curr, pos)$$
$$\land |res(C)| = 1 \land |clm(C)| = 0$$
$$\land cclm' = cclm \oplus \{C \mapsto \emptyset\} \tag{8}$$
$$\land cres' = cres \oplus \{C \mapsto cclm(C)\}$$

$$TS \xrightarrow{\text{wd } rc(C)} TS' \iff TS' = (res, clm, cres', cclm, pth, curr, pos)$$
$$\land cres' = cres \oplus \{C \mapsto \emptyset\}. \tag{9}$$

In (1) we allow for the passage of time, where cars move along their paths in the urban road network. We assume this movement to be continuous but give no concrete formula here, because of the idea of separating spatial reasoning from car dynamics. We reuse the transitions (2)–(5) from [1] for creating (resp. removing) a claim or reservation for a neighbouring lane. These transitions are only allowed on road segments between two intersections. To (2) and (4) we add the further restriction that we are not allowed to make new claims (resp. reservations) while on a crossing segment. With (6)–(9) we add four transitions which create (resp. remove) claims and reservations for crossing segments to pass through an intersection.

2.3 Bended View and Virtual Lanes

For proving safety we restrict ourselves to finite parts of the traffic snapshot TS. The intuition is that the safety of a car depends only on its immediate surroundings. We encode this by introducing a structure called *view*, which only contains the part of lanes and crossing segments that lie within some distance h called *horizon*.

In previous works [1,2] covering highway and country road traffic, the set of lanes L in a view was obtained by taking a subinterval of the global set of parallel lanes $\mathbb{L}$. This is no longer possible for the urban traffic scenario, since taking an arbitrary subset of lanes can yield a set of lanes which are not neighbouring. We therefore have to construct a view from the urban road network, the current traffic snapshot, a given real-valued interval $X = [a, b]$ and a car E which we call *owner of the view*. We can adapt the definition from [1] as follows.

Definition 4 (View). *The view $V = (L, X, E)$ of car E contains*

- *a set of virtual lanes $L \subseteq \mathcal{P}(V^{\mathbb{Z}})$ and*
- *an interval of space along the lanes $X = [a, b] \subseteq \mathbb{R}$ relative to E's position.*

If an intersection is within the horizon h, we deal with a *bended view* as cars are allowed to turn in any possible direction at the crossing (cf. $V(E)$ in Fig. 1). To allow for our spatial reasoning with the logic UMLSL, we construct straight and adjacent *virtual lanes* from the urban road network and the path of car E. For each of these virtual lanes, we map the nodes of E's path to the appropriate segments of the virtual lane, such that traversing along the virtual lanes corresponds to moving through the nodes of the urban road network along the path of E.

To build our virtual lanes, we first need to find the crossings and road segments E traverses. These can be obtained from the path $pth(E)$ by building a second path called *track* of the car, $track(E)$, which is more coarse-grained. Every element of E's path is abstracted to its connected component id_{cs} (id_{rs} resp.). Subsequent occurrences of the same component are recorded only once.

Example (track of a car). Let us recall the situation in Fig. 1 and the corresponding Urban Road Network from Fig. 2. Furthermore, assume that car E's path contains the subpath $\langle 7, c_0, c_1, c_2, 4 \rangle$. From Fig. 1 we can see, that $6, 7$ form road segment r_3 and $4, 5$ are road segment r_2. We additionally name the strongly connected component composed of $c_0, \ldots, c_3$ with cr_0. We then obtain $track(E) = \langle \ldots, r_3, cr_0, r_2, \ldots \rangle$. $\quad\square$

From these tracks we now construct two neighbouring *infinite lanes*, $\overrightarrow{\pi_E}$ for the lane in E's driving direction and $\overleftarrow{\pi_E}$ for the opposing traffic. We find one path in the urban road network, which traverses the crossing and road segments in $track(E)$ in the same order along the directed edges ($\overrightarrow{\pi_E}$) and one traversing them in opposite order along the edges ($\overleftarrow{\pi_E}$).

Example (infinite lanes). For car E with $track(E) = \langle \ldots, r_3, cr_0, r_2, \ldots \rangle$ of the previous example, we need to find a π_i with $\pi_i \in r_3$ and there is a directed edge from π_i to some node in $cr_0 = \{c_0, \ldots, c_4\}$. This is only satisfied for the choice of 7 and c_0. Continuing this way, we obtain the infinite partial lane $\langle \ldots, 7, c_0, c_1, c_2, 4, \ldots \rangle$. Similarly we obtain $\langle \ldots, 6, c_3, 5, \ldots \rangle$ as second lane. $\quad\square$

We have now obtained a set of infinite lanes. Since views are just subsets of lanes the last step for finding our virtual lanes is to obtain the finite infix of path $\overrightarrow{\pi_E}$ which is visible in the view. We start at $pth(E)_{curr(E)}$ and examine the next elements $\pi_n \in \overrightarrow{\pi_E}$ and sum up all weights of nodes between $pth(E)_{curr(E)}$ and π_n. If this sum is in the interval $X = [a, b]$ the path element is visible in

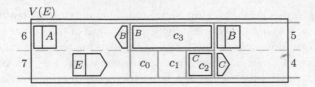

Fig. 3. Virtual lanes with cars for view $V(E)$ in Fig. 1.

the view and therefore part of the finite virtual lane $\overrightarrow{\pi_E}^v$. $\overleftarrow{\pi_E}^v$ is constructed symmetrically.

Example (virtual lanes). Previously, for car E we determined the infinite lane $\langle\ldots,7,c_0,c_1,c_2,4,\ldots\rangle$. We construct the finite virtual lane $\langle 7,c_0,c_1,c_2,4\rangle$. The second virtual lane is determined symmetrically and we get the set of virtual lanes depicted in Fig. 3 for car E.   □

Definition 5 (Constructed View). *For a car E and $X = [a,b]$ the constructed view of E is defined by $V_C = (\{\overrightarrow{\pi_E}^v, \overleftarrow{\pi_E}^v\}, [a,b], E)$, where $\overrightarrow{\pi_E}^v$ and $\overleftarrow{\pi_E}^v$ are virtual lanes.*

We adapt the chopping operations of views from [11] to the new notion of views. Note that while horizontal chopping is analogous to ITL [3], vertical chopping is defined concretely for two lanes.

Definition 6 (Operations on Views). *Let V_1, V_2 and V be views of a traffic snapshot TS. Then $V = V_1 \ominus V_2$ iff $V = (L,X,E)$, $V_1 = (L_1,X,E)$ and $V_2 = (L_2,X,E)$ and $L_1 = \emptyset$ or $L_1 = \{\overleftarrow{\pi_E}\}$ or $L_1 = \{\overrightarrow{\pi_E},\overleftarrow{\pi_E}\}$. Furthermore, $V = V_1 \oslash V_2$ iff $V = (L,[r,t],E)$ and $\exists s \in [r,t]$ with $V_1 = V_{[r,s]}$ and $V_2 = V_{[s,t]}$.*

We call a view of the form $V_S(E,TS) = (\{\overrightarrow{\pi_E}^v, \overleftarrow{\pi_E}^v\}, [-h,+h], E)$ a *standard view* of E, where h is a sufficiently large horizon, such that any car at maximum velocity can come to a complete standstill within this distance.

To model sensor capabilities of a car E, we introduce a *sensor function* Ω_E. This function returns the size of a car as perceived by the sensors of E. A possible implementation is, e.g., a car may calculate its own braking distance, while it can only perceive the physical size of all other cars. In the remainder of this paper we will consider an implementation, where every car perceives the breaking distances of all other cars.

Sensor Function. The car dependent sensor function $\Omega_E : \mathbb{I} \times TS \to \mathbb{R}$ yields, given an arbitrary car C and a traffic snapshot TS, the size of a car C as perceived by E's sensors.

Visible Segments of Cars. For both virtual lanes, we need to find all segments $seg_V(C)$ which are (partially) occupied by a car C and visible in the view of E. Therefore, we first add the value of $\Omega_E(C,TS)$ to the position $pos(C)$. Since this distance may span more than one lane segment (crossing segment resp.) we have to split this distance whenever the weight $\omega(s)$ of a segment along C's path is exceeded. Finally, we make the position of the segment relative to the position of the car E by adding the distance ($dist$) between the positions of E and C along the path of E.

2.4 UMLSL Syntax and Semantics

For the specification of the *Multi-lane Spatial Logic* for urban traffic *UMLSL*, we start with the set of car variables CVar ranging over car identifiers with typical elements c and d. RVar is used for variables ranging over the real numbers, with typical elements x, y. We assume RVar $\cap$ CVar $= \emptyset$. The set of all *variables* Var $=$ CVar $\cup$ RVar $\cup$ {ego} has typical elements u, v. For this paper we restrict ourselves to real-valued constants for the values of RVar, for a more complete discussion including real-valued terms see [2].

Formulas of UMLSL are built from atoms, Boolean connectors and first-order quantifiers. Furthermore, we use two *chop operations*, one for a horizontal chop, denoted by $\frown$ like for interval temporal logic and a vertical chop operator given by the vertical arrangement of formulas. Intuitively, a formula $\phi_1 \frown \phi_2$ holds if we can split the view V horizontally into two views V_1 and V_2 such that on V_1 ϕ_1 holds and V_2 satisfies ϕ_2. Similarly a formula $\frac{\phi_2}{\phi_1}$ is satisfied by V, if V can be chopped at a lane into two subviews, V_1 and V_2, where V_i satisfies ϕ_i for $i = 1, 2$.

Besides the atom *free* which represents free space on a lane, UMLSL extends MLSL by the atom cs to represent crossing segments. Hereby, we can, e.g., state that car E occupies a crossing segment ($cs \wedge re(\mathrm{ego})$) or that a crossing segment is free ($cs \wedge free$).

Definition 7 (Syntax). *The syntax of the* Urban Multi-lane Spatial Logic UMLSL *is defined as follows.*

$$\phi ::= true \mid u = v \mid free \mid cs \mid re(c) \mid cl(c) \mid \neg\phi \mid \phi_1 \wedge \phi_2 \mid \exists c \bullet \phi_1 \mid \phi_1 \frown \phi_2 \mid \tfrac{\phi_2}{\phi_1},$$

where $c \in$ CVar$\cup$*{ego} and* $u, v \in$ Var. *We denote the set of all UMLSL formulas by* $\Phi_\mathbb{U}$.

Definition 8 (Valuation and Modification). *A* valuation *is a function* $\nu: \text{Var} \to \mathbb{I} \cup \mathbb{R}$ *respecting the types of variables. To modify the valuation we use the overriding notation* $\nu \oplus \{v \mapsto \alpha\}$, *where the value of* v *is modified to* α.

While the syntax of UMLSL is not very different from previous versions of MLSL [1,2], the semantics of the atoms change drastically, to accommodate for bended views. Note that for the following definition of the semantics of UMLSL, the semantics of $\neg$, $\wedge$ and $\exists$ are defined as usual and thus not given here.

Definition 9 (Semantics). *The* satisfaction *of formulas with respect to a traffic snapshot* $\mathcal{TS}$, *a view* $V = (L, X, E)$ *and a valuation of variables* ν *is defined inductively as follows:*

$$\mathcal{TS}, V, \nu \models true \qquad for\ all\ \mathcal{TS}, V, \nu$$

$$\mathcal{TS}, V, \nu \models u = v \quad \Leftrightarrow \nu(u) = \nu(v)$$

$$\mathcal{TS}, V, \nu \models free \quad \Leftrightarrow |L| = 1\ and\ \|X\| > 0\ and$$
$$\forall i \in \mathbb{I}: seg_V(i) = \emptyset$$

$$\mathcal{TS}, V, \nu \models cs \quad \Leftrightarrow \{\pi_L\} = L\ and\ \|X\| > 0\ and$$
$$\forall \pi \in \pi_L: \pi \in cs$$

$$\mathcal{TS}, V, \nu \models re(c) \quad \Leftrightarrow \{\pi_L\} = L\ and\ \|X\| > 0\ and$$
$$\forall \pi \in \pi_L: \exists X': (\pi, X') \in seg_V(\nu(c))\ and$$
$$X \subseteq \bigcup_{\{(s,X')|s\in seg_V(\nu(c))\ and\ s\in \pi_L\ and\ s\notin cclm(\nu(c))\}} X'$$

$$\mathcal{TS}, V, \nu \models cl(c) \quad \Leftrightarrow \{\pi_L\} = L\ and\ \|X\| > 0\ and$$
$$(\forall \pi \in \pi_L: \pi \in \mathbb{CS}\ and\ \pi \in cclm(\nu(c))\ or$$
$$(\{\pi\} = \pi_L\ and\ \pi \notin \mathbb{CS}\ and\ seg_V(\nu(c)) = \{(\pi, X')\}$$
$$and\ X \subseteq X')$$

$$\mathcal{TS}, V, \nu \models \ell = x \quad \Leftrightarrow \|X\| = \nu(x)$$

$$\mathcal{TS}, V, \nu \models \phi_1 \frown \phi_2 \quad \Leftrightarrow \exists V_1, V_2 \bullet V = V_1 \oplus V_2\ and$$
$$\mathcal{TS}, V_1, \nu \models \phi_1\ and\ \mathcal{TS}, V_2, \nu \models \phi_2$$

$$\mathcal{TS}, V, \nu \models {\phi_1 \atop \phi_2} \quad \Leftrightarrow \exists V_1, V_2 \bullet V = V_1 \ominus V_2\ and$$
$$\mathcal{TS}, V_1, \nu \models \phi_1\ and\ \mathcal{TS}, V_2, \nu \models \phi_2$$

For $re(c)$ to hold the whole view has to be occupied by a reservation of the car C. This is the case if the view consists of only one virtual lane π_L and has a size larger than 0 ($\|X\| > 0$). If all of these conditions are satisfied, we finally have to check that all segments in this view are completely occupied by C.

Abbreviation. In the following we will often use the abbreviation $\langle \phi \rangle$ to state that a formula ϕ holds *somewhere* in the considered view of car E.

3   Controllers for Safe Crossing Manoeuvres

For urban traffic manoeuvres, we have three types of controllers: A *lane-change controller* to cover lane-change manoeuvres on road segments, a *crossing controller* to handle crossing manoeuvres and a *distance controller* that maintains the braking distance to a car in front of an occupied crossing. For every car, each equipped with these three controllers, we will show safety in Sect. 4.

Our controllers are defined as timed automata [9] extended with UMLSL formulas as guards, data variables for cars and lanes and controller actions to, e.g., change lanes or reserve crossing segments. We call these extended timed automata *automotive-controlling timed automata* (ACTA).

3.1   Syntax of Automotive-Controlling Timed Automata

To describe traffic situations we use *data variables* $\mathbb{D}_{\mathbb{L}}$ ranging over the set of lanes $\mathbb{L}$ and *clock variables* ranging over $\mathbb{R}_+$. For data and clock variables, we use *data constraints* $\varphi_{\mathbb{D}} \in \Phi_{\mathbb{D}}$, similar to guards and invariants in UPPAAL [12] and build up our guards and invariants from these data constraints and UMLSL formulas.

Definition 10 (Guards and Invariants). *With the set of all data constraints* $\varphi_{\mathbb{D}} \in \Phi_{\mathbb{D}}$ *and the set of all UMLSL formulas* $\varphi_{\mathbb{U}} \in \Phi_{\mathbb{U}}$ *the set of guards and invariants* $\varphi \in \Phi$ *is inductively defined by*

$$\varphi ::= \varphi_{\mathbb{D}} \mid \varphi_{\mathbb{U}} \mid \varphi_1 \wedge \varphi_2 \mid true.$$

Example. The guard (resp. invariant) $\phi \equiv \exists c, d : \langle re(c) \wedge re(d) \rangle \vee x > t$ consists of the MLSL formula $\exists c, d : \langle re(c) \wedge re(d) \rangle$ and the data constraint $x > t$. It states that somewhere exists a collision between cars C and D or the clock x exceeded t.   □

For modifications $\nu_{act} \in \mathcal{V}_{Act}$ of data and clock variables, we again refer to UPPAAL. We express possible driving manoeuvres by *controller actions*, which may occur at the transitions of ACTA. For this, we introduce the set of *car variables* $\mathbb{D}_{\mathbb{I}}$ that map to car identifiers $\mathbb{I}$. Controller actions enable a car to set or withdraw a (crossing) claim or a (crossing) reservation.

Definition 11 (Controller Actions). *With* $c \in \mathbb{D}_{\mathbb{I}}$, *the set of all controller actions* $Ctrl_{Act}$ *is defined by*

$$c_{act} ::= c(c, \psi_{\mathbb{D}}) \mid wd\, c(c) \mid cc(c) \mid wd\, cc(c) \mid r(c) \mid wd\, r(c, \psi_{\mathbb{D}})$$
$$\mid rc(c) \mid wd\, rc(c) \mid \tau,$$

where $\psi_{\mathbb{D}} ::= k \mid d_1 \mid d_1 + d_2 \mid d_1 - d_2$ *with* $k \in \mathbb{N}$, $d_1, d_2 \in \mathbb{D}_{\mathbb{L}}$ *and* $r \in \mathbb{R}$.

Note that in contrast to $c(c, \psi_{\mathbb{D}})$, the action $cc(c)$ gets along without a second parameter, because the path through the crossing will automatically be claimed by the traffic snapshot (cf. Sect. 2.2). The case for $wd\, rc(c)$ is analogous.

Since we are considering at least three controllers (distance controller, lane-change controller, crossing controller) in one single car we need a way to identify the car in which an ACTA is located. For this purpose we introduce an *identifying tuple* $I \in \mathbb{D}_{\mathbb{I}} \cup \{ego\} \times \mathbb{I}$ whereby for a controller action $r(c)$ with $I = (c, C)$ the traffic snapshot will recognise a reservation for car C.

We will use the special constant ego to identify the *car under consideration* E with $I = (ego, E)$. This car is the *owner of the view* $V(E)$.

Definition 12 (Automotive-Controlling Timed Automaton). *An ACTA is defined by a tuple* $\mathcal{A} = (Q, \mathbb{X}, \mathbb{D}, \mathcal{I}, T, q_{ini}, I)$, *where*

- Q *is a finite set of states* $q_0, q_1, q_2, \ldots \in Q$,
- $\mathbb{X}$ *is the set of clocks* $x, y, z, \ldots$,
- $\mathbb{D} = \mathbb{D}_{\mathbb{L}} \cup \mathbb{D}_{\mathbb{I}}$ *is the set of data variables* $d_1, d_2, d_3, \ldots$,

- $\mathcal{I} : Q \to \Phi$ *assigns an invariant* $\mathcal{I}(q)$ *to every state* q,
- $T \subseteq Q \times \Phi \times Ctrl_{Act} \times \mathcal{V}_{Act} \times Q$ *is the set of all directed edges, where an element* $(q, \varphi, c_{act}, \nu_{act}, q') \in T$ *is an edge from states* q *to* q' *labelled with a guard* φ, *a controller action* c_{act} *and a set of variable modifications* ν_{act},
- $q_{ini} \in Q$ *is the initial state and*
- $I \in \mathbb{D}_\mathbb{I} \cup \{ego\} \times \mathbb{I}$ *identifies the car for which the controller works.*

An example for an automotive-controlling timed automaton is given later by the crossing controller in Sect. 3.3 Fig. 4.

3.2   Semantics of Automotive-Controlling Timed Automata

The formal semantics of an automotive-controlling timed automaton $ACTA = (Q, \mathbb{X}, \mathbb{D}, \mathcal{I}, T, q_{ini}, I)$ is defined by a transition system

$$\mathcal{T}(ACTA) = (Conf(ACTA), Ctrl_{Act} \cup Time, \{ \xrightarrow{\lambda} \mid \lambda \in Ctrl_{Act} \cup Time \}, \mathcal{C}_{ini}),$$

where a configuration $\mathcal{C} \in Conf(ACTA)$ consists of a traffic snapshot $\mathcal{TS}$, a valuation ν for all clock and data variables and a state q. The set of all configurations is given by $Conf(ACTA) = \{ \langle \mathcal{TS}, \nu, q \rangle \mid q \in Q \land \mathcal{TS}, V_s(E, \mathcal{TS}), \nu \models \mathcal{I}(q) \}$.

Furthermore there exists an initial configuration $\mathcal{C}_{ini} \in Conf(ACTA)$, where $\mathcal{C}_{ini} = \langle \mathcal{TS}_{ini}, \nu_{ini}, q_{ini} \rangle$ with $\nu_{ini}(x) = 0$ for all clocks $x \in \mathbb{X}$ and $\nu_{ini} \models \mathcal{I}(q_{ini})$.

For $\sim \in \{cc, wd\ c, wd\ cc, r, rc, wd\ rc\}$ exists $\langle \mathcal{TS}, \nu, q \rangle \xrightarrow{\sim(c)} \langle \mathcal{TS}', \nu', q' \rangle$, iff with respective transitions $\mathcal{TS} \xrightarrow{\sim(C)} \mathcal{TS}'$ and $q \xrightarrow{\varphi/\sim(c);\nu_{act}} q'$, where $\mathcal{TS}, V_s(E, \mathcal{TS}), \nu \oplus I \models \varphi$ and $\mathcal{TS}', V_s(E, \mathcal{TS}'), \nu' \oplus I \models \mathcal{I}(q')$ hold.

With $\sim \in \{c, wd\ r\}$ and $n \in \mathbb{L}$, a transition $\langle \mathcal{TS}, \nu, q \rangle \xrightarrow{\sim(c,n)} \langle \mathcal{TS}', \nu', q' \rangle$ exists, iff there are transitions $\mathcal{TS} \xrightarrow{\sim(C,n)} \mathcal{TS}'$ and $q \xrightarrow{\varphi/\sim(c,l);\nu_{act}} q'$, where $\mathcal{TS}, V_s(E, \mathcal{TS}), \nu \oplus I \models \varphi$ and with $\nu'(l) = n$ furthermore $\mathcal{TS}', V_s(E, \mathcal{TS}'), \nu' \oplus I \models \mathcal{I}(q')$ holds.

For $\lambda \in Time$, we refer to the semantics of timed automata, where additionally guards and invariants must hold for $V_s(E, \mathcal{TS})$ and traffic snapshot $\mathcal{TS}$.

3.3   Controller Construction

We now construct the crossing controller $\mathcal{A}_{cc}$ for turning manoeuvres on crossings and we adapt the lane change controller for two-way traffic from [2] to a road controller $\mathcal{A}_{rc}$ for manoeuvres on road segments between intersections. Finally, we need a distance controller $\mathcal{A}_{dc}$ to preserve the safety envelope of an arbitrary car and prevent it from driving onto an intersection without a reservation. For such a distance controller we refer to [13].

As mentioned in the introduction, we separate our controllers from the car dynamics. That is, our controllers, e.g., decide how and whether a lane change or a crossing manoeuvre is conducted. Setting inputs for the actuators is delegated to a lower level of controllers. This approach allows for a purely spatial reasoning in our safety proof in Sect. 4. However, a good example for controllers on the dynamics level is given by Damm et al. in [14], where the authors introduce a velocity and a steering controller.

Crossing Controller. This controller is based on the idea of the lane change controller from [1]. Therefore, we first *claim* an area we want to enter and *reserve* it only if no collision is detected. We assume a crossing manoeuvre to take at most t_{cr} time to finish. At this point we would like to recall the constant ego with valuation $\nu(\text{ego}) = E$.

As we want to prevent different reservations from overlapping, we introduce a *collision check* for the actor E expressed by the UMLSL formula

$$col(\text{ego}) \equiv \exists c : c \neq \text{ego} \wedge \langle re(\text{ego}) \wedge re(c) \rangle.$$

We assume $\neg col(\text{ego})$ to hold in the initial state of our crossing controller. Next we need to detect whether a car approaches a crossing. To that end, we formalise a *crossing ahead check* for the actor E by the formula

$$ca(\text{ego}) \equiv \langle re(\text{ego}) \frown \neg re(\text{ego}) \wedge \neg \langle cs \rangle \wedge \ell \geq d_c \frown cs \rangle,$$

where d_c is a constant, whose length is at most the size of the safety envelope of the fastest car with the weakest brakes. In order to enter a crossing, a car first needs to claim a path through the crossing and check if there are any overlaps of other cars' claims or reservations, formalised by the *potential collision check*

$$pc(c) \equiv c \neq \text{ego} \wedge \langle cl(\text{ego}) \wedge re(c) \vee cl(c) \rangle.$$

If a potential collision is detected, the car must withdraw its claim. Further on, we want to exclude that the actor is entering an intersection while changing lanes. Therefore we introduce the *lane change check*

$$lc(\text{ego}) \equiv \left\langle \begin{matrix} re(\text{ego}) \\ re(\text{ego}) \end{matrix} \right\rangle.$$

When $lc(\text{ego})$ does not hold, the actor reserves the claimed path and starts the crossing manoeuvre. To prevent deadlocks, we set a time bound t_c for the time that may pass between claiming and reserving crossing segments. If the actor reserves any crossing segments, the *on crossing check*

$$oc(\text{ego}) \equiv \langle re(\text{ego}) \wedge cs \rangle$$

holds. When the actor has left the last crossing segment and is driving on a normal lane, the crossing manoeuvre is finished. The reservation of actor E is then reduced to the lane which is the next segment in $pth(E)$. The constructed crossing controller is depicted in Fig. 4

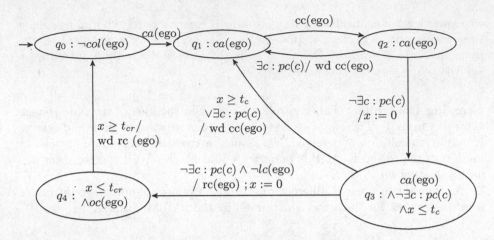

Fig. 4. Crossing controller $\mathcal{A}_{cc}$

Road Controller. This controller is responsible for overtaking manoeuvres on road segments between intersections. As these road segments are structurally comparable to country roads, we refer to [2], where a lane change controller for these types of roads was presented. We only modify this controller by the requirement, that as soon as ca(ego) holds, any claim must be withdrawn immediately and no new claim or reservation might be created until the crossing is passed. However, the car may finish an already begun overtaking manoeuvre, wherefore we make sure the distance d_c used in ca(ego) is big enough to do so.

4   Safe Crossing Manoeuvres

The desired safety property is that at any moment the spaces reserved by different cars are disjoint. This property is formalised by the formula

$$Safe \equiv \forall c, d : c \neq d \rightarrow \neg \langle re(c) \wedge re(d) \rangle,$$

which defines that in any lane the reservations of two arbitrary different cars are not overlapping. A traffic snapshot $\mathcal{TS}$ is called *safe*, if $\mathcal{TS} \models Safe$ holds. For the following proof sketch, we assume that every car perceives the safety envelope of all other cars in its view. Furthermore, we rely on the following assumptions for the overall safety property to hold.

Assumption A1. The initial traffic snapshot $\mathcal{TS}_0$ is *safe*.

Assumption A2. Every car is equipped with a distance controller $\mathcal{A}_{dc}$, a road controller $\mathcal{A}_{rc}$ and a crossing controller $\mathcal{A}_{cc}$ as introduced in Sect. 3.

Theorem 1 (Urban Traffic Safety). *If Assumptions A1 and A2 hold, every traffic snapshot $\mathcal{TS}$ that is reachable from $\mathcal{TS}_0$ by time transitions and transitions allowed by the road controller and the crossing controller is safe.*

Proof. We prove safety from the perspective of an arbitrary car E, because all cars behave similarly. Therefore, we show, that all traffic snapshot $\mathcal{TS}$ reachable from $\mathcal{TS}_0$, for all subviews V of E and all valuations ν with $\nu(\text{ego}) = E$ are safe

$$\mathcal{TS}, V, \nu \models Safe_{\text{ego}} \text{ , where } Safe_{\text{ego}} \equiv \neg\exists c \neq \text{ego} \wedge \langle re(\text{ego}) \wedge re(c)\rangle. \quad (10)$$

Our approach is a proof by induction over the number of transitions needed to reach a traffic snapshot from $\mathcal{TS}_0$. The induction basis holds, as $\mathcal{TS}_0$ is already safe by Assumption A1. Now assume that that $\mathcal{TS}_k$ is reachable from $\mathcal{TS}_0$ in k steps and that $\mathcal{TS}_k$ is safe. For the induction step we show for $k \to k+1$ that the traffic snapshot $\mathcal{TS}_{k+1}$, reachable from $\mathcal{TS}_k$ by one further transition, is safe as well.

Observe that overlaps can only occur if either the road controller $\mathcal{A}_{rc}$ creates a reservation on a lane, the crossing controller creates a crossing reservation for a crossing segment or time passes.

Time passes. For time transitions $\mathcal{TS}_k \xrightarrow{t} \mathcal{TS}_{k+1}$ the distance controller $\mathcal{A}_{dc}$ assures that the distance to a car ahead or an intersection remains positive. Thus if E reaches a crossing and $\mathcal{A}_{cc}$ is not permitted to commit a crossing reservation timely, $\mathcal{A}_{dc}$ will force E to decelerate and probably stop. In this case no automatic crossing reservation is reserved for E by the traffic snapshot and the safety property is not violated.

Reservations on lanes. The road controller $\mathcal{A}_{rc}$ corresponds to the lane change controller proven safe in [2] except for one additional feature; As provided in Sect. 3.3, $\mathcal{A}_{rc}$ will withdraw a committed claim as soon as a crossing is ahead and will not create a new claim or reservation, but E may finish an already begun overtaking manoeuvre. [2] specifies lane change manoeuvres to take at most t_{lc} time to finish, wherefore we simply make sure the distance d_c used in the crossing ahead check is large enough to let E finish the overtaking manoeuvre.

Crossing reservations. For the crossing controller $\mathcal{A}_{cc}$ the only possibility of a crossing reservation is the transition from state q_3 to q_4. At this transition, the guard $\neg lc(\text{ego})$ again keeps the car from triggering a crossing reservation with an active lane change manoeuvre. If more than t_c time passes and $lc(\text{ego})$ still holds or a potential collision is detected, $\mathcal{I}(q_3)$ and the outgoing transitions from state q_3 force $\mathcal{A}_{cc}$ to change back to q_1 and withdraw its crossing claim. Observe that the invariant $ca(\text{ego})$ holds until E is granted a crossing reservation and $\mathcal{A}_{cc}$ therefore changed to state q_4. While all crossing segments necessary for E's crossing manoeuvre are reserved exclusively for E, no other car can create a crossing reservation for the reserved segments until the on crossing check $oc(\text{ego})$ does not hold anymore and therefore $cres(E) = \emptyset$.

We analysed all cases where a reservation for a lane or a crossing reservation is created or time passes and ensured that the safety property (10) is not violated in any case. $\square$

5 Conclusion

The innovation of our approach is the level of abstraction from car dynamics to merely spatial reasoning for safety properties in urban traffic scenarios. In [1,2] this was shown for motorways and country roads. We were able to build on these settings, by adding new topological features (urban road network, paths) to the abstract model and a representation of crossing segments to the Multi-lane Spatial Logic. Furthermore, we defined a crossing controller with new controller actions for crossing manoeuvres (crossing claim and reservation). With the syntax and semantics of automotive-controlling timed automata we presented a new type of automaton to formally implement our crossing controller as well as the lane-change controllers from [1,2]. Finally, we proved safety of our crossing controller.

More on related work. While we detached spatial aspects from the car dynamics, it is of high interest to relate our spatial reasoning to car dynamics and thus link our work to hybrid systems. In [15] Olderog et al. propose a concrete dynamic model suitable for our abstract model. There the authors refine the spatial atoms of MLSL to distance measures.

In [16], Linker shows that the spatial fragment of the original logic MLSL is undecidable. As UMLSL is an extension of MLSL, this undecidability result applies for UMLSL too. Fortunately, in [17] Fränzle et al. prove that MLSL is decidable, when considering only a bounded *scope* around the cars. This is a reduction motivated by reality, because actual autonomous cars can only process state information of finitely many environmental cars in real-time.

Future work. So far, we considered a concept of perfect knowledge, where every car knows the safety envelopes of all other cars, including their braking distances. To build a more realistic model, we will extend this approach to imperfect knowledge, where every car only perceives the physical size of other cars, as given by its sensors. To this end, we plan to integrate a concept of communication via broadcast channels. In [18], Alrahman et al. provide a calculus for attribute-based broadcast communication between dynamic components, that could be of use for our purposes.

We intend to formalise our safety proof by extending and applying the proof system defined for highway manoeuvres by Linker [16].

In addition, we would like to investigate our abstract model for further applications, for example an extension to roundabout traffic and crossings with more or less than four intersecting road segments. Furthermore, one could consider cars driving onto crossings while being in an overtaking manoeuvre or a crossing controller that is able to dynamically change its path at an oncoming crossing.

Finally, an interesting question is to extend our controllers to consider not only safety, but also liveness and fairness properties. To this end we would first need to extend the syntax and semantics of UMLSL by operators from temporal logic [19]. With this we could express, that a car that desires so *finally* ($\Diamond$) passes an intersection:

$$Life \equiv \forall c : (ca(c) \land pth(c)_{next(c)} \in \mathbb{CS} \rightarrow \Diamond oc(c)).$$

References

1. Hilscher, M., Linker, S., Olderog, E.-R., Ravn, A.P.: An abstract model for proving safety of multi-lane traffic manoeuvres. In: Qin, S., Qiu, Z. (eds.) ICFEM 2011. LNCS, vol. 6991, pp. 404–419. Springer, Heidelberg (2011). doi:10.1007/978-3-642-24559-6_28
2. Hilscher, M., Linker, S., Olderog, E.-R.: Proving safety of traffic manoeuvres on country roads. In: Liu, Z., Woodcock, J., Zhu, H. (eds.) Theories of Programming and Formal Methods. LNCS, vol. 8051, pp. 196–212. Springer, Heidelberg (2013). doi:10.1007/978-3-642-39698-4_12
3. Moszkowski, B.: A temporal logic for multilevel reasoning about hardware. Computer 18, 10–19 (1985)
4. Zhou, C., Hoare, C., Ravn, A.: A calculus of durations. Inf. Process. Lett. 40, 269–276 (1991)
5. Schäfer, A.: A calculus for shapes in time and space. In: Liu, Z., Araki, K. (eds.) ICTAC 2004. LNCS, vol. 3407, pp. 463–477. Springer, Heidelberg (2005). doi:10.1007/978-3-540-31862-0_33
6. Loos, S.M., Platzer, A.: Safe intersections: at the crossing of hybrid systems and verification. In: Yi, K. (ed.) Intelligent Transportation Systems (ITSC), pp. 1181–1186. Springer, Heidelberg (2011)
7. Werling, M., Gindele, T., Jagszent, D., Gröll, L.: A robust algorithm for handling moving traffic in urban scenarios. In: Proceedings of IEEE Intelligent Vehicles Symposium, Eindhoven, The Netherlands, pp. 168–173 (2008)
8. Colombo, A., Del Vecchio, D.: Efficient algorithms for collision avoidance at intersections. In: Proceedings of the 15th ACM International Conference on Hybrid Systems: Computation and Control, HSCC 2012, pp. 145–154. ACM, New York (2012)
9. Alur, R., Dill, D.: A theory of timed automata. TCS 126, 183–235 (1994)
10. Woodcock, J., Davies, J.: Using Z - Specification, Refinement, and Proof. Prentice Hall, Upper Saddle River (1996)
11. Linker, S., Hilscher, M.: Proof theory of a multi-lane spatial logic. In: Liu, Z., Woodcock, J., Zhu, H. (eds.) ICTAC 2013. LNCS, vol. 8049, pp. 231–248. Springer, Heidelberg (2013). doi:10.1007/978-3-642-39718-9_14
12. Behrmann, G., David, A., Larsen, K.G.: A tutorial on UPPAAL. In: Bernardo, M., Corradini, F. (eds.) SFM-RT 2004. LNCS, vol. 3185, pp. 200–236. Springer, Heidelberg (2004). doi:10.1007/978-3-540-30080-9_7
13. Damm, W., Hungar, H., Olderog, E.R.: Verification of cooperating traffic agents. Int. J. Control 79, 395–421 (2006)
14. Damm, W., Möhlmann, E., Rakow, A.: Component based design of hybrid systems: a case study on concurrency and coupling. In: Proceedings of the 17th International Conference on Hybrid Systems: Computation and Control, pp. 145–150. ACM (2014)
15. Olderog, E., Ravn, A.P., Wisniewski, R.: Linking spatial and dynamic models for traffic maneuvers. In: 54th IEEE Conference on Decision and Control, CDC 2015, Osaka, Japan, pp. 6809–6816, 15–18 December 2015
16. Linker, S.: Proofs for Traffic Safety - Combining Diagrams and Logic. Ph.d thesis, University of Oldenburg (2015)
17. Fränzle, M., Hansen, M.R., Ody, H.: No need knowing numerous neighbours. In: Meyer, R., Platzer, A., Wehrheim, H. (eds.) Correct System Design. LNCS, vol. 9360, pp. 152–171. Springer, Heidelberg (2015). doi:10.1007/978-3-319-23506-6_11

18. Alrahman, Y.A., De Nicola, R., Loreti, M., Tiezzi, F., Vigo, R.:A calculus for attribute-based communication. In: Proceedings of the 30th Annual ACM Symposium on Applied Computing, SAC 2015, pp. 1840–1845. ACM, New York (2015)
19. Pnueli, A.: The temporal logic of programs. In: Proceedings of the 18th Annual Symposium on Foundations of Computer Science, SFCS 1977, Washington, DC, USA, pp. 46–57. IEEE Computer Society (1977)

Composition and Transformation

Unifying Heterogeneous State-Spaces
with Lenses

Simon Foster[1(✉)], Frank Zeyda[2], and Jim Woodcock[1]

[1] Department of Computer Science, University of York, York YO10 5GH, UK
{simon.foster,jim.woodcock}@york.ac.uk
[2] School of Computing, Teesside University, Middlesbrough TS1 3BA, UK
f.zeyda@tees.ac.uk

Abstract. Most verification approaches embed a model of program state into their semantic treatment. Though a variety of heterogeneous state-space models exists, they all possess common theoretical properties one would like to capture abstractly, such as the common algebraic laws of programming. In this paper, we propose lenses as a universal state-space modelling solution. Lenses provide an abstract interface for manipulating data types through spatially-separated views. We define a lens algebra that enables their composition and comparison, and apply it to formally model variables and alphabets in Hoare and He's Unifying Theories of Programming (UTP). The combination of lenses and relational algebra gives rise to a model for UTP in which its fundamental laws can be verified. Moreover, we illustrate how lenses can be used to model more complex state notions such as memory stores and parallel states. We provide a mechanisation in Isabelle/HOL that validates our theory, and facilitates its use in program verification.

1 Introduction

Predicative programming [17] is a unification technique that uses predicates to describe abstract program behaviour and executable code alike. Programs are denoted as logical predicates that characterise the observable behaviours as mappings between the state before and after execution. Thus one can apply predicate calculus to reason about programs, as well as prove the algebraic laws of programming themselves [20]. These laws can then be applied to construct semantic presentations for the purpose of verification, such as operational semantics, Hoare calculi, separation logic, and refinement calculi, to name a few [2,8]. This further enables the application of automated theorem provers to build program verification tools, an approach which has seen multiple successes [1,23].

Modelling the state space of a program and manipulation of its variables is a key problem to be solved when building verification tools [27]. Whilst relation algebra, Kleene algebra, quantales, and related algebraic structures provide excellent models for point-free laws of programming [3,14], when one considers point-wise laws for operators that manipulate state, like assignment, additional behavioural semantics is needed. State spaces can be heterogeneous — that is

© Springer International Publishing AG 2016
A. Sampaio and F. Wang (Eds.): ICTAC 2016, LNCS 9965, pp. 295–314, 2016.
DOI: 10.1007/978-3-319-46750-4_17

consisting of different representations of state and variables. For example, separation logic [6] considers both the store, a static mapping from names to values, and the heap, a dynamic mapping from addresses to values. Nevertheless, one would like a uniform interface for different variable models to facilitate the definition and use of generic laws of programming. When considering parallel programs [21], one also needs to consider subdivision of the state space into non-interfering regions for concurrent threads, and their eventual reconciliation post execution. Moreover, we have the overarching need for meta-logical operators on state, like variable substitution and freshness, that are often considered informally but are vital to express and mechanise many laws of programming [17,20,21].

In this paper, we propose *lenses* [12] as a unifying solution to state-space modelling. Lenses provide a solution to the view-update problem in database theory [13], and are similarly applied to manipulation of data structures in functional programming [11]. They employ well-behaved *get* and *put* functions to identify a particular view of a source data structure, and allow one to perform transformations on it independently of the wider context.

Our contribution is an extension of the theory of lenses that allows their use in modelling variables as abstract views on program state spaces with a uniform semantic interface. We define a novel lens algebra for manipulation of variables and state spaces, including separation-algebra-style operators [6] such as state (de)composition, that enable abstract reasoning about program operators that modify state spaces in sophisticated ways. Our algebra has been mechanised in Isabelle/HOL [24] and includes a repository of verified lens laws.

We apply the lens algebra to model heterogeneous state space models within the context of Hoare and He's Unifying Theories of Programming [7,21] (UTP), a predicative programming framework with an incremental and modular approach to denotational model construction. Therein, we use lenses to semantically model UTP variables and the predicate calculus' meta-logical functions, with no need for explicit abstract syntax, and thence provide a purely algebraic basis for the meta-logical laws, predicate calculus laws, and the laws of programming. We have further used Isabelle/HOL to mechanise a large repository of UTP laws; this both validates the soundness of our lens-based UTP framework and, importantly, paves the way for future program verification tools[1].

The structure of our paper is as follows. In Sect. 2, we provide background material and related work. In Sect. 3, we present a mechanised theory of lenses, in the form of an algebraic hierarchy, concrete instantiations, and algebraic operators, including a useful equivalence relation. This theory is standalone, and we believe has further applications beyond modelling state. Crucially, all the constructions we describe require only a first-order polymorphic type system which makes it suitable for Isabelle/HOL. In Sect. 4, we apply the theory of lenses to show how different state abstractions can be given a unified treatment. For this, we construct the UTP's relational calculus, associated meta-logical operators, and prove various laws of programming. Along the way, we show how our model satisfies various

---

[1] For supporting Isabelle theories, including mechanised proofs for all laws in this paper, see http://cs.york.ac.uk/~simonf/ictac2016.

important algebraic structures to validate its adequacy. We also use lenses to give an account to parallel state in Sect. 4.4. Finally, in Sect. 5, we conclude.

2   Background and Related Work

2.1   Unifying Theories of Programming

The UTP [21] is a framework for defining denotational semantic models based on an alphabetised predicate calculus. A program is denoted as a set of possible observations. In the relational calculus, imperative programs are in view and thus observations consist of before variables x and after variables x'. This allows operators like assignment, sequential composition, if-then-else, and iteration to be denoted as predicates over these variables, as illustrated in Table 1. From these denotations, algebraic laws of programming can be proved, such as those in Table 2, and more specialised semantic models developed for reasoning about programs, such as Hoare calculi and operational semantics. UTP also supports more sophisticated modelling constructs; for example concurrency is treated in [21, Chap. 7] via the *parallel-by-merge* construct $P\|_M Q$, a general scheme for parallel composition that creates two copies of the state space, executes P and Q in parallel on them, and then merges the results through the merge predicate M. This is then applied to UTP theory of communication in Chap. 8, and henceforth to give a UTP semantics to the process calculus CSP [7,19].

Mechanisation of the UTP for the purpose of verification necessitates a model for the predicate and relational calculi [16,29] that must satisfy laws such as those in Table 2. LP1 and LP2 are point-free laws, and can readily be derived from algebras like relation algebra or Kleene algebra [14]. The remaining laws, however, are point-wise in the sense that they rely on the predicate variables. Whilst law LP3 can be modelled with KAT [2] (Kleene Algebra with Tests) by considering b to be a test, the rest explicitly reference variables. LP4 and LP5 require

Table 1. Imperative programming in the alphabetised relational calculus

$$x := v \ \triangleq\ x' = v \wedge y' = y \qquad P\,;\,Q \ \triangleq\ \exists x_0 \bullet P[x_0/x'] \wedge Q[x_0/x]$$
$$(P \lhd b \rhd Q) \ \triangleq\ (b \wedge P) \vee (\neg b \wedge Q) \qquad P^* \ \triangleq\ \nu X \bullet P\,;\,X$$

Table 2. Typical laws of programming

$$(P\,;\,Q)\,;\,R \ =\ P\,;\,(Q\,;\,R) \tag{LP1}$$

$$P\,;\,\textbf{false} \ =\ \textbf{false}\,;\,P \ =\ \textbf{false} \tag{LP2}$$

$$\textbf{while}\,b\,\textbf{do}\,P \ =\ (P\,;\,\textbf{while}\,b\,\textbf{do}\,P) \lhd b \rhd \boldsymbol{I} \qquad \text{if } \forall x \bullet x' \notin \mathsf{fv}(b) \tag{LP3}$$

$$P\,;\,Q \ =\ \exists x_0 \bullet P[x_0/x]\,;\,Q[x_0/x'] \tag{LP4}$$

$$x := e\,;\,P \ =\ P[e/x] \tag{LP5}$$

$$(x := e\,;\,y := f) \ =\ (y := f\,;\,x := e) \qquad \text{if } x \neq y, x \notin \mathsf{fv}(f), y \notin \mathsf{fv}(e) \tag{LP6}$$

that we support quantifiers and substitution. LP6 additionally requires we can specify free variables. Thus, to truly provide a generic algebraic foundation for the UTP, a more expressive model supporting these operators is needed.

2.2 Isabelle/HOL

Isabelle/HOL [24] is a proof assistant for Higher Order Logic. It includes a functional specification language, a proof language for discharging specified goals in terms of proven theorems, and tactics that help automate proof. Its type system supports first-order parametric polymorphism, meaning types can carry variables – e.g. α list for type variable α. Built-in types include total functions $\alpha \Rightarrow \beta$, tuples $\alpha \times \beta$, booleans bool, and natural numbers nat. Isabelle also includes partial function maps $\alpha \rightharpoonup \beta$, which are represented as $\alpha \Rightarrow \beta$ option, where β option can either take the value Some $(v : \beta)$ or None. Function dom(f) gives the domain of f, $f(k \mapsto v)$ updates a key k with value v, and function the $: \alpha$ option $\Rightarrow \alpha$ extracts the valuation from a Some constructor, or returns an underdetermined value if None is present.

Record types can be created using **record** $\mathcal{R} = f_1 : \tau_1 \cdots f_n : \tau_n$, where $f_i : \tau_i$ is a field. Each field f_i yields a query function $f_i : \mathcal{R} \Rightarrow \tau_i$, and update function $f_i\text{-}upd : (\tau_i \Rightarrow \tau_i) \Rightarrow (\mathcal{R} \Rightarrow \mathcal{R})$ with which to transform $\mathcal{R}$. Moreover Isabelle provides simplification theorems for record instances $(\!| f_1 = v_1 \cdots f_n = v_n |\!)$:

$$f_i(\!| \cdots f_i = v \cdots |\!) = v \qquad f_i\text{-}upd\ g\ (\!| \cdots f_i = v \cdots |\!) = (\!| \cdots f_i = g(v) \cdots |\!)$$

The HOL logic includes an equality relation $\_ = \_ : \alpha \Rightarrow \alpha \Rightarrow$ bool that equates values of the same type α. In terms of tactics, Isabelle provides an equational simplifier simp, generalised deduction tactics blast and auto, and integration of external automated provers using the sledgehammer tool [5].

Our paper does not rely on detailed knowledge of Isabelle, as we present our definitions and theorems mathematically, though with an Isabelle feel. Technically, we make use of the lifting and transfer packages [22] that allow us to lift definitions and associated theorems from super-types to sub-types. We also make use of Isabelle's **locale** mechanism to model algebraic hierarchies as in [14].

2.3 Mechanised State Spaces

Several mechanisations of the UTP in Isabelle exist [9,10,16,29] that take a variety of approaches to modelling state; for a detailed survey see [29]. A general comparison of approaches to modelling state was made in [27] which identifies four models of state, namely state as functions, tuples, records, and abstract types, of which the first and third seem the most prevalent.

The first approach models state as a function Var $\Rightarrow$ Val, for suitable value and variable types. This approach is taken by [8,16,25,29], and requires a deep model of variables and values, in which concepts such as typing are first-class. This provides a highly expressive model with few limitations on possible manipulations [16]. However, [27] highlights two obstacles: (1) the machinery required

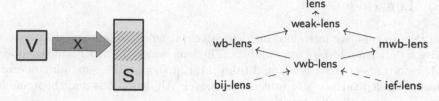

Fig. 1. Visualisation of a simple lens   **Fig. 2.** Lens algebraic hierarchy

for deep reasoning about program values is heavy and *a priori* limits possible constructions, and (2) explicit variable naming requires one to consider issues like α-renaming. Whilst our previous work [29] effectively mitigates (1), at the expense of introducing axioms, the complexities associated with (2) remain. Nevertheless, the approach seems necessary to model dynamic creation of variables, as required, for example, in modelling memory heaps in separation logic [6,8].

The alternative approach uses records to model state; a technique often used by verification tools in Isabelle [1,2,9,10]. In particular, [9] uses this approach to create a shallow embedding of the UTP and library of laws[2] which, along with [25], our work is inspired by. A variable in this kind of model is abstractly represented by pairing the field-query and update functions, f_i and f_i-*upd*, yielding a nameless representation. As shown in [2,9,10], this approach greatly simplifies automation of program verification in comparison to the former functional approach through directly harnessing the polymorphic type system and automated proof tactics. However, the expense is a loss of flexibility compared to the functional approach, particularly in regards to decomposition of state spaces and handling of extension as required for local variables [27]. Moreover, those employing records seldom provide general support for meta-logical concepts like substitution, and do not abstractly characterise the behaviour of variables.

Our approach generalises all these models by abstractly characterising the behaviour of state and variables using lenses. Lenses were created as an abstraction for bidirectional programming and solving the view-update problem [12]. They abstract different views on a data space, and allow their manipulation independently of the context. A lens consists of two functions: *get* that extracts a view from a larger source, and *put* that puts back an updated view. [11] gives a detailed study of the algebraic lens laws for these functions. Combinators are also provided for composing lenses [12,13]. They have been practically applied in the *Boomerang* language[3] for transformations on textual data structures.

Our lens approach is indeed related to the state-space solution in [27] of using Isabelle locales to characterise a state type abstractly and polymorphically. A difference though is the use of explicit names, where our lenses are nameless. Moreover, the core lens laws [11] bear a striking resemblance to Back's variable laws [4], which he uses to form the basis for the meta-logical operators of substitution, freshness, and specification of procedures.

---

[2] See archive of formal proofs: https://www.isa-afp.org/entries/Circus.shtml.
[3] Boomerang home page: http://www.seas.upenn.edu/~harmony/.

3 Lenses

In this section, we introduce our lens algebra, which is later used in Sect. 4 to give a uniform interface for variables. The lens laws in Sect. 3.1 and composition operator of Sect. 3.3 are adapted from [11,12], though the remaining operators, such as independence and sublens, are novel. All definitions and theorems have been mechanically validated[1].

3.1 Lens Laws

A lens $X : V \implies S$, for source type S and view type V, identifies V with a subregion of S, as illustrated in Fig. 1. The arrow denotes X and the hatched area denotes the subregion V it characterises. Transformations on V can be performed without affecting the parts of S outside the hatched area. The lens signature consists of a pair of total functions[4] $get_X : S \Rightarrow V$ that extracts a view from a source, and $put_X : S \Rightarrow V \Rightarrow S$ that updates a view within a given source. When speaking about a particular lens we omit the subscript name. The behaviour of a lens is constrained by one or more of the following laws [11].

$$get\ (put\ s\ v) = v \qquad \text{(PutGet)}$$
$$put\ (put\ s\ v')\ v = put\ s\ v \qquad \text{(PutPut)}$$
$$put\ s\ (get\ s) = s \qquad \text{(GetPut)}$$

PutGet states that if we update the view in s to v, then extracting the view yields v. PutPut states that if we make two updates, then the first update is overwritten. GetPut states that extracting the view and then putting it back yields the original source. These laws are often grouped into two classes [12]: *well-behaved lenses* that satisfy PutGet and GetPut, and *very well-behaved lenses* that additionally satisfy PutPut. We also identify *weak lenses* that satisfy only PutGet, and *mainly well-behaved lenses* that satisfy PutGet and PutPut but not GetPut. These weaker classes prove useful in certain contexts, notably in the map lens implementation (see Sect. 3.2). Moreover [11,12] also identify the class of *bijective lenses* that satisfy PutGet and also the following law.

$$put\ s\ (get\ s') = s' \qquad \text{(StrongGetPut)}$$

StrongGetPut states that updating the view completely overwrites the state, and thus the source and view are, in some sense, equivalent. Finally we have the class of *ineffectual lenses* whose views do not effect the source. Our complete algebraic hierarchy of lenses is illustrated in Fig. 2, where the arrows are implicative.

---

[4] Partial functions are sometimes used in the literature, e.g. [13]. We prefer total functions, as these circumvent undefinedness issues and are at the core of Isabelle/HOL.

3.2   Concrete Lenses

We introduce lenses that exemplify the above laws and are applicable to modelling different kinds of state spaces. The function lens (fl) can represent total variable state functions $\mathsf{Var} \Rightarrow \mathsf{Val}$ [16], whilst the map lens (ml) can represent heaps [8]. The record lens (rl) can represent static variables [2,10].

Definition 1 (Function, Map, and Record lenses)

$$
\begin{aligned}
get_{\mathsf{fl}(k)} &\triangleq \lambda f.\, f(k) & put_{\mathsf{fl}(k)} &\triangleq \lambda f\ v.\, f(k := v) \\
get_{\mathsf{ml}(k)} &\triangleq \lambda f.\, \mathsf{the}(f(k)) & put_{\mathsf{ml}(k)} &\triangleq \lambda f\ v.\, f(k \mapsto v) \\
get_{\mathsf{rl}(f_i)} &\triangleq f_i & get_{\mathsf{rl}(f_i)} &\triangleq \lambda r\ v.\, f_i\text{-}upd\ (\lambda x.\, v)\ r
\end{aligned}
$$

The (total) function lens fl(k) focusses on a specific output associated with input k. The *get* function applies the function to k, and the *put* function updates the valuation of k to v. It is a very well-behaved lens:

Theorem 1 (The function lens is very well-behaved)

Proof. Included in our mechanised Isabelle theories[1].

The map lens $\mathsf{ml}(k)$ likewise focusses on the valuation associated with a given key k. If no value is present at k then *get* returns an arbitrary value. The map lens is therefore not a well-behaved lens since it does not satisfy GetPut, as $f(k \mapsto \mathsf{the}(f(k))) \neq f$ when $k \notin \mathsf{dom}(f)$ since the maps have different domains.

Theorem 2 (The map lens is mainly well-behaved)

Finally, we consider the record lens $\mathsf{rec}(f_i)$. As mentioned in Sect. 2.3, each record field yields a pair of functions f_i and f_i-upd, and associated simplifications for record instances. Together these can be used to prove the following theorem:

Theorem 3 (Record lens). *Each $f_i : \mathcal{R} \Rightarrow \tau_i$ yields a very well-behaved lens.*

This must be proved on a case-by-case basis for each field in each newly defined record; however the required proof obligations can be discharged automatically.

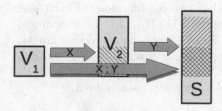

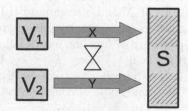

Fig. 3. Lens composition visualised   **Fig. 4.** Lens independence visualised

3.3 Lens Algebraic Operators

Lens composition $X \,\mathring{\,} Y : V_1 \Longrightarrow S$, for $X : V_1 \Longrightarrow V_2$ and $Y : V_2 \Longrightarrow S$ allows one to focus on regions within larger regions. The intuition in Fig. 3 shows how composition of X and Y yields a lens that focuses on the V_1 subregion of S. For example, if a record has a field which is itself a record, then lens composition allows one to focus on the inner fields by composing the lenses for the outer with those of the inner record. The definition is given below.

Definition 2 (Lens composition)

$$put_{X \,\mathring{\,} Y} \triangleq \lambda s \, v. \; put_Y \; s \; (put_X \; (get_Y \; s) \; v) \qquad get_{X \,\mathring{\,} Y} \triangleq get_X \circ get_Y$$

The *put* operator of lens composition first extracts view V_2 from source S, puts $v : V_1$ into this, and finally puts the combined view. The *get* operator simply composes the respective *get* functions. Lens composition is closed under all lens classes ($\{\text{weak}, \text{wb}, \text{mwb}, \text{vwb}\}$-lens). We next define the unit lens, $\mathbf{0} : \text{unit} \Longrightarrow S$, and identity lens, $\mathbf{1} : S \Longrightarrow S$.

Definition 3 (Unit and identity lenses)

$$put_0 \triangleq \lambda s \, v.s \qquad get_0 \triangleq \lambda s.() \qquad put_1 \triangleq \lambda s \, v.v \qquad get_1 \triangleq \lambda s.s$$

The unit lens view is the singleton type unit. Its *put* has no effect on the source, and *get* returns the single element (). It is thus an ineffectual lens. The identity lens identifies the view with the source, and it is thus a bijective lens. Lens composition and identity form a monoid. We now consider operators for comparing lenses which may have different view types, beginning with lens independence.

Definition 4 (Lens independence). *Lenses* $X : V_1 \Longrightarrow S$ *and* $Y : V_2 \Longrightarrow S$ *are independent, written* $X \bowtie Y$, *provided they satisfy the following laws:*

$$put_X(put_Y \; s \; v) \; u \; = \; put_Y(put_X \; s \; u) \; v \tag{LI1}$$
$$get_X(put_Y \; s \; v) \; = \; get_X \; s \tag{LI2}$$
$$get_Y(put_X \; s \; u) \; = \; get_Y \; s \tag{LI3}$$

Intuitively, two lenses are independent if they identify disjoint regions of the source as illustrated in Fig. 4. We characterise this by requiring that the *put* functions of X and Y commute (LI1), and that the *put* functions of each lens has no effect on the result of the *get* function of the other (LI2, LI3). For example, independence of function lenses follows from inequality of the respective inputs, i.e. $\text{fl}(k_1) \bowtie \text{fl}(k_2) \iff k_1 \neq k_2$. Lens independence is a symmetric relation, and it is also irreflexive ($\neg(X \bowtie X)$), unless X is ineffectual.

The second type of comparison between two lenses is containment.

Definition 5 (Sublens relation). *Lens* $X : V_1 \Longrightarrow S$ *is a sublens of* $Y : V_2 \Longrightarrow S$, *written* $X \preceq Y$, *if the equation below is satisfied.*

$$X \preceq Y \triangleq \exists Z : V_1 \Longrightarrow V_2. \; Z \in \textit{wb-lens} \; \wedge \; X = Z \,\mathring{\,} Y$$

The intuition of sublens is simply that the source region of X is contained within that of Y. The definition is explained by the following commuting diagram:

$$X \xrightarrow{\quad} S \xleftarrow{\quad} Y$$
$$V_1 \dashrightarrow \underset{Z}{\quad} \dashrightarrow V_2$$

Intuitively, Z is a "shim" lens that identifies V_1 with a subregion of V_2. Focusing on region V_1 in V_2, followed by V_2 in S is the same as focusing on V_1 in S. The sublens relation is transitive and reflexive, and thus a preorder. Moreover 0 is the least element ($0 \preceq X$), and 1 is the greatest element ($X \preceq 1$), provided X is well-behaved. Sublens orders lenses by the proportion of the source captured. We have also proved the following theorem relating independence to sublens:

Theorem 4 (Sublens preserves independence)

If $X \preceq Y$ and $Y \bowtie Z$ then also $X \bowtie Z$.

We use sublens to induce an equivalence relation $X \approx Y \triangleq X \preceq Y \wedge Y \preceq X$. It is a weaker notion than homogeneous HOL equality $=$ between lenses as it allows the comparison of lenses with differently-typed views. We next prove two correspondences between bijective and ineffectual lenses.

Theorem 5 (Bijective and ineffectual lenses equality equivalence)

$$X \in \textit{ief-lens} \Leftrightarrow X \approx 0 \qquad X \in \textit{bij-lens} \Leftrightarrow X \approx 1$$

The first law states that ineffectual lenses are equivalent to 0, and the second that bijective lenses are equivalent to 1. Showing that a lens is bijective thus entails demonstrating that it characterises the whole state space, though potentially with a different view type. We lastly describe lens summation.

Definition 6 (Lens sum)

$$\textit{put}_{X \oplus Y} \triangleq \lambda s \, (u, v). \, \textit{put}_X \, (\textit{put}_Y \, s \, v) \, u \qquad \textit{get}_{X \oplus Y} \triangleq \lambda s.(\textit{get}_X \, s, \textit{get}_Y \, s)$$

The intuition is given in Fig. 5. Given independent lenses $X : V_1 \Longrightarrow S$ and $Y : V_2 \Longrightarrow S$, their sum yields a lens $V_1 \times V_2 \Longrightarrow S$ that characterises both subregions. The combined *put* function executes the *put* functions sequentially, whilst the *get* extracts both values simultaneously. A notable application is to

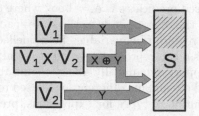

Fig. 5. Lens sum visualised

define when a source can be divided into two disjoint views $X \bowtie Y$, a situation we can describe with the formula $X \oplus Y \approx \mathbf{1}$, or equivalently $X \oplus Y \in$ bij-lens, which can be applied to framing or division of a state space for parallel programs (see Sect. 4.4). Lens sum is closed under all lens classes. We also introduce two related lenses for viewing the left and right of a product source-type, respectively.

Definition 7 (First and second lenses)

$$put_{\mathsf{fst}} \triangleq (\lambda(s,t)u.(u,t)) \qquad get_{\mathsf{fst}} \triangleq fst$$
$$put_{\mathsf{snd}} \triangleq (\lambda(s,t)u.(s,u)) \qquad get_{\mathsf{snd}} \triangleq snd$$

We then prove the following lens sum laws:

Theorem 6 (Sum laws). *Assuming* $X \bowtie Y$, $X \bowtie Z$, *and* $Y \bowtie Z$:

$$
\begin{array}{rclcrcl}
X \oplus Y & \approx & Y \oplus X & \qquad & X \oplus (Y \oplus Z) & \approx & (X \oplus Y) \oplus Z \\
X \oplus 0 & \approx & X & \qquad & (X \oplus Y) \,;\, Z & = & (X \,;\, Z) \oplus (Y \,;\, Z) \\
X & \preceq & X \oplus Y & \qquad & \mathsf{fst} \oplus \mathsf{snd} & = & \mathbf{1} \\
X \oplus Y & \bowtie & Z \quad \text{if } X \bowtie Z \text{ and } Y \bowtie Z & & & &
\end{array}
$$

Lens sum is commutative, associative, has **0** as its identity, and distributes through lens composition. Naturally, each summand is a sublens of the whole, and it preserves independence as the next law demonstrates. The remaining law demonstrates that a product is fully viewed by its first and second component.

4 Unifying State-Space Abstractions

In this section, we apply our lens theory to modelling state spaces in the context of the UTP's predicate calculus. We construct the core calculus (Sect. 4.1), meta-logical operators (Sect. 4.2), apply these to the relational laws of programming (Sect. 4.3), and finally give an algebraic basis to parallel-by-merge (Sect. 4.4). We also show that our model satisfies various important algebras, and thus justify its adequacy.

4.1 Alphabetised Predicate Calculus

Our model of alphabetised predicates is $\alpha \Rightarrow$ bool, where α is a suitable type for modelling the alphabet, that corresponds to the state space. We do not constrain the structure of α, but require that variables be modelled as lenses into it. For example, the record lens rl can represent a typed static alphabet [2,9,10], whilst the map lens ml can support dynamically allocated variables [8]. Moreover, lens composition can be used to combine different lens-based representations of state. We begin with the definition of types for expressions, predicates, and variables.

Definition 8 (UTP types)

$$(\tau, \alpha)\,\mathsf{uexpr} \triangleq (\alpha \Rightarrow \tau) \qquad\qquad \alpha\,\mathsf{upred} \triangleq (\mathsf{bool}, \alpha)\,\mathsf{uexpr}$$

$$(\alpha, \beta)\,\mathsf{urel} \triangleq (\alpha \times \beta)\,\mathsf{upred} \qquad (\tau, \alpha)\,\mathsf{uvar} \triangleq (\tau \Longrightarrow \alpha)$$

All types are parametric over alphabet type α. An expression $(\tau, \alpha)\,\mathsf{uexpr}$ is a query function mapping a state α to a given value in τ. A predicate $\alpha\,\mathsf{upred}$ is a boolean-valued expression. A (heterogeneous) relation is a predicate whose alphabet is $\alpha \times \beta$. A variable $x : (\tau, \alpha)\,\mathsf{uvar}$ is a lens that views a particular subregion of type τ in α, which affords a very general state model. We already have meta-logical functions for variables, in the form of lens equivalence $\approx$ and lens independence $\bowtie$. Moreover, we can construct variable sets using operators **0** which corresponds to $\emptyset$, $\oplus$ which corresponds to $\cup$, **1** which corresponds to the whole alphabet, and $\preceq$ that can model set membership $x \in A$. Theorem 6 justifies these interpretations. We define several core expression constructs for literals, variables, and operators, from which most other operators can be built.

Definition 9 (UTP expression constructs)

$$\mathsf{lit} : \tau \Rightarrow \tau\,\mathsf{uexpr} \qquad \mathsf{var} : (\tau, \alpha)\,\mathsf{uvar} \Rightarrow (\tau, \alpha)\,\mathsf{uexpr}$$
$$\mathsf{lit}\ k \triangleq \lambda s.\,k \qquad\qquad \mathsf{var}\ x \triangleq \lambda s.\,\mathsf{get}_x\ s$$
$$\mathsf{uop} : (\tau \Rightarrow \phi) \Rightarrow (\tau, \alpha)\,\mathsf{uexpr} \Rightarrow (\phi, \alpha)\,\mathsf{uexpr}$$
$$\mathsf{uop}\ f\ v \triangleq \lambda s.\,f\ (v(s))$$
$$\mathsf{bop} : (\tau \Rightarrow \phi \Rightarrow \psi) \Rightarrow (\tau, \alpha)\,\mathsf{uexpr} \Rightarrow (\phi, \alpha)\,\mathsf{uexpr} \Rightarrow (\psi, \alpha)\,\mathsf{uexpr}$$
$$\mathsf{bop}\ f\ u\ v \triangleq \lambda s.\,f\ (u(s))\ (v(s))$$

A literal lit lifts a HOL value to an expression via a constant λ-abstraction, so it yields the same value for any state. A variable expression var takes a lens and applies the *get* function on the state space s. Constructs uop and bop lift functions to unary and binary operators, respectively. These lifting operators enable a proof tactic for predicate calculus we call pred-tac [16] that uses the transfer package [22] to compile UTP expressions and predicates to HOL predicates, and afterwards apply auto or sledgehammer to discharge the resulting conjecture. Unless otherwise stated, all theorems below are proved in this manner.

The predicate calculus' boolean connectives and equality are obtained by lifting the corresponding HOL functions, leading to the following theorem:

Theorem 7 (Boolean Algebra). *UTP predicates form a Boolean Algebra*

We define the refinement order on predicates $P \sqsubseteq Q$, as usual, as universally closed reverse implication $[Q \Rightarrow P]$, and use it to prove the following theorem.

Theorem 8 (Complete Lattice). *UTP predicates form a Complete Lattice*

This provides suprema ($\bigsqcup$), infima ($\bigsqcap$), and fixed points (μ, ν) which allow us to express recursion. The bottom of the lattice is **true**, the most non-deterministic specification, and the top is **false**, the miraculous program. Next we define the existential and universal quantifiers using the lens operation *put*:

Definition 10 (Existential and universal quantifiers)

$$\exists x \bullet P \triangleq (\lambda s.\exists v.P(put_x\ s\ v)) \qquad \forall x \bullet P \triangleq (\lambda s.\forall v.P(put_x\ s\ v))$$

The quantifiers on the right-hand side are HOL quantifiers. Existential quantification ($\exists x \bullet P$) states that there is a valuation for x in state s such that P holds, specified using *put*. Universal quantification is defined similarly and satisfies ($\forall x \bullet P$) = ($\neg\exists x \bullet \neg P$). We derive universal closure $[P] \triangleq \forall 1 \bullet P$, that quantifies all variables in the alphabet (**1**). Alphabetised predicates then form a Cylindric Algebra [18], which axiomatises the quantifiers of first-order logic.

Theorem 9 (Cylindric Algebra). *UTP predicates form a Cylindric Algebra; the following laws are satisfied for well-behaved lenses x, y, and z:*

$$(\exists x \bullet \mathbf{false}) \iff \mathbf{false} \tag{C1}$$

$$P \Rightarrow (\exists x \bullet P) \tag{C2}$$

$$(\exists x \bullet (P \land (\exists x \bullet Q))) \iff ((\exists x \bullet P) \land (\exists x \bullet Q)) \tag{C3}$$

$$(\exists x \bullet \exists y \bullet P) \iff (\exists y \bullet \exists x \bullet P) \tag{C4}$$

$$(x = x) \iff \mathbf{true} \tag{C5}$$

$$(y = z) \iff (\exists x \bullet y = x \land x = z) \qquad if\ x \bowtie y, x \bowtie z \tag{C6}$$

$$\mathbf{false} \iff \begin{pmatrix} (\exists x \bullet x = y \land P)\land \\ (\exists x \bullet x = y \land \neg P) \end{pmatrix} \qquad if\ x \bowtie y \tag{C7}$$

Proof. Most proofs are automatic, the one complexity being C4 which we have to split into cases for (1) $x \bowtie y$, when x and y are different, and (2) $x \approx y$, when they're the same. We thus implicitly assume that variables cannot overlap, though lenses can. C6 and C7 similarly require independence assumptions.

From this algebra, the usual laws of quantification can be derived [18], even for nameless variables. Since lenses can also represent variable sets, we can also model quantification over multiple variables such as $\exists x, y, z \bullet P$, which is represented as $\exists x \oplus y \oplus z \bullet P$, and then prove the following laws.

Theorem 10 (Existential quantifier laws)

$$(\exists A \oplus B \bullet P) = (\exists A \bullet \exists B \bullet P) \tag{Ex1}$$

$$(\exists B \bullet \exists A \bullet P) = (\exists A \bullet P) \qquad if\ B \preceq A \tag{Ex2}$$

$$(\exists x \bullet P) = (\exists y \bullet Q) \qquad if\ x \approx y \tag{Ex3}$$

Ex1 shows that quantifying over two disjoint sets or variables equates to quantification over both. Ex2 shows that quantification over a larger lens subsumes a smaller lens. Finally Ex3 shows that if we quantify over two lenses that identify the same subregion then those two quantifications are equal.

In addition to quantifiers for UTP variables we also provide quantifiers for HOL variables in UTP expressions, $\exists x \bullet P$ and $\forall x \bullet P$, that bind x in a closed λ-term. These are needed to quantify logical meta-variables, which are often useful in proof. This completes the specification of the predicate calculus.

4.2 Meta-logical Operators

We next move onto the meta-logical operators, first considering fresh variables, which we model by a weaker semantic property known as *unrestriction* [16,25].

Definition 11 (Unrestriction)

$$x \,\sharp\, P \;\Leftrightarrow\; (\forall s, v \bullet P(\mathit{put}_x \; s \; v) = P(s))$$

Intuitively, lens x is unrestricted in P, written $x \,\sharp\, P$, provided that P's valuation does not depend on x. Specifically, the effect of P evaluated under state s is the same if we change the value of x. It is thus a sufficient notion to formalise the meta-logical provisos for the laws of programming. Unrestriction can alternatively be characterised as predicates whose satisfy the fixed point $P = (\exists x \bullet P)$ for very well-behaved lens x. We now show some of the key unrestriction laws.

Theorem 11 (Unrestriction laws)

$$\text{U1} \; \frac{-}{\mathbf{0} \,\sharp\, P} \qquad \text{U2} \; \frac{x \preceq y \quad y \,\sharp\, P}{x \,\sharp\, P} \qquad \text{U3} \; \frac{x \,\sharp\, P \quad y \,\sharp\, P \quad x \bowtie y}{(x \oplus y) \,\sharp\, P}$$

$$\text{U4} \; \frac{-}{x \,\sharp\, \mathbf{true}} \qquad \text{U5} \; \frac{x \,\sharp\, P \quad x \,\sharp\, Q}{x \,\sharp\, P \wedge Q} \qquad \text{U6} \; \frac{x \,\sharp\, P \quad x \,\sharp\, Q}{x \,\sharp\, (P = Q)} \qquad \text{U7} \; \frac{x \,\sharp\, P}{x \,\sharp\, \neg P}$$

$$\text{U8} \; \frac{x \bowtie y}{x \,\sharp\, y} \qquad \text{U9} \; \frac{x \in \mathsf{mwb\text{-}lens}}{x \,\sharp\, (\exists x \bullet P)} \qquad \text{U10} \; \frac{x \bowtie y \quad x \,\sharp\, P}{x \,\sharp\, (\exists y \bullet P)} \qquad \text{U11} \; \frac{-}{x \,\sharp\, [P]}$$

Laws U1–U3 correspond to unrestriction of multiple variables using the lens operations; for example U2 states that sublens preserves unrestriction. Laws U4–U7 show that unrestriction distributes through the logical connectives. Laws U8–U11 show the behaviour of unrestriction with respect to variables. U8 states that x is unrestricted in variable expression y if x and y are independent. U9 and U10 relate to unrestriction over quantifiers; the proviso $x \in \mathsf{mwb\text{-}lens}$ means, for example, that a law is applicable to variables modelled by maps. Finally U11 states that all variables are unrestricted in a universal closure.

We next introduce substitution $P[v/x]$, which is also encoded semantically using homogeneous substitution functions $\sigma : \alpha \Rightarrow \alpha$ over state space α. We define functions for application, update, and querying of substitutions:

Definition 12 (Substitution functions)

$$\sigma \dagger P \;\triangleq\; \lambda s. P(\sigma(s))$$

$$\sigma(x \mapsto_s e) \;\triangleq\; (\lambda s. \mathit{put}_x \; (e(s)) \; (\sigma(s)))$$

$$\langle \sigma \rangle_s \, x \;\triangleq\; (\lambda s. \mathit{get}_x \; (\sigma(s)))$$

Substitution application $\sigma \dagger P$ takes the state, applies σ to it, and evaluates P under this updated state. The simplest substitution, $\mathsf{id} \triangleq \lambda x.\, x$, effectively maps all variables to their present value. Substitution lookup $\langle \sigma \rangle_s \, x$ extracts the expression associated with variable x from σ. Substitution update $\sigma(x \mapsto_s e)$

assigns the expression e to variable x in σ. It evaluates e under the incoming state s and then puts the result into the state updated with the original substitution σ applied. We also introduce the short-hand $[x_1 \mapsto_s e_1, \cdots, x_n \mapsto_s e_n] = \mathsf{id}(x_1 \mapsto_s e_1, \cdots, x_n \mapsto_s e_n)$. A substitution $P[e_1, \cdots, e_n/x_1, \cdots, x_n]$ of n expressions to corresponding variables is then expressed as $[x_1 \mapsto_s e_1, \cdots, x_n \mapsto_s e_n] \dagger P$.

Theorem 12 (Substitution query laws)

$$\langle \sigma(x \mapsto_s e) \rangle_s x = e \qquad \qquad \text{(SQ1)}$$
$$\langle \sigma(y \mapsto_s e) \rangle_s x = \langle \sigma \rangle_s x \qquad \qquad if\ x \bowtie y \qquad \text{(SQ2)}$$
$$\sigma(x \mapsto_s e, y \mapsto_s f) = \sigma(y \mapsto_s f) \qquad \qquad if\ x \preceq y \qquad \text{(SQ3)}$$
$$\sigma(x \mapsto_s e, y \mapsto_s f) = \sigma(y \mapsto_s f, x \mapsto_s e) \qquad \qquad if\ x \bowtie y \qquad \text{(SQ4)}$$

SQ1 and SQ2 show how substitution lookup is evaluated. SQ3 shows that an assignment to a larger lens overrides a previous assignment to a small lens and SQ4 shows that independent lens assignments can commute. We next prove the laws of substitution application.

Theorem 13 (Substitution application laws)

$$\sigma \dagger x = \langle \sigma \rangle_s x \qquad \qquad \text{(SA1)}$$
$$\sigma(x \mapsto_s e) \dagger P = \sigma \dagger e \qquad \qquad if\ x \sharp P \qquad \text{(SA2)}$$
$$\sigma \dagger \mathsf{uop}\ f\ v = \mathsf{uop}\ f\ (\sigma \dagger v) \qquad \qquad \text{(SA3)}$$
$$\sigma \dagger \mathsf{bop}\ f\ u\ v = \mathsf{bop}\ f\ (\sigma \dagger u)\ (\sigma \dagger v) \qquad \qquad \text{(SA4)}$$
$$(\exists y \bullet P)[e/x] = (\exists y \bullet P[e/x]) \qquad \qquad if\ x \bowtie y, y \sharp e \qquad \text{(SA5)}$$

These laws effectively subsume the usual syntactic substitution laws, for an arbitrary number of variables, many of which simply show how substitution distributes through expression and predicate operators. SA2 shows that a substitution of an unrestricted variable has no effect. SA5 captures when a substitution can pass through a quantifier. The variables x and y must be independent, and furthermore the expression e must not mention y such that no variable capture can occur. Finally, we will use unrestriction and substitution to prove the one-point law of predicate calculus [17, Sect. 3.1].

Theorem 14 (One-point)

$$(\exists x \bullet P \wedge x = e) = P[e/x] \qquad \qquad if\ x \in \mathsf{mwb\text{-}lens},\ x \sharp e$$

Proof. By predicate calculus with pred-tac.

The one-point law states that a quantification can be eliminated if precisely one value for the quantified variable is specified. We state the requirement "x does not appear in e" with unrestriction. Thus we have now constructed a set of metalogical operators and laws which can be applied to the laws of programming, all the while remaining within our algebraic lens framework and mechanised model. Indeed, all our operators are deeply encoded first-class entities in Isabelle/HOL.

4.3   Relational Laws of Programming

We now show how lenses can be applied to prove the common laws of programming within the relational calculus, by augmenting the alphabetised predicate calculus with relational variables and operators. Recall that a relation is simply a predicate over a product state: $(\alpha \times \beta)$ upred. Input and output variables can thus be specified as lenses that focus on the before and after state, respectively.

Definition 13 (Relational variables)

$$[\![x]\!] \;=\; x \,\text{\small ⸵}\, \mathsf{fst} \qquad\qquad [\![x']\!] \;=\; x \,\text{\small ⸵}\, \mathsf{snd}$$

A variable x is lifted to a input variable x by composing it with fst, or to an output variable x' by composing it with snd. We can then proceed to define the operators of the relational calculus.

Definition 14 (Relational operators)

$$P \,;Q \;\triangleq\; \exists v \bullet P[v/1'] \wedge Q[v/1] \qquad\qquad \mathbb{I} \;\triangleq\; (1'=1)$$
$$P \lhd b \rhd Q \;\triangleq\; (b \wedge P) \vee (\neg b \wedge Q) \qquad\qquad x := v \;\triangleq\; \mathbb{I}\,[v/x]$$

The definition of sequential composition is similar to the standard UTP presentation [21], but we use 1 and $1'$ to represent the input and output alphabets of Q and P, respectively. Skip ($\mathbb{I}$) similarly uses 1 to state that the before state is the same as the after state. We then combine $\mathbb{I}$ with substitution to define the assignment operator. Note that because x is a lens, and v could be a product expression, this operator can be used to represent multiple assignments. We also describe the if-then-else conditional operator $P \lhd b \rhd Q$. Sequential composition and skip, combined with the already defined predicate operators, provide us with the facilities for describing point-free while programs [2], which we illustrate by proving that alphabetised relations form a quantale.

Theorem 15 (Unital quantale). *UTP relations form a unital quantale; that is they form a complete lattice and in addition satisfy the following laws:*

$$(P \,;Q)\,;R = P \,;(Q \,;R) \qquad\qquad P \,;\mathbb{I} = P = \mathbb{I}\,;P$$

$$P \,;\left(\bigsqcap_{Q \in \mathcal{Q}} Q\right) = \bigsqcap_{Q \in \mathcal{Q}} (P \,;Q) \qquad \left(\bigsqcap_{P \in \mathcal{P}} P\right)\,;Q = \bigsqcap_{P \in \mathcal{P}} (P \,;Q)$$

This is proved in the context of Armstrong's Regular Algebra library [2], which also derives a proof that UTP relations form a Kleene algebra. This in turn allows definition of iteration using **while** b **do** $P \triangleq (b \wedge P)^\star \wedge (\neg b')$, where b' denotes relational converse of b, and thence to prove the usual laws of loops. We next describe the laws of assignment.

Theorem 16 (Assignment laws)

$$x := e \,;P \;=\; P[e/x] \tag{ASN1}$$

$$x := e \; ; x := f \;\; = \;\; x := f \hspace{4cm} if\, x \,\sharp\, f \quad (\text{ASN2})$$

$$x := e \; ; y := f \;\; = \;\; y := f \; ; x := e \hspace{2cm} if\, x \bowtie y, x \,\sharp\, f, y \,\sharp\, e \quad (\text{ASN3})$$

$$\dot{x} := e \; ; (P \lhd b \rhd Q) \;\; = \;\; (x := e \; ; P) \lhd b[e/x] \rhd$$
$$(x := e \; ; Q) \hspace{3.5cm} if\, 1' \,\sharp\, b \quad (\text{ASN4})$$

We focus on ASN3 that demonstrates when assignments to x and y commute, and models law LP6 on p. 3. Thus we have illustrated how lenses provide a general setting in which the laws of programming can be proved, including those that require meta-logical assumptions.

4.4 Parallel-by-merge

We further illustrate the flexibility of our model by implementing one of the more complex UTP operators: parallel-by-merge. Parallel-by-merge is a general schema for parallel composition as described in [21, Chap. 7]. It enables the expression of sophisticated forms of parallelism that merge the output of two programs into a single consistent after state. It is illustrated in Fig. 6 for two programs P and Q acting on variables x and y. The input values are fed into P and Q, and their output values are fed into predicates U0 and U1. The latter two rename the variables so that the outputs from both programs can be distinguished by the merge predicate M. M takes as input the variable values before P and Q were executed, and the respective outputs. It then implements a specific mechanism for reconciling these outputs depending on the semantic model of the target language. For example, if P and Q both yield event traces as in CSP [7,19], then only those traces that are consistent will be permitted.

Lenses can be used to define the merge predicate and post-state renamings U0 and U1. The merge predicate takes as input three copies of the state: the outputs from P and Q, and the before state of the entire computation. Thus if the state has type A then $M : ((A \times A) \times A, A)$ urel, and similarly U0, U1 : $(A, (A \times A) \times A)$ urel. We thus give syntax to refer to indexed variables $n.x$, and prior variables $_{<}x$, that give the input values, using the following lens compositions:

Definition 15 (Separated and prior variables)

$$[\![0.x]\!] \;\; = \;\; x \mathbin{\fatsemi} \mathsf{fst} \mathbin{\fatsemi} \mathsf{fst} \hspace{1cm} [\![1.x]\!] \;\; = \;\; x \mathbin{\fatsemi} \mathsf{snd} \mathbin{\fatsemi} \mathsf{fst} \hspace{1cm} [\![_{<}x]\!] \;\; = \;\; x \mathbin{\fatsemi} \mathsf{snd}$$

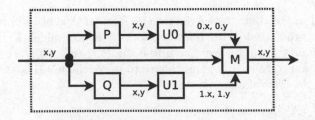

Fig. 6. Pictorial representation of parallel-by-merge $P \,\|_M\, Q$

Lenses $0.x$ and $1.x$ focus on the first and second elements of the tuple's first element, and $_<x$ focusses on the second element. We now define U0 and U1:

Definition 16 (Separating simulations)

$$\mathsf{U0} \triangleq 0.1' = 1 \wedge {_<1}' = 1 \qquad\qquad \mathsf{U1} \triangleq 1.1' = 1 \wedge {_<1}' = 1$$

U0 and U1 copy the before value of the whole state into both their respective indexed variables, and also the prior state. We can now describe parallel-by-merge, given a suitable basic parallel composition operator $\|$ which could, for example, be plain conjunction or design parallel composition (see [21, Chap. 3]):

Definition 17 (Parallel-by-merge)

$$P\|_M Q \triangleq ((P \,;\, \mathsf{U0})\|(Q \,;\, \mathsf{U1})) \,;\, M$$

We also define predicate $\mathsf{swap}_m \triangleq 0.x, 1.x := 1.x, 0.x$ that swaps the left and right copies, and then prove the following generalised commutativity theorem:

Theorem 17 (Commutativity of parallel-by-merge). *If* $M;\mathsf{swap}_m = M$ *then* $P\|_M Q = Q\|_M P$.

This theorem states that if a merge predicate is symmetric, the resulting parallel composition is commutative. In the future we will also show the other properties of parallel composition [21], such as associativity and units. Nevertheless, we have shown that lenses enable a fully algebraic treatment of parallelism.

5   Conclusions

We have presented an enriched theory of lenses, with algebraic operators and lens comparators, and shown how it can be applied to generically modelling the state space of programs in predicative semantic frameworks. We showed how lenses characterise variables, express meta-logical properties, and enrich and validate the laws of programming. The theory of lenses is general, and we believe it has many applications beyond program semantics, such as verifying bidirectional transformations [12]. We have also defined various other useful lens operations, such as lens quotient which is dual to composition. Space has not allowed us to cover this, but we claim this is useful for expressing the contraction of state spaces. Further study of the algebraic properties of these operators is in progress.

Overall, lenses have proven to be a useful abstraction for reasoning about state, in terms of properties like independence and combination. We have used our model to prove several hundred laws of predicate and relational calculus from the UTP book [21] and other sources [7,17,26]. We have also mechanised the Hoare calculus and a weakest precondition calculus that support practical program verification. Although details were omitted for brevity, lenses enable definition of operators like alphabet extension and restriction, through the description of alphabet coercion lenses that are used to represent local variables and methods. We are currently exploring links with Back's variable calculus [4].

In future work we will to apply lenses to additional theories of programming, such as hybrid systems [15] and separation logic [28], especially since our lens algebra resembles a separation algebra. Moreover, we will use our UTP theorem prover[5] to apply our database of programming laws to build practical verification tools for a variety of semantically rich languages [26], in particular for the purpose of analysing heterogeneous Cyber-Physical Systems [15]. We also plan to integrate our work with the existing *Isabelle/Circus* [10] library[2] to further improve verification support for concurrent and reactive systems.

Acknowledgements. This work is partly supported by EU H2020 project *INTO-CPS*, grant agreement 644047. http://into-cps.au.dk/. We also thank Prof. Burkhart Wolff for his generous and helpful comments on our work.

References

1. Alkassar, E., Hillebrand, M.A., Leinenbach, D., Schirmer, N.W., Starostin, A.: The Verisoft approach to systems verification. In: Shankar, N., Woodcock, J. (eds.) VSTTE 2008. LNCS, vol. 5295, pp. 209–224. Springer, Heidelberg (2008). doi:10.1007/978-3-540-87873-5_18

2. Armstrong, A., Gomes, V., Struth, G.: Building program construction and verification tools from algebraic principles. Formal Aspects Comput. **28**(2), 265–293 (2015)

3. Armstrong, A., Struth, G., Weber, T.: Program analysis and verification based on Kleene algebra in Isabelle/HOL. In: Blazy, S., Paulin-Mohring, C., Pichardie, D. (eds.) ITP 2013. LNCS, vol. 7998, pp. 197–212. Springer, Heidelberg (2013). doi:10.1007/978-3-642-39634-2_16

4. Back, R.-J., Preoteasa, V.: Reasoning about recursive procedures with parameters. In: Proceedings of the Workshop on Mechanized Reasoning About Languages with Variable Binding, MERLIN 2003, pp. 1–7. ACM (2003)

5. Blanchette, J.C., Bulwahn, L., Nipkow, T.: Automatic proof and disproof in Isabelle/HOL. In: Tinelli, C., Sofronie-Stokkermans, V. (eds.) FroCoS 2011. LNCS (LNAI), vol. 6989, pp. 12–27. Springer, Heidelberg (2011). doi:10.1007/978-3-642-24364-6_2

6. Calcagno, C., O'Hearn, P., Yang, H.: Local action and abstract separation logic. In: LICS, pp. 366–378. IEEE, July 2007

7. Cavalcanti, A., Woodcock, J.: A tutorial introduction to CSP in *Unifying Theories of Programming*. In: Cavalcanti, A., Sampaio, A., Woodcock, J. (eds.) PSSE 2004. LNCS, vol. 3167, pp. 220–268. Springer, Heidelberg (2006). doi:10.1007/11889229_6

8. Dongol, B., Gomes, V.B.F., Struth, G.: A program construction and verification tool for separation logic. In: Hinze, R., Voigtländer, J. (eds.) MPC 2015. LNCS, vol. 9129, pp. 137–158. Springer, Heidelberg (2015). doi:10.1007/978-3-319-19797-5_7

9. Feliachi, A., Gaudel, M.-C., Wolff, B.: Unifying theories in Isabelle/HOL. In: Qin, S. (ed.) UTP 2010. LNCS, vol. 6445, pp. 188–206. Springer, Heidelberg (2010). doi:10.1007/978-3-642-16690-7_9

---

[5] See our repository at github.com/isabelle-utp/utp-main/tree/shallow.2016.

10. Feliachi, A., Gaudel, M.-C., Wolff, B.: Isabelle/Circus: a process specification and verification environment. In: Joshi, R., Müller, P., Podelski, A. (eds.) VSTTE 2012. LNCS, vol. 7152, pp. 243–260. Springer, Heidelberg (2012). doi:10.1007/978-3-642-27705-4_20

11. Fischer, S., Hu, Z., Pacheco, H.: A clear picture of lens laws. In: Hinze, R., Voigtländer, J. (eds.) MPC 2015. LNCS, vol. 9129, pp. 215–223. Springer, Heidelberg (2015). doi:10.1007/978-3-319-19797-5_10

12. Foster, J.: Bidirectional programming languages. Ph.D. thesis, University of Pennsylvania (2009)

13. Foster, J., Greenwald, M., Moore, J., Pierce, B., Schmitt, A.: Combinators for bidirectional tree transformations: a linguistic approach to the view-update problem. ACM Trans. Program. Lang. Syst. **29**(3), 17 (2007). doi:10.1145/1232420.1232424

14. Foster, S., Struth, G., Weber, T.: Automated engineering of relational and algebraic methods in Isabelle/HOL. In: Swart, H. (ed.) RAMICS 2011. LNCS, vol. 6663, pp. 52–67. Springer, Heidelberg (2011). doi:10.1007/978-3-642-21070-9_5

15. Foster, S., Thiele, B., Cavalcanti, A., Woodcock, J.: Towards a UTP semantics for Modelica. In Proceedings of the 6th International Symposium on Unifying Theories of Programming, June 2016. To appear

16. Foster, S., Zeyda, F., Woodcock, J.: Isabelle/UTP: a mechanised theory engineering framework. In: Naumann, D. (ed.) UTP 2014. LNCS, vol. 8963, pp. 21–41. Springer, Heidelberg (2015). doi:10.1007/978-3-319-14806-9_2

17. Hehner, E.C.R.: A Practical Theory of Programming. Texts and Monographs in Computer Science. Springer, New York (1993)

18. Henkin, L., Monk, J., Tarski, A.: Cylindric Algebras, Part I. North-Holland, Amsterdam (1971)

19. Hoare, T.: Communicating Sequential Processes. Prentice-Hall, London (1985)

20. Hoare, T., Hayes, I., He, J., Morgan, C., Roscoe, A., Sanders, J., Sørensen, I., Spivey, J., Sufrin, B.: The laws of programming. Commun. ACM **30**(8), 672–687 (1987)

21. Hoare, T., He, J.: Unifying Theories of Programming. Prentice-Hall, Englewood Cliffs (1998)

22. Huffman, B., Kunčar, O.: Lifting and transfer: a modular design for quotients in Isabelle/HOL. In: Gonthier, G., Norrish, M. (eds.) CPP 2013. LNCS, vol. 8307, pp. 131–146. Springer, Heidelberg (2013). doi:10.1007/978-3-319-03545-1_9

23. Klein, G., et al.: seL4: Formal verification of an OS kernel. In: Proceedings of the 22nd Symposium on Operating Systems Principles (SOSP), pp. 207–220. ACM (2009)

24. Nipkow, T., Wenzel, M., Paulson, L.C. (eds.): Isabelle/HOL: A Proof Assistant for Higher-Order Logic. LNCS, vol. 2283. Springer, Heidelberg (2002)

25. Oliveira, M., Cavalcanti, A., Woodcock, J.: Unifying theories in ProofPower-Z. In: Dunne, S., Stoddart, B. (eds.) UTP 2006. LNCS, vol. 4010, pp. 123–140. Springer, Heidelberg (2006). doi:10.1007/11768173_8

26. Oliveira, M., Cavalcanti, A., Woodcock, J.: A UTP semantics for circus. Formal Aspects Comput. **21**, 3–32 (2009)

27. Schirmer, N., Wenzel, M.: State spaces - the locale way. In: SSV 2009. ENTCS, vol. 254, pp. 161–179 (2009)

28. Woodcock, J., Foster, S., Butterfield, A.: Heterogeneous semantics and unifying theories. In: 7th International Symposium on Leveraging Applications of Formal Methods, Verification, and Validation (ISoLA) (2016). To appear

29. Zeyda, F., Foster, S., Freitas, L.: An axiomatic value model for Isabelle/UTP. In: Proceedings of the 6th International Symposium on Unifying Theories of Programming (2016). To appear

Ensuring Correctness of Model Transformations While Remaining Decidable

Jon Haël Brenas[1], Rachid Echahed[1(✉)], and Martin Strecker[2]

[1] CNRS and Université de Grenoble Alpes, Grenoble, France
echahed@imag.fr
[2] Université de Toulouse / IRIT, Toulouse, France

Abstract. This paper is concerned with the interplay of the expressiveness of model and graph transformation languages, of assertion formalisms making correctness statements about transformations, and the decidability of the resulting verification problems. We put a particular focus on transformations arising in graph-based knowledge bases and model-driven engineering. We then identify requirements that should be satisfied by logics dedicated to reasoning about model transformations, and investigate two promising instances which are decidable fragments of first-order logic.

Keywords: Graph transformation · Model transformation · Program verification · Classical logic · Modal logic

1 Introduction

We tackle the problem of model transformations and their correctness, where transformations are specified with the aid of rules and correctness properties are stated as logical formulas. By model we intend a graph structure enriched with logical formulas which label either nodes or edges. In our approach, a rule is composed of a left-hand side which is a graph annotated with logical formulas, and a right-hand side which is a sequence of actions. The shape of the graph and the formulas yield an applicability condition of the rule at a matching subgraph of the model; the right-hand side transforms this subgraph with actions such as creation, deletion or cloning of nodes or insertion and deletion of arcs.

Rewrite systems come with a specification in the form of pre- and postconditions, and we aim at full deductive verification, ascertaining that any model satisfying the precondition is transformed into a model satisfying the postcondition.

The correctness of model transformations has attracted some attention in the last years. One prominent approach is model checking, such as implemented by the Groove tool [13]. The idea is to carry out a symbolic exploration of the state space, starting from a given model, in order to find out whether certain invariants

---

This research has been supported by the *Climt* project (ANR-11-BS02-016).

A. Sampaio and F. Wang (Eds.): ICTAC 2016, LNCS 9965, pp. 315–332, 2016.
DOI: 10.1007/978-3-319-46750-4_18

are maintained or certain states (*i.e.*, model configurations) are reachable. The Viatra tool has similar model checking capabilities [25] and in addition allows the verification of elaborate well-formedness constraints imposed on models [23]. Well-formedness is within the realm of our approach (and amounts to checking the consistency of a formula), but is not the primary goal of this paper which is on the dynamics of models.

The Alloy analyser [17] uses bounded model checking for exploring relational designs and transformations (see for example [5] for an application in graph transformations). Counter-examples are presented in graphical form. All the aforementioned techniques use powerful SAT- or SMT-solvers, but do not carry out a complete deductive verification. In our paper, we aim at full-fledged verification of transformations.

General-purpose program verification with systems such as AutoProof [24] and Dafny [18] becomes increasingly automated and thus interesting as push-button technology for model transformations. In this context, fragments of first-order logic have been proposed that are decidable and are useful for dealing with pointer structures [16].

The question explored in this paper is: which requirements does a logic have to fulfill in order to allow for such a verification technique to succeed?

Several different logics have been proposed over the years to tackle the problem of graph transformation verification. Among the most prominent approaches figure nested conditions [15,20] that are explicitly created to describe graph properties. Another widely used logic in graph transformation verification is monadic second-order logic [10,21] that allows to go beyond first-order definable properties. [4] introduces a logic closer to modal logic that allows to express both graph properties and the transformations at the same time.

Nonetheless, these approaches are not flawless. They are all undecidable in general and thus either cannot be used to prove correctness of graph transformations in an automated way or only work on limited classes of graphs. Starting from the other side of the logical spectrum, one could consider using Description Logics to describe graph properties [1,6] that are decidable. Another choice could be the use of modal logics as they are suited to reason about programs. Obviously, this comes at a cost in term of expressiveness.

Separation logic [22] is another choice that is worth considering when dealing with transformations of graphs. It has been developed especially to be able to talk about pointers in conventional programming languages.

In this paper, we proceed in an orthogonal direction. Instead of introducing a logic and advising users to tailor their problem so that it is expressible in our logic and that its models comply with the restrictions so that the verification is actually possible, we aim at providing a means for the users to decide whether the logic they have used to represent their problem will actually allow them to prove their transformations correct or whether they have to use several different systems in parallel.

We are in particular interested in decidable logics, and so we instantiate our general framework with two decidable logics: Two-variable logic with counting (in Sect. 5.1) and logics with exists-forall-prefix (in Sect. 5.2). The fragment of effectively propositional logic [19], that is implemented by the Z3 prover [11] and is closely related to the logical fragment we discuss in Sect. 5.2, has been known for a long time to be decidable [8]. The use of two-variable logics [14] for the verification of model transformation is relatively novel even though it contains all Description Logics without role inclusions. Once more the goal is not to advocate the use of any logic but to give the user the ability to decide if the logics that are planned to be used satisfy some minimal conditions so that the verification can be carried out effectively.

The rest of the paper is structured as follows: we start with an example, in Sect. 2, motivating our model transformation approach, which we then make more formal in Sect. 3. In Sect. 4, we propose general principles that a logic has to fulfill to be usable for verifying model transformations. Then, in Sect. 5, we illustrate our proposal through the two aforementioned logics. Concluding remarks are provided in Sect. 6.

2   Motivating Example

In order to better illustrate our purpose, an example modelling a sample of the information system of a hospital is introduced. Figure 1 is the UML model of this sample.

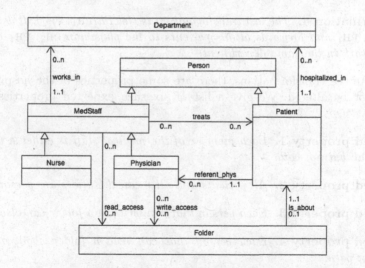

Fig. 1. A sample UML model for the hospital example

We consider persons (shortened to PE). Some of them work in the hospital and form the medical staff (MS) and others are patients (PA). The medical staff is partitioned into physicians (PH) and nurses (NU). In addition, the hospital is split into several departments (DE) or services. Documents pertaining to patients are stored in folders (FO).

Each member of the medical staff is assigned (denoted by *works_in*) to a department. The same way, each patient is hospitalized (*hospitalized_in*) in one of the departments. There may be several members of the medical staff that may collaborate to treat (*treats*) a patient at a given time but one of them is considered as the referent physician (*referent_phys*), that is to say she is in charge of the patient. Part of the medical staff can access the folder containing the documents about (*is_about*) a patient either to read (*read_access*) or to write (*write_access*) information.

The fact is the hospital is bound to evolve: new patients arrive to be cured and others leave, new medical staffers are hired and others move out. To illustrate our purpose, four possible transformations are specified below.

Transformation 1. *The first transformation is New_Ph(*PH$_1$*, D$_1$*). It creates a new physician to which is associated an identifier* PH$_1$*. This physician will be working in the department identified with* D$_1$*.*

Transformation 2. *The second transformation is New_Pa(*PA$_1$*,* PH$_1$*,* FO$_1$*). It adds a new patient. The patient* PA$_1$ *is created alongside his folder* FO$_1$*. She is then assigned* PH$_1$ *as referent physician.*

Transformation 3. *The third transformation is Del_Pa(*PA$_1$*). It modifies data so that patient* PA$_1$ *is no more hospitalized.*

Transformation 4. *The last transformation is Del_Ph(*PH$_1$*,* PH$_2$*). It deletes the physician* PH$_1$ *and forwards all his patients to the physician* PH$_2$*.* PH$_1$ *and* PH$_2$ *have to work in the same department.*

Despite the transformations, there are some properties of the hospital that should not be altered. We give a list of six such expected properties in the following.

Expected property 1. *Each member of the medical staff is either a nurse or a physician but not both.*

Expected property 2. *All patients and all medical staffers are persons.*

Expected property 3. *Each person that can write in a folder can also read it.*

Expected property 4. *Each person that can read a folder about a patient treats that patient.*

Expected property 5. *Only medical staffers can treat persons and only patients can be treated.*

Expected property 6. *Every patient has exactly one referent physician.*

3   A Model Transformation Framework

In this section, a framework used to describe models as well as their transformations is introduced. A model is considered hereafter as a graph, labeled by logical formulae. The logic in which these formulae are expressed is considered as a parameter, say $\mathcal{L}$, of the proposed framework. Required features of such a logic are discussed in the next section. Nevertheless, we assume in this section that the logic $\mathcal{L}$ is endowed with a relation $\models$ over its formulae. That is to say, $n \models B$ (resp. $e \models B$) should be understood as "formula B is satisfied at node n (resp. edge e)".

Definition 1 (Graph). *Let $\mathcal{L}$ be a logic. A graph G is a tuple $(N, E, \mathcal{C}, \mathcal{R}, \phi_N, \phi_E, s, t)$ where N is a set of nodes, E is a set of edges, $\mathcal{C}$ is a set of (node) formulae (of $\mathcal{L}$) or concepts, $\mathcal{R}$ is the set of edge formulae (of $\mathcal{L}$) or roles, ϕ_N is the node labeling function, $\phi_N : N \to \mathcal{P}(\mathcal{C})$, ϕ_E is the edge labeling function, $\phi_E : E \to \mathcal{R}$, s is the source function $s : E \to N$ and t is the target function $t : E \to N$.*

Labeling a graph with logical formulae is quite usual in Kripke structures. In this paper, labeling formulae will play a role either in the transformation process or in the generation of proof obligations for the properties intended to be proved.

Transformations of models are performed by means of graph rewrite systems. These rewrite systems are extensions of those defined in [12] where graphs are labeled with formulae. Thus, the left-hand sides of the rules are labeled graphs as defined in Definition 1, whereas the right-hand sides are defined as sequences of elementary actions. Elementary actions constitute a set of basic transformations used in graph transformation processes. They are given in the following definition.

Definition 2 (Elementary action, action). *An elementary action, say a, has one of the following forms:*

- *a concept assignment $c := i$ where i is a node and c is an atomic formula (a unary predicate). It sets the valuation of c such that the only node labeled by c is i.*
- *a concept addition $c := c + i$ (resp. concept deletion $c := c - i$) where i is a node and c is an atomic formula (a unary predicate). It adds the node i to (resp. removes the node i from) the valuation of the formula c.*
- *a role addition $r := r + (i, j)$ (resp. role deletion $r := r - (i, j)$) where i and j are nodes and r is an atomic role (a binary predicate). It adds the pair (i, j) to (resp. removes the pair (i, j) from) the valuation of the role r.*
- *a node addition $new(i)$ (resp. node deletion $del_I(i)$) where i is a new node (resp. an existing node). It creates the node i. i has no incoming nor outgoing edge and there is no atomic formula such that i belongs to its valuation (resp. it deletes i and all its incoming and outgoing edges).*
- *a global incoming edge redirection $i \gg^{in} j$ where i and j are nodes. It redirects all incoming edges of i towards j.*

– a global outgoing edge redirection $i \gg^{out} j$ where i and j are nodes. It redefines the source of all outgoing edges of i as j.

– a node cloning $clone(i, i')$ where i is a node, i' is a node that does not exist yet. It creates a new node i' that has the same labels as i and the same incoming and outgoing edges[1] (see Fig. 3).

The result of performing the elementary action α on a graph $G = (N^G, E^G, \mathcal{C}^G, \mathcal{R}^G, \phi_N^G, \phi_E^G, s^G, t^G)$ produces the graph $G' = (N^{G'}, E^{G'}, \mathcal{C}^{G'}, \mathcal{R}^{G'}, \phi_N^{G'}, \phi_E^{G'} s^{G'}, t^{G'})$ as defined in Fig. 2 and write $G' = G[\alpha]$ or $G \Rightarrow_\alpha G'$. An action, say α, is a sequence of elementary actions of the form $\alpha = a_1; a_2; \ldots; a_n$. The result of performing α on a graph G is written $G[\alpha]$. $G[a; \alpha] = (G[a])[\alpha]$ and $G[\epsilon] = G$, ϵ being the empty sequence.

Definition 3 (Rule, Graph Rewrite Systems). A rule $\rho[\boldsymbol{n}]$ is a pair (lhs, α) where $\boldsymbol{n}$ is a vector of concept variables. These variables are instantiated by means of actual concepts when a rule is applied. lhs, called the left-hand side, is a graph and α, called the right-hand side, is an action. Rules are usually written $\rho[\boldsymbol{n}] : lhs \rightarrow \alpha$. Concept variables n_i in $\boldsymbol{n}$ may appear both in lhs and in α. A graph rewrite system is a set of rules.

Notice that a rule $\rho[\boldsymbol{n}] : lhs \rightarrow \alpha$ may be considered as a generic rule which yields an actual rewrite rule for every instance of the variables $\boldsymbol{n}$. We write $\rho[\boldsymbol{c}]$ to denote the rule obtained from $\rho[\boldsymbol{n}] : lhs \rightarrow \alpha$ by replacing every variable concept n_i appearing either in lhs or in α by the actual concept c_i. Now let us define when a rule can be applied to a graph.

Definition 4 (Match). Let $\rho[\boldsymbol{n}] : lhs \rightarrow \alpha$ be a rule and G be a graph. Let $\rho[\boldsymbol{c}]$ be an instance of rule $\rho[\boldsymbol{n}]$ and inst be the instance function defined as $inst(n_i) = c_i$ for $i \in \{0, \ldots, k\}$. We say that the instance $\rho[\boldsymbol{c}]$ matches the graph G via the match $h = (h_N, h_E)$, where $h_N : N^{lhs} \rightarrow N^G$ and $h_E : E^{lhs} \rightarrow E^G$ if the following conditions hold:

1. $\forall n \in N^{lhs}, \forall d \in \phi_{N_{lhs}}(n), h_N(n) \models inst(d)$
2. $\forall e \in E^{lhs}, \forall r \in \phi_{E_{lhs}}(e), h_E(e) \models inst(r)$[2]
3. $\forall e \in E^{lhs}, s_G(h_E(e)) = h_N(s_{lhs}(e))$
4. $\forall e \in E^{lhs}, t_G(h_E(e)) = h_N(t_{lhs}(e))$

The third and the fourth conditions are classical and say that the source and target functions and the match have to agree. The first condition says that for every node n of the left-hand side, the node to which it is associated, $h_N(n)$, in G has to satisfy every concept that n satisfies. This condition clearly expresses additional negative and positive conditions which are added to the "structural" pattern matching. The second condition expresses the same conditions on the edges.

---

[1] This action has the same effect as the one defined by means of sesquipushout [9].

[2] $inst(r)$ (resp. $inst(d)$) replaces in r (resp. in d) every occurrence of a concept variable n_i by its instance c_i. The formal definition of the function $inst$ depends on the structure of the considered concepts and roles.

If $\alpha = c := i$ then:

$N^{G'} = N^G, E^{G'} = E^G, \mathcal{C}^{G'} = \mathcal{C}^G, \mathcal{R}^{G'} = \mathcal{R}^G$

$\phi_N^{G'}(n) = \begin{cases} \phi_N^G(n) \cup \{c\} & \text{if } n = i \\ \phi_N^G(n)\backslash\{c\} & \text{if } n \neq i \end{cases}, \phi_E^{G'} = \phi_E^G,$

$s^{G'} = s^G, \, t^{G'} = t^G$

If $\alpha = c := c + i$ then:

$N^{G'} = N^G, E^{G'} = E^G, \mathcal{C}^{G'} = \mathcal{C}^G, \mathcal{R}^{G'} = \mathcal{R}^G,$

$\phi_E^{G'} = \phi_E^G, \phi_N^{G'}(n) = \begin{cases} \phi_N^G(n) \cup \{c\} & \text{if } n = i \\ \phi_N^G(n) & \text{if } n \neq i \end{cases}$

$s^{G'} = s^G, \, t^{G'} = t^G$

If $\alpha = c := c - i$ then:

$N^{G'} = N^G, E^{G'} = E^G, \mathcal{C}^{G'} = \mathcal{C}^G, \mathcal{R}^{G'} = \mathcal{R}^G,$

$\phi_E^{G'} = \phi_E^G, \, \phi_N^{G'}(n) = \begin{cases} \phi_N^G(n)\backslash\{c\} & \text{if } n = i \\ \phi_N^G(n) & \text{if } n \neq i \end{cases}$

$s^{G'} = s^G, \, t^{G'} = t^G$

If $\alpha = r := r + (i,j)$ then :

$N^{G'} = N^G, \mathcal{C}^{G'} = \mathcal{C}^G, \mathcal{R}^{G'} = \mathcal{R}^G,$

$E^{G'} = E^G \cup \{e\}$ where e is a new element

$\phi_N^{G'} = \phi_N^G, \, \phi_E^{G'}(e') = \begin{cases} r & \text{if } e' = e \\ \phi_E^G(e') & \text{if } e' \neq e \end{cases}$

$s^{G'}(e') = \begin{cases} i & \text{if } e' = e \\ s^G(e') & \text{if } e' \neq e \end{cases}$

$t^{G'}(e') = \begin{cases} j & \text{if } e' = e \\ t^G(e') & \text{if } e' \neq e \end{cases}$

If $\alpha = r := r - (i,j)$ then:

$N^{G'} = N^G, \mathcal{C}^{G'} = \mathcal{C}^G, \mathcal{R}^{G'} = \mathcal{R}^G$

$E^{G'} = E^G \backslash r_{i,j}$, where

$r_{i,j} = \{e \in E^G | s^G(e) = i \wedge t^G(e) = j \wedge \phi_E^G(e) = r\}$

$\phi_N^{G'} = \phi_N^G, \phi_E^{G'}$ is the restriction of ϕ_E^G to $E^{G'}$

$s^{G'}$ is the restriction of s^G to $E^{G'}$

$t^{G'}$ is the restriction of t^G to $E^{G'}$

If $\alpha = new(i)$ then:

$N^{G'} = N^G \cup \{i\}$ where i is a new node,

$E^{G'} = E^G, \mathcal{C}^{G'} = \mathcal{C}^G, \mathcal{R}^{G'} = \mathcal{R}^G,$

$\phi_N^{G'}(n) = \begin{cases} \emptyset & \text{if } n = i \\ \phi_N^G(n') & \text{if } n \neq i \end{cases}$

$\phi_E^{G'} = \phi_E^G, \, s^{G'} = s^G, \, t^{G'} = t^G$

If $\alpha = del(i)$ then:

$N^{G'} = N^G \backslash\{i\}, \mathcal{C}^{G'} = \mathcal{C}^G, \mathcal{R}^{G'} = \mathcal{R}^G,$

$E^{G'} = E^G \backslash \{e | s^G(e) = i \vee t^G(e) = i\}$

$\phi_N^{G'}$ is the restriction of ϕ_N^G to $N^{G'}$

$\phi_E^{G'}$ is the restriction of ϕ_E^G to $E^{G'}$

$s^{G'}$ is the restriction of s^G to $E^{G'}$

$t^{G'}$ is the restriction of t^G to $E^{G'}$

If $\alpha = i \gg^{in} j$ then :

$N^{G'} = N^G, E^{G'} = E^G, \mathcal{C}^{G'} = \mathcal{C}^G,$

$\mathcal{R}^{G'} = \mathcal{R}^G, \phi_N^{G'} = \phi_N^G, \phi_E^{G'} = \phi_E^G,$

$s^{G'} = s^G, t^{G'}(e) = \begin{cases} j & \text{if } t^G(e) = i \\ t^G(e) & \text{if } t^G(e) \neq i \end{cases}$

If $\alpha = i \gg^{out} j$ then:

$N^{G'} = N^G, E^{G'} = E^G, \mathcal{C}^{G'} = \mathcal{C}^G,$

$\mathcal{R}^{G'} = \mathcal{R}^G, \phi_N^{G'} = \phi_N^G, \phi_E^{G'} = \phi_E^G,$

$\phi_N^{G'} = \phi_N^G, t^{G'} = t^G,$

$s^{G'}(e) = \begin{cases} j & \text{if } s^G(e) = i \\ s^G(e) & \text{if } s^G(e) \neq i \end{cases}$

If $\alpha = clone(i, i')$ then:

$\mathcal{C}^{G'} = \mathcal{C}^G, \mathcal{R}^{G'} = \mathcal{R}^G$

$N^{G'} = N^G \cup \{i'\}, E^{G'} = E^G \cup E_i'$ where

$E_i' = E_i^{in} \cup E_i^{out} \cup E_i^{loop}$ with

$E_i^{in} = \{e^{in} | \exists e \in E^G, t^G(e) = i\}$

$E_i^{out} = \{e^{out} | \exists e \in E^G, s^G(e) = i\}$

$E_i^{loop} = \{e^{loop} | \exists e \in E^G, s^G(e) = t^G(e) = i\}$

$\phi_N^{G'}(n) = \begin{cases} \phi_N^G(n) & \text{if } n \neq i' \\ \phi_N^G(i) & \text{otherwise} \end{cases}$

$\phi_E^{G'}(e) = \begin{cases} \phi_E^G(e) & \text{if } e \notin E_i' \\ \phi_E^G(co(e)) & \text{otherwise} \end{cases}$

$t^{G'}(e) = \begin{cases} t^G(e) & \text{if } e \notin E_i' \\ t^G(co(e)) & \text{if } e \in E_i^{out} \\ i' & \text{if } e \in E_i^{in} \cup E_i^{loop} \end{cases}$

$s^{G'}(e) = \begin{cases} s^G(e) & \text{if } e \notin E_i' \\ s^G(co(e)) & \text{if } e \in E_i^{in} \\ i' & \text{if } e \in E_i^{out} \cup E_i^{loop} \end{cases}$

where for $e' \in E'$, $co(e')$ is the edge e that e' is a copy of.

Fig. 2. Summary of the effects of atomic actions

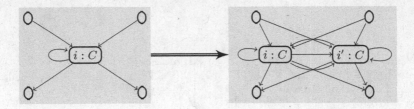

Fig. 3. Example of node cloning. The action $clone(i, i')$ is performed.

Definition 5 (Rule application). *We define the applicability condition as:* $App(\rho[c], G)$ *iff there exists a match* h *from the instance* $\rho[c]$ *to* G. *A graph* G *rewrites to graph* G' *using a rule* $\rho[c] : lhs \to \alpha$ *iff* $App(\rho[c], G)$ *holds and* G' *is obtained from* G *by performing actions in* $h(\alpha)$[3]. *Formally,* $G' = G[h(\alpha)]$. *We write* $G \to_{\rho[c]} G'$ *or* $G \to_{\rho[c],h} G'$.

Example 1. Let us consider again the example given in Sect. 2. We provide in Fig. 4, for every transformation already presented informally, a corresponding rewrite rule.

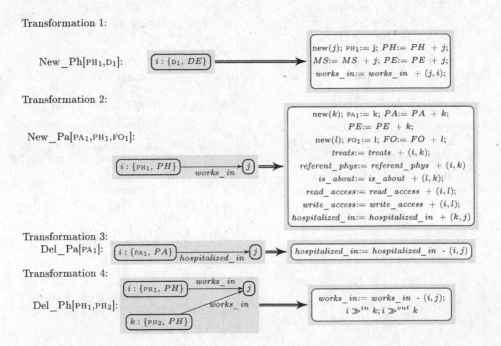

Fig. 4. Transformation rules for the sample hospital model

---

[3] $h(\alpha)$ is obtained from α by replacing every node name, n, of lhs by $h(n)$.

Very often, transforming models by means of rewrite rules necessitates the use of the notion of strategies. Informally, a strategy acts as a recipe indicating in which order the rules are applied.

Definition 6 (Strategy). *Given a graph rewriting system $\mathcal{R}$, a strategy is a word of the following language defined by s:*

$$s := \rho[c_0, \ldots, c_k] \ \text{(Rule application)} \mid s^* \quad \text{(Closure)}$$
$$ s; s \qquad \qquad \text{(Composition)} \mid s \oplus s \ \text{(Choice)} \quad \text{where } \rho[c_0, \ldots, c_k] \text{ is an}$$

instance of a rule in $\mathcal{R}$.

We write $G \Rightarrow_S G'$ when G rewrites to G' following the rules given by the strategy S.

Informally, the strategy "$\rho_1; \rho_2$" means that rule ρ_1 should be applied first, followed by the application of rule ρ_2. Notice that the strategies as defined above allow one to define infinite derivations from a given graph G because we have included the Kleene star construct s^* as a constructor of strategies. Handling the Kleene star does not introduce much more difficulties but requires the use of the notion of invariants in the verification procedures, as it is the case for while loops in imperative languages. It also requires us to extend the notion of applicability from rules to strategies:

$$\text{App}(s^*, G) \qquad = \text{true} \qquad\qquad \text{App}(s_0; s_1, G) = \text{App}(s_0, G)$$
$$\text{App}(s_0 \oplus s_1, G) = \text{App}(s_0, G) \vee \text{App}(s_1, G)$$

In Fig. 5, we provide the rules that specify how strategies are used to rewrite a model (graph). Notice that a closure free strategy is always terminating while a choice free strategy is always confluent.

(RULE APPLICATION)	(CHOICE LEFT)	(CHOICE RIGHT)
$\dfrac{G \to_{\rho[c]} G'}{G \Rightarrow_{\rho[c]} G'}$	$\dfrac{G \Rightarrow_{s_0} G'}{G \Rightarrow_{s_0 \oplus s_1} G'}$	$\dfrac{G \Rightarrow_{s_1} G'}{G \Rightarrow_{s_0 \oplus s_1} G'}$
(COMPOSITION)	(CLOSURE APPLICABLE)	(CLOSURE INAPPLICABLE)
$\dfrac{G \Rightarrow_{s_0} G'' \quad G'' \Rightarrow_{s_1} G'}{G \Rightarrow_{s_0;s_1} G'}$	$\dfrac{G \Rightarrow_s G'' \quad G'' \Rightarrow_{s*} G' \quad \text{App}(s, G)}{G \Rightarrow_{s*} G'}$	$\dfrac{\neg \text{App}(s, G)}{G \Rightarrow_{s*} G}$

Fig. 5. Strategy application rules

To end this section we define the notion of a specification which consists in providing *Pre* and *Post* conditions that one may want to ensure for a given strategy. More precisely, we propose the following definitions.

Definition 7 (Program, Specification). *A program is a tuple $(\mathcal{R}, \mathcal{S})$ where $\mathcal{R}$ is a graph rewrite system and $\mathcal{S}$ is a strategy. A specification SP is a tuple $(Pre, Post, \mathcal{P})$ where Pre and Post are formulae and $\mathcal{P}$ is a program.*

Notice that *Pre* and *Post* are supposed to be formulae of a given logic. We do not specify such a logic in the above definition. We provide actual examples in Sect. 5. A specification (*Pre*, *Post*, $\mathcal{P}$) asserts that for all models G that satisfies the formula *Pre*, all models G' obtained after rewriting G according to strategy $\mathcal{S}$ of program $\mathcal{P} = (\mathcal{R}, \mathcal{S})$, (i.e. $G \Rightarrow_S G'$), G' satisfies formula *Post*.

4   General Logical Framework

Our aim in this section is to discuss general requirements for a logic, say $\mathcal{L}$, that might be considered either to specify pre and post conditions of specifications or to label models.

Let $SP = (Pre, \ Post, \ \mathcal{P})$ be a specification. If SP is correct, then if a model G satisfies *Pre* ($G \models Pre$) and G rewrites to model G' via a strategy $\mathcal{S}$ of a program $\mathcal{P} = (\mathcal{R}, \mathcal{S})$ ($G \Rightarrow_S G'$), then G' satisfies *Post* ($G' \models Post$). In addition to the general requirements for logics $\mathcal{L}$, a Hoare-like calculus dedicated to prove the correctness of specifications is also discussed in this section.

The first, and most obvious, requirements for a logic, $\mathcal{L}$, is that it can express the labeling of models with formulae which specify nodes and edges.

Requirement 1. *Node formulae (concepts in $\mathcal{C}$) should be adequate to the notion of nodes. That is to say, nodes might be candidates to interpret node formulae.*

Requirement 2. *Edge formulae (roles in $\mathcal{R}$) should be adequate to the notion of edges. That is to say, edges might be candidates to interpret edge formulae.*

The conditions *Pre* and *Post* are properties of models. Thus, we have the following requirement.

Requirement 3. *Assertions Pre and Post should be adequate to the notion of graphs (i.e. models). That is to say, models might be candidates to interpret Pre and Post assertions.*

The main ingredient of the verification calculus consists in computing weakest preconditions of postconditions (see function wp defined in Fig. 6). The basic cases of the computations of weakest precondition deal with elementary actions. For that, to every elementary action is associated a so called substitution. Such substitutions are the elementary building blocks allowing the verification of a program.

Definition 8. *Let a be an elementary action, as defined in Definition 2. The substitution $[a]$ associated to the elementary action a is the formula constructor which associates, to each formula ϕ of $\mathcal{L}$, the formula $\phi[a]$. Given a model $\mathcal{M}$, $\phi[a]$ is defined such that $\mathcal{M} \models \phi[a] \Leftrightarrow$ for all models $\mathcal{M}', \mathcal{M} \Rightarrow_a \mathcal{M}'$ implies $\mathcal{M}' \models \phi$.*
A logic $\mathcal{L}'$ is said to be closed under substitutions if for each action a, for each formula ϕ of $\mathcal{L}'$, $\phi[a]$ is also a formula of $\mathcal{L}'$.

$$wp(\rho[\boldsymbol{c}],\ Q) = App(tag(\rho[\boldsymbol{c}])) \Rightarrow wp(tag(\alpha_{\rho[\boldsymbol{c}]}), Q)$$
$$wp(s_0; s_1,\ Q) = wp(s_0,\ wp(s_1,\ Q)) \qquad\qquad wp(s^*) = inv_s$$
$$wp(s_0 \oplus s_1,\ Q) = wp(s_0, Q) \wedge wp(s_1, Q)$$

Fig. 6. Weakest preconditions for strategies.

$$vc(\rho[\boldsymbol{c}],\ Q) = \text{true} \qquad\qquad vc(s_0; s_1,\ Q) = vc(s_0,\ wp(s_1,\ Q)) \wedge vc(s_1, Q)$$
$$vc(s_0 \oplus s_1,\ Q) = vc(s_0, Q) \wedge vc(s_1, Q)$$
$$vc(s^*,\ Q) = (inv_s \wedge App(s) \Rightarrow wp(s, inv_s)) \wedge (inv_s \wedge \neg App(s) \Rightarrow Q)$$
$$\wedge vc(s, inv_s) \wedge vc(s_1, Q)$$

Fig. 7. Verification conditions for strategies.

Weakest preconditions for actions come in two flavors: for elementary actions a, we have $wp(a, Q) = Q[a]$, and for composite actions, $wp(a; \alpha,\ Q) = wp(a,\ wp(\alpha,\ Q))$. On this basis, weakest preconditions for strategies can be easily computed as depicted in Fig. 6. These preconditions follow the principles of Hoare Logic calculi except for the one dedicated to rules, viz. $wp(\rho[\boldsymbol{c}],\ Q)$. This latter corresponds essentially to an "if-then" structure in imperative programs. Put it simply, it checks three properties that are required for the application of a rule to be correct. Up to now, App depended on G. However, correctness proofs should hold for all possible models (graphs). That is way we modify App to be dependent only on the rules and strategies. First, App is a function which applies to a rule $\rho[\boldsymbol{c}]$ and returns a formula of $\mathcal{L}$ stating that there exists a match from the left-hand side of $\rho[\boldsymbol{c}]$ to a potential graph. If the formula $App(\rho[\boldsymbol{c}])$ is satisfied, the rule can be performed. Second, whenever the formula $App(\rho[\boldsymbol{c}]) \Rightarrow wp(\alpha_{\rho[\boldsymbol{c}]}, Q)$ is valid, then if there exists a match, the conditions, viz. $wp(\alpha_{\rho[\boldsymbol{c}]}, Q)$, which ensure the postcondition to be satisfied, are satisfied too. This corresponds to the usual weakest-precondition in Hoare Logic.

There is one additional issue which deserves to be handled carefully. Actually, one same rule can be fired several times during the execution of a program. It is thus mandatory to keep track of where each occurence of the rule is applied. To be more precise, App introduces a condition that uses the names of the nodes in the left-hand sides of rules. As these names uniquely define nodes and edges, if a same rule were used several times with the same names of nodes and edges, the rule would be applied to the exact same nodes and edges. This issue is solved by renaming the individuals (i.e., nodes and edges) each time the rule is fired. This is done through the function tag. That is why $wp(\rho[\boldsymbol{c}], Q) = App(tag(\rho[\boldsymbol{c}])) \Rightarrow wp(tag(\alpha_{\rho[\boldsymbol{c}]}), Q)$.

Finally, the closure of a strategy, s^*, which is close to while structures in imperative programs, needs the definition of an invariant, inv_s, and the introduction of verification conditions, $vc(s^*, Q)$, shown in Fig. 7. Basically, the idea is that a closure is considered as a subprogram whose correctness is proven on the side. The verification condition checks that the specification of this subprogram whose pre and post conditions are the invariant.

From the discussion above, we come to a new requirement about the logic $\mathcal{L}$, regarding the use of substitutions within weakest preconditions.

Requirement 4. $\mathcal{L}$ *must be closed under substitutions.*

If this last requirement is not satisfied, the computation of weakest preconditions may lead to formulas not expressible in $\mathcal{L}$. In this case, the verification of the correctness of specifications would need new proof procedures different from those of $\mathcal{L}$.

In addition, $App(\rho[c])$ must be definable in $\mathcal{L}$. Obviously, this depends mainly on the rules one wants to use. It is thus possible, for a given problem, to use one logic that may not be powerful enough for other problems. Nonetheless, one of the requirements this entails on $\mathcal{L}$ is that it must allow some kind of existential quantification so that the graph can be traversed to look for a match. Obviously, the $\exists$-quantifier of first-order logic is a prime candidate but some other mechanisms like individual assertions $a : C$ in Description Logics [3] or the @ operator of hybrid logic [2] can be used.

Requirement 5. $\mathcal{L}$ *must be able to express $App(\rho[c])$ for all rules $\rho[c]$ of the graph rewrite system under study.*

Theorem 1 (Soundness). *Let $\mathcal{L}$ be a logic satisfying requirements 1 to 5. Let $SP = (Pre, Post, (\mathcal{R}, \mathcal{S}))$ be a specification. If $(Pre \Rightarrow wp(\mathcal{S}, Post)) \wedge vc(\mathcal{S}, Post)$ is valid in $\mathcal{L}$, then for all graphs G, G' such that $G \Rightarrow_\mathcal{S} G'$, $G \models Pre$ implies $G' \models Post$.*

Proof (Sketch). The proof of this theorem is quite straightforward. One just has to check for every atomic strategy s that if $Pre \Rightarrow wp(s, Post)$ and $G \models Pre$ then $G' \models Post$. We give the proof for the rule application which is the most complex.

Assume $\mathcal{S} = \rho[c]$ where $\rho[c]$ is a rule of $\mathcal{R}$. Let us assume $Pre \Rightarrow wp(\rho[c], Post)$ is valid. Because $wp(\rho[c], Post) = App(tag(\rho[c])) \Rightarrow wp(tag(\alpha_{\rho[c]}), Post)$, also $(Pre \wedge App(tag(\rho[c]))) \Rightarrow wp(tag(\alpha_{\rho[c]}), Post)$ is valid. Let G be a graph. If $G \models App(\rho[c])$, there is a match h. Let G' be such that $G \Rightarrow_{\rho[c],h} G'$. By definition of the substitutions, $G \Rightarrow_{\rho[c],h} G'$ and $G \models wp(tag(\alpha_{\rho[c]}), Post)$ implies $G' \models Post$. On the other hand, if $G \not\models App(\rho[c])$, there does not exist any G' such that $G \Rightarrow_{\rho[c]} G'$ and thus the program fails. Thus $G \models Pre$ implies that $G' \models Post$ $\square$.

After performing the calculus, one gets a formula $vc(\mathcal{S}, Post) \wedge (Pre \Rightarrow wp(\mathcal{S}, Post))$. Obviously, in order to be able to decide whether or not a program is correct, one has to prove that the obtained formula is valid. Hence the following requirement.

Requirement 6. *The validity problem for $\mathcal{L}$ is decidable.*

Nevertheless, this last requirement could be optional if interactive theorem provers are preferred.

5   Instances of the Example

Hereafter, we illustrate the general logical framework proposed in the previous section through the Hospital example by providing logics which fulfill the six proposed requirements. In [7] another instance is proposed using an extension of propositional dynamic logic is proposed.

First, let us observe that all of the invariants that we defined can be expressed in first-order logic (Formulae on the right).

Property 1:

$MS = NU \uplus PH$ $\rightsquigarrow \forall x. MS(x) \Leftrightarrow (NU(x) \wedge \neg PH(x)) \vee$
$(\neg NU(x) \wedge PH(x))$

Property 2:

$PA \cup MS \subseteq PE$ $\rightsquigarrow \forall x. PA(x) \vee MS(x) \Rightarrow PE(x)$

Property 3:

$write\_access \subseteq read\_access$ $\rightsquigarrow \forall x, y. write\_access(x, y) \Rightarrow$
$read\_access(x, y)$

Property 4:

$read\_access \circ is\_about \subseteq treats$ $\rightsquigarrow \forall x, y, z. read\_access(x, y) \wedge is\_about(y, z)$
$\Rightarrow treats(x, z)$

Property 5:

$treats \subseteq MS \times PA$ $\rightsquigarrow \forall x, y. treats(x, y) \Rightarrow MS(x) \wedge PA(y)$

Property 6:

$PA \Rightarrow \exists^{=1} referent\_phys$ $\rightsquigarrow \forall x. PA(x) \Rightarrow (\exists y.\ referent\_phys(x, y) \wedge$
$\forall z. referent\_phys(x, z) \Rightarrow z = y)$

First-order logic is not decidable though, and thus one may want to use a different logic in order to be able to decide the correctness of the considered properties. In the following, we use the 2-variable fragment of first-order logic with counting ($\mathcal{C}^2$) [14] and $\exists^* \forall^*$, the fragment of first-order logic whose formula in prenex form are of the form $\exists i_0, \ldots, i_k. \forall j_0, \ldots, j_l. A(i_0, \ldots, i_k, j_0, \ldots, j_l)$.

In order to be able to distinguish between nodes of a model (active nodes) and those which are not part of a given model, we add to the signature of the logic a unary predicate $Active$ which ranges over nodes and edges. Creating a new node becomes adding it to the $Active$ nodes. This also requires to add that $\forall x, y. \neg Active(x) \Rightarrow (\bigwedge_{\psi \text{ an atomic unary predicate}} \neg \psi(x) \wedge \bigwedge_{r \text{ an atomic binary predicate}} \neg r(x, y) \wedge \neg r(y, x))$. I.e., non active nodes are not assumed to satisfy any property.

Let SPH be the specification $(Pre, Post, \mathcal{P})$ associated to the hospital example. Assume the strategy is $\mathcal{S} = New\_Ph[\text{NPH}, \text{NEONAT}]; Del\_Pa[\text{OPA}]$ while the considered rewrite system $\mathcal{R}$ is the one from Fig. 4. This program $\mathcal{P}$ creates a new physician NPH and lets the patient OPA leave the hospital. Let inv denote the conjunction of the expected properties. Let the precondition Pre be $inv \wedge$ $\exists x. (\text{NEONAT}(x) \wedge DE(x)) \wedge \exists x. (\text{OPA}(x) \wedge PA(x)) \wedge \forall x. \neg \text{NPH}(x)$. Let the postcondition $Post$ be $inv \wedge \exists x, y. (\text{NPH}(x) \wedge PH(x) \wedge works\_in(x, y) \wedge \text{NEONAT}(y) \wedge DE(y))$. Proving the correctness of SPH amounts to proving that $Pre \Rightarrow wp(\mathcal{S}, Post)$ is valid. This is a formula in first-order logic. In the following two subsections, this specification is proven to be correct using two different decidable logics that are able to express parts of Pre and $Post$.

5.1 Two-Variable Logic with Counting : $\mathcal{C}^2$

$\mathcal{C}^2$ is the two-variable fragment of first-order logic with counting. Its formulas are those of first-order logic than can be expressed with only two variables and using the counting quantifier constructor $\exists^{<n} x.P$ expressing that there are less than n values x that satisfy P. In our case, this constructor will mostly be used to express that there exist less than n different r-successors of a given node.

Definition 9. *Let $\mathcal{U}$ be a set of unary predicates, $u \in \mathcal{U}$, $\mathcal{B}$ be a set of binary predicates, $b \in \mathcal{B}$, n an integer. A formula ϕ of $\mathcal{C}^2$ is defined as:*

$$\phi \quad := \top \mid \phi \wedge \phi \mid \neg\phi \mid \exists^{<n} x.\phi_x \mid \exists^{<n} y.\phi_y$$
$$\phi_x := \phi \mid u(x) \mid b(x,x) \mid \phi_x \wedge \phi_x \mid \neg\phi_x \mid \exists^{<n} x.\phi_x \mid \exists^{<n} y.\phi_{x,y}$$
$$\phi_y := \phi \mid u(y) \mid b(y,y) \mid \phi_y \wedge \phi_y \mid \neg\phi_y \mid \exists^{<n} y.\phi_y \mid \exists^{<n} x.\phi_{x,y}$$
$$\phi_{x,y} := \phi_x \mid \phi_y \mid b(x,y) \mid b(y,x) \mid \phi_{x,y} \wedge \phi_{x,y} \mid \neg\phi_{x,y} \mid \exists^{<n} x.\phi_{x,y} \mid \exists^{<n} y.\phi_{x,y}$$

As usual, $\perp$ means $\neg\top$, $\phi \vee \psi$ means $\neg(\neg\phi \wedge \neg\psi)$, $\phi \Rightarrow \psi$ means $\neg\phi \vee \psi$, $\exists^{\geq n} v.\phi$ means $\neg\exists^{<n} v.\phi$, $\exists v.\phi$ means $\exists^{\geq 1} v.\phi$, $\forall v.\phi$ means $\neg\exists v.\neg\phi$.

Definition 10. *Let $\mathcal{G} = (N, E, \mathcal{C}, \mathcal{R}, \phi_N, \phi_E, s, t)$ be a graph. We define the valuation of formulae as follows:*

$\top^I \qquad\quad = true$

$(\phi \wedge \psi)^I \quad = \phi^I$ and ψ^I

$(\neg\phi)^I \qquad = not\ \phi^I$

$$(\exists^{<n} x.\phi_x)^I = \begin{cases} true & \text{if there does not exist } n \text{ nodes } m_1, \ldots, m_n, \\ & \quad m_i \neq m_j \text{ for } 0 < i < j \leq n \text{ such that } m_i \models \phi_x \\ false & \text{otherwise} \end{cases}$$

$(\exists^{<n} y.\phi_y)^I$ *is defined the same as* $(\exists^{<n} x.\phi_x)^I$ *but replacing x's with y's*

Let us now focus on $m \models \phi_x$:

$m \models \phi \qquad\quad iff\ \phi^I$

$m \models u(x) \qquad iff\ u \in \phi_N(m)$

$m \models b(x,x) \qquad iff\ there\ exists\ e \in E.s(e) = m, t(e) = m\ and\ b = \phi_E(e)$

$m \models (\phi_x \wedge \psi_x)\ iff\ m \models \phi_x\ and\ m \models \psi_x$

$m \models \neg\phi_x \qquad iff\ m \not\models \phi_x$

$m \models \exists^{<n} x.\phi_x\ \ iff\ there\ does\ not\ exist\ n\ nodes\ m'_1, \ldots, m'_n,$
$\qquad\qquad\qquad\qquad m_i \neq m_j\ for\ 0 < i < j \leq n\ such\ that\ m'_i \models \phi_x$

$m \models \exists^{<n} y.\phi_{x,y}\ iff\ there\ does\ not\ exist\ n\ nodes\ w_1, \ldots, w_n,$
$\qquad\qquad\qquad\qquad w_i \neq w_j\ for\ 0 < i < j \leq n\ such\ that\ (m, w_i) \models \phi_{x,y}$

$m \models \phi_y$ *is defined the same way but swapping the x's and the y's. Let us now focus on $(m, m') \models \phi_{x,y}$:*

$(m, m') \models \phi_x \qquad\qquad iff\ m \models \phi_x$

$(m, m') \models \phi_y \qquad\qquad iff\ m' \models \phi_y$

$(m, m') \models b(x,y) \qquad\quad iff\ there\ exists\ e \in E.s(e) = m, t(e) = m'\ and\ b = \phi_E(e)$

$(m, m') \models b(y,x) \qquad\quad iff\ there\ exists\ e \in E.s(e) = m', t(e) = m\ and\ b = \phi_E(e)$

$(m, m') \models (\phi_{x,y} \wedge \psi_{x,y})\ iff\ (m, m') \models \phi_{x,y}\ and\ (m, m') \models \psi_{x,y}$

$(m, m') \models \neg\phi_{x,y} \qquad\ \ iff\ (m, m') \not\models \phi_{x,y}$

$(m, m') \models \exists^{<n} x.\phi_{x,y} \quad iff\ there\ does\ not\ exist\ n\ nodes\ m_1, \ldots, m_n, m_i \neq m_j$
$\qquad\qquad\qquad\qquad\qquad for\ all\ 0 < i < j \leq n\ such\ that\ (m_i, m') \models \phi_{x,y}$

$(m, m') \models \exists^{<n} y.\phi_{x,y} \quad iff\ there\ does\ not\ exist\ n\ nodes\ m'_1, \ldots, m'_n, m'_i \neq m'_j$
$\qquad\qquad\qquad\qquad\qquad for\ all\ 0 < i < j \leq n\ such\ that\ (m, m'_i) \models \phi_{x,y}$

Theorem 2 ([14]). *The validity problem of $\mathcal{C}^2$ is decidable.*

Let us now check the six requirements of the previous section. $\mathcal{C}^2$ contains unary predicates that are interpreted on nodes and binary predicates that are interpreted on edges. *Pre* and *Post* are interpreted on graphs.

Theorem 3. $\mathcal{C}^2$ *is closed under substitutions.*

The proof relies on the fact that first-order logic is closed under substitution. The proof provides a system of rewrite rules that removes substitutions. As it does not introduce new variables, it also works for $\mathcal{C}^2$. We give three example rules to understand better how does it work:

- $(\phi \wedge \psi)[\sigma] \rightsquigarrow \phi[\sigma] \wedge \psi[\sigma]$ as if $\phi \wedge \psi$ is satisfied after performing σ, so must be ϕ and ψ and the other way round.
- $r(x,y)[r := r + (i,j)] \rightsquigarrow r(x,y) \vee (i(x) \wedge j(y))$ as $r^{I'}$ is $r^I \cup (i^I, j^I)$.
- $r(x,y)[clone(i,i')] \rightsquigarrow r(x,y) \vee (i'(x) \wedge \exists x.(i(x) \wedge r(x,y))) \vee (i'(y) \wedge \exists y.(i(y) \wedge r(x,y))) \vee (i'(x) \wedge i'(y) \wedge \exists x.(i(x) \wedge r(x,x)))$.

Example 2. $\mathcal{C}^2$ can express all the predicates $App(\rho)$ for the rules of the considered example (see Fig. 4):

- $App(New\_Ph[\text{PH}_1, \text{D}_1]) = \exists x.(\text{D}_1(x) \wedge \text{DE}(x)) \wedge \exists x.(\neg Active(x) \wedge \text{PH}_1(x))$
- $App(New\_Pa[\text{PA}_1, \text{PH}_1, \text{FO}_1]) = \exists x,y.(\text{PH}_1(x) \wedge \text{PH}(x) \wedge works\_in(x,y)) \wedge \exists x.(\neg Active(x) \wedge \text{PA}_1(x)) \wedge \exists x.(\neg Active(x) \wedge \text{FO}_1(x))$
- $App(Del\_Pa[\text{PA}_1]) = \exists x,y.(\text{PA}_1(x) \wedge \text{PA}(x) \wedge hospitalized\_in(x,y))$
- $App(Del\_Ph[\text{PH}_1, \text{PH}_2]) = \exists x,y.(\text{PH}_1(x) \wedge \text{PH}(x) \wedge works\_in(x,y) \wedge \exists x.(\text{PH}_2(x) \wedge \text{PH}(x) \wedge works\_in(x,y)))$

One should also be interested in the ability of the logic to express the properties to be verified.

Example 3. $\mathcal{C}^2$ is not able to express Property 4: $read\_access \circ is\_about \subseteq treats$ as one would need to keep track of three variables at a time. On the other hand, Property 6: $\forall x.\text{PA}(x) \Rightarrow \exists^{=1} referent\_phys.\top$ is a formula of $\mathcal{C}^2$.

5.2   Exist-Forall-Prefix

The logic $\exists^* \forall^*$ is the fragment of first-order logic such that its prefix in prenex normal form is composed of a sequence of existential quantifiers and then a sequence of universal quantifiers.

Definition 11. *Let $\mathcal{U}$ be a set of unary predicates, $u \in \mathcal{U}$ and $\mathcal{B}$ a set of binary predicates, $b \in \mathcal{B}$. Let $x_1, \ldots, x_k, a_1, \ldots, a_l$ be variables and v, w denote two of them. A formula ϕ of $\exists^* \forall^*$ is defined as:*
$$\phi := \exists x_0, \ldots, x_k, \forall a_0, \ldots, a_l.\psi(x_1, \ldots, x_k, a_1, \ldots, a_l)$$
$$\psi := \top \mid \psi \wedge \psi \mid \neg \phi \mid u(v) \mid b(v,w)$$

As usual, $\bot$ means $\neg \top$, $\phi \vee \psi$ means $\neg(\neg \phi \wedge \neg \psi)$, $\phi \Rightarrow \psi$ means $\neg \phi \vee \psi$.

Definition 12. Let $\mathcal{G} = (N, E, \mathcal{C}, \mathcal{R}, \phi_N, \phi_E, s, t)$ be a graph. We defined the valuation of formulae: $(\exists x_1, \ldots, x_k, \forall a_1, \ldots, a_l . \psi(x_0, \ldots, x_k, a_0, \ldots, a_l))^I = N$ iff there exist k nodes $(x_1, \ldots, x_k)$ such that for all choices of l nodes $(a_1, \ldots, a_l)$, $(x_1, \ldots, x_k, a_1, \ldots, a_l) \models \psi$. Let us define $(x_1, \ldots, x_k, a_1, \ldots, a_l) \models \psi$:

$(x_1, \ldots, a_l) \models \top$

$(x_1, \ldots, a_l) \models (\phi \wedge \psi)$ iff $(x_1, \ldots, a_l) \models \phi$ and $(x_1, \ldots, a_l) \models \psi$

$(x_1, \ldots, a_l) \models (\neg \phi)$ iff $(x_1, \ldots, a_l) \not\models \phi$

$(x_1, \ldots, a_l) \models u(v)$ iff $u \in \phi_N(v)$

$(x_1, \ldots, a_l) \models b(v, w)$ iff there exists $e \in E$. $s(e) = v$, $t(e) = w$ and $b = \phi_E(e)$

Theorem 4. The validity problem of $\exists^* \forall^*$ is decidable.

This is a well-known result ([8], Chap. 6).

The six requirements of the previous section clearly hold for this logic. $\exists^* \forall^*$ contains unary predicates that are interpreted on nodes and binary predicates that are interpreted on edges.

Theorem 5. $\exists^* \forall^*$ is closed under substitutions.

The proof is exactly the same as the one for $\mathcal{C}^2$ and $\mathcal{FO}$. One needs to be careful though as additional quantifiers are introduced. They are always of the form $\exists x.(i(x) \wedge c(x))$ or $\exists x.(i(x) \wedge r(x, y))$ that can be rewritten as $\forall x.(\neg i(x) \vee c(x))$ or $\forall x.(\neg i(x) \vee r(x, y))$. Thus one can consider that only universal quantifiers are introduced.

Example 4. $\exists^* \forall^*$ can express all the predicates $App(\rho)$ for the rules of the considered example (see Fig. 4):

- $App(New\_Ph[\text{PH}_1, \text{D}_1]) = \exists x.(\text{D}_1(x) \wedge \text{DE}(x)) \wedge \exists x.(\neg Active(x) \wedge \text{PH}_1(x))$
- $App(New\_Pa[\text{PA}_1, \text{PH}_1, \text{FO}_1]) = \exists x, y.(\text{PH}_1(x) \wedge \text{PH}(x) \wedge works\_in(x, y)) \wedge \exists x.(\neg Active(x) \wedge \text{PA}_1(x)) \wedge \exists x.(\neg Active(x) \wedge \text{FO}_1(x))$
- $App(Del\_Pa[\text{PA}_1]) = \exists x, y.(\text{PA}_1(x) \wedge \text{PA}(x) \wedge hospitalized\_in(x, y))$
- $App(Del\_Ph[\text{PH}_1, \text{PH}_2]) = \exists x, y, z.(\text{PH}_1(x) \wedge \text{PH}(x) \wedge works\_in(x, y) \wedge \text{PH}_2(z) \wedge \text{PH}(z) \wedge works\_in(z, y))$

It is worth noting that the definition of $App(\rho)$ introduces new existential quantifiers as it checks for the existence of a match. This could seem to lead to a problem as the formula no longer is in $\exists^* \forall^*$. Actually, as the existentially quantified variables do not depend on the previously defined universally quantified variables, it is possible to move them at the beginning thus yielding a formula in $\exists^* \forall^*$.

Once more one has to check whether all properties can be expressed in the chosen logic.

Example 5. $\exists^* \forall^*$ is not able to express Property 6: $PA \Rightarrow \exists^{=1} referent\_phys$ as it needs an existential quantifier after the universal ones to express the existence of an edge labeled with $referent\_phys$. On the other hand, Property 4: $\forall x, y, z . read\_access(x, y) \wedge is\_about(y, z) \Rightarrow treats(x, z)$ is part of $\exists^* \forall^*$.

6 Conclusions

We considered the verification problem of model/graph transformations. We introduced a notion of specification consisting of pre- and postcondition which specify the correctness of the run of rewrite rules performed according to a given rewrite strategy.

Deciding the correctness of a given specification is not an easy and decidable task in general. We proposed some criteria which may be helpful to choose the most appropriate logics one can use to express proof obligations related to the correctness problem. We illustrated our proposal by considering a running example for which two decidable logics have been used to prove its correctness.

Even in the relatively simple considered example, none of the investigated logics is expressive enough to be able to deal with all the discussed properties. This is a deliberate choice. Our point is that one has to select for each problem one or several logics that are relevant and we proposed some criteria that help to select such logics.

References

1. Ahmetaj, S., Calvanese, D., Ortiz, M., Simkus, M.: Managing change in graph-structured data using description logics. In: Proceedings of the Twenty-Eighth AAAI Conference on Artificial Intelligence, Québec City, Québec, Canada, 27–31 July 2014, pp. 966–973 (2014)
2. Areces, C., Blackburn, P., Marx, M.: Hybrid logics: characterization, interpolation and complexity. J. Symb. Log. **66**(3), 977–1010 (2001)
3. Baader, F., Calvanese, D., McGuinness, D.L., Nardi, D., Patel-Schneider, P.F. (eds.): The Description Logic Handbook: Theory, Implementation, and Applications. Cambridge University Press, Cambridge (2003)
4. Balbiani, P., Echahed, R., Herzig, A.: A dynamic logic for termgraph rewriting. In: Ehrig, H., Rensink, A., Rozenberg, G., Schürr, A. (eds.) ICGT 2010. LNCS, vol. 6372, pp. 59–74. Springer, Heidelberg (2010). doi:10.1007/978-3-642-15928-2_5
5. Baresi, L., Spoletini, P.: On the use of alloy to analyze graph transformation systems. In: Corradini, A., Ehrig, H., Montanari, U., Ribeiro, L., Rozenberg, G. (eds.) ICGT 2006. LNCS, vol. 4178, pp. 306–320. Springer, Heidelberg (2006). doi:10.1007/11841883_22
6. Brenas, J.H., Echahed, R., Strecker, M.: On the closure of description logics under substitutions. In: Proceedings of the 29th International Workshop on Description Logics, Cape Town, South Africa, 22–25 April 2016
7. Brenas, J.H., Echahed, R., Strecker, M.: Proving correctness of logically decorated graph rewriting systems. In: 1st International Conference on Formal Structures for Computation and Deduction, FSCD 2016, Porto, Portugal, 22–26 June 2016, pp. 14:1–14:15 (2016)
8. Börger, E., Grädel, E., Gurevich, Y.: The Classical Decision Problem. Springer, New York (2000)
9. Corradini, A., Heindel, T., Hermann, F., König, B.: Sesqui-pushout rewriting. In: Corradini, A., Ehrig, H., Montanari, U., Ribeiro, L., Rozenberg, G. (eds.) ICGT 2006. LNCS, vol. 4178, pp. 30–45. Springer, Heidelberg (2006). doi:10.1007/11841883_4

10. Courcelle, B.: The monadic second-order logic of graphs. I. Recognizable sets of finite graphs. Inf. Comput. **85**(1), 12–75 (1990)
11. de Moura, L., Bjørner, N.: Z3: an efficient SMT solver. In: Ramakrishnan, C.R., Rehof, J. (eds.) TACAS 2008. LNCS, vol. 4963, pp. 337–340. Springer, Heidelberg (2008). doi:10.1007/978-3-540-78800-3_24
12. Echahed, R.: Inductively sequential term-graph rewrite systems. In: Ehrig, H., Heckel, R., Rozenberg, G., Taentzer, G. (eds.) ICGT 2008. LNCS, vol. 5214, pp. 84–98. Springer, Heidelberg (2008). doi:10.1007/978-3-540-87405-8_7
13. Ghamarian, A.H., de Mol, M., Rensink, A., Zambon, E., Zimakova, M.: Modelling and analysis using GROOVE. STTT **14**(1), 15–40 (2012)
14. Grädel, E., Otto, M., Rosen, E.: Two-variable logic with counting is decidable. In: Proceedings of 12th IEEE Symposium on Logic in Computer Science, LICS 1997, Warschau (1997)
15. Habel, A., Pennemann, K.: Correctness of high-level transformation systems relative to nested conditions. Math. Struct. Comput. Sci. **19**(2), 245–296 (2009)
16. Itzhaky, S., Banerjee, A., Immerman, N., Nanevski, A., Sagiv, M.: Effectively-propositional reasoning about reachability in linked data structures. In: Sharygina, N., Veith, H. (eds.) CAV 2013. LNCS, vol. 8044, pp. 756–772. Springer, Heidelberg (2013). doi:10.1007/978-3-642-39799-8_53
17. Jackson, D.: Software Abstractions. MIT Press, Cambridge (2011)
18. Leino, K.R.M.: Dafny: an automatic program verifier for functional correctness. In: Clarke, E.M., Voronkov, A. (eds.) LPAR 2010. LNCS (LNAI), vol. 6355, pp. 348–370. Springer, Heidelberg (2010). doi:10.1007/978-3-642-17511-4_20
19. Piskac, R., de Moura, L.M., Bjørner, N.: Deciding effectively propositional logic using DPLL and substitution sets. J. Autom. Reason. **44**(4), 401–424 (2010)
20. Poskitt, C.M., Plump, D.: A Hoare calculus for graph programs. In: Ehrig, H., Rensink, A., Rozenberg, G., Schürr, A. (eds.) ICGT 2010. LNCS, vol. 6372, pp. 139–154. Springer, Heidelberg (2010). doi:10.1007/978-3-642-15928-2_10
21. Poskitt, C.M., Plump, D.: Verifying monadic second-order properties of graph programs. In: Giese, H., König, B. (eds.) ICGT 2014. LNCS, vol. 8571, pp. 33–48. Springer, Heidelberg (2014). doi:10.1007/978-3-319-09108-2_3
22. Reynolds, J.C.: An overview of separation logic. In: Meyer, B., Woodcock, J. (eds.) VSTTE 2005. LNCS, vol. 4171, pp. 460–469. Springer, Heidelberg (2008). doi:10.1007/978-3-540-69149-5_49
23. Semeráth, O., Barta, Á., Szatmári, Z., Horváth, Á., Varró, D.: Formal validation of domain-specific languages with derived features and well-formedness constraints. Int. J. Softw. Syst. Model., July 2015
24. Tschannen, J., Furia, C.A., Nordio, M., Polikarpova, N.: AutoProof: auto-active functional verification of object-oriented programs. In: Baier, C., Tinelli, C. (eds.) TACAS 2015. LNCS, vol. 9035, pp. 566–580. Springer, Heidelberg (2015). doi:10.1007/978-3-662-46681-0_53
25. Varró, D.: Automated formal verification of visual modeling languages by model checking. Softw. Syst. Model. **3**(2), 85–113 (2004)

ProofScript: Proof Scripting for the Masses

Steven Obua[✉], Phil Scott, and Jacques Fleuriot

School of Informatics, Edinburgh University,
10 Crichton Street, Edinburgh EH8 9AB, Scotland, UK
steven.obua@gmail.com
http://www.proofpeer.net

Abstract. The goal of the *ProofPeer* project is to make *collaborative theorem proving* a reality. An important part of our plan to make this happen is *ProofScript*, a language designed to be the main user interface of ProofPeer. Of foremost importance in the design of ProofScript is its fit within a collaborative theorem proving environment. By this we mean that it needs to fit into an environment where peers who are not necessarily part of the current theorem proving and programming language communities work independently from but collaboratively with each other to produce formal definitions and proofs. All aspects of ProofScript are shaped by this design principle. In this paper we will discuss ProofScript's most important aspect of being an integrated language both for interactive proof and for proof scripting.

1 Introduction

Interactive theorem proving (ITP) has come a long way since its inception in the seventies. We have argued elsewhere [2] that *collaborative theorem proving* (CTP) represents its natural evolution, where we defined CTP as the social machine of ITP. The goal of the *ProofPeer* [1] project is to make CTP a reality both by developing CTP fundamentals and by building a practical CTP system.

An important component of ProofPeer is the language that peers use to formulate theorems and proofs, which we call *ProofScript*. In many respects our role model and arguably the state of the art for a structured proof language is Isabelle/Isar [6]. The Isabelle/Isar system consists of a complicated stack in which the programming language Standard ML powers the Isabelle kernel and its programmatic extensions. Isar is the most important of these extensions. That means that in order to add automation and scripting to Isar, one needs to be familiar with the low-level ML fundamentals on which Isar itself is built. In acknowledgment of this, limited capabilities for proof automation called *Eisbach* have recently been added to Isar so that proof methods can be formulated within the Isar language itself [9]. Apart from the fact that this only makes possible proof automation of limited scope, we think that the general situation has become even more complex by this addition, as depicted in Fig. 1.

Some say that complexity crushes the human mind, others say that one should design things as simple as possible, but not simpler. Not wanting to ignore

© Springer International Publishing AG 2016
A. Sampaio and F. Wang (Eds.): ICTAC 2016, LNCS 9965, pp. 333–348, 2016.
DOI: 10.1007/978-3-319-46750-4_19

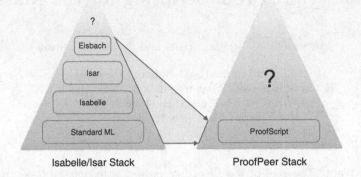

Fig. 1. Surface complexity of Isabelle/Isar vs. ProofPeer

any of these truths, with ProofScript we are trying to condense the capabilities, ideas, and experience embodied by the Isabelle/Isar stack into a single language which serves us as a fresh starting point for exploring collaborative theorem proving. This is not unsimilar to how ITP started out, when ML was designed as a language to explore the design space of ITP [10] before extension pyramids like Isabelle/Isar were built on top of it.

Our goal is to minimize the *surface complexity* the user of our system perceives, thus opening up the system to a wide audience of mathematicians, engineers, basically anyone in need of designing correct virtual or physical artifacts. We do not expect proofpeers to be familiar with type theory or to possess other knowledge usually common only in the programming language community. ProofScript provides us with a unified layer of abstraction which we hope is easy to learn and become productive in. Picking up ProofScript should not be harder than, for example, learning Javascript. Having this abstraction layer also stabilizes ProofPeer as an environment, as changes to how theories are stored, how they are interpreted/compiled, how their execution is distributed in our cloud computing environment etc. can to a large extent happen under the hood.

In this paper we describe the current state of the ProofScript language. Although capability wise we are nowhere near yet what the whole Isabelle/Isar stack can do, we have made important steps towards this goal and we think that the promise of our approach is now apparent.

We will first give an overview of the programming language aspects of ProofScript in Sect. 2. After describing the logical foundations of ProofScript in Sect. 3, we describe in Sect. 4 how we integrated them with the programming language into an approach we call *structured proof scripting*. We conclude in Sect. 5.

For further information about ProofScript, and to experiment with our current implementation, please consult the ProofScript documentation [3].

2   The Programming Language

In this section we give an overview of the ProofScript programming language, with a focus on *why* we designed the language as we did.

2.1   Theories and Namespaces

ProofScript is written in units called *theories*. All theories except the `root` theory [8] have one or more parent theories which they extend, thus forming a directed, acyclic and connected theory extension graph. Before a theory is executed, all of its parents must have been executed. The result of executing a theory is basically a binding of names to computed values, which is usually persisted after execution.

Each theory has its own unique *namespace*. Like Java packages, namespaces are organized as a directory-like structure. We use them to limit both name conflicts in theories and user access privileges to those theories. The namespace tree is orthogonal to the extension graph, and thus one way for peers to collaborate with each other is by extending each other's theories. Within a theory, namespace *aliases* can be defined.

2.2   Types and Patterns

ProofScript is dynamically typed. This means types are checked during the execution of a theory, and not statically, i.e. at some time between the writing and the execution of the theory. This choice is somewhat unorthodox, given that we have drawn earlier parallels with Standard ML, which is famous for the invention of static type checking via the Hindley-Milner type-system. The main reason for our choice is that a static type system that approximates the flexibility of dynamical typing would need to include advanced concepts like functors, type classes, and so on. This would presume a depth of programming knowledge which is simply not realistic for our intended target audience of proofpeers. To further strengthen our argument, languages like Python and Julia have shown that dynamic typing is well-received in the scientific community.

Another more subtle reason is that ProofScript's logic is based on set theory, which in some sense is closer to dynamic typing than static typing. Of course there is no technical reason why there should be any affinity between scripting language and logic, but from an anthropological point of view there seems to be an advantage in fostering these affinities.

Unlike set theory though, being a vehicle for theorem proving, ProofScript has strong types. In particular this means that values of type `Theorem` can only be created by truth preserving operations sanctioned by the logical kernel. Similarly, any other value obeys the invariants established by its type.

There are four built-in logic related types: `Context`, `Theorem`, `Term` and `Type`. These are covered in Sect. 3. Furthermore, there are currently eight built-in non-logical types, shown in Fig. 2.

Type	Example Value
Nil	nil
Integer	-18446744073709551617
String	"ProofScript"
Boolean	false
Tuple	(7, "seven", ("two", 2))
Set	{2, 3, 5, 7, "prime"}
Map	{7 → "seven", 3 → "three", "four" → 4}
Function	x ↦ x * x

Fig. 2. Built-in non-logical types of ProofScript

ProofScript has pattern matching. The matching for logical terms is treated in Sect. 4. Examples for other patterns are shown in Fig. 3. As shown in the examples, patterns can be used to perform simple type checks.

Pattern Example	Explanation
x : Integer	matches any integer
(x, _, y) if x == y	matches any triple that has equal first and third elements
x <+ _ +> y if x == y	matches any tuple with at least two elements where the first and the last element are equal
None	matches a value equal to the custom constructor None
Some x	matches any value created by applying the custom constructor Some
f x	matches any value which is a constructor application
_ : Option	matches any value that has the custom type Option

Fig. 3. Examples of non-logical patterns

It is possible in ProofScript to define *custom datatypes*. An example is shown in Fig. 4 (left hand side) where the custom type Option is defined, together with two constructors None and Some. Some takes an argument and None does not. An example for an expression yielding a value of type Option is Some (1, None). Examples for using pattern matching with this custom type are shown in Fig. 3. Note that type and constructor names must start with an uppercase letter. Mostly this is just a convention we like to enforce, but for constructor names it also makes it easier to distinguish in patterns between constructors and variables.

```
datatype Option          datatype List
   None                      Nil
   Some x                    Cons (head, tail : List)
```

Fig. 4. Custom types Option and List

The argument of a datatype constructor can be constrained by an arbitrary pattern. A datatype modelling heterogeneous lists could therefore be defined as shown in Fig. 4 (right hand side). While expressions like `Cons(1, Cons(2, Cons (3, Nil)))` would evaluate successfully to a value of type `List`, an expression like `Cons(1, 2)` would lead to a runtime error.

2.3 Purely Functional Structured Programming

ProofScript is a purely functional programming language with strict evaluation. By this we mean that functions are first-class, there are no side-effects, and arguments are evaluated *before* they are passed to a function. This makes it possible for peers to write concise, expressive, modular, yet predictable code.

The absence of side-effects, besides its other obvious advantages, also immensely simplifies the semantics and economy of how theories are managed and shared. Imagine theory A being extended by n theories $T_1, \ldots, T_n$. In ProofPeer, after executing a theory, its resulting state is persisted and reused, immediately or at a later time. In our example, that would mean that after executing all of the aforementioned theories, $n + 1$ theory states would have been persisted. If instead A were to contain a function which depended on and mutated some state, then calling this function from T_i could lead to a changed state of A which then would need separate persisting. Assuming none of the T_i extend each other, after executing all theories, up to $2n + 1$ theory states would have been persisted. Now assume that the T_i contained mutable state themselves and depended on each other, say T_j also extends T_i for all $i < j$. In the worst possible case, that would lead to $\frac{(n+1)(n+2)}{2}$ persisted theory states. Our intention is to develop technologies for collaborative theorem proving which scale, so avoiding such quadratic growth by enforcing the absence of side-effects is an obvious thing to do.

Despite its advantages, purely functional programming is not mainstream. Imperative programming is a much more popular style to program in. A major reason for this is that structured programming, which is one of the pillars on which imperative programming rests, is just much more readable to most people than a program written by composing higher-order functions.

We want the advantages of purely functional programming, but we also want to appeal to the mainstream. Fortunately, purely functional programming and structured programming are not at odds with each other at all, but can easily be combined in an approach called *purely functional structured programming* [11]. We have adopted this approach for ProofScript, and therefore it is possible to program in ProofScript *both* by composing functions *and* by writing block-structured code, including **for**-loops, **while**-loops, etc.

As a simple example, consider computing the greatest common divisor of two integers in ProofScript as shown in Fig. 5. On the left hand side, typical functional code for computing the `gcd` is presented. The right hand side displays an alternative formulation of `gcd` using structured programming. Both styles fit within the paradigm of purely functional structured programming, and both versions of `gcd` are side-effect free.

```
def gcd (a, b) =                    def gcd (a, b) =
  if a < 0 then gcd (-a, b)           if a < 0 then a = -a
  else                                if b < 0 then b = -b
    if b < 0 then gcd (a, -b)         while b > 0 do
    else                                (a, b) = (b, a mod b)
      if b == 0 then a               return a
      else gcd (b, a mod b)
```

Fig. 5. Greatest common divisor in ProofScript

2.4   Layout-Sensitive Syntax

Rules that endow indentation and layout with meaning have been adopted by a diverse range of computer languages, from programming languages like Python, Haskell or Scala to markup languages like Markdown. It seems that having explicit rules for layout is especially helpful for novices [12] by removing clutter in the form of curly braces, semicolons, and the like, making errors more obvious by making intended, indented and actual structure synonymous.

That is why we have created ProofScript from the start as a language where indentation and layout is meaningful. ProofScript's syntax is defined by a context-free grammar with added explicit annotations which constrain the possible shapes of parsed text. The technical details of this have been developed over the past three years and are actually still evolving. It was our hope from the beginning that semantic layout would not only make it easier directly for the users themselves to read and write code, but also that the error recovery mechanisms of the parser would profit from it – this is important for the interactive experience we are striving for. As it turns out, this is indeed the case, and has lead us to the discovery of a general way of combining lexing and parsing which we call local lexing [4] and which will be described in a forthcoming separate paper.

The main use of indentation in ProofScript is to delineate the *block structure* of code, making it an ideal companion of our paradigm of purely functional structured programming. An example is the use of indentation to resolve the *dangling else* conflict, as might already be apparent from Fig. 5.

It is often argued that while semantic layout might be beneficial for novices, in a professional setting meaningful layout is detrimental to productivity. A simple example is that usually text editors can be configured to equate a tab character with a fixed number of space characters, often this number is 2 or 4. Differing editor configurations can thus lead to differing meanings of the same program code, which is clearly undesirable. Our solution to this particular problem is to simply disallow the tab character in legal ProofScript. Additionally, we hope to avoid this problem entirely by letting peers interact with ProofPeer in a web environment under our control.

Another example is that of the naive use of Landin's *offside rule* [13] to associate indentation with meaning, which leads to code that might easily break or change its meaning when common automatic refactorings like changing the name of an identifier are applied to it. The offside rule states that *"The southeast*

```
do*                                do*
  val x = 10                         val x = 10
  def f x = do* x + 1                def somefunction x = do* x + 1
              x = x * x                              x = x * x
              x * x                                  x * x
  f x                                somefunction x
```

Fig. 6. Instabilities in Landin's offside rule under refactorings

```
do*                                do*
  val x = 10                         val x = 10
  def f x =                          def somefunction x = do* x + 1
  do*                                x = x * x
    x + 1                            x * x
    x = x * x                        somefunction x
    x * x
  f x
```

Fig. 7. Corrected versions of the invalid code in Fig. 6

quadrant that just contains the phrase's first symbol must contain the entire phrase [...]." If we apply this rule to the hypothetical code snippets in Fig. 6, then it would make sense for the left hand side to evaluate to [(11, 10000)] and for the right hand side to yield [(10000, [101])] (the **do*** control flow statement evaluates all expressions in its argument block and returns them as a tuple). But the right hand side is just a refactoring of the left hand side where we replaced f with somefunction! To combat this problem, neither of the two code snippets shown in Fig. 6 are valid ProofScript. We achieve this by only using layout rules which only compare distances made up entirely of spaces, as opposed to comparing the lengths of texts made up of arbitrary characters. For example, we cannot compare the length (in pixels or centimeters) of the text "**def f x = do***" which consists of 13 characters with the length of 13 spaces. This also solves related problems, such as the fact that a different variable-width font could possibly change the meaning of a program. The legal ProofScript versions of the code shown in Fig. 6 are presented in Fig. 7. Clearly, their meaning is stable under the refactorings f ↦ somefunction and somefunction ↦ f, respectively.

3 Logical Foundations

Most people who use math do not work in a formal setting. They have been taught a naive form of set theory, and usually they know that more rigorous and (hopefully) paradox-free versions exist which are very similar to naive set theory, like Zermelo Fraenkel set theory (ZF).

To accommodate all of these people, ProofScript's logic is based on Zermelo Fraenkel set theory. Isabelle/ZF [14, 15] has pioneered how to embed Zermelo Fraenkel set theory with choice (ZFC) in *intuitionistic* higher-order logic. Our approach differs from the Isabelle/ZF approach in that we embed ZF in *classical* higher-order logic. We call this logic *ZFH* [17] (the H stands both for higher-order logic and Hilbert choice). The reason for this is that we want to be able to draw on the wealth of experience and tools which have been developed in the realm of classical higher-order logic (HOL).

ZFH is almost identical to the logic of HOL-ST [7] and Isabelle/HOLZF [16], which embed ZFC within HOL via a special type representing the universe of ZF sets that comes with the constants and axioms which make up ZF (note that the axiom of choice is implied by the properties of the Hilbert choice operator in HOL). A dilemma that comes up in both HOL-ST and Isabelle/HOLZF is that it is often not clear how to choose between ZF and HOL. Natural numbers for example could be formalised as a ZF set, but they could also be defined as a type in HOL. We resolve this dilemma in ZFH by not having any facilities for defining new HOL types, and by disallowing type variables in terms. This restriction makes it clear that HOL is to be used as the "meta logic", whereas ZF is the logic where all the real work gets done. Because ZFH is a restricted version of HOL-ST and Isabelle/HOLZF, it is consistent if they are.

In the following we will explain the various parts that constitute ZFH and how ProofScript's logical kernel manages them.

3.1 Types and Terms

A *type* τ is either the universal type of ZF sets $\mathcal{U}$, the propositional/boolean type $\mathbb{P}$, or a function type $\tau_1 \rightarrow \tau_2$:

$$\tau ::= \mathcal{U} \mid \mathbb{P} \mid \tau_1 \rightarrow \tau_2.$$

A *term* t is either a constant c, a polymorphic constant $p[\tau]$, a higher-order function $x : \tau_1 \mapsto t$, a bound variable x or a higher-order application $t_1 t_2$:

$$t ::= c \mid p[\tau] \mid x : \tau \mapsto t \mid x \mid t_1 t_2.$$

We have chosen the notation $x : \tau \mapsto t$ over the notation $\lambda x : \tau. t$ because the former is more familiar to a wider audience than the latter.

There are only three polymorphic constants p: equality $=$, universal quantification $\forall$ and existential quantification $\exists$. All other constants are monomorphic constants c; this means in particular that all user-defined constants are monomorphic.

Terms on their own do not have any type because we do not know the types of the monomorphic constants c which appear in a term. The type of a term can only be determined relative to a *context* $\mathcal{C}$. Contexts are the topic of Sect. 3.3. For our purposes here we can simply view a context $\mathcal{C}$ as a partial function from constants c to types $\mathcal{C}(c)$. We can then define the type $\Gamma_{\mathcal{C}}(t)$ of a term t relative to a context $\mathcal{C}$ as shown in Fig. 8. $\Gamma_{\mathcal{C}}$ is a partial function, and we call t *valid* (in $\mathcal{C}$) if $\Gamma_{\mathcal{C}}$ is defined at t, i.e. if $\Gamma_{\mathcal{C}}(t) = \tau$ for some type τ. Note that a valid term has *no free variables*: in valid ZFH terms, all variables appearing in the term must be bound to the argument of an enclosing higher-order function.

$$\Gamma_{\mathcal{C}}(t) = \Gamma_{\mathcal{C},\emptyset}$$
$$\Gamma_{\mathcal{C},\mathcal{V}}(c) = \mathcal{C}(c) \quad \text{if } \mathcal{C} \text{ is defined at } c$$
$$\Gamma_{\mathcal{C},\mathcal{V}}(x) = \mathcal{V}(x) \quad \text{if } \mathcal{V} \text{ is defined at } x$$
$$\Gamma_{\mathcal{C},\mathcal{V}}(p[\tau]) = \begin{cases} (\tau \to \mathbb{P}) \to \mathbb{P} & \text{if } p \in \{\forall, \exists\} \\ \tau \to \tau \to \mathbb{P} & \text{if } p \in \{=\} \end{cases}$$
$$\Gamma_{\mathcal{C},\mathcal{V}}(x : \tau \mapsto t) = \tau \to \rho \quad \text{if } \Gamma_{\mathcal{C},\mathcal{V}[x:=\tau]} = \rho$$
$$\Gamma_{\mathcal{C},\mathcal{V}}(t_1\, t_2) = \rho \quad \text{if } \Gamma_{\mathcal{C},\mathcal{V}}(t_1) = \tau \to \rho \text{ and } \Gamma_{\mathcal{C},\mathcal{V}}(t_2) = \tau$$

Fig. 8. The type $\Gamma_{\mathcal{C}}(t)$ of a term t relative to a context $\mathcal{C}$

Because a term without a context is usually useless, the kernel only operates on *certified terms*, which are basically pairs $(\mathcal{C}, t)$ of a context $\mathcal{C}$ and a term t such that t is valid in $\mathcal{C}$.

3.2 Type Inference

Of course when actually writing down concrete terms in ProofScript, most of the time there is no need to explicitly provide the type τ in the terms $p[\tau]$ and $x : \tau \mapsto t$ as ProofScript performs fully automatic type inference in the spirit of Hindley-Milner. This can be done by allowing type variables for the purposes of the internal type inference algorithm only. There is a caveat though: given that there are no type variables in ZFH terms, what do we do if the result of the type inference still contains type variables?

A simple solution to this problem is to replace all type variables simply by $\mathcal{U}$, the universe of ZF sets. This makes sense as our focus is on set theory anyway, so among the infinitely many possible instantiations this is the most likely one.

This is *almost* the solution we chose; our actual solution is slightly more involved but allows the overloading of syntactic function application with both higher-order function application and set theoretic function application. This is described in detail in [17]. The resulting type inference can be described as being basically Hindley-Milner, but preferring set-theoretic function application over higher-order function application and the type $\mathcal{U}$ over all other types.

3.3 Contexts and Theorems

We have already pointed out that in ProofScript terms only make sense relative to a context $\mathcal{C}$. The context is responsible for maintaining a record of which constants have been defined or introduced so far, and which axioms have been assumed. Contexts in ProofScript are not derived constructs but axiomatic and built-in, and the kernel is responsible for their creation and maintenance. The specification for the concrete syntax of terms is also associated with contexts but not maintained by the kernel.

A context can be thought of as unifying two concepts which in other HOL systems are separate: that of the global logical state of the kernel (or theory context in a system that supports theories), and that of the local logical context.

At the start of a theory, a new context is created based on the contexts of all theories that the theory extends. From then on, there are basically four kernel operations to construct new contexts from existing ones: Introduce, Assume, Define and Choose. Each of them creates a new context of the kind indicated by the operation, and each of these new contexts maintains a backpointer to the context it was created from, its *parent context*. During the execution of a Proof-Script theory, a tree of contexts is created, the root being the context created at the start of the theory.

Theorems are basically certified terms $(\mathcal{C}, t)$ such that $\Gamma_{\mathcal{C}}(t) = \mathbb{P}$ and such that t has been proven to represent a true proposition in context $\mathcal{C}$. There are three different ways to create theorems: (1) Theorems can be created via built-in functions. Examples are `reflexive`, which returns theorems of the form $(\mathcal{C}, t = t)$, or the function `instantiate`, which allows the instantiation of (some of) the universally quantified variables of an existing theorem. (2) Theorems can be created by lifting them between contexts. This is described later. (3) Theorems are created as byproducts of constructing new contexts. This is described in the following where we discuss the four basic kernel operations for constructing new contexts.

All of the following kernel operations are being applied to an existing context $\mathcal{C}$. The newly created context $\mathcal{D}$ has $\mathcal{C}$ as its parent; the context $\mathcal{D}$ also stores which operation using what parameters created it. Here are the operations:

Introduce(n, τ) takes a name n and a type τ and creates a new context $\mathcal{D}$ from $\mathcal{C}$ with an additional constant d with name n and of type τ in $\mathcal{D}$. The name n is not allowed to belong to any constant c in $\mathcal{C}$.

Assume(t) takes a certified term t of type $\mathbb{P}$ and autolifts it into context $\mathcal{C}$ (see Sect. 3.5), resulting in the certified term $(\mathcal{C}, t')$. It then creates a new context $\mathcal{D}$ from $\mathcal{C}$ and returns the theorem $(\mathcal{D}, t')$.

Define(n, t) takes a name n and a certified term t. The name n is not allowed to belong to any constant in $\mathcal{C}$. It first autolifts t into context $\mathcal{C}$, resulting in the certified term $(\mathcal{C}, t')$. After creating a new context $\mathcal{D}$ from $\mathcal{C}$ with an additional constant d with name n and of type $\Gamma_{\mathcal{C}}(t')$, it returns the theorem $(\mathcal{D}, d = t')$.

Choose(n, t) takes a name n and a theorem t. The name n is not allowed to belong to any constant in $\mathcal{C}$, and the theorem must have the form

$$(\mathcal{C}, \forall x_1 : \tau_1. \ldots \forall x_k : \tau_k. \exists y : \tau. t') \quad \text{for some } k \geq 0.$$

We assume that all x_i are different from each other (otherwise we would enforce this via automatic α-conversion).

The operation first creates the context $\mathcal{D}$ from $\mathcal{C}$ with an additional constant d with name n and of type $\tau_1 \rightarrow \ldots \rightarrow \tau_k \rightarrow \tau$ and then returns the theorem

$$(\mathcal{D}, \forall x_1 : \tau_1. \ldots \forall x_k : \tau_k. t'[d\, x_1 \ldots x_k/y]).$$

3.4   Theories and Namespaces

When a theory has been executed, one of the leaves of its context tree becomes the *completed context* of that theory. In principle, it would be possible to select any of the leaves *after* the creation of the complete context tree, but instead we have chosen that during the creation of the context tree of a theory it is always clear which one of the current leaves of the tree is the one which will lead to the completed context. The sequence of these designated leaves forms the *main context thread* of the theory, which at the end of execution will be the same as the path from the context at the start of the theory to the final completed context. The kernel manages the thread by maintaining for each executing theory a pointer to the current leaf of the main context thread; each time a new context is created whose parent is on the main context thread (the only exception is SpawnThread, introduced below), it asks the kernel to put it on the main context thread as well. The kernel will oblige if the pointer it maintains points to the parent of the new context, and will adjust it to point now to the new context instead; otherwise the creation of the new context will fail. Figure 9 illustrates the main context thread. Note that there are two additional kinds of contexts for managing the main context thread:

Complete() demands that context $\mathcal{C}$ is on the main context thread and creates a
new context $\mathcal{D}$ on it which has all unqualified constants removed from it and
which is not allowed to have any children.
SpawnThread() demands that context $\mathcal{C}$ is on the main context thread and creates a new context $\mathcal{D}$ which is *not* on the main context thread.

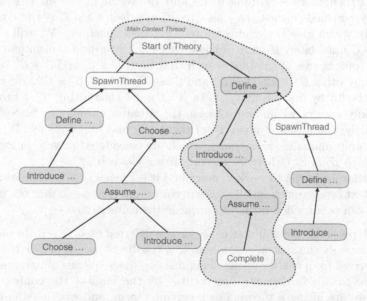

Fig. 9. Context tree and main context thread

The names of logical constants may include a namespace qualification. Only constants which are introduced on the main context thread may be qualified; the namespace they are qualified with is of course the namespace of the theory the main context thread belongs to. Unqualified constants are treated as *private* constants which cannot be referred to by extending theories. This is reflected in how the context at the start of the theory is formed: it contains all and only the constants of the completed contexts of the theories it extends – and these constants are all qualified.

Note that it is not allowed to have an Assume context on the main context thread of any theory except theory root which introduces all axioms of ZFH. This means that all theories are conservative extensions of the root theory.

3.5   Lifting Between Contexts

We will consider in this section how the kernel lifts theorems and certified terms between contexts.

Let us first introduce some shorthand notation. For two theories A and B, let us write $B \rhd A$ if B directly or transitively extends A. Furthermore, for any context C let us write $\mathsf{theory}(C)$ for the theory that C belongs to. Finally, for two contexts C and D we write $D \succ C$ if $\mathsf{theory}(C) = \mathsf{theory}(D)$ and C is a direct or transitive parent of D.

Consider a theorem or certified term (D, t) which we wish to lift into context C. In practice, there are only two relevant situations, and the kernel only supports these: we have either $\mathsf{theory}(C) \rhd \mathsf{theory}(D)$ or $\mathsf{theory}(C) = \mathsf{theory}(D)$.

We will first consider the situation where $\mathsf{theory}(C) = \mathsf{theory}(D)$. If $C = D$ or $C \succ D$ then there is nothing to do and the result of the lift is simply the theorem / certified term (C, t) – an exception arises when C is the completed context, in which case t may only contain *qualified* constants. We will defer the case $D \succ C$ until later. If none of these cases hold, we find the unique context C' in the context tree of the theory such that both $D \succ C'$ and $C \succ C'$, and such that for any other C'' with $D \succ C''$ and $C \succ C''$ we have $C' \succ C''$. We can now perform the lift by first lifting (D, t) to (C', t') and then lifting (C', t') to (C, t').

Secondly, assume we find ourselves in the situation $\mathsf{theory}(C) \rhd \mathsf{theory}(D)$. We then find the completed context D' of $\mathsf{theory}(D)$ and lift (D, t) to (D', t'). If t' would contain unqualified constants, which are considered private in $\mathsf{theory}(D)$, the lifting to D' fails. Otherwise the full lifting yields (C, t').

Note that for lifting to work as described it is crucial that a name introduced via a context is different from all names in the parent context so that no "constant capture" can occur when lifting downwards the context tree.

Lifting Upwards. We will now consider the deferred case $D \succ C$. In particular we consider only the special case where C is the direct parent of D – the general case is derived from this by lifting along a successive sequence of parents leading from D to C. We furthermore distinguish by the kind of the context D and whether we are lifting a theorem or a certified term, and whether the lifting is done in a *canonical* or a *structure preserving* way. Even without the cases for the context kinds SpawnThread and Complete (which are trivial) this leaves us with a total of 16 cases. Therefore we treat here only the cases for Introduce.

So assume that we want to lift a theorem $(\mathcal{D}, t)$ from an Introduce(n, τ) context to its parent in a structure preserving way to obtain a theorem $(\mathcal{C}, t')$. Let c be the constant with name n that has been introduced by the context. Then we set $t' = \forall x : \tau.t[x/c]$ where x is a fresh variable not occuring as a bound variable anywhere in t. If instead we lift the canonical way, we arrive at the same t' if c actually appears in t; otherwise we just set $t' = t$.

Lifting a certified term $(\mathcal{D}, t)$ works similarly. When lifting in a structure preserving way we set $t' = x : \tau \mapsto t[x/c]$, when lifting canonically we simply set $t' = t$ if c does not occur in t and $t' = x : \tau \mapsto t[x/c]$ otherwise.

Auto Lifting. We have previously used the phrase of *autolifting* a certified term $(\mathcal{D}, t)$ to context $\mathcal{C}$. By this we simply mean that we lift $(\mathcal{D}, t)$ to $(\mathcal{C}, t')$ the canonical way; but in addition the autolift only succeeds if t and t' are identical.

Correctness. Introducing contexts as a first-class concept into the kernel and providing axiomatic lifting between contexts is an interesting and we think promising new approach to building HOL systems. Contexts correspond closely to how mathematicians actually reason and it seems to be clear that they are reducible to ordinary logic; therefore we believe that our implementation of them is correct. Obviously it would be better to have a formal proof for this which we do not have yet. But then again, there are only very few theorem proving systems around where the kernel has actually been proven to be correct.

4   Structured Proof Scripting

Structured proof and purely functional structured programming are such a good fit, one might almost think that one was made for the other. Both are side-effect free, and both use a block structure notation which can be nested. Contexts as introduced in Sect. 3.3 make it easy to marry the two as we will outline in this section.

To turn purely functional structured programming into structured proof scripting, we augment the program state with two additional components, the *current* context, and the *literal* context. Literal context and current context are the same, except during the execution of a function call: then the literal context is the same as the literal context *at the point of the function definition*. The reason for this is that we need both a dynamically scoped context, which is the current context, and a lexically scoped one, which is the literal context.

There are several built-in statements which manipulate the current context. One of them is the **let**-statement which introduces either an Introduce or a Define context, depending on its argument which is a *term literal*. For example,

```
let 'x'
```

takes the current context, applies Introduce(x, $\mathcal{U}$) to it, and makes the result the new current context, whereas

```
theorem t: '∀ p. p → p'          val tm = '∀ p. p → p'
  let 'p : ℙ'                      def prf () =
  assume prop: 'p'                   let p: 'p : ℙ'
  prop                               assume prop: p
                                     prop
                                   theorem t: tm
                                     prf ()
```

Fig. 10. A simple theorem

```
let x_def: 'x = d'
```

makes Define(x, 'd') the new current context $\mathcal{D}$ and binds x_def to the theorem
$(\mathcal{D}, 'x = d')$.

Another logical statement is the **theorem**-statement. Assume the statement
shown in Fig. 10 (left hand side) is issued in the current context $\mathcal{C}$. What is
happening here is that the block starting at **let** is evaluated to yield the theorem
$(\mathcal{D}, 'p')$ where $\mathcal{D}$ results from $\mathcal{C}$ by first applying to it SpawnThread (if $\mathcal{C}$ is on
the main context thread) and then Define(p, ℙ) and Assume('p'). The theorem
$(\mathcal{D}, 'p')$ is lifted back into context $\mathcal{C}$ both in a canonical and in a structure
preserving way, and the results are compared with $(\mathcal{C}, '∀ p. p → p')$; if at
least one of them is equal (modulo $\alpha/\beta/\eta$-conversion), and in this case both are,
the lifted theorem is bound to the identifier t (after a possible $\alpha/\beta/\eta$-conversion
to match the theorem statement).

Logical statements and non-logical statements can be freely mixed as shown
in the example in Fig. 10 (right hand side), which proves the same theorem as
the previous example. Note how the current context is dynamically passed along
during the function call prf().

In general it is not a good idea to introduce a logical constant with a fixed
name p within a function because that name might already have been taken
within the context that the function is called with. A more general way to write
prf which avoids name clashes is shown on the left hand side of Fig. 11. The
expression fresh "p" returns a constant name similar to p which has not been
used yet within the current context. This new name is then inserted into the
term literal argument of **let** via the *quote* ‹p›. As a syntactic convenience, this
can be written simply as displayed on the right hand side of Fig. 11.

```
def prf () =                     def prf () =
  val p : String = fresh "p"       let '‹val p› : ℙ'
  let '‹p› : ℙ'                     assume prop: '‹p›'
  val p : Term = '‹p›'             prop
  assume prop: '‹p›'
  prop
```

Fig. 11. Avoiding name clashes with fresh

Quotes are useful not only in the above situation, but whenever one wishes to insert a string or a term into a term literal. Furthermore, quotes are used within *term literal patterns* to designate the holes inside the term literal pattern which are expected to be filled by the pattern match. For example here is how one would define a function which takes apart an implication into premise and conclusion and which fails if the argument term is not an implication:

$$\text{def dest\_imp '⟨p⟩} \to \text{⟨c⟩' = (p, c)}$$

The allowed term patterns are those of the higher-order pattern fragment identified by Dale Miller [18,19].

To conclude this outline of structured proof scripting, let us return to the issue of current and literal context. Our examples so far have shown how the *current* context is used and evolved. But what is the *literal* context good for? The answer is that the literal context is used for parsing term literals. Imagine for example you had implemented a theory `Geometry` in which you define the logical constant `sin`. You also provide a helper function which determines if a term represents the `sin` function as shown in Fig. 12. Now imagine that another theory A also shown in Figure 12 extends theory `Geometry` and defines its own `sin` constant. What would you expect the **show**-statement to print, **true** or **false**? It will print **false**, as the term literal `'sin'` in theory `Geometry` refers to the logical constant `\Geometry\sin`, and in theory A refers to `\A\sin` instead. To implement this behaviour, term literals are parsed by the literal context, and not by the current context. Note however that quotes within term literals are evaluated still within the current context, not the literal one.

```
theory Geometry              theory A extends Geometry
let 'sin = ...'              let 'sin = ...'
def                          show is_sin 'sin'
  is_sin 'sin' = true
  is_sin _ = false
```

Fig. 12. Theories `Geometry` and A

5 Conclusion

We have presented ProofScript, a language for structured proof scripting. We believe that this language can become a firm and simple foundation of ProofPeer, our system in the making for collaborative theorem proving. ProofScript today still needs to grow as a language and as an environment. Many features are still being developed, among them: extensible syntax, theories with assumptions, execution of/code generation for functions defined in logic, automation, and an actual interactive user interface. Yet, it is already a functioning theorem proving system with promise – we have bootstrapped the system to a point where it can convert certificates from an automated theorem prover for first-order logic into valid ProofScript proofs, which is described in a forthcoming paper [5].

References

1. ProofPeer. http://www.proofpeer.net
2. Obua, S., Fleuriot, J., Scott, P., Aspinall, D.: ProofPeer: Collaborative Theorem Proving. arXiv: 1404.6186 (2013)
3. ProofScript. http://proofpeer.net/topics/proofscript
4. Obua, S., Scott, P., Fleuriot, J.: Local Lexing. http://proofpeer.net/papers/locallexing
5. Scott, P., Obua, S., Fleuriot, J.: Bootstrapping LCF Declarative Proofs. http://proofpeer.net/papers/bootstrapping
6. Wenzel, M.: Isar — a generic interpretative approach to readable formal proof documents. In: Bertot, Y., Dowek, G., Théry, L., Hirschowitz, A., Paulin, C. (eds.) TPHOLs 1999. LNCS, vol. 1690, pp. 167–183. Springer, Heidelberg (1999). doi:10.1007/3-540-48256-3_12
7. Agerholm, S., Gordon, M.: Experiments with ZF set theory in HOL and Isabelle. In: Thomas Schubert, E., Windley, P.J., Alves-Foss, J. (eds.) TPHOLs 1995. LNCS, vol. 971, pp. 32–45. Springer, Heidelberg (1995). doi:10.1007/3-540-60275-5_55
8. ProofPeer Root Theory. http://proofpeer.net/repository?root.thy
9. Matichuk, D., Wenzel, M., Murray, T.: An Isabelle proof method language. In: Klein, G., Gamboa, R. (eds.) ITP 2014. LNCS, vol. 8558, pp. 390–405. Springer, Heidelberg (2014). doi:10.1007/978-3-319-08970-6_25
10. Gordon, M., Milner, A., Wadsworth, C.: Edinburgh LCF. LNCS, vol. 78. Springer, Heidelberg (1979). doi:10.1007/3-540-09724-4
11. Obua, S.: Purely Functional Structured Programming. arXiv:1007.3023 (2010)
12. Okasaki, C.: In praise of mandatory indentation for novice programmers, February 2008. http://okasaki.blogspot.co.uk/2008/02/in-praise-of-mandatory-indentation-for.html
13. Landin, P.: The Next 700 Programming Languages (1966). doi:10.1145/365230.365257
14. Paulson, L.: Set theory for verification: I. From foundations to functions. J. Autom. Reason. 11(3), 353–389 (1993). doi:10.1007/BF00881873
15. Paulson, L.: Set theory for verification: II. Induction and recursion. J. Autom. Reason. 15, 167–215 (1995). doi:10.1007/BF00881916
16. Obua, S.: Partizan games in Isabelle/HOLZF. In: Barkaoui, K., Cavalcanti, A., Cerone, A. (eds.) ICTAC 2006. LNCS, vol. 4281, pp. 272–286. Springer, Heidelberg (2006). doi:10.1007/11921240_19
17. Obua, S., Fleuriot, J., Scott, P., Aspinall, D.: Type inference for ZFH. In: Kerber, M., Carette, J., Kaliszyk, C., Rabe, F., Sorge, V. (eds.) CICM 2015. LNCS (LNAI), vol. 9150, pp. 87–101. Springer, Heidelberg (2015). doi:10.1007/978-3-319-20615-8_6
18. Miller, D.: A logic programming language with lambda-abstraction, function variables, and simple unification. In: Schroeder-Heister, P. (ed.) ELP 1989. LNCS, vol. 475, pp. 253–281. Springer, Heidelberg (1991). doi:10.1007/BFb0038698
19. Nipkow, T.: Functional Unification of Higher-Order Patterns (1993). doi:10.1109/LICS.1993.287599

Automata

Derived-Term Automata for Extended Weighted Rational Expressions

Akim Demaille[✉]

EPITA Research and Development Laboratory (LRDE),
14-16, rue Voltaire, 94276 Le Kremlin-Bicêtre, France
akim@lrde.epita.fr

Abstract. We present an algorithm to build an automaton from a rational expression. This approach introduces support for extended weighted expressions. Inspired by derived-term based algorithms, its core relies on a different construct, *rational expansions*. We introduce an inductive algorithm to compute the expansion of an expression from which the automaton follows. This algorithm is independent of the size of the alphabet, and actually even supports infinite alphabets. It can easily be accommodated to generate deterministic (weighted) automata. These constructs are implemented in Vcsn, a free-software platform dedicated to weighted automata and rational expressions.

1 Introduction

Foundational to Automata Theory, the Kleene Theorem (and its weighted extension, the Kleene–Schützenberger Theorem) states the equivalence of *recognizability*—the property of being accepted by an automaton—and *rationality*—the property of being defined by a *rational*, or *regular*, expression. Numerous constructive proofs (read *algorithms*) have been proposed to go from rational expressions to automata, and vice versa. This paper focuses on building an automaton from an expression.

In 1961 Glushkov [12] provides an algorithm to build a nondeterministic automaton (without spontaneous transitions) now often called the standard (or position, or Glushkov) automaton. Earlier (1960), McNaughton and Yamada [15] proposed the same construct for *extended* rational expressions (i.e., including intersection and complement operators), but performed the now usual subset-automaton construction on-the-fly, thus yielding a deterministic automaton. A key ingredient of these algorithms is that they build an automaton whose states represent positions in the rational expression, and computations on these automata actually represent "executions" of the rational expression.

In 1964 Brzozowski [4] shows that *extended* expressions can be used directly as acceptors: following a transition corresponds to computing the left-quotient of the current expression by the current letter. With a proper equivalence relation between expressions (namely ACI: associativity, commutativity, and idempotence of the addition), Brzozowski shows that there is a finite number of

© Springer International Publishing AG 2016
A. Sampaio and F. Wang (Eds.): ICTAC 2016, LNCS 9965, pp. 351–369, 2016.
DOI: 10.1007/978-3-319-46750-4_20

equivalence classes of such quotients, called *derivatives*. This leads to a very natural construction of a *deterministic* automaton whose states are these derivatives. A rather discreet sentence [4, last line of p. 484] introduces the concept of *expansion*, which is not further developed.

In 1996 Antimirov [3] introduces a novel idea: do not apply ACI equivalence globally; rather, when computing the derivative of an expression which is a sum, split it in a set of *partial derivatives* (or *derived terms*) — which amounts to limiting ACI to the sums that are at the very upper level of the expression. A key feature of the built automaton is that it is non-deterministic; as a result the worst-case size of the resulting automaton (i.e., its number of states) is linear in the size of the expression, instead of exponential with Brzozowski's construct. Antimirov also suggests *not* to rely on derivation in implementations, but on so called *linear forms*, which are closely related to Brzozowski's expansions; derivation is used to prove correctness.

In 2005 Lombardy and Sakarovitch [14] generalize the computation of the derivation and derived-term automaton to support weights. Since, as is well-known, not all weighted non-deterministic automata can be determinized, their construction relies on a generalization of Antimirov's derived-term that generates a *non-deterministic* automaton. In their formalization, Antimirov's sets of derived terms naturally turn into *weighted* sets—each term is associated with a weight—that they name *polynomials* (of expressions). However, linear forms completely disappear, and the construction of the derived-term automaton relies on derivatives. Independently Rutten [18] proposes a similar construction.

In 2011, Caron et al. [5] complete Antimirov's construct to support extended expressions. This is at the price of a new definition of derivatives: sets of sets of expressions, interpreted as disjunctions of conjunctions of expressions.

The contributions of this paper are threefold. Firstly, we introduce *expansions*, which generalize Brzozowski's expansions and Antimirov's linear forms to support weighted expressions; they bind together the derivatives, the constant terms and the *firsts* of an expression (letters with which words of the language/series start). They make the computation of the derived-term automaton independent of the size of the alphabet, and actually completely eliminate the need for the alphabet to be finite. Secondly, we provide support for extended weighted rational expressions, which generalizes both Lombardy and Sakarovitch [14] and Caron et al. [5]. And thirdly, we introduce a variation of this algorithm to build *deterministic* (weighted) automata.

We first settle the notations in Sect. 2, provide an algorithm to compute the expansion of an expression in Sect. 3, which is used in Sect. 4 to propose an alternative construction of the derived-term automaton. In Sect. 5 we expose related work and conclude in Sect. 6.

The concepts introduced here are implemented in Vcsn. Vcsn is a free-software platform dedicated to weighted automata and rational expressions [10]. It supports both derivations and expansions, as exposed in this paper, and the corresponding constructions of the derived-term automaton[1].

2 Notations

Our purpose is to define, compute, and use *rational expansions*. They intend to be to the differentiation (derivation) of rational expressions what differential forms are to the differentiation of functions. Defining expansions requires several concepts, defined bottom-up in this section. The following figure should help understanding these different entities, how they relate to each other, and where we are heading to: given a weighted rational expression $\mathsf{E}_1 = \langle 5 \rangle 1 + \langle 2 \rangle\, ace + \langle 6 \rangle\, bce + \langle 4 \rangle\, ade + \langle 3 \rangle\, bde$ (weights are written in angle brackets), compute its expansion:

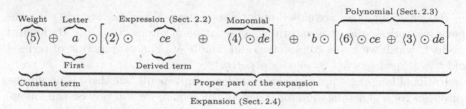

It is helpful to think of expansions as a normal form for expressions.

2.1 Rational Series

Series are to weighted automata what languages are to Boolean automata. Not all languages are rational (denoted by an expression), and similarly, not all series are rational (denoted by a weighted expression). We follow Sakarovitch [19].

Let A be a (finite) alphabet, and $\langle \mathbb{K}, +, \cdot, 0_{\mathbb{K}}, 1_{\mathbb{K}} \rangle$ a semiring whose (possibly non commutative) multiplication will be denoted by implicit concatenation. A (formal power) *series* over A^* with *weights* (or *multiplicities*) in $\mathbb{K}$ is any map from A^* to $\mathbb{K}$. The weight of a word m in a series s is denoted $s(m)$. The *support* of a series s is the language of words that have a non-zero weight in s. The *empty* series, $m \mapsto 0_{\mathbb{K}}$, is denoted 0; for any word u (including ε), u denotes the series $m \mapsto 1_{\mathbb{K}}$ if $m = u, 0_{\mathbb{K}}$ otherwise. Equipped with the pointwise addition ($s + t := m \mapsto s(m) + t(m)$) and the Cauchy product ($s \cdot t := m \mapsto \sum_{u,v \in A^* | u \cdot v = m} s(u) \cdot t(v)$) as multiplication, the set of these series forms a semiring denoted $\langle \mathbb{K}\langle\!\langle A^* \rangle\!\rangle, +, \cdot, 0, \varepsilon \rangle$.

The *constant term* of a series s, denoted s_ε, is $s(\varepsilon)$, the weight of the empty word. A series s is *proper* if $s_\varepsilon = 0_{\mathbb{K}}$. The *proper part* of s, denoted s_p, is the proper series which coincides with s on non empty words: $s = s_\varepsilon + s_p$.

---

[1] See the interactive environment, http://vcsn-sandbox.lrde.epita.fr, or http://vcsn.lrde.epita.fr/dload/2.3/notebooks/expression.derived_term.html, its documentation, or this paper's companion notebook, http://vcsn.lrde.epita.fr/dload/2.3/notebooks/ICTAC-2016.html.

A weight $k \in \mathbb{K}$ is *starrable* if its *star*, $k^* := \sum_{n \in \mathbb{N}} k^n$, is defined. To ensure semantic soundness, we suppose that $\mathbb{K}$ is a *topological semiring*, i.e., it is equipped with a topology, and both addition and multiplication are continuous. Besides, it is supposed to be *strong*, i.e., the product of two summable families is summable. This ensures that $\mathbb{K}\langle\langle A^* \rangle\rangle$, equipped with the product topology derived from the topology on $\mathbb{K}$, is also a strong topological semiring. The *star* of a series is an infinite sum: $s^* := \sum_{n \in \mathbb{N}} s^n$.

Proposition 1. *Let $\mathbb{K}$ be a strong topological semiring. Let $s \in \mathbb{K}\langle\langle A^* \rangle\rangle$, s^* is defined iff s_ε^* is defined and then $s^* = s_\varepsilon^* + s_\varepsilon^* s_p s^*$.*

Proof. By [19, Proposition 2.6, p. 396] s^* is defined iff s_ε^* is defined and then $s^* = (s_\varepsilon^* s_p)^* s_\varepsilon^* = s_\varepsilon^* (s_p s_\varepsilon^*)^*$. The result then follows directly from $s^* = \varepsilon + ss^*$: $s^* = s_\varepsilon^* (s_p s_\varepsilon^*)^* = s_\varepsilon^* (\varepsilon + (s_p s_\varepsilon^*)(s_p s_\varepsilon^*)^*) = s_\varepsilon^* + s_\varepsilon^* s_p (s_\varepsilon^* (s_p s_\varepsilon^*)^*) = s_\varepsilon^* + s_\varepsilon^* s_p s^*$. □

Rational languages are closed under intersection. When the semiring is commutative, series support a natural generalization of intersection, the Hadamard product, which we name *conjunction* and denote &. The conjunction of series s and t is defined as $s \& t := m \mapsto s(m) \cdot t(m)$.

Rational languages are also closed under complement, but there is no unique consensus for a generalization for series. In the sequel, we will rely on the following definition: "s^c is the characteristic series of the complement of the support of s." More precisely, $s^c(m) := s(m)^c$ where $\forall k \in \mathbb{K}, k^c := 1_{\mathbb{K}}$ if $k = 0_{\mathbb{K}}$, $0_{\mathbb{K}}$ otherwise.

Proposition 2. *For series $s, s', t, t', s_a, t_a \in \mathbb{K}\langle\langle A^* \rangle\rangle$ with $a \in A$, for $S, T \subseteq A$, and weights $k, h, s_\varepsilon, t_\varepsilon \in \mathbb{K}$:*

$$(s + s') \& t = s \& t + s' \& t \quad s \& (t + t') = s \& t + s \& t' \tag{1}$$

$$(ks) \& (ht) = (kh)(s \& t) \quad \text{when } \mathbb{K} \text{ is commutative} \tag{2}$$

$$\left(s_\varepsilon + \sum_{a \in S} a \cdot s_a\right) \& \left(t_\varepsilon + \sum_{a \in T} a \cdot t_a\right) = s_\varepsilon t_\varepsilon + \sum_{a \in S \cap T} a \cdot (s_a \& t_a) \tag{3}$$

$$\left(s_\varepsilon + \sum_{a \in S} a \cdot s_a\right)^c = s_\varepsilon^c + \sum_{a \in S} a \cdot s_a^c + \sum_{a \in A \setminus S} a \cdot 0^c \tag{4}$$

From now on, when conjunction is used, we implicitly assume that the semiring is commutative.

2.2 Extended Weighted Rational Expressions

Several definitions of the weighted rational expressions compete, for instance (i) depending where the weights are expressions by themselves [8], or only appear as left- and right-exterior products [1,14], and (ii) on the representation of the empty word: as a special expression "1" [14], or as a simple label "ε" belonging to $\Sigma \cup \{\varepsilon\}$ [13, Sect. 3.1], or finally, as a simple instance of a weight-as-expression for $1_{\mathbb{K}}$ [8]. We follow Sakarovitch [19, Definition III.2.3 p. 399].

Definition 1 (Extended Weighted Rational Expression). *A rational (or regular) expression* E *is a term built from the following grammar, where* $a \in A$ *is a letter, and* $k \in \mathbb{K}$ *a weight:* E $::= 0 \mid 1 \mid a \mid$ E $+$ E $\mid \langle k \rangle$ E $\mid$ E $\langle k \rangle \mid$ E $\cdot$ E $\mid$ E$^* \mid$ E $\&$ E $\mid$ E^c.

Since the product of $\mathbb{K}$ does not need to be commutative (unless conjunction is used) there are two exterior products: $\langle k \rangle$ E and E $\langle k \rangle$. The *size* (aka *length*) of an expression E, $|$E$|$, is its number of symbols, excluding parentheses; its *width* (aka *literal length*), $\|$E$\|$, is the number of occurrences of letters.

Rational expressions are syntactic objects; they provide a finite notation for (some) series, which are semantic objects.

Definition 2 (Series Denoted by an Expression). *Let* E *be an expression. The series denoted by* E*, noted* $[\![$E$]\!]$*, is defined by induction on* E*:*

$$[\![0]\!] := 0 \qquad [\![1]\!] := \varepsilon \qquad [\![a]\!] := a$$

$$[\![\text{E} + \text{F}]\!] := [\![\text{E}]\!] + [\![\text{F}]\!] \qquad [\![\langle k \rangle \text{E}]\!] := k[\![\text{E}]\!] \qquad [\![\text{E} \langle k \rangle]\!] := [\![\text{E}]\!]k$$

$$[\![\text{E} \cdot \text{F}]\!] := [\![\text{E}]\!] \cdot [\![\text{F}]\!] \quad [\![\text{E}^*]\!] := [\![\text{E}]\!]^* \quad [\![\text{E} \& \text{F}]\!] := [\![\text{E}]\!] \& [\![\text{F}]\!] \quad [\![\text{E}^c]\!] := [\![\text{E}]\!]^c$$

An expression is *valid* if it denotes a series. More specifically, this requires that $[\![\text{F}]\!]^*$ is well defined for each subexpression of the form F*, i.e., that the constant term of $[\![\text{F}]\!]$ is *starrable* in $\mathbb{K}$ (Proposition 1). So for instance, $1_{\mathbb{K}}^*$ and $(a^*)^*$ are valid in $\mathbb{B}$, but invalid in $\mathbb{Q}$. This definition of validity, which involves series (semantics) to define a property of expressions (syntax), will be made effective (syntactic) with the appropriate definition of the constant term $c(\text{E})$ *of an expression* E (Definition 8).

Two expressions E and F are *equivalent* iff $[\![\text{E}]\!] = [\![\text{F}]\!]$. Some expressions are "trivially equivalent"; any candidate expression will be rewritten via the following *trivial identities*. Any subexpression of a form listed to the left of a '$\Rightarrow$' is rewritten as indicated on the right.

$$\text{E} + 0 \Rightarrow \text{E} \qquad 0 + \text{E} \Rightarrow \text{E}$$

$$\langle 0_{\mathbb{K}} \rangle \text{E} \Rightarrow 0 \quad \langle 1_{\mathbb{K}} \rangle \text{E} \Rightarrow \text{E} \quad \langle k \rangle 0 \Rightarrow 0 \quad \langle k \rangle \langle h \rangle \text{E} \Rightarrow \langle kh \rangle \text{E}$$

$$\text{E} \langle 0_{\mathbb{K}} \rangle \Rightarrow 0 \quad \text{E} \langle 1_{\mathbb{K}} \rangle \Rightarrow \text{E} \quad 0 \langle k \rangle \Rightarrow 0 \quad \text{E} \langle k \rangle \langle h \rangle \Rightarrow \text{E} \langle kh \rangle$$

$$(\langle k \rangle \text{E}) \langle h \rangle \Rightarrow \langle k \rangle (\text{E} \langle h \rangle) \qquad \ell \langle k \rangle \Rightarrow \langle k \rangle \ell$$

$$\text{E} \cdot 0 \Rightarrow 0 \qquad 0 \cdot \text{E} \Rightarrow 0$$

$$(\langle k \rangle^? 1) \cdot \text{E} \Rightarrow \langle k \rangle \text{E} \qquad \text{E} \cdot (\langle k \rangle^? 1) \Rightarrow \text{E} \langle k \rangle$$

$$0^* \Rightarrow 1$$

$$\text{E} \& 0 \Rightarrow 0 \qquad 0 \& \text{E} \Rightarrow 0 \qquad \text{E} \& 0^c \Rightarrow \text{E} \qquad 0^c \& \text{E} \Rightarrow \text{E}$$

$$\langle k \rangle^? \ell \& \langle h \rangle^? \ell \Rightarrow \langle kh \rangle \ell \qquad \langle k \rangle^? \ell \& \langle h \rangle^? \ell' \Rightarrow 0$$

$$(\langle k \rangle \text{E})^c \Rightarrow \text{E}^c \qquad (\text{E} \langle k \rangle)^c \Rightarrow \text{E}^c$$

where E stands for a rational expression, $a \in A$ is a letter, $\ell, \ell' \in A \cup \{1\}$ denote two different labels, $k, h \in \mathbb{K}$ are weights, and $\langle k \rangle^? \ell$ denotes either $\langle k \rangle \ell$, or ℓ in which case $k = 1_{\mathbb{K}}$ in the right-hand side of $\Rightarrow$. The choice of these identities is beyond the scope of this paper (see Lombardy and Sakarovitch [14, p. 149]), however note that, with the exception of the last line, they are limited to trivial properties; in particular *linearity* ("weighted ACI": associativity, commutativity, and $\langle k \rangle \, \mathsf{E} + \langle h \rangle \, \mathsf{E} \Rightarrow \langle k + h \rangle \, \mathsf{E}$) is not enforced — polynomials will take care of it (Sect. 2.3). In practice, additional identities help reducing the number of derived terms [17], hence the final automaton size. The last two rules, about complement, will be discussed in Sect. 4.3; they are disabled when $\mathbb{K}$ has zero divisors for cases such as $(\langle x \rangle \, (a + \langle y \rangle \, b))^c$ with $xy = 0_{\mathbb{K}}$.

Example 1. Conjunction and complement can be combined to define new operators which are convenient syntactic sugar. For instance, $\mathsf{E} \lessdot \mathsf{F} := \mathsf{E} + (\mathsf{E}^c \, \& \, \mathsf{F})$ allows to define a left-biased $+$ operator: $[\![\mathsf{E} \lessdot \mathsf{F}]\!] \, (u) = [\![\mathsf{E}]\!] \, (u)$ if $[\![\mathsf{E}]\!] \, (u) \neq 0_{\mathbb{K}}$, $[\![\mathsf{F}]\!] \, (u)$ otherwise. The following example mocks Lex-like scanners: identifiers are non-empty sequences of letters of $\{a, b\}$ that are not reserved keywords. The expression $\mathsf{E}_3 := \langle 2 \rangle \, ab \lessdot \langle 3 \rangle \, (a + b)^+$, with weights in $\mathbb{Z}$, maps the "keyword" ab to 2, and "identifiers" to 3. Once desugared and simplified by the trivial identities, we have $\mathsf{E}_3 = \langle 2 \rangle \, ab + ((ab)^c \, \& \, \langle 3 \rangle \, ((a + b)(a + b)^*))$.

2.3 Rational Polynomials

At the core of the idea of "partial derivatives" introduced by Antimirov [3], is that of *sets* of rational expressions, later generalized in *weighted sets* by Lombardy and Sakarovitch [14], i.e., functions (partial, with finite domain) from the set of rational expressions into $\mathbb{K} \setminus \{0_{\mathbb{K}}\}$. It proves useful to view such structures as "polynomials of rational expressions". In essence, they capture the linearity of addition.

Definition 3 (Rational Polynomial). *A polynomial (of rational expressions) is a finite (left) linear combination of rational expressions. Syntactically it is represented by a term built from the grammar* $\mathsf{P} ::= 0 \mid \langle k_1 \rangle \odot \mathsf{E}_1 \oplus \cdots \oplus \langle k_n \rangle \odot \mathsf{E}_n$ *where* $k_i \in \mathbb{K} \setminus \{0_{\mathbb{K}}\}$ *denote non-null weights, and* E_i *denote non-null expressions. Expressions may not appear more than once in a polynomial. A monomial is a pair* $\langle k_i \rangle \odot \mathsf{E}_i$. *The terms of* P *is the set* $\mathsf{exprs} \, (\mathsf{P}) := \{\mathsf{E}_1, \ldots, \mathsf{E}_n\}$.

We use specific symbols ($\odot$ and $\oplus$) to clearly separate the outer polynomial layer from the inner expression layer. A polynomial P of expressions can be "projected" as a rational expression $\mathsf{expr} \, (\mathsf{P})$ by mapping its sum and left-multiplication by a weight onto the corresponding operators on rational expressions. This operation is performed on a canonical form of the polynomial (expressions are sorted in a well defined order). Polynomials denote series: $[\![\mathsf{P}]\!] := [\![\mathsf{expr} \, (\mathsf{P})]\!]$.

Example 2. Let $\mathsf{E}_1 := \langle 5 \rangle \, 1 + \langle 2 \rangle \, ace + \langle 6 \rangle \, bce + \langle 4 \rangle \, ade + \langle 3 \rangle \, bde$. The polynomial '$\mathsf{P}_{1a} := \langle 2 \rangle \odot ce \oplus \langle 4 \rangle \odot de$' has two monomials: '$\langle 2 \rangle \odot ce$' and '$\langle 4 \rangle \odot de$'. It denotes the (left) quotient of $[\![\mathsf{E}_1]\!]$ by a, and '$\mathsf{P}_{1b} := \langle 6 \rangle \odot ce \oplus \langle 3 \rangle \odot de$' the quotient by b.

Let $\mathsf{P} = \langle k_1 \rangle \odot \mathsf{E}_1 \oplus \cdots \oplus \langle k_n \rangle \odot \mathsf{E}_n$ be a polynomial, k a weight (possibly null) and F an expression (possibly null), we introduce the following operations:

$$\mathsf{P} \cdot \mathsf{F} := \langle k_1 \rangle \odot (\mathsf{E}_1 \cdot \mathsf{F}) \oplus \cdots \oplus \langle k_n \rangle \odot (\mathsf{E}_n \cdot \mathsf{F})$$

$$\langle k \rangle \, \mathsf{P} := \langle kk_1 \rangle \odot \mathsf{E}_1 \oplus \cdots \oplus \langle kk_n \rangle \odot \mathsf{E}_n$$

$$\mathsf{P} \langle k \rangle := \langle k_1 \rangle \odot (\mathsf{E}_1 \langle k \rangle) \oplus \cdots \oplus \langle k_n \rangle \odot (\mathsf{E}_n \langle k \rangle)$$

$$\mathsf{P}_1 \,\&\, \mathsf{P}_2 := \bigoplus_{\substack{\langle k_1 \rangle \odot \mathsf{E}_1 \in \mathsf{P}_1 \\ \langle k_2 \rangle \odot \mathsf{E}_2 \in \mathsf{P}_2}} \langle k_1 k_2 \rangle \odot (\mathsf{E}_1 \,\&\, \mathsf{E}_2) \qquad \mathsf{P}^c := \langle 1_{\mathbb{K}} \rangle \odot \mathsf{expr}\,(\mathsf{P})^c \qquad (5)$$

Trivial identities might simplify the result, e.g., $(\langle 1_{\mathbb{K}} \rangle \odot a) \,\&\, (\langle 1_{\mathbb{K}} \rangle \odot b) = \langle 1_{\mathbb{K}} \rangle \odot (a \,\&\, b) = 0$. Note the asymmetry between left and right exterior products. The addition of polynomials is commutative, multiplication by zero (be it an expression or a weight) evaluates to the null polynomial, and the left-multiplication by a weight is distributive.

Lemma 1. $\llbracket \mathsf{P} \cdot \mathsf{F} \rrbracket = \llbracket \mathsf{P} \rrbracket \cdot \llbracket \mathsf{F} \rrbracket$ $\qquad \llbracket \langle k \rangle \, \mathsf{P} \rrbracket = \langle k \rangle \, \llbracket \mathsf{P} \rrbracket$ $\qquad \llbracket \mathsf{P} \langle k \rangle \rrbracket = \llbracket \mathsf{P} \rrbracket \langle k \rangle$
$\llbracket \mathsf{P}_1 \,\&\, \mathsf{P}_2 \rrbracket = \llbracket \mathsf{P}_1 \rrbracket \,\&\, \llbracket \mathsf{P}_2 \rrbracket$ $\qquad \llbracket \mathsf{P}^c \rrbracket = \llbracket \mathsf{P} \rrbracket^c$.

Proof. The first three are trivial. The case of & follows from (1) and (2). Complement follows from its definition: $\llbracket \mathsf{P}^c \rrbracket = \llbracket \langle 1_{\mathbb{K}} \rangle \odot \mathsf{expr}\,(\mathsf{P})^c \rrbracket = \llbracket \mathsf{expr}\,(\mathsf{P})^c \rrbracket = \llbracket \mathsf{expr}\,(\mathsf{P}) \rrbracket^c = \llbracket \mathsf{P} \rrbracket^c$. $\qquad\square$

2.4 Rational Expansions

Expansions group together a distinguished weight, and, for each letter, its associated polynomial. Let $[n]$ denote $\{1, \ldots, n\}$.

Definition 4 (Rational Expansion). *A rational expansion X is a term built from the grammar* $\mathsf{X} ::= \langle k \rangle \oplus a_1 \odot [\mathsf{P}_1] \oplus \cdots \oplus a_n \odot [\mathsf{P}_n]$ *where* $k \in \mathbb{K}$ *is a weight (possibly null),* $a_i \in A$ *letters (occurring at most once), and* P_i *non-null polynomials. The constant term is* k*, the proper part is* $a_1 \odot [\mathsf{P}_1] \oplus \cdots \oplus a_n \odot [\mathsf{P}_n]$*, the firsts is* $\{a_1, \ldots, a_n\}$ *(possibly empty), and the terms are* $\mathsf{exprs}\,(\mathsf{X}) := \bigcup_{i \in [n]} \mathsf{exprs}\,(\mathsf{P}_i)$.

To ease reading, polynomials are written in square brackets. Contrary to expressions and polynomials, there is no specific term for the empty expansion: it is represented by $\langle 0_{\mathbb{K}} \rangle$, the null weight. Except for this case, null constant terms are left implicit. Besides their support for weights, expansions differ from Antimirov's linear forms in that they integrate the constant term, which gives them a flavor of series. Given an expansion X, we denote by X_ε (or $\mathsf{X}(\varepsilon)$) its constant term, by $f(\mathsf{X})$ its firsts, by X_p its proper part, and by X_a (or $\mathsf{X}(a)$) the polynomial corresponding to a in X, or the null polynomial if $a \notin f(\mathsf{X})$. Expansions will thus be written: $\mathsf{X} = \langle \mathsf{X}_\varepsilon \rangle \oplus \bigoplus_{a \in f(\mathsf{X})} a \odot [\mathsf{X}_a]$.

An expansion whose polynomials are monomials is said to be *deterministic*. An expansion X can be "projected" as a rational expression $\mathsf{expr}\,(\mathsf{X})$ by mapping weights, letters and polynomials to their corresponding rational expressions, and $\oplus/\odot$ to the sum/concatenation of rational expressions. Again, this is performed on a canonical form of the expansion: letters and polynomials are sorted. Expansions also denote series: $\llbracket \mathsf{X} \rrbracket := \llbracket \mathsf{expr}\,(\mathsf{X}) \rrbracket$. An expansion X is said to be *equivalent* to an expression E iff $\llbracket \mathsf{X} \rrbracket = \llbracket \mathsf{E} \rrbracket$.

Example 3 (Example 2 continued). Expansion $X_1 := \langle 5 \rangle \oplus a \odot [P_{1a}] \oplus b \odot [P_{1b}]$ has $X_1(\varepsilon) = \langle 5 \rangle$ as constant term, and maps the letter a (resp. b) to the polynomial $X_1(a) = P_{1a}$ (resp. $X_1(b) = P_{1b}$). X_1 can be proved to be equivalent to E_1.

Let X, Y be expansions, k a weight, and E an expression (all possibly null):

$$X \oplus Y := \langle X_\varepsilon + Y_\varepsilon \rangle \oplus \bigoplus_{a \in f(X) \cup f(Y)} a \odot [X_a \oplus Y_a] \tag{6}$$

$$\langle k \rangle X := \langle kX_\varepsilon \rangle \oplus \bigoplus_{a \in f(X)} a \odot [\langle k \rangle X_a] \qquad X \langle k \rangle := \langle X_\varepsilon k \rangle \oplus \bigoplus_{a \in f(X)} a \odot [X_a \langle k \rangle] \tag{7}$$

$$X \cdot E := \bigoplus_{a \in f(X)} a \odot [X_a \cdot E] \qquad \text{with } X \text{ proper: } X_\varepsilon = 0_\mathbb{K} \tag{8}$$

$$X \,\&\, Y := \langle X_\varepsilon Y_\varepsilon \rangle \oplus \bigoplus_{a \in f(X) \cap f(Y)} a \odot [X_a \,\&\, Y_a] \tag{9}$$

$$X^c := \langle X_\varepsilon^c \rangle \oplus \bigoplus_{a \in f(X)} a \odot [X_a^c] \oplus \bigoplus_{a \in A \setminus f(X)} a \odot [0^c] \tag{10}$$

Since by definition expansions never map to null polynomials, some firsts might be smaller sets than suggested by these equations. For instance in $\mathbb{Z}$ the sum of $\langle 1 \rangle \oplus a \odot [\langle 1 \rangle \odot b]$ and $\langle 1 \rangle \oplus a \odot [\langle -1 \rangle \odot b]$ is $\langle 2 \rangle$, and $(a \odot [\langle 1 \rangle \odot b]) \,\&\, (a \odot [\langle 1 \rangle \odot c])$ is $\langle 0 \rangle$ since $b \,\&\, c \Rightarrow 0$. Note that X^c is a deterministic expansion.

The following lemma is simple to establish: lift semantic equivalences, such as those of Proposition 2, to syntax, using Lemma 1.

Lemma 2. $[\![X \oplus Y]\!] = [\![X]\!] + [\![Y]\!] \qquad [\![\langle k \rangle X]\!] = \langle k \rangle [\![X]\!] \qquad [\![X \langle k \rangle]\!] = [\![X]\!] \langle k \rangle$
$[\![X \cdot E]\!] = [\![X]\!] \cdot [\![E]\!] \qquad [\![X \,\&\, Y]\!] = [\![X]\!] \,\&\, [\![Y]\!] \qquad [\![X^c]\!] = [\![X]\!]^c$.

2.5   Weighted Automata

Definition 5 (Automaton). *A weighted automaton $\mathcal{A}$ is a tuple $\langle A, \mathbb{K}, Q, E, I, T \rangle$ where:*

- *A (the set of labels) is an alphabet (usually finite), $\mathbb{K}$ (the set of weights) is a semiring,*
- *Q is a set of states, I and T are the initial and final functions from Q into $\mathbb{K}$,*
- *E is a (partial) function from $Q \times A \times Q$ into $\mathbb{K} \setminus \{0_\mathbb{K}\}$;*
  its domain represents the transitions: $\langle source, label, destination \rangle$.

An automaton is *locally finite* if each state has a finite number of outgoing transitions ($\forall s \in Q, \{s\} \times A \times Q \cap E$ is finite). A *finite automaton* has a finite number of states. A *path* p in an automaton is a sequence of transitions $(q_0, a_0, q_1)(q_1, a_1, q_2) \cdots (q_n, a_n, q_{n+1})$ where the source of each is the destination of the previous one; its *label* is the word $a_0 a_1 \cdots a_n$, its *weight* is $I(q_0) \otimes E(q_0, a_0, q_1) \otimes \cdots \otimes E(q_n, a_n, q_{n+1}) \otimes T(q_{n+1})$. The *evaluation* of word

u by a locally finite automaton $\mathcal{A}$, $\mathcal{A}(u)$, is the (finite) sum of the weights of all the paths labeled by u, or $0_{\mathbb{K}}$ if there are no such path. The *behavior* of such an automaton $\mathcal{A}$ is the series $[\![\mathcal{A}]\!] := u \mapsto \mathcal{A}(u)$. A state q is *initial* if $I(q) \neq 0_{\mathbb{K}}$. A state q is *accessible* if there is a path from an initial state to q. The *accessible* part of an automaton $\mathcal{A}$ is the subautomaton whose states are the accessible states of $\mathcal{A}$. The size of a finite automaton, $|\mathcal{A}|$, is its number of states.

Definition 6 (Semantics of a State). *Given a weighted automaton $\mathcal{A} = \langle A, \mathbb{K}, Q, E, I, T \rangle$, inductively[2] define a semantic mapping $[\![-]\!] : Q \to \mathbb{K}\langle\!\langle A^* \rangle\!\rangle$ as follows:*

- *For all $q \in Q$, $[\![q]\!](\varepsilon) := T(q)$.*
- *For all $q \in Q$, $a \in A$, and $u \in A^*$, $[\![q]\!](au) := \sum_{q' \in Q} E(q, a, q')\left([\![q']\!](u)\right)$.*

It follows by a simple inductive proof from this definition, that for all $u = a_1 \ldots a_n$,

$$[\![q]\!](u) = \sum_{q_1 \in Q} \cdots \sum_{q_{n+1} \in Q} E(q, a_0, q_1) \cdots E(q_n, a_n, q_{n+1}) T(q_{n+1})$$

and from here it follows directly from the definition of $[\![\mathcal{A}]\!]$ that:

$$[\![\mathcal{A}]\!](u) = \sum_{q \in Q} I(q)\left([\![q]\!](u)\right) \tag{11}$$

We are interested, given an expression E, in an algorithm to compute an automaton $\mathcal{A}_\mathsf{E}$ such that $[\![\mathcal{A}_\mathsf{E}]\!] = [\![\mathsf{E}]\!]$ (Sect. 4). To this end, we first introduce a simple recursive procedure to compute *the* expansion of an expression.

3   Computing Expansions of Expressions

3.1   Expansion of a Rational Expression

Definition 7 (Expansion of a Rational Expression). *The expansion of a rational expression E, written $d(\mathsf{E})$, is the expansion defined inductively as follows:*

$$d(0) := \langle 0_{\mathbb{K}} \rangle \qquad d(1) := \langle 1_{\mathbb{K}} \rangle \qquad d(a) := a \odot [\langle 1_{\mathbb{K}} \rangle \odot 1] \tag{12}$$

$$d(\mathsf{E} + \mathsf{F}) := d(\mathsf{E}) \oplus d(\mathsf{F}) \qquad d(\langle k \rangle \mathsf{E}) := \langle k \rangle\, d(\mathsf{E}) \qquad d(\mathsf{E}\langle k \rangle) := d(\mathsf{E})\langle k \rangle \tag{13}$$

$$d(\mathsf{E} \cdot \mathsf{F}) := d_p(\mathsf{E}) \cdot \mathsf{F} \oplus \langle d_\varepsilon(\mathsf{E}) \rangle\, d(\mathsf{F}) \tag{14}$$

$$d(\mathsf{E}^*) := \langle d_\varepsilon(\mathsf{E})^* \rangle \oplus \langle d_\varepsilon(\mathsf{E})^* \rangle\, d_p(\mathsf{E}) \cdot \mathsf{E}^* \tag{15}$$

$$d(\mathsf{E} \,\&\, \mathsf{F}) := d(\mathsf{E}) \,\&\, d(\mathsf{F}) \tag{16}$$

$$d(\mathsf{E}^c) := d(\mathsf{E})^c \tag{17}$$

where $d_\varepsilon(\mathsf{E}) := d(\mathsf{E})_\varepsilon, d_p(\mathsf{E}) := d(\mathsf{E})_p$ are the constant term/proper part of $d(\mathsf{E})$.

---

[2] The induction is on the length of the word u in $[\![q]\!](u)$, which is defined for all q and all words of the given length simultaneously.

The right-hand sides are indeed expansions. The computation trivially terminates: induction is performed on strictly smaller subexpressions. These formulas
are enough to compute the expansion of an expression; there is no secondary
process for the firsts — indeed $d(a) := a \odot [\langle 1_{\mathbb{K}} \rangle \odot 1]$ suffices and every other
case simply propagates or assembles the firsts — or the constant terms. In an
implementation a single recursive call to $d(\mathsf{E})$ is performed for (14) and (15),
from which $d_\varepsilon(\mathsf{E})$ and $d_p(\mathsf{E})$ are obtained. So for instance (15) should rather be
written: $d(\mathsf{E}^*) := \text{let } \mathsf{X} = d(\mathsf{E}) \text{ in } \langle \mathsf{X}_\varepsilon^* \rangle \oplus \langle \mathsf{X}_\varepsilon^* \rangle \mathsf{X}_p \cdot \mathsf{E}^*$. Besides, existing expressions should be referenced to, not duplicated: in the previous piece of code, E^*
is not built again, the input argument is reused.

Lemma 3. *For any expression* E, $[\![d(\mathsf{E})]\!] = [\![\mathsf{E}]\!]$.

Proof. Proved by induction over E. The proof is straightforward for (12), (13),
(16) and (17), using Lemma 2. The case of multiplication, (14), follows from:

$$[\![d(\mathsf{E} \cdot \mathsf{F})]\!] = \left[\!\left[d_p(\mathsf{E}) \cdot \mathsf{F} \oplus \langle d_\varepsilon(\mathsf{E}) \rangle \cdot d(\mathsf{F}) \right]\!\right] = [\![d_p(\mathsf{E})]\!] \cdot [\![\mathsf{F}]\!] + \langle d_\varepsilon(\mathsf{E}) \rangle \cdot [\![d(\mathsf{F})]\!]$$

$$= [\![d_p(\mathsf{E})]\!] \cdot [\![\mathsf{F}]\!] + \langle d_\varepsilon(\mathsf{E}) \rangle \cdot [\![\mathsf{F}]\!] \quad \text{(by inductive hypothesis)}$$

$$= \left([\![\langle d_\varepsilon(\mathsf{E}) \rangle]\!] + [\![d_p(\mathsf{E})]\!] \right) \cdot [\![\mathsf{F}]\!] = \left[\!\left[\langle d_\varepsilon(\mathsf{E}) \rangle + d_p(\mathsf{E}) \right]\!\right] \cdot [\![\mathsf{F}]\!]$$

$$= [\![d(\mathsf{E})]\!] \cdot [\![\mathsf{F}]\!] = [\![\mathsf{E}]\!] \cdot [\![\mathsf{F}]\!] \quad \text{(by inductive hypothesis)} = [\![\mathsf{E} \cdot \mathsf{F}]\!]$$

It might seem more natural to exchange the two terms (i.e., $\langle d_\varepsilon(\mathsf{E}) \rangle \cdot d(\mathsf{F}) \oplus$
$d_p(\mathsf{E}) \cdot \mathsf{F}$), but an implementation first computes $d(\mathsf{E})$ and then computes $d(\mathsf{F})$
only if $d_\varepsilon(\mathsf{E}) \neq 0_{\mathbb{K}}$. The case of Kleene star, (15), follows from Proposition 1. $\square$

By Lemma 3, given an expression E and its expansion $d(\mathsf{E}) = \langle k \rangle \oplus a_1 \odot [\mathsf{P}_1] \oplus$
$\cdots \oplus a_n \odot [\mathsf{P}_n]$, we have $[\![\mathsf{E}]\!] = [\![\langle k \rangle \oplus a_1 \odot [\mathsf{P}_1] \oplus \cdots \oplus a_n \odot [\mathsf{P}_n]]\!]$ and thus, by
the definition of the semantics of an expansion we have: $[\![\mathsf{E}]\!] = k + a_1 \cdot [\![\mathsf{P}_1]\!] +$
$\ldots + a_n \cdot [\![\mathsf{P}_n]\!]$. Now, the following facts on the constant term and left quotients
of $[\![\mathsf{E}]\!]$ immediately follow:

$$[\![\mathsf{E}]\!]_\varepsilon = k \qquad\qquad a_i^{-1} [\![\mathsf{E}]\!] = [\![\mathsf{P}_i]\!] \qquad \text{for } 1 \leq i \leq n \qquad (18)$$

3.2 Connection with Derivatives

We reproduce here the definition of constant terms and derivatives from
Lombardy et al. [14, p. 148 and Definition 2], with our notations and covering extended expressions. To facilitate reading, weights such as the constant
term are written in angle brackets, although so far this was reserved to syntactic
constructs.

Definition 8 (Constant Term and Derivative).

$$c(0) := \langle 0_{\mathbb{K}} \rangle, \quad c(1) := \langle 1_{\mathbb{K}} \rangle, \qquad \partial_a 0 := 0, \quad \partial_a 1 := 0, \qquad (19)$$

$$c(a) := \langle 0_{\mathbb{K}} \rangle, \forall a \in A, \qquad\qquad \partial_a b := 1 \text{ if } b = a, \text{ 0 } \textit{otherwise}, \quad (20)$$

$$c(\mathsf{E} + \mathsf{F}) := c(\mathsf{E}) + c(\mathsf{F}), \qquad \partial_a(\mathsf{E} + \mathsf{F}) := \partial_a \mathsf{E} \oplus \partial_a \mathsf{F}, \tag{21}$$

$$c(\langle k \rangle \, \mathsf{E}) := \langle k \rangle \, c(\mathsf{E}), \qquad \partial_a(\langle k \rangle \, \mathsf{E}) := \langle k \rangle \, (\partial_a \mathsf{E}), \tag{22}$$

$$c(\mathsf{E} \, \langle k \rangle) := c(\mathsf{E}) \, \langle k \rangle, \qquad \partial_a(\mathsf{E} \, \langle k \rangle) := (\partial_a \mathsf{E}) \, \langle k \rangle, \tag{23}$$

$$c(\mathsf{E} \cdot \mathsf{F}) := c(\mathsf{E}) \cdot c(\mathsf{F}), \qquad \partial_a(\mathsf{E} \cdot \mathsf{F}) := (\partial_a \mathsf{E}) \cdot \mathsf{F} \oplus \langle c(\mathsf{E}) \rangle \, \partial_a \mathsf{F}, \tag{24}$$

$$c(\mathsf{E}^*) := c(\mathsf{E})^*, \qquad \partial_a \mathsf{E}^* := \langle c(\mathsf{E})^* \rangle \, (\partial_a \mathsf{E}) \cdot \mathsf{E}^* \tag{25}$$

$$c(\mathsf{E} \,\&\, \mathsf{F}) := c(\mathsf{E}) \cdot c(\mathsf{F}), \qquad \partial_a(\mathsf{E} \,\&\, \mathsf{F}) := \partial_a \mathsf{E} \,\&\, \partial_a \mathsf{F}, \tag{26}$$

$$c(\mathsf{E}^c) := c(\mathsf{E})^c, \qquad \partial_a \mathsf{E}^c := (\partial_a \mathsf{E})^c \tag{27}$$

where (25) *applies iff* $c(\mathsf{E})^*$ *is defined in* $\mathbb{K}$.

The reader is invited to compare Definitions 7 and 8, which does not include the computation of the firsts.

The following lemma is a syntactic version of (18).

Lemma 4. *For any expression* E, $d(\mathsf{E})(\varepsilon) = c(\mathsf{E})$, *and* $d(\mathsf{E})(a) = \partial_a \mathsf{E}$.

Proof. A straightforward induction on E. The cases of constants and letters are immediate consequences of (19) and (20) on the one hand, and (12) on the other hand. Equation (13) matches (21) to (23). Multiplication (concatenation) is again barely a change of notation between (14) and (24), and likewise for the Kleene star ((15) and (25)). Conjunction, (26), follows from (9) and (16), and complement, (27), from (17) and (10). □

Lemma 4 states that expansions, like Antimirov's linear forms, offer a different means to compute the expression derivatives. However expansions seem to better capture the essence of the process, where the computations of constant terms are tightly coupled with that of the derivations. The formulas are more concise. Expansions are also "more complete" than derivations, viz., the expansion of an expression can be seen as a normal-form of this expression: $\mathsf{E} \equiv \mathsf{expr}\,(d(\mathsf{E}))$ and $d(\mathsf{E}) = d(\mathsf{expr}\,(d(\mathsf{E})))$. Expansions are more efficient to perform effective calculations, such as building an automaton (Sect. 4.4).

4 Expansion-Based Derived-Term Automaton

Definition 9 (Derived-Term Automaton). *The derived-term automaton of an expression* E *is the accessible part of the automaton* $\mathcal{A}_\mathsf{E} := \langle A, \mathbb{K}, Q, E, I, T \rangle$ *defined as follows:*

- Q *is the set of rational expressions on alphabet* A *with weights in* $\mathbb{K}$,
- $E(\mathsf{F}, a, \mathsf{F}') = k$ *iff* $a \in f(d(\mathsf{F}))$ *and* $\langle k \rangle \, \mathsf{F}' \in d(\mathsf{F})(a)$,
- $I = \mathsf{E} \mapsto 1_\mathbb{K}$, $T(\mathsf{F}) = k$ *iff* $\langle k \rangle = d(\mathsf{F})(\varepsilon)$.

The resulting automaton is locally finite, and not necessarily deterministic.

Input   : E, a rational expression
Output : $\langle E, I, T \rangle$ an automaton (simplified notation)

$I(\mathsf{E}) := 1_{\mathbb{K}}$;   // Unique initial state
$Q := \mathrm{Queue}(\mathsf{E})$;   // A work list (queue) loaded with E
while Q *is not empty* **do**
   | $\mathsf{E} := \mathrm{pop}(Q)$;   // A new state/expression to complete
   | $\mathsf{X} := d(\mathsf{E})$;   // The expansion of E
   | $T(\mathsf{E}) := \mathsf{X}(\varepsilon)$;   // Final weight: the constant term
   | **foreach** $a \odot [\mathsf{P}_a] \in \mathsf{X}$ **do**   // For each first/polynomial in X
      | **foreach** $\langle k \rangle \odot \mathsf{F} \in \mathsf{P}_a$ **do**   // For each monomial of $\mathsf{P}_a = \mathsf{X}(a)$
       | $E(\mathsf{E}, a, \mathsf{F}) := k$;   // New transition
       | **if** $\mathsf{F} \notin Q$ **then**
        | $\mathrm{push}(Q, \mathsf{F})$;   // F is a new state, to complete later
      | **end**
    | **end**
  | **end**
end

Algorithm 1. Building the derived-term automaton. The set of states is implicitly grown when transitions are added.

Example 4 (Examples 2 and 3 continued). We have $d(\mathsf{E}_1) = \mathsf{X}_1$. $\mathcal{A}_{\mathsf{E}_1}$ is:

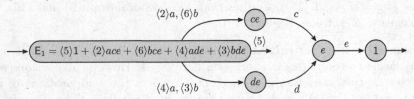

It is straightforward to extract an algorithm from Definition 9, using a work-list of states whose outgoing transitions to compute (see Algorithm 1). This approach admits a natural lazy implementation: the whole automaton is not computed at once, but rather, states and transitions are computed on-the-fly, on demand, for instance when evaluating a word.

Theorem 1. *Any (valid) expression* E *and its expansion-based derived-term automaton* $\mathcal{A}_\mathsf{E}$ *denote the same series, i.e.,* $[\![\mathcal{A}_\mathsf{E}]\!] = [\![\mathsf{E}]\!]$.

This theorem is a straightforward consequence of the following lemma. To make sure we do not get confused between, on the one hand, the semantics $[\![\mathsf{E}]\!]$ of expressions, and on the other hand the (as above defined) semantics of states in the derived term-automaton (which are also given by expressions), let the state associated with an expression E be denoted by q_E. Thus, the semantics of the state corresponding to E in the derived term automaton is written as $[\![q_\mathsf{E}]\!]$.

Lemma 5. *For all rational expressions* E, $[\![\mathsf{E}]\!] = [\![q_\mathsf{E}]\!]$.

Proof. We show by induction on the length of words u, that for all expressions E and all words u, $[\![\mathsf{E}]\!](u) = [\![q_\mathsf{E}]\!](u)$, from which the desired result follows.

Base case. Say we are given an expression E, with $d(\mathsf{E}) = \langle k \rangle \oplus a_1 \odot [\mathsf{P}_1] \oplus \cdots \oplus a_n \odot [\mathsf{P}_n]$. We have $[\![\mathsf{E}]\!](\varepsilon) = k$ by (18); similarly, by the definition of semantics of a state in an automaton and the definition of the derived term automaton, we have $[\![q_\mathsf{E}]\!](\varepsilon) = T(q_\mathsf{E}) = k$, as $k = d(\mathsf{E})(\varepsilon)$.

Inductive case. Assume the statement holds for all words u with $|u| < n$ and all expressions E. Now let $v = au$ be a word with $|v| = n$, and say, we have an expression E, again with $d(\mathsf{E}) = \langle k \rangle \oplus a_1 \odot [\mathsf{P}_1] \oplus \cdots \oplus a_n \odot [\mathsf{P}_n]$.

If $a = a_i$ for some i with $1 \leq i \leq n$, we then have $[\![\mathsf{E}]\!](au) = (a^{-1}[\![\mathsf{E}]\!])(u) = [\![\mathsf{P}_i]\!](u)$. Now let P_i be of the form

$$\mathsf{P}_i = \langle k_1 \rangle \odot \mathsf{F}_1 \oplus \ldots \oplus \langle k_m \rangle \odot \mathsf{F}_m, \tag{28}$$

giving:

$$
\begin{aligned}
[\![\mathsf{E}]\!](au) &= [\![\mathsf{P}_i]\!](u) \\
&= [\![\langle k_1 \rangle \odot \mathsf{F}_1 \oplus \ldots \oplus \langle k_m \rangle \odot \mathsf{F}_m]\!](u) && \text{(def. of } \mathsf{P}_i, \text{ (28))} \\
&= (k_1 [\![\mathsf{F}_1]\!] + \ldots + k_m [\![\mathsf{F}_m]\!])(u) && \text{(def. of expr)} \\
&= k_1 [\![\mathsf{F}_1]\!](u) + \ldots + k_m [\![\mathsf{F}_m]\!](u) && \text{(general fact about power series)} \\
&= k_1 [\![q_{\mathsf{F}_1}]\!](u) + \ldots + k_m [\![q_{\mathsf{F}_m}]\!](u) && \text{(inductive hypothesis)}
\end{aligned}
$$

We also have (again assuming that $a = a_i$, and thus that $a \in f(d(\mathsf{E}))$)

$$[\![q_\mathsf{E}]\!](au) = \sum_{q' \in Q} E(q_\mathsf{E}, a, q') \left([\![q']\!](u) \right) \qquad \text{(Definition 6)}$$

and, by the definition of E in the derived term automaton, we have $E(q_\mathsf{E}, a, q_\mathsf{F}) = k$ iff $\langle k \rangle \mathsf{F} \in d(\mathsf{E})(a) = \mathsf{P}_i$. This gives (using our expansion (28) of P_i)

$$[\![q_\mathsf{E}]\!](au) = k_1 [\![q_{\mathsf{F}_1}]\!](u) + \ldots + k_m [\![q_{\mathsf{F}_m}]\!](u)$$

and completes the inductive case whenever $a = a_i$ for some i.

Finally, if $a \neq a_i$ for all i, it is easy to see that $[\![\mathsf{E}]\!](au) = 0 = [\![q_\mathsf{E}]\!](au)$, and the inductive step is now complete. $\qquad \square$

Proof (Theorem 1). Follows from Lemma 5, and the fact, resulting from the definition of I in Definition 9 in combination with (11), that $[\![\mathcal{A}_\mathsf{E}]\!] = [\![q_\mathsf{E}]\!]$. $\qquad \square$

4.1 Derived-Term Automaton Size

The smallness of the derived-term automaton for basic operators ($|\mathcal{A}_\mathsf{E}| \leq \|\mathsf{E}\|+1$) [14, Theorem 2]) no longer applies with extended operators. Let m and n be coprime integers, $\mathsf{E} := (a^m)^* \,\&\, (a^n)^*$ has width $\|\mathsf{E}\| = m + n$; it is easy to see that $|\mathcal{A}_\mathsf{E}| = mn$. It is also a classical result that the minimal (trim) automaton to recognize the language of $\mathsf{F}_n := (a + b)^* a (a + b)^n$ has 2^{n+1} states; so $\|\mathsf{F}_n^c\| = 2n + 3$, but $|\mathcal{A}_{\mathsf{F}_n^c}| = 2^{n+1} + 1$ (the additional state is the sink state needed to get a *complete* deterministic automaton before complement). Actually, when complement is used on an infinite semiring, it is not even guaranteed that the automaton is finite (see Sect. 4.3).

Theorem 2. *If $\mathbb{K}$ is finite, or if E has no complement, then $\mathcal{A}_E$ is finite.*

Proof. The proof goes in several steps. First introduce the *proper derived terms* of E, a set of expressions noted $PD(E)$, and the *derived terms* of E, $D(E) :=$ $PD(E) \cup \{E\}$. $PD(E)$ is defined inductively as in [14, Definition 3], to which we add

$$PD(E \& F) := \{E_i \& F_j \mid E_i \in PD(E), F_j \in PD(F)\}$$
$$PD(E^c) := \{((\langle k_1 \rangle E_1 + \cdots + \langle k_n \rangle E_n)^c \mid k_1, \ldots, k_n \in \mathbb{K}, E_1, \ldots, E_n \in PD(E)\}$$

Second, verify that $PD(E)$ is finite (under the proper assumptions). Third, prove that $D(E)$ is "stable by expansion", i.e., $\forall F \in D(E)$, $\mathsf{exprs}\,(d(F)) \subseteq D(E)$. Finally, observe that the states of $\mathcal{A}_E$ are therefore members of $D(E)$, which is finite. $\square$

4.2 Deterministic Automata

The exposed approach can be used to generate *deterministic* automata by *determinizing* the expansions: $\mathsf{det}(X) := \bigoplus_{a \in f(X)} \langle 1_{\mathbb{K}} \rangle \odot \mathsf{expr}\,(X_a)$. The expr operator "consolidates" a polynomial into an expression that ensures this determinism. For instance the expansion $a \odot [\langle 1_{\mathbb{K}} \rangle \odot b \oplus \langle 1_{\mathbb{K}} \rangle \odot c]$, which would yield two transitions labeled by a (one to b and the other to c) is determinized into $a \odot [\langle 1_{\mathbb{K}} \rangle \odot (b + c)]$, yielding a single transition (to $b + c$).

It is well known that some nondeterministic *weighted* automata have no deterministic equivalent, in which case determinization loops. Our construct is subject to the same condition. The expression $E := a^* + (\langle 2 \rangle a)^*$ on the alphabet $\{a\}$ admits an infinite number of derivatives: $\partial_{a^n}(E) = a^* \oplus \langle 2^n \rangle (\langle 2 \rangle a)^*$. Therefore our construction of *deterministic* automata would not terminate: the automaton is locally finite but infinite (and there is no finite deterministic automaton equivalent to E). However, a lazy implementation as available in Vcsn (see footnote 1) would uncover the automaton on demand, for instance when evaluating a word.

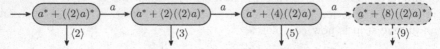

To improve determinizability, when $\mathbb{K}$ features a left-division $\backslash$, we apply the usual technique used in the weighted determinization of automata: normalize the results to keep a unique representative of colinear polynomials. Concretely, when determinizing expansions, polynomials are first normalized: $\mathsf{det}(X) :=$ $\bigoplus_{a \in f(X)} \langle |X_a| \rangle \odot \mathsf{expr}\,(|X_a| \backslash X_a)$ where, for a polynomial $P = \bigoplus_{i \in I} \langle k_i \rangle \odot E_i$, and a weight k, $k \backslash P := \bigoplus_{i \in I} \langle k \backslash k_i \rangle \odot E_i$, and the weight $|P|$ denotes some "norm" of (the coefficients of) P. For instance $|P|$ can be the GCD of the k_i (so that the coefficients are coprime), or, in the case of a field, the first non null k_i (so that the first non null coefficient is $1_{\mathbb{K}}$), or the sum of the k_i provided it's not null (so that the sum of the coefficients is $1_{\mathbb{K}}$), etc.

Example 5 (Examples 2 to 4 cont.). The deterministic derived-term automaton of E_1 using GCD-normalization is:

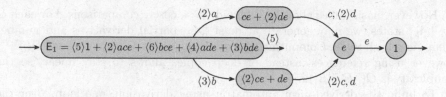

4.3 The Case of Complement

The classical algorithm to complement an (unweighted) automaton, by complementation of the set of the final states, requires a deterministic and complete automaton (which can lead to an exponential number of states). In our case "local" determinism (i.e., restricted to complemented subexpressions) is ensured by expr in the definition of the complement of an expansion in (5) and (10).

In the case of weighted expressions, we hit the same problems—and apply the same techniques—as in Sect. 4.2: not all expressions generate finite automata. A strict (non-lazy) implementation would not terminate on $\left(a^* + (\langle 2 \rangle\, a)^*\right)^c$; a lazy implementation would uncover finite portions of the automaton, on demand. However, although $\mathsf{F} := (\langle 2 \rangle\, a)^* + (\langle 4 \rangle\, aa)^*$ admits an infinite number of derivatives, F^c features only two: $\left(\langle 2 \rangle\, (\langle 2 \rangle\, a)^* + \langle 4 \rangle\, (a(\langle 4 \rangle\, aa)^*)\right)^c \Rightarrow \left(((\langle 2 \rangle\, a)^* + \langle 2 \rangle\, (a(\langle 4 \rangle\, aa)^*))\right)^c$ and itself. It is the trivial identity $(\langle k \rangle\, \mathsf{E})^c \Rightarrow \mathsf{E}^c$ that eliminates the common factor.

Example 6 (Example 1 continued). We have:

$$d(\mathsf{E}_3) = a \odot [\langle 2 \rangle \odot b \oplus \langle 3 \rangle \odot (b^c\, \&\, (a+b)^*)] \oplus b \odot [\langle 3 \rangle \odot (a+b)^*]$$

The lower part of $\mathcal{A}_{\mathsf{E}_3}$ is characteristic of the complement of a complete deterministic automaton:

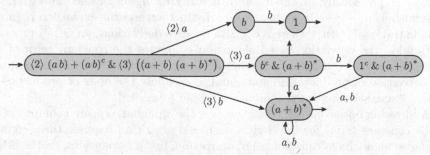

4.4 Complexity and Performances

We focus on basic expressions. Obviously, $\|\mathsf{E}\| \leq |\mathsf{E}|$, and we know $|\mathcal{A}_\mathsf{E}| \leq \|\mathsf{E}\|+1$.

The complexity of Antimirov's algorithm is $O(\|E\|^3 |E|^2)$ [7]: for each of the $|\mathcal{A}_\mathsf{E}|$ states, we may generate at most $|\mathcal{A}_\mathsf{E}|$ partial derivatives, each one to compare to the $|\mathcal{A}_\mathsf{E}|$ derived-terms. That's $O(|\mathcal{A}_\mathsf{E}|^3)$ comparisons to perform on objects of size $O(|\mathsf{E}|^2)$.

However, hash tables allow us to avoid these costly comparisons. For each of the $|\mathcal{A}_E|$ states, we may generate at most $|\mathcal{A}_E|$ partial derivatives and number them via a hash table. Computing an expansion builds an object of size $O(|E|^2)$, however using references instead of deep copies allows to stay linear, so the complexity is $O(\|E\|^2|E|)$.

To build the derived-term automaton using derivation, one loops over the alphabet for each derived term. This incurs a performance penalty with large alphabets. Let a_i, b_i, for $i \in [m]$, be distinct letters. The following table reports the duration of the process, in milliseconds, for $E_n^m := \sum_{i=1}^m (a_i + b_i)^* a_i (a_i + b_i)^n$ (right associative) by Vcsn[3]. The parameter m controls the alphabet size: 2, 128, and 254 (Vcsn reserves two `chars`) corresponding to $m = 1, 64$, and 127. The parameter n makes the expression arbitrarily long.

		n					
	m	1	10	50	100	500	1000
derivation	1	0.07	0.16	0.93	2.7	48.4	199
	64	17.1	84.0	548	1,471	23,608	87,137
	127	64.4	319	2,041	5,407	85,611	323,652
expansion	1	0.04	0.11	0.48	1.3	17.8	76
	64	1.82	6.40	39.19	100	1,368	4,951
	127	3.62	12.92	75.70	205	2,741	10,045

The expansion-based algorithm always performs better than the derivation-based one, dramatically on large alphabets on this benchmark.

One can optimize the derivation-based algorithm by computing the firsts globally [17] or locally, on-the-fly, and then derivating on this set. However, on sums such as $a_1 + \cdots + a_n$ (where a_i are distinct letters) the expansion requires a single traversal ($O(n)$) whereas one still needs n derivations, an $O(n^2)$ process.

Besides, the derivation-based algorithm computes the constant term of an expression several times: to check whether the current state is final, to compute the derivation of products and stars, and to compute the firsts of products. To fix this issue, these repeated computations can be cached.

Addressing both concerns (iteration over the alphabet, repeated computation of the constant term) for the derivation-based algorithm requires three tightly entangled algorithms (constant term, derivation, first). Expansions, on the other hand, keep them together, in a single construct, computed in a single traversal of the expression.

---

[3] Vcsn 2.2 as of 2016-05-16, compiled with Clang 3.6 with options `-O3 -DNDEBUG`, and run on a Mac OS X 10.11.4, Intel Core i7 2.9GHz, 8GB of RAM. Best run out of five.

5 Related Work

Compared to Brzozowski [4] we introduced *weighted* expansions, and their direct computation, making them the core computation of the algorithm. This was partly done for basic Boolean expressions by Antimirov [3] as "linear forms."

Our work is deeply influenced by Lombardy and Sakarovitch [14] and shares many similarities. However, by carefully avoiding the derivatives with respect to words, we get simpler proofs for Theorems 1 and 2. Besides, these proofs free us from the requirement of a yielding a finite automaton, which is needed to establish the correctness of Vcsn's on-the-fly computation of infinite derived-term automata. We also introduced the construction of deterministic (weighted) derived-term automata. Fischer et al. [11] also make profit from laziness to address non rational languages (such as $a^n b^n c^n$); however their approach is quite different: their core construction is the Glushkov automaton, and they use named rational expressions to build "infinite rational expressions", gracefully handled thanks to Haskell's laziness.

There is a striking similarity between (non-weighted) expansions, and auxiliary tools used by Mirkin to define a "prebase" of an expression [16]. See for instance the proof of Proposition 1 as reproduced by Champarnaud and Ziadi [6].

Aside from our support for weighted expressions, our approach of extended operators is comparable to that of Caron et al. [5], but, we believe, using a simpler framework. Basically, their sets of sets of expressions correspond to polynomials of conjunctions: their $\{\{E, F\}, \{G, H\}\}$ is our $E \,\&\, F \oplus G \,\&\, H$. Using our framework, the automaton of Fig. 3 [5] has one state less, since $\{E, F\}$ and $\{E \cap F\}$ both are $E \,\&\, F$. Actually, the main point of sets of sets of expressions is captured by our *distributive* definition of the conjunction of polynomials, (5), which matches that of their $\odot$ operator; indeed what they call the "natural extension" [5, Sect. 3.1] would correspond to $P_1 \,\&\, P_2 := \mathsf{expr}\,(P_1) \,\&\, \mathsf{expr}\,(P_2)$. Additional properties of $\&$ (e.g., associativity), can be enabled via new trivial identities. Like us, their $\ominus$ operator ensures that complemented expressions generate deterministic automata.

For basic (weighted) expressions, completely different approaches build the derived-term automaton with a quadratic complexity [1,8]. However, the expansion-based algorithm features some unique properties. It supports a simple and natural on-the-fly implementation. It provides insight on the built automata by labeling states with the language/series they denote (e.g., Vcsn renders derived-term automata as in Examples 4 to 6). It is a flexible framework in which new operators can be easily supported (e.g., the shuffle and infiltration operators in Vcsn), and even multitape rational expressions to generate transducers [9]. It supports the direct construction of deterministic automata. And it copes easily with alternative derivation schemes, such as the "broken derived-terms" [2].

6   Conclusion

The construction of the derived-term automaton from a weighted rational expression is a powerful technique: states have a natural interpretation (they are identified by their future: the series they compute), extended rational expressions are easily supported, determinism can be requested, and it even offers a natural lazy, on-the-fly, implementation to handle infinite automata.

To build the derived-term automaton, we generalized Brzozowski's expansions to weighted expressions, and an inductive algorithm to compute the expansion of a rational expression. The formulas on which this algorithm is built reunite as a unique entity three facets that were kept separated in previous works: constant term, firsts, and derivatives. This results in a simpler set of equations, a proof that does not require the resulting automaton to be finite, and an implementation whose complexity is independent of the size of the alphabet and even applies when it is infinite (e.g., when labels are strings, integers, etc.). Building the derived-term automaton using expansions is straightforward, even when required to be deterministic. We have also shown that using proper techniques, the complexity of the algorithm is much better that previously reported.

The computation of expansions and derivations are implemented in Vcsn (see footnote 1), together with their automaton construction procedures (possibly lazy, possibly deterministic). Our implementation actually supports for additional operators on rational expressions (e.g., shuffle and infiltration).

Acknowledgments. Interactions with A. Duret-Lutz, S. Lombardy, L. Saiu and J. Sakarovitch resulted in this work. Anonymous reviewers made very helpful comments. In particular, an anonymous reviewer of ICALP 2016 contributed the proof of Theorem 1, much simpler than the original one (which was still based on derivatives), and proposed the benchmark of Sect. 4.4.

References

1. Allauzen, C., Mohri, M.: A unified construction of the Glushkov, follow, and Antimirov automata. In: Královič, R., Urzyczyn, P. (eds.) MFCS 2006. LNCS, vol. 4162, pp. 110–121. Springer, Heidelberg (2006). doi:10.1007/11821069_10
2. Angrand, P.-Y., Lombardy, S., Sakarovitch, J.: On the number of broken derived terms of a rational expression. J. Automata Lang. Comb. **15**(1/2), 27–51 (2010)
3. Antimirov, V.: Partial derivatives of regular expressions and finite automaton constructions. TCS **155**(2), 291–319 (1996)
4. Brzozowski, J.A.: Derivatives of regular expressions. J. ACM **11**(4), 481–494 (1964)
5. Caron, P., Champarnaud, J.-M., Mignot, L.: Partial derivatives of an extended regular expression. In: Dediu, A.-H., Inenaga, S., Martín-Vide, C. (eds.) LATA 2011. LNCS, vol. 6638, pp. 179–191. Springer, Heidelberg (2011). doi:10.1007/978-3-642-21254-3_13
6. Champarnaud, J.-M., Ziadi, D.: From Mirkin's prebases to Antimirov's word partial derivatives. Fundam. Inf. **45**(3), 195–205 (2001)
7. Champarnaud, J.-M., Ziadi, D.: Canonical derivatives, partial derivatives and finite automaton constructions. TCS **289**(1), 137–163 (2002)

8. Champarnaud, J.-M., Ouardi, F., Ziadi, D.: An efficient computation of the equation $\mathbb{K}$-automaton of a regular $\mathbb{K}$-expression. In: Harju, T., Karhumäki, J., Lepistö, A. (eds.) DLT 2007. LNCS, vol. 4588, pp. 145–156. Springer, Heidelberg (2007). doi:10.1007/978-3-540-73208-2_16
9. Demaille, A.: Derived-term automata of multitape rational expressions. In: Han, Y.-S., Salomaa, K. (eds.) CIAA 2016. LNCS, vol. 9705, pp. 51–63. Springer, Heidelberg (2016). doi:10.1007/978-3-319-40946-7_5
10. Demaille, A., Duret-Lutz, A., Lombardy, S., Sakarovitch, J.: Implementation concepts in Vaucanson 2. In: Konstantinidis, S. (ed.) CIAA 2013. LNCS, vol. 7982, pp. 122–133. Springer, Heidelberg (2013). doi:10.1007/978-3-642-39274-0_12
11. Fischer, S., Huch, F., Wilke, T.: A play on regular expressions: functional pearl. In: Proceedings of the 15th ACM SIGPLAN International Conference on Functional Programming, ICFP 2010, pp. 357–368. ACM (2010)
12. Glushkov, V.M.: The abstract theory of automata. Russ. Math. Surv. **16**, 1–53 (1961)
13. Kaplan, R.M., Kay, M.: Regular models of phonological rule systems. Comput. Linguist. **20**(3), 331–378 (1994)
14. Lombardy, S., Sakarovitch, J.: Derivatives of rational expressions with multiplicity. TCS **332**(1–3), 141–177 (2005)
15. McNaughton, R., Yamada, H.: Regular expressions and state graphs for automata. IEEE Trans. Electron. Comput. **9**, 39–47 (1960)
16. Mirkin, B.G.: An algorithm for constructing a base in a language of regular expressions. Eng. Cybern. **5**, 110–116 (1966)
17. Owens, S., Reppy, J., Turon, A.: Regular-expression derivatives re-examined. J. Funct. Program. **19**(2), 173–190 (2009)
18. Rutten, J.J.M.M.: Behavioural differential equations: a coinductive calculus of streams, automata, and power series. TCS **308**(1–3), 1–53 (2003)
19. Sakarovitch, J.: Elements of Automata Theory. Cambridge University Press (2009). Corrected English translation of Éléments de théorie des automates, Vuibert, 2003

Weighted Register Automata and Weighted Logic on Data Words

Parvaneh Babari$^{(\boxtimes)}$, Manfred Droste, and Vitaly Perevoshchikov

Institut für Informatik, Universität Leipzig, 04109 Leipzig, Germany
{babari,droste,perev}@informatik.uni-leipzig.de

Abstract. In this paper, we investigate automata models for quantitative aspects of systems with infinite data domains, e.g., the costs of storing data on a remote server or the consumption of resources (e.g., memory, energy, time) during a data analysis. We introduce weighted register automata on data words and investigate their closure properties. In our main result, we give a logical characterization of weighted register automata by means of weighted existential monadic second-order logic; for the proof we employ a new class of determinizable visibly register automata.

1 Introduction

In the areas of static analysis of databases and software verification, there is much interest in processing information taken from an infinite domain. The notion of data words is a well-known concept for the modelling of these situations. A data word can be considered as a sequence of pairs where the first element is taken from a finite alphabet (as in classical words) and the second element is taken from an infinite data domain. Register automata introduced by Kaminski and Francez [18] provide a widely studied model for reasoning on data words. These automata can be considered as classical nondeterministic finite automata equipped with a finite set of registers which are used to store data in order to compare it with some data in the future. This enables them to handle parameters like user names, passwords, identifiers of connections, etc., in a fashion similar to, and slightly more expressive than, the class of *data-independent* systems. This model served as a basis for the study of various automata models and logics on data words and trees [6,9,16,17, 19]. Classical *timed automata* of Alur and Dill [2] for real-time systems are another example of automata on data (timed) words.

For many Computer Science applications, quantitative properties of systems such as costs, probabilities, vagueness and uncertainty of a statement, consumption of memory and energy are of significant importance. Weighted automata (cf. [13] for surveys) form a well-known model for quantitative aspects which has been extended to various different settings (cf., e.g., weighted timed automata [5,20] for the quantitative analysis of real-time systems). In this paper, we introduce *weighted register automata* for quantitative reasoning on data words. Motivated by the seminal Büchi-Elgot-Trakhtenbrot theorem [11] about the expressive

P. Babari—Supported by DFG Graduiertenkolleg 1763 (QuantLA).

A. Sampaio and F. Wang (Eds.): ICTAC 2016, LNCS 9965, pp. 370–384, 2016.
DOI: 10.1007/978-3-319-46750-4_21

equivalence of finite automata and monadic second-order (MSO) logic and by the weighted MSO logic of Droste and Gastin [12], we introduce *weighted MSO logic on data words* and give a logical characterization of weighted register automata.

To the best of our knowledge, quantitative extensions of register automata have not been studied yet. However, there is a plenty of quantitative models which are quite natural in the context of register automata. We give several examples. For instance, since register automata process data taken from an infinite domain, there arises a natural question about the efficiency of the processing of big data which can invoke big costs for data storing and data comparison. As another example, one may ask about the maximal size of data which will be stored in registers along a computation of a register automaton. Thus, our goal is to develop a model which will reflect this kind of data-dependent costs. Now we give a summary of our results.

We introduce weighted register automata over *commutative data semirings* equipped with a collection of binary data functions in the spirit of the classical theory of weighted automata [13]. Whereas in the models of register automata known from the literature data are usually compared with respect to equality or a linear order, here we allow data comparison by means of an arbitrary collection of binary data relations. This approach permits easily to incorporate timed automata [2] and weighted timed automata [5,20] into our framework. Moreover, this approach gives rise to the further investigations of the models for data processing, e.g., data comparison with an approximation error.

We introduce semiring-weighted existential MSO logic on data words equipped with binary data functions. Recall that weighted MSO logic on classical words is introduced and investigated in [12]. In order to model data comparison, we use the data predicates in the spirit of relative distance predicates of Wilke [21]. Our goal is to prove the expressive equivalence of this new logic with weighted register automata. To reach this goal, we faced the following difficulties.

- The unrestricted use of binary data functions and weighted universal quantifiers goes beyond recognizability even for very simple formulas.
- Register automata are neither determinizable nor closed under complement.

In order to overcome these problems, we obtain the suitable fragment of our weighted existential MSO logic by restricting the use of the weighted universal quantifier to formulas without weighted quantifiers and by restricting the use of data functions to an intuitively defined logical operator. In our main result, we state that this restricted weighted EMSO-logic is equivalent to weighted register automata. The existing proof techniques and ideas for weighted logic cannot be applied in our setting, since register automata are not closed under complement and we need a new construction to deal with binary data functions. For this purpose, we introduce a determinizable class of unweighted register automata, called visibly register automata. In visibly register automata, the transition label determines which registers will be updated when taking a transition. This determinizable class of register automata could be also of independent interest. Moreover, we introduce a new normalization technique for the binary data functions, in order to provide a translation of formulas with weighted universal quantifiers into weighted register automata. With this, we achieve our goal; moreover, our construction of weighted register automata equivalent to a given weighted MSO formula is effective.

2 Register Automata

Register automata are nondeterministic finite state automata on data words equipped with a finite set of registers for storing data (from an infinite data domain) and comparison of new data instances with the already stored data. Whereas in the original definition of register automata [18] it is only possible to check the equality of data, the later models augment a data domain with a linear order and allow to compare data with respect to this linear order. However, more complicated situations of data comparison are reasonable as well (cf., e.g., timed automata [2] or Example 2.2). In this paper, we consider the model of register automata where we augment a data domain with some arbitrary binary relations which will be used for data comparison. Notice that our register automata also incorporate timed automata [2].

For a set X, let $\mathcal{P}(X) = \{Y \mid Y \subseteq X\}$, the *powerset* of X. A *data structure* is a pair $\mathbb{D} = \langle D, \mathcal{R} \rangle$ where D is an arbitrary set called a *data domain* and $\mathcal{R}$ is a set of binary relations on D called *data relations*. For the convenience of presentation, we assume throughout all of this paper that D contains a designated initial data value $\perp \in D$ which will be the initial data value of all registers. An *alphabet* is a non-empty finite set. Let Σ be an alphabet and $\mathbb{D} = \langle D, \mathcal{R} \rangle$ a data structure. A *data word* over Σ and $\mathbb{D}$ is a finite sequence $w = (a_1, d_1)...(a_n, d_n)$ where $a_1, ..., a_n \in \Sigma$ and $d_1, ..., d_n \in D$. Let $\mathbb{D}\Sigma^+ = (\Sigma \times D)^+$ stand for the set of all data words over Σ and $\mathbb{D}$. Any subset $\mathcal{L} \subseteq \mathbb{D}\Sigma^+$ is called a *data language*.

Let Reg be a finite set of *registers* which take values from the data domain D of the data structure $\mathbb{D}$. A *register valuation* over Reg and $\mathbb{D}$ is any mapping $\vartheta : \text{Reg} \rightarrow D$. Let $\text{Val}(\text{Reg}, \mathbb{D})$ denote the collection of all register valuations over Reg and $\mathbb{D}$. For a register valuation $\vartheta \in \text{Val}(\text{Reg}, \mathbb{D})$, a subset $\Lambda \subseteq \text{Reg}$ and a data value $d \in D$, the *update* $\vartheta[\Lambda := d]$ is the register valuation in $\text{Val}(\text{Reg}, \mathbb{D})$ defined for all registers r by $\vartheta[\Lambda := d](r) = d$ if $r \in \Lambda$ and $\vartheta[\Lambda := d](r) = \vartheta(r)$ otherwise. The set $\text{Guard}(\text{Reg}, \mathbb{D})$ of *register guards* over Reg and $\mathbb{D}$ is defined by the grammar

$$\phi ::= \text{True} \mid Rr \mid \phi \wedge \phi \mid \neg\phi$$

where $r \in \text{Reg}$ and R is a data relation in D. Given a register valuation $\vartheta \in \text{Val}(\text{Reg}, \mathbb{D})$, a data value $d \in D$ and a register guard $\phi \in \text{Guard}(\text{Reg}, \mathbb{D})$, the *satisfaction relation* $(\vartheta, d) \models \phi$ is defined inductively on the structure of ϕ as follows: $(\vartheta, d) \models \text{True}$ always holds; $(\vartheta, d) \models Rr$ iff $(\vartheta(r), d) \in R$; $(\vartheta, d) \models \phi_1 \wedge \phi_2$ iff $(\vartheta, d) \models \phi_1$ and $(\vartheta, d) \models \phi_2$; $(\vartheta, d) \models \neg\phi$ iff $(\vartheta, d) \models \phi$ does not hold.

Definition 2.1. Let Σ be an alphabet and $\mathbb{D}$ a data structure. A *register automaton* over Σ and $\mathbb{D}$ is a tuple $\mathcal{A} = (Q, \text{Reg}, I, T, F)$ where Q is a finite set of *states*, Reg is a finite set of *registers*, $I, F \subseteq Q$ are sets of *initial* resp. *final* states, and $T \subseteq Q \times \Sigma \times \text{Guard}(\text{Reg}, \mathbb{D}) \times 2^{\text{Reg}} \times Q$ is a finite set of *transitions*.

We denote a transition $t = (q, a, \phi, \Lambda, q') \in T$ by $q \xrightarrow[\phi,\Lambda]{a} q'$. Let $\mathsf{label}(t) = a \in \Sigma$, the *label* of t. A *configuration* of $\mathcal{A}$ is a pair $c = \langle q, \vartheta \rangle$ consisting of a state $q \in Q$ and a register valuation $\vartheta \in \mathsf{Val}(\mathsf{Reg}, \mathbb{D})$. We say that c is *initial* if $q \in I$ and $\vartheta(r) = \bot$ for all $r \in \mathsf{Reg}$. We call c *final* if $q \in F$. Let $c = \langle q, \vartheta \rangle$ and $c' = \langle q', \vartheta' \rangle$ be configurations of $\mathcal{A}$, $t \in T$ a transition of the form $p \xrightarrow[\phi,\Lambda]{a} p'$, and $d \in D$ a data value. We say that $c \vdash_{t,d} c'$ is a *switch* from c to c' via the transition t and the data value d if $p = q$, $p' = q'$, $(\vartheta, d) \models \phi$ and $\vartheta' = \vartheta[\Lambda := d]$. Note that c' is uniquely determined by c, t and d. A *run* ρ of $\mathcal{A}$ is a non-empty sequence of switches between configurations starting in an initial configuration and ending in a final configuration. Formally, ρ is a sequence of the form $c_0 \vdash_{t_1,d_1} c_1 \vdash_{t_2,d_2} \cdots \vdash_{t_n,d_n} c_n$ where $n \geq 1$, c_0 is an initial configuration, $c_1, ..., c_{n-1}$ are configurations, c_n is a final configuration, $t_1, ..., t_n \in T$ and $d_1, ..., d_n \in D$. Let $\mathsf{label}(\rho) = (\mathsf{label}(t_1), d_1)...(\mathsf{label}(t_n), d_n) \in \mathbb{D}\Sigma^+$ be the *label* of ρ. For any data word $w \in \mathbb{D}\Sigma^+$, let $\mathsf{Run}_\mathcal{A}(w)$ denote the set of all runs of $\mathcal{A}$ with label w. Let $\mathcal{L}(\mathcal{A}) = \{w \in \mathbb{D}\Sigma^+ \mid \mathsf{Run}_\mathcal{A}(w) \neq \emptyset\}$, the data language *recognized* by $\mathcal{A}$.

Example 2.2. Now we give some examples of data structures for register automata. For any data domain D, let $(=_D) \subseteq D \times D$ denote the data relation $\{(d, d) \mid d \in D\}$.

(a) Let D be any non-empty set. Then, $\mathbb{D} = (D, \{=_D\})$ is a data structure which corresponds to the original model of register automata [18]. Note that the register automata of [18] are also equipped with an initial vector of data values. In order to model this feature in our setting, we can extend the set of data relations of $\mathbb{D}$ with the set $\{R_d \mid d \in D\}$ where $R_d = \{(d', d) \mid d' \in D\}$ for all $d \in D$.

(b) Let $(D, <_D)$ be a linear order with a non-empty set D. Then, $(D, \{=_D, <_D\})$ is a data structure for the register automata considered in [17].

(c) In various situations, exact data values are not known and we deal with their approximated values, e.g., obtained from some experiments. Therefore, it can be reasonable to compare data values with respect to a given approximation error. For instance, let the data domain D be the set of all rational numbers. For any nonnegative rational number ε, let $R_\varepsilon = \{(q, q') \mid q, q' \in D \text{ and } |q - q'| \leq \varepsilon\}$. Note that R_0 is equal to $=_D$. Then, $(D, \{R_\varepsilon \mid \varepsilon \geq 0\})$ is a data structure for register automata with approximated values.

(d) In this example we show that *timed automata* [2] are also included in our model. The idea is that instead of clock resets and clock constraints of the form $x \bowtie k$ (where x is a clock, $\bowtie \in \{<, =, >\}$ and $k \in \mathbb{N}$) we can record instants when the clock x is reset in a register r_x and replace $x \bowtie k$ by the constraint $t - r_x \bowtie k$ where t is the current time moment. Let D be the set of non-negative real numbers. For $\bowtie \in \{<, =, >\}$ and $k \in \mathbb{N}$, let $R_{\bowtie k} = \{(t, t') \mid t, t' \in D \text{ and } t' - t \bowtie k\}$. Note that $R_{=0}$ is equal to $=_D$. Then, $\mathbb{D}^{\mathsf{Timed}} = (D, \{R_{\bowtie k} \mid \bowtie \in \{<, =, >\} \text{ and } k \in \mathbb{N}\})$ is a data structure for timed automata; here we take $\bot = 0$ as the initial data value. Note that register automata over $\mathbb{D}^{\mathsf{Timed}}$ can accept non-monotonic data sequences (which do not correspond to timed words). These difficulties can be avoided if we add a special register $\tilde{r}$ which will control the monotonicity of a data word. The register $\tilde{r}$ is updated after taking every transition and the constraint $(R_{\geq 0})\tilde{r}$ is added to every transition.

3 Weighted Register Automata

In this section, we introduce *weighted register automata* as a quantitative model for reasoning about data words. They extend the qualitative register automata of the previous section with weights and reflect the following quantitative information:

- As in classical weighted automata [13], transitions of our model also carry weights which do not depend on data.
- As opposed to weighted automata, weighted register automata must be able to process data taken from an infinite data domain. Therefore, the size of data can be very large and processing of such data can be expensive. Our weighted model takes into account the costs of data processing.

Note that our weighted register automata model is different from the cost register automata model of Alur et al. [1], because cost register automata run on words over a finite alphabet and the registers are used to compute the weight of a run and can be updated depending on the previous register values. In contrast, in our model we deal with data words over an infinite alphabet and the registers can be updated only depending on the current input data value.

In order to be able to reflect various quantitative settings, we will consider a general structure for weighted register automata. For this purpose, we adopt the structure of semirings (as in the classical weighted automata [13]) to our new setting of data words.

For any sets X, Y, let Y^X denote the collection of all mappings $f : X \to Y$. For $y \in Y$, let y^X denote the mapping $y^X : X \to \{y\}$. A *data semiring* over a data structure $\mathbb{D} = \langle D, \mathcal{R} \rangle$ is a pair $\mathbb{S} = \langle \mathcal{S}, \mathcal{F} \rangle$ where $\mathcal{S} = (S, +, \cdot, 0, 1)$ is a semiring and $\mathcal{F} \subseteq S^{D \times D}$ such that $1^{D \times D} \in \mathcal{F}$. We call $\mathbb{S}$ *commutative* if $\mathcal{S}$ is a commutative semiring, i.e., $s \cdot s' = s' \cdot s$ for all $s, s' \in S$.

Definition 3.1. Let Σ be an alphabet, $\mathbb{D}$ a data structure and $\mathbb{S} = \langle (S, +, \cdot, 0, 1), \mathcal{F} \rangle$ a data semiring over $\mathbb{D}$. A *weighted register automaton (WRA)* over Σ, $\mathbb{D}$ and $\mathbb{S}$ is a tuple $\mathcal{A} = (Q, \mathsf{Reg}, I, T, F, \mathsf{wt})$ where $(Q, \mathsf{Reg}, I, T, F)$ is a register automaton over Σ and $\mathbb{D}$, and $\mathsf{wt} = \langle \mathsf{wt}_{\mathsf{trans}}, \mathsf{wt}_{\mathsf{data}} \rangle$ is a pair of weight functions such that $\mathsf{wt}_{\mathsf{trans}} : T \to S$ and $\mathsf{wt}_{\mathsf{data}} : (T \times \mathsf{Reg}) \to \mathcal{F}$.

Note that $\mathsf{wt}_{\mathsf{trans}} : T \to S$ can be considered as a weight function of the classical weighted automata [13] and describes data-independent costs for transitions. The weight function $\mathsf{wt}_{\mathsf{data}}$ assigns to every transition $t \in T$, every register $r \in \mathsf{Reg}$, every data value d stored in r and every new data value $d' \in D$ the cost $\mathsf{wt}_{\mathsf{data}}(t, r)(d, d') \in S$ of data processing in the register r which includes the cost of comparison d with d' and, if necessary, the cost of storing d' in the register r.

Let $c = \langle q, \vartheta \rangle$ and $c' = \langle q', \vartheta' \rangle$ be configurations of $\mathcal{A}$ and $c \vdash_{t,d} c'$ a switch between them. Then, its *weight* is defined as the product of the costs of data processing in the registers (defined by $\mathsf{wt_{data}}$) and the transition cost of $\mathsf{wt_{trans}}(t)$. Formally, we let $\mathsf{wt}(c \vdash_{t,d} c') = \prod_{r \in \mathsf{Reg}} \mathsf{wt_{data}}(t, r)(\vartheta(r), d) \cdot \mathsf{wt_{trans}}(t)$. Now let $\rho = (c_0 \vdash_{t_1,d_1} c_1 \vdash_{t_2,d_2} \cdots \vdash_{t_n,d_n} c_n)$ be a run of $\mathcal{A}$. Then, the *weight* of ρ is defined as the product of the weights of all switches of ρ. Formally, we let $\mathsf{wt}(\rho) = \prod_{i=1}^{n} \mathsf{wt}(c_{i-1} \vdash_{t_i,d_i} c_i)$. Then, the *behavior* of $\mathcal{A}$ is the mapping $[\![\mathcal{A}]\!] : \mathbb{D}\Sigma^+ \to S$ defined for all $w \in \mathbb{D}\Sigma^+$ by $[\![\mathcal{A}]\!](w) = \sum (\mathsf{wt}(\rho) \mid \rho \in \mathsf{Run}_{\mathcal{A}}(w))$. We will call any mapping $\mathbb{L} : \mathbb{D}\Sigma^+ \to S$ a *data series* over Σ, $\mathbb{D}$ and $\mathbb{S}$. Let $\mathbb{S}\langle\!\langle \mathbb{D}\Sigma^+ \rangle\!\rangle$ denote the collection of all data series over Σ, $\mathbb{D}$ and $\mathbb{S}$. We say that $\mathbb{L} \in \mathbb{S}\langle\!\langle \mathbb{D}\Sigma^+ \rangle\!\rangle$ is *recognizable* if there exists a WRA $\mathcal{A}$ over Σ, $\mathbb{D}$ and $\mathbb{S}$ such that $[\![\mathcal{A}]\!] = \mathbb{L}$.

Example 3.2.

(a) Consider the *arctic semiring* $\mathsf{Arc} = (\mathbb{N} \cup \{-\infty\}, \max, +, -\infty, 0)$ of natural numbers. Let $\mathbb{D} = \langle D, \mathcal{R} \rangle$ be a data structure augmented with a *size function* $\mathsf{size} : D \to \mathbb{N}$ (e.g., length of a data string or number of bits of an integer). Let $\mathcal{F}_1$ be the collection of functions $f_c : D \times D \to \mathbb{N}$ with $c \in \mathbb{N}$ and $f_c(d, d') = c \cdot \mathsf{size}(d')$ for all $d, d' \in D$. The collection $\mathcal{F}_1$ can be useful, e.g., for the cases where we need to estimate the costs of checking the equality of data or to update a register. Then $\langle \mathsf{Arc}, \mathcal{F}_1 \rangle$ is a data semiring. Alternatively, we can consider the collection $\mathcal{F}_2$ of functions $g_{k,l} : D \times D \to \mathbb{N}$ with $k, l \in \mathbb{N}$ and $g_{k,l}(d, d') = k \cdot \mathsf{size}(d) + l \cdot \mathsf{size}(d')$ for all $d, d' \in D$. The collection $\mathcal{F}_2$ can be useful, e.g., for the cases where we need to estimate the costs of finding a pattern in a data string (e.g., using the well-known Knuth-Morris-Pratt algorithm). Then $\langle \mathsf{Arc}, \mathcal{F}_2 \rangle$ is a data semiring.

(b) A semiring *weighted timed automata* (WTA) model of [5] was investigated in [20]. We show that WTA can be simulated by our WRA model. Let $\mathcal{S} = (S, +, \cdot, 0, 1)$ be an arbitrary semiring. As discussed in Example 2.2(d), we take the data structure $\mathbb{D}^{\mathsf{Timed}}$ and use a designated register $\tilde{r}$ which will be updated after taking every transition. Note that WTA of [5] have location weight functions from a family $\mathcal{G} \subseteq S^{\mathbb{R}_{\geq 0}}$. In the case of WRA, we can simulate a unary function $g \in \mathcal{G}$ by, e.g., the binary function $f_g \in S^{\mathbb{R}_{\geq 0} \times \mathbb{R}_{\geq 0}}$ such that, for all $t, t' \in \mathbb{R}_{\geq 0}$, $f_g(t, t') = g(t' - t)$ if $t \leq t'$ and $f_g(t, t') = 1$ otherwise. Note that $\mathbb{S} = \langle \mathcal{S}, \{f_g \mid g \in \mathcal{G}\} \rangle$ is a data semiring. Then, the location weight functions g can be reflected as $\mathsf{wt_{data}}(t, \tilde{r}) = f_g$ where t is a transition of a WRA over Timed and $\mathbb{S}$.

Example 3.3. Consider the data structure $\langle D, \{=_D\} \rangle$ of Example 2.2 (a) and the alphabet $\Sigma = \{a\}$. For $w \in \mathbb{D}\Sigma^+$ and $d \in D$, let $|w|_d \in \mathbb{N}$ be the number of d's in w. Let $\mathbb{L} : \mathbb{D}\Sigma^+ \to \mathbb{N}$ be a data series defined by $\mathbb{L}(w) = \max_{d \in D} |w|_d$. Consider the data semiring $\mathbb{S} = \langle \mathsf{Arc}, \{1^{D \times D}\} \rangle$ and the WRA $\mathcal{A}$, with a single register r, over Σ, $\mathbb{D}$ and $\mathbb{S}$ depicted in Fig. 1. Moreover, for all transitions t of $\mathcal{A}$, we put $\mathsf{wt_{data}}(t, r) = 1^{D \times D}$. Then $[\![\mathcal{A}]\!] = \mathbb{L}$. Note that in Fig. 1 we omit the transition label a, register guard True and the empty register update; $\mathsf{upd}(r)$ means that the register r is updated.

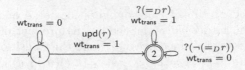

Fig. 1. The WRA $\mathcal{A}$ of Example 3.3

We establish some basic closure properties for the class of recognizable data series. We will apply them in the proof of our logical characterization result. Note that the class of timed series recognizable by weighted timed automata of [20] is not stable under the Hadamard product even in the case of commutative semirings (cf. Example 5 of [20]). Interestingly, our model of WRA extends weighted timed automata and the class of recognizable data languages is closed under the Hadamard product in the case of commutative semiring. This is due to the fact that in WRA we assign a data-dependent weight to every register.

Let Σ be an alphabet, $\mathbb{D} = \langle \mathcal{D}, \mathcal{R} \rangle$ a data structure and $\mathbb{S} = \langle (S, +, \cdot, 0, 1), \mathcal{F} \rangle$ a commutative data semiring over $\mathbb{D}$. Let $\mathbb{L}_1, \mathbb{L}_2 \in \mathbb{S}\langle\!\langle \mathbb{D}\Sigma^+ \rangle\!\rangle$ be data series. The *sum* $\mathbb{L}_1 + \mathbb{L}_2 \in \mathbb{S}\langle\!\langle \mathbb{D}\Sigma^+ \rangle\!\rangle$ and the *Hadamard product* $\mathbb{L}_1 \odot \mathbb{L}_2 \in \mathbb{S}\langle\!\langle \mathbb{D}\Sigma^+ \rangle\!\rangle$ are defined by $(\mathbb{L}_1 + \mathbb{L}_2)(w) = \mathbb{L}_1(w) + \mathbb{L}_2(w)$ respectively $(\mathbb{L}_1 \odot \mathbb{L}_2)(w) = \mathbb{L}_1(w) \cdot \mathbb{L}_2(w)$ for all $w \in \mathbb{D}\Sigma^+$. Let Γ be an alphabet and $h : \Gamma \to \Sigma$ a mapping called henceforth a *renaming*. For a data word $u = (a_1, d_1)...(a_n, d_n) \in \mathbb{D}\Gamma^+$, let $h(u) = (h(a_1), d_1)...(h(a_n), d_n) \in \mathbb{D}\Sigma^+$. For a data series $\mathbb{L} \in \mathbb{S}\langle\!\langle \mathbb{D}\Gamma^+ \rangle\!\rangle$, the *renaming* $h(\mathbb{L}) \in \mathbb{S}\langle\!\langle \mathbb{D}\Sigma^+ \rangle\!\rangle$ is defined for all $w \in \mathbb{D}\Sigma^+$ by $h(\mathbb{L})(w) = \sum (\mathbb{L}(u) \mid u \in \mathbb{D}\Gamma^+$ and $h(u) = w)$. For a data series $\mathbb{L} \in \mathbb{S}\langle\!\langle \mathbb{D}\Sigma^+ \rangle\!\rangle$, the *inverse renaming* $h^{-1}(\mathbb{L}) \in \mathbb{S}\langle\!\langle \mathbb{D}\Gamma^+ \rangle\!\rangle$ is defined for all $u \in \mathbb{D}\Gamma^+ \to S$ by $h^{-1}(\mathbb{L})(u) = \mathbb{L}(h(u))$.

Lemma 3.4. *The class of data series recognizable over Σ, $\mathbb{D}$ and $\mathbb{S}$ is closed under the sum, Hadamard product, renaming and inverse renaming.*

4   Weighted Existential MSO Logic for Data Words

In this section we introduce weighted existential monadic second-order (wEMSO) logic over data semirings for data words augmented with binary data functions. Then we show that a suitable fragment of our weighted logic and our weighted register automata model are expressively equivalent. As in [8], in order to describe easily boolean properties, we introduce two levels of formulas: boolean and weighted. We operate with the boolean formulas as in the usual logic. On the weighted level, we add weights and binary functions from a data semiring and extend the logical operations by computations in the data semiring.

4.1   Weighted Existential MSO Logic

Let V_1 and V_2 be countable pairwise disjoint sets of first-order and second-order variables. Let $V = V_1 \cup V_2$. Let Σ be an alphabet, $\mathbb{D} = (D, \mathcal{R})$ a data structure with the initial data value $\perp \in D$ and $\mathbb{S} = \langle (S, +, \cdot, 0, 1), \mathcal{F} \rangle$ a data semiring over $\mathbb{D}$. *Weighted first-order logic* $\mathsf{wFO}(\Sigma, \mathbb{D}, \mathbb{S})$ over Σ, $\mathbb{D}$ and $\mathbb{S}$ is defined by the grammar

$$\beta ::= P_a(x) \mid x \leq y \mid x \in X \mid R(X, x) \mid \beta \vee \beta \mid \neg \beta \mid \exists x.\beta$$
$$\varphi ::= \beta \mid s \mid f(x, y) \mid f(\bot, y) \mid \varphi \oplus \varphi \mid \varphi \otimes \varphi \mid \bigoplus x.\varphi \mid \bigotimes x.\varphi$$

where $a \in \Sigma$, $x, y \in V_1$, $X \in V_2$, $R \in \mathcal{R}$, $s \in S$ and $f \in \mathcal{F}$. The formulas β are called *boolean* over Σ and $\mathbb{D}$. Let $\mathsf{Bool}(\Sigma, \mathbb{D})$ denote the set of all boolean formulas. Note that a formula $R(X, x)$ relates to a relative distance formula of Wilke [21] and reflects the performance of a register r: here the second-order variable X keeps track of positions where r is updated; moreover, at the position x the register guard Rr is checked. Using boolean formulas, we define the boolean formulas $x < y$, $x = y$, $x \notin X$, $\beta_1 \wedge \beta_2$, $\forall x.\beta$, $\beta_1 \rightarrow \beta_2$ and $\beta_1 \leftrightarrow \beta_2$ as usual.

Weighted existential MSO logic $\mathsf{wEMSO}(\Sigma, \mathbb{D}, \mathbb{S})$ over Σ, $\mathbb{D}$ and $\mathbb{S}$ is defined to be the set of all formulas of the form $\bigoplus X_1 ... \bigoplus X_n.\varphi$ where $n \geq 0$, $X_1, ..., X_n$ are second-order variables and $\varphi \in \mathsf{wFO}(\Sigma, \mathbb{D}, \mathbb{S})$. Given a formula $\psi \in \mathsf{wEMSO}(\Sigma, \mathbb{D}, \mathbb{S})$, the set $\mathsf{Free}(\psi) \subseteq V$ of *free variables* of ψ is defined as usual. We say that ψ is a *sentence* if $\mathsf{Free}(\psi) = \emptyset$.

Let $w = (a_1, d_1)...(a_n, d_n) \in \mathbb{D}\Sigma^+$ be a data word. Let $\mathsf{dom}(w) = \{1, ..., n\}$, the *domain* of w. A *w-assignment* is a mapping $\sigma : V \rightarrow \mathsf{dom}(w) \cup \mathcal{P}(\mathsf{dom}(w))$ which maps first-order variables to elements in $\mathsf{dom}(w)$ and second-order variables to subsets of $\mathsf{dom}(w)$. For a first-order variable x and a position $i \in \mathsf{dom}(w)$, the w-assignment $\sigma[x/i]$ is defined on $V \setminus \{x\}$ as σ, and we let $\sigma[x/i](x) = i$. We also let $\sigma[x/i] \upharpoonright_{V \setminus \{x\}} = \sigma \upharpoonright_{V \setminus \{x\}}$. For a second-order variable X and $I \subseteq \mathsf{dom}(w)$, the w-assignment $\sigma[X/I]$ is defined similarly. Given a formula $\beta \in \mathsf{Bool}(\Sigma, \mathbb{D})$ and a w-assignment σ, the satisfaction relation $(w, \sigma) \models \beta$ is defined by induction on the structure of β as usual where, for new formulas of the form $R(X, x)$, we let $(w, \sigma) \models R(X, x)$ iff, letting $d_0 = \bot$, for the greatest position $i \in \sigma(X) \cup \{0\}$ with $i < \sigma(x)$ we have $(d_i, d_{\sigma(x)}) \in R$. Since the satisfaction relation depends only on values of free variables, we will abuse notation and also write $(w, \sigma|_U) \models \beta$ for any $U \subseteq V$ with $\mathsf{Free}(\beta) \subseteq U$.

Let $\mathbb{D}\Sigma_V^+$ denote the set of all pairs (w, σ) where $w \in \mathbb{D}\Sigma^+$ and σ is a w-assignment. Given a formula $\psi \in \mathsf{wEMSO}(\Sigma, \mathbb{D}, \mathbb{S})$, the *semantics* of ψ is the mapping $[\![\psi]\!]_V : \mathbb{D}\Sigma_V^+ \rightarrow S$ defined for all $(w, \sigma) \in \mathbb{D}\Sigma_V^+$ with $w = (a_1, d_1)...(a_n, d_n)$ as shown in Table 1. If ψ is a sentence, then we can ignore the w-assignments in the definition of the semantics and consider it as the data series $[\![\psi]\!] : \mathbb{D}\Sigma^+ \rightarrow S$.

Table 1. The semantics of wEMSO-formulas

$$[\![\beta]\!](w, \sigma) = \begin{cases} 1, & \text{if } (w, \sigma) \models \beta, \\ 0, & \text{otherwise} \end{cases}$$

$$[\![s]\!](w, \sigma) = s$$

$$[\![f(x, y)]\!](w, \sigma) = f(d_{\sigma(x)}, d_{\sigma(y)})$$

$$[\![f(\bot, y)]\!](w, \sigma) = f(\bot, d_{\sigma(y)})$$

$$[\![\varphi_1 \oplus \varphi_2]\!](w, \sigma) = [\![\varphi_1]\!](w, \sigma) + [\![\varphi_2]\!](w, \sigma)$$

$$[\![\varphi_1 \otimes \varphi_2]\!](w, \sigma) = [\![\varphi_1]\!](w, \sigma) \cdot [\![\varphi_2]\!](w, \sigma)$$

$$[\![\bigoplus x.\varphi]\!](w, \sigma) = \sum_{i \in \mathsf{dom}(w)} [\![\varphi]\!](w, \sigma[x/i])$$

$$[\![\bigotimes x.\varphi]\!](w, \sigma) = \prod_{i \in \mathsf{dom}(w)} [\![\varphi]\!](w, \sigma[x/i])$$

$$[\![\bigoplus X.\varphi]\!](w, \sigma) = \sum_{I \subseteq \mathsf{dom}(w)} [\![\varphi]\!](w, \sigma[X/I])$$

Example 4.1. Consider the data series $\mathbb{L}$ of Example 3.3. Note that $\mathbb{L}$ is definable by the wEMSO$(\Sigma, \mathbb{D}, \mathbb{S})$-sentence $\bigoplus X.\bigoplus x.[(X = \{x\}) \otimes \bigotimes (y > x).([R(X, y) \otimes 1] \oplus \neg R(X, y))]$ where $X = \{x\}$ is an abbreviation for the formula $\forall z.(z \in X \leftrightarrow z = x)$, $\bigotimes (y > x).\varphi$ abbreviates the formula $\bigotimes y.([(y > x) \otimes \varphi] \oplus [y \leq x])$, and $R = (=_D)$.

4.2 Restricted wEMSO

Our goal is to study the connection between weighted register automata and our new weighted logic on data words. Similarly to the result of [12], the unrestricted use of formulas of the form $\bigotimes x.\varphi$ leads to unrecognizable data series. Below we give a further example of unrecognizability which is specific for wEMSO on data words.

Example 4.2. Let $\Sigma = \{a\}$ be a singleton alphabet and $\mathbb{D}$ be a data structure with the data domain $\mathbb{N}$. Consider the data semiring $\mathbb{S} = \langle \text{Arc}, \mathcal{F} \rangle$ where Arc is the arctic semiring of Example 3.2 (a) and $\mathcal{F} = \{0^{\mathbb{N} \times \mathbb{N}}, f\}$ where $f : \mathbb{N} \times \mathbb{N} \to \mathbb{N}$ is defined by $f(n, n') = n$ for all $n, n' \in \mathbb{N}$. Consider the sentence $\varphi \in$ wEMSO$(\Sigma, \mathbb{D}, \mathbb{S})$ defined by $\varphi = \bigoplus x.\bigoplus y.f(x, y)$. Note that, for every $n \in \mathbb{N}$ and the data word $w_n = (a, n) \in \mathbb{D}\Sigma^+$ of length 1, we have $[\![\varphi]\!](w_n) = n$. Now suppose that there exists a WRA $\mathcal{A}$ over Σ, $\mathbb{D}$ and $\mathbb{S}$ with $[\![\mathcal{A}]\!] = [\![\varphi]\!]$. Then there exists a constant $M \in \mathbb{N}$ such that $[\![\mathcal{A}]\!](w_n) \leq M$ for all $n \in \mathbb{N}$. A contradiction.

Now we investigate a fragment of wEMSO which is expressively equivalent to WRA. We follow the approach of Wilke [21] for a logical characterization of timed automata where the use of every expressive time distance predicate dist$(x, y) \bowtie k$ with $x, y \in V_1$, $\bowtie \in \{<, =, >\}$ was restricted to relative time distance predicates dist$(X, y) \bowtie k$ with $X \in V_2$ (note that the relative time distance predicates correspond to the formulas $R(X, y)$ in our logic Bool$(\Sigma, \mathbb{D})$). We replace the formulas $f(x, y)$ by the formulas $f(X, y)$ whose semantics is defined in a similar manner as for $R(X, y)$ as follows. For $f \in \mathcal{F}$, a first-order variable x and a second-order variable X, let $f(X, x)$ denote the wFO$(\Sigma, \mathbb{S}, \mathbb{D})$-formula $\bigoplus y.(\beta(X, x, y) \otimes f(y, x)) \oplus (\beta'(X, x) \otimes f(\perp, x))$ where $\beta(X, x, y)$ is the boolean formula $y \in X \wedge y < x \wedge \forall z.([y < z \wedge z < x] \to z \notin X)$ and $\beta'(X, x)$ is the boolean formula $\forall y.(y < x \to y \notin X)$. Then, for all $(w, \sigma) \in \mathbb{D}\Sigma^+$ with $w = (a_1, d_1)...(a_n, d_n)$, we have $[\![f(X, x)]\!](w, \sigma) = f(d_i, d_{\sigma(x)})$ where $i \in \sigma(X) \cup \{0\}$ is the greatest position with $i < \sigma(x)$ and $d_0 = \perp$.

The following example shows that the use of our new logical operator $f(X, x)$ in the scope of a weighted quantifier $\bigotimes y$ with $y \neq x$ also goes beyond recognizability by WRA.

Example 4.3. Let $\Sigma = \{a\}$ be an alphabet and $\mathbb{D}$ be a data structure with the data domain $\mathbb{N}$. Consider the data semiring $\mathbb{S} = \langle \text{Arc}, \mathcal{F} \rangle$ with $\mathcal{F} = \{0^{\mathbb{N} \times \mathbb{N}}, f\}$ where f is defined for all $n, n' \in \mathbb{N}$ by $f(n, n') = n'$. Consider the sentence $\varphi \in$ wEMSO$(\Sigma, \mathbb{D}, \mathbb{S})$ defined by $\varphi = \bigoplus X.\bigoplus x.(\beta(X, x) \otimes \bigotimes y.f(X, x))$ where $y \neq x$ and the boolean formula $\beta(X, x)$ is defined as $\forall y.y \leq x \wedge \forall y.(y \in X \leftrightarrow \forall z.y \leq z)$. Note that $\beta(X, x)$ describes that x is the last position of a data word and X is the set containing only the first position of a data word. For any $n \geq 2$ and the data word $w_n = (a,0)^{n-1}(a,n) \in \mathbb{D}\Sigma^+$, we have $\llbracket \varphi \rrbracket(w_n) = n^2$. Suppose that there exists a WRA $\mathcal{A}$ over Σ, $\mathbb{D}$ and $\mathcal{S}$ with $\llbracket \mathcal{A} \rrbracket = \llbracket \varphi \rrbracket$. Then, there exists a constant $M \in \mathbb{N}$ such that $\llbracket \mathcal{A} \rrbracket(w_n) \leq M \cdot n$ for all $n \geq 2$. A contradiction.

Now, based on the explanations above, we will define the desired fragment of wEMSO for WRA. Similarly to [12], we must restrict the use of $\bigotimes x$ to simplified formulas without weighted quantifiers. Let x be a first-order variable. We say that a formula $\gamma \in \mathsf{wFO}(\Sigma, \mathbb{D}, \mathbb{S})$ is *almost boolean over x* if it is derived by the grammar

$$\gamma ::= \beta \mid s \mid f(X,x) \mid \gamma \oplus \gamma \mid \gamma \otimes \gamma$$

where $\beta \in \mathsf{Bool}(\Sigma, \mathbb{D})$, $s \in S$, $f \in \mathcal{F}$ and X is a second-order variable. Let $\mathsf{aBool}[x](\Sigma, \mathbb{D}, \mathbb{S})$ denote the set of all almost boolean formulas over x. Then, *restricted weighted first-order logic* $\mathsf{wFO}^{\mathsf{res}}(\Sigma, \mathbb{D}, \mathbb{S}) \subseteq \mathsf{wFO}(\Sigma, \mathbb{D}, \mathbb{S})$ is defined by the grammar

$$\varphi ::= \beta \mid s \mid f(X,x) \mid \varphi \oplus \varphi \mid \varphi \otimes \varphi \mid \bigoplus x.\varphi \mid \bigotimes x.\gamma$$

where $\beta \in \mathsf{Bool}(\Sigma, \mathbb{D})$, $s \in S$, $f \in \mathcal{F}$, x is a first-order variable, X is a second-order variable and $\gamma \in \mathsf{aBool}[x](\Sigma, \mathbb{D}, \mathbb{S})$. *Restricted weighted existential MSO logic* $\mathsf{wEMSO}^{\mathsf{res}}(\Sigma, \mathbb{D}, \mathbb{S}) \subseteq \mathsf{wEMSO}(\Sigma, \mathbb{D}, \mathbb{S})$ is defined to be the set of all formulas of the form $\bigoplus X_1 ... \bigoplus X_n.\varphi$ where $n \geq 0$, $X_1, ..., X_n$ are second-order variables and $\varphi \in \mathsf{wFO}^{\mathsf{res}}(\Sigma, \mathbb{D}, \mathbb{S})$.

We say that a fragment $\mathsf{Frag} \subseteq \mathsf{wEMSO}(\Sigma, \mathbb{D}, \mathbb{S})$ is *expressively equivalent to WRA* if, for every data series $\mathbb{L} : \mathbb{D}\Sigma^+ \to S$, $\mathbb{L}$ is recognizable by a WRA over Σ, $\mathbb{D}$ and $\mathbb{S}$ iff $\mathbb{L}$ is definable by a sentence in Frag.

Theorem 4.4. *Let Σ be an alphabet, $\mathbb{D}$ a data structure, and $\mathbb{S}$ a commutative data semiring over $\mathbb{D}$. Then $\mathsf{wEMSO}^{\mathsf{res}}(\Sigma, \mathbb{D}, \mathbb{S})$ is expressively equivalent to WRA.*

In the following two sections we will represent the results which are the main ingredients to prove Theorem 4.4.

5 Determinizable Class of Register Automata

Two of the main difficulties of the proof of Theorem 4.4 are that register automata are neither determinizable nor closed under complement. The goal of this section is to investigate a subclass of register automata which can be applied in the proof of our logical characterization result. Our determinizable subclass could be also of independent interest.

The idea is to make the register updates visible in transition labels. Note that a similar idea was applied in *event-clock automata* [3] and *visibly pushdown automata* [4]. However, the class of languages recognizable by event-clock automata is not closed under renamings of input symbols [3], and thus, this model is not suitable for the translation of logical formulas. Recall that in event-clock automata, the input alphabet is arbitrary and with every letter a clock is associated. In contrast, in our model we take an arbitrary set of registers and an input alphabet is defined depending on this set of registers.

Throughout all of this section, we fix an alphabet Σ, a data structure $\mathbb{D} = (\mathcal{D}, \mathcal{R})$ and a finite set of *registers* Reg. Let $\Sigma^{\langle \text{Reg} \rangle}$ denote the alphabet $\Sigma \times \{0,1\}^{\text{Reg}}$. A *visibly register automaton* over Σ, $\mathbb{D}$ and Reg is a register automaton $\mathcal{A}$ over $\Sigma^{\langle \text{Reg} \rangle}$ and $\mathbb{D}$ with the set of registers Reg such that, for every transition $q \xrightarrow[\phi, \Lambda]{(a,\theta)} q'$ of $\mathcal{A}$ where q, q' are states, $a \in \Sigma$, $\theta \in \{0,1\}^{\text{Reg}}$, $\Lambda \subseteq$ Reg and ϕ is a register guard, we have $\Lambda = \{r \in \text{Reg} \mid \theta(r) = 1\}$. Note that $\mathcal{A}$ recognizes the language $\mathcal{L}(\mathcal{A}) \subseteq \mathbb{D}(\Sigma^{\langle \text{Reg} \rangle})^{+}$. Note also that visibly register automata form a subclass of register automata. We say that a register automaton $\mathcal{A}$ over Σ and $\mathbb{D}$ is *deterministic* if it has a single initial state and whenever $p \xrightarrow[\phi, \Lambda]{a} q$ and $p \xrightarrow[\phi', \Lambda']{a} q'$ are two distinct transitions of $\mathcal{A}$, then ϕ and ϕ' are mutually exclusive, i.e., for all registers valuations ϑ and all data values $d \in \mathcal{D}$, we have $(\vartheta, d) \nvDash \phi \wedge \phi'$. We call $\mathcal{A}$ *complete* if for all states p of $\mathcal{A}$, all letters $a \in \Sigma$, all register valuations ϑ and all data values $d \in \mathcal{D}$, there exists a transition $p \xrightarrow[\phi, \Lambda]{a} q$ of $\mathcal{A}$ with $(\vartheta, d) \models \phi$.

Theorem 5.1. *Let $\mathcal{A}$ be a visibly register automaton over Σ, $\mathbb{D}$ and Reg. Then, there exists a deterministic and complete visibly register automaton $\mathcal{A}'$ over Σ, $\mathbb{D}$ and Reg such that $\mathcal{L}(\mathcal{A}) = \mathcal{L}(\mathcal{A}')$.*

Proof (Sketch). The proof follows a similar idea as the proof of Theorem 1 of [3] about determinization of event-clock automata. Let $\mathcal{A} = (Q, \text{Reg}, I, T, F)$ be a visibly register automaton over the alphabet Σ recognizing the data language $\mathcal{L}(\mathcal{A})$. From $\mathcal{A}$ we construct a visibly register automaton $\mathcal{A}' = (Q', \text{Reg}, I', T', F')$ with $\mathcal{L}(\mathcal{A}') = \mathcal{L}(\mathcal{A})$, where $Q' = \mathcal{P}(Q)$, $I' = \{I\}$, $F' = \{U \subseteq Q \mid U \cap F \neq \emptyset\}$. T' is defined as follows. Suppose that $U \in \mathcal{P}(Q)$ and $(a, \theta) \in \Sigma^{\langle \text{Reg} \rangle}$. Let $(t_i)_{i \in \{1,...,m\}}$ be an enumeration of the set of all transitions in T with label (a, θ) starting in a state from U. For each $i \in \{1, ..., m\}$ let $t_i = (p_i \xrightarrow[\phi_i, \Lambda]{(a,\theta)} q_i)$ with $\Lambda = \{r \in \text{Reg} \mid \theta(r) = 1\}$. For any subset $J \subseteq \{1, ..., m\}$, we add to T' the transition $t' = (U \xrightarrow[\phi_J, \Lambda]{(a,\theta)} U')$ where $U' = \{q_i \mid i \in J\}$ with $\phi_J = \bigwedge_{i \in J} \phi_i \wedge \bigwedge_{i \notin J} \neg \phi_i$. With this construction, we can show that $\mathcal{A}'$ is deterministic and complete and $\mathcal{L}(\mathcal{A}') = \mathcal{L}(\mathcal{A})$.

Let Γ, Δ be alphabets and $h : \Gamma \to \Delta$ a renaming. We say that a mapping $h : \mathbb{D}(\Gamma^{\langle \text{Reg} \rangle})^{+} \to \mathbb{D}(\Delta^{\langle \text{Reg} \rangle})^{+}$ is a Reg-*independent renaming* if it is induced by a mapping $\tilde{h} : \Gamma \to \Delta$. Then, for data languages $\mathcal{L} \subseteq \mathbb{D}(\Gamma^{\langle \text{Reg} \rangle})^{+}$ and $\mathcal{L}' \subseteq \mathbb{D}(\Delta^{\langle \text{Reg} \rangle})^{+}$, the Reg-*independent renaming* $h(\mathcal{L}) \subseteq \mathbb{D}(\Delta^{\langle \text{Reg} \rangle})^{+}$ and Reg-*independent inverse renaming* $h^{-1}(\mathcal{L}) \subseteq \mathbb{D}(\Gamma^{\langle \text{Reg} \rangle})^{+}$ are defined as usual. Using

Theorem 5.1 for the complement, it is not difficult to verify the closure properties for visibly register automata stated in the next lemma.

Lemma 5.2. *The class of data languages recognizable by visibly register automata over an arbitrary alphabet, $\mathbb{D}$ and Reg is closed under union, intersection and complement, Reg-independent renaming and Reg-independent inverse renaming.*

Now let $\beta \in \mathsf{Bool}(\Sigma, \mathbb{D})$ be a formula and Reg the set of all second-order variables X occurring in a subformula of β of the form $R(X, x)$. Using the standard encoding of free variables, we encode the set of all pairs $(w, \sigma|_{\mathsf{Free}(\beta)})$ such that $(w, \sigma) \in \mathbb{D}\Sigma^+$ and $(w, \sigma) \models \beta$ as the data language $\mathcal{L}(\varphi) \subseteq \mathbb{D}(\Gamma^{\langle\mathsf{Reg}\rangle})^+$ where $\Gamma = \Sigma \times \{0,1\}^{\mathsf{Free}(\beta)\backslash\mathsf{Reg}}$. Using Lemma 5.2 and Theorem 5.1, one can show by induction the following theorem.

Theorem 5.3. *Let $\beta \in \mathsf{Bool}(\Sigma, \mathbb{D})$ be a formula and Reg the set of all second-order variables X occurring in a subformula of β of the form $R(X, x)$. Then, there exists a deterministic visibly register automaton $\mathcal{A}$ over Γ, $\mathbb{D}$ and Reg such that $\mathcal{L}(\mathcal{A}) = \mathcal{L}(\varphi)$.*

6 Definability Equals Recognizability

The proof of the fact that recognizability implies definability relies on a similar construction as the proof of Theorem 29 of [20]:

Theorem 6.1. *Let $\mathcal{A}$ be a WRA over Σ, $\mathbb{D}$ and $\mathbb{S}$. Then there exists a sentence $\varphi \in \mathsf{wEMSO}^{\mathsf{res}}(\Sigma, \mathbb{D}, \mathbb{S})$ such that $[\![\varphi]\!] = [\![\mathcal{A}]\!]$.*

Now we turn to the converse direction of Theorem 4.4. Our proof will follow a similar strategy as the proof of the corresponding theorem in [12], i.e., we proceed by induction on the structure of the formula, encode the values of variables as letters of an extended alphabet and apply closure properties stated in Lemma 3.4 of our paper. A crucial problem occurs with the $\bigotimes x$-quantifiers and this case requires a new proof technique, since unweighted register automata are not determinizable and our almost boolean formulas contain functions of the form $f(X, x)$ where f is taken from $\mathcal{F}$ which is not necessarily closed under $+$ and $\cdot$. We solve this problem by translating a $\mathsf{wEMSO}^{\mathsf{res}}$-sentence into a sentence where $\bigotimes x$-quantifiers are applied to formulas of the simplified form. Then, using our Theorem 5.3, we can construct a WRA for $\bigotimes x$-formulas.

For simplicity, we denote the triple $(\Sigma, \mathbb{D}, \mathbb{S})$ by Υ. Let $x \in V_1$ be a first-order variable. We say that a formula $\kappa \in \mathsf{wFO}(\Upsilon)$ is a *semi-granular weight formula* over Υ and x if it is of the form $s \otimes f_1(X_1, x) \otimes \ldots \otimes f_r(X_r, x)$ where $s \in S$, $r \geq 0$, $f_1, \ldots, f_r \in \mathcal{F}$ and $X_1, \ldots, X_r$ are second-order variables. If $X_1, \ldots, X_r$ are pairwise distinct, then κ is called a *granular weight formula*. Let $\mathsf{Gran}[x](\Upsilon)$ denote the set of all granular weight formulas over Υ and x. We say that a formula $\gamma \in \mathsf{aBool}[x](\Upsilon)$ is a *simple almost boolean formula* over Υ and x if it is of the form

$\bigoplus_{i=1}^n (\beta_i \otimes \kappa_i)$ where $n \geq 1$, $\kappa_1, ..., \kappa_n \in \mathsf{Gran}[x](\Upsilon)$ and $\beta_1, ..., \beta_n \in \mathsf{Bool}(\Sigma, \mathbb{D})$ are boolean formulas (not necessarily mutually exclusive). We say that a formula $\psi \in \mathsf{wEMSO}^{\mathsf{res}}(\Upsilon)$ is *canonical* over Υ if whenever it contains a subformula of the form $\bigotimes x.\gamma$, then γ is a simple almost boolean formula over Υ and x. Now we show that each sentence $\psi \in \mathsf{wEMSO}^{\mathsf{res}}(\Upsilon)$ can be translated into a canonical sentence over Υ:

Lemma 6.2. *Let $\psi \in \mathsf{wEMSO}^{\mathsf{res}}(\Upsilon)$ be a sentence. Then, there exists a canonical sentence ζ over Υ such that $[\![\zeta]\!] = [\![\psi]\!]$.*

Proof (Sketch). First, using the commutativity and distributivity of the data semiring $\mathbb{S}$, we can replace every almost boolean formula $\gamma \in \mathsf{aBool}[x](\Upsilon)$ occurring in ψ by a formula $\gamma' = \bigoplus_{i=1}^n (\beta_i \otimes \kappa_i)$ where $\beta_1, ..., \beta_n \in \mathsf{Bool}(\Sigma, \mathbb{D})$ and each κ_i is a semi-granular weight formula over Υ and x which is of the form $s_i \otimes \bigotimes_{k=1}^r f_{ik}(Y_k, x)$. Let $\eta \in \mathsf{wEMSO}^{\mathsf{res}}(\Upsilon)$ be the sentence obtained after these replacements. Second, we replace semi-granular weight formulas in η by granular weight formulas. The idea is the following. Assume that $\eta = \bigoplus X_1, ..., X_k.\varphi$ with $\varphi \in \mathsf{wFO}^{\mathsf{res}}(\Upsilon)$. In the case when φ contains a semi-granular formula $\kappa = s \otimes f_1(Y_1, x) \otimes ... \otimes f_i(Y_i, x) \otimes ... \otimes f_j(Y_j, x) \otimes ... \otimes f_n(Y_n, x)$ with $i \neq j$ and $Y_i = Y_j$, then we take a fresh second-order variable Z and replace η by the sentence $\bigoplus X_1, ..., X_k, Z.([\forall z.(z \in Z \leftrightarrow z \in Y_i)] \otimes \tilde{\varphi})$ where $\tilde{\varphi}$ is obtained from φ by replacing the variable Y_j in κ by the fresh variable Z. Following this process, in finitely many steps we can get rid of all repeating second-order variables in semi-granular weight formulas and obtain the desired canonical sentence ζ with $[\![\zeta]\!] = [\![\psi]\!]$.

Theorem 6.3. *Let $\psi \in \mathsf{wEMSO}^{\mathsf{res}}(\Sigma, \mathbb{S}, \mathbb{D})$ be a sentence. Then there exists a WRA $\mathcal{A}$ over Σ, $\mathbb{S}$ and $\mathbb{D}$ such that $[\![\mathcal{A}]\!] = [\![\psi]\!]$.*

Proof (Sketch). By Lemma 6.2, we may assume that ψ is canonical. We proceed by induction on the structure of a subformula ζ of ψ. As usual, whenever ζ contains free variables, ζ will be translated into a WRA over the extended alphabet $\Sigma \times \{0,1\}^{\mathsf{Free}(\zeta)}$. We restrict ourselves to the most interesting case $\zeta = \bigotimes x.\gamma$. Since ψ is canonical, γ is a simple almost boolean formula. Let Reg be the set of all second-order variables Y such that γ has a subformula of the form $R(Y, x)$ with $R \in \mathcal{R}$ or $f(Y, x)$ with $f \in \mathcal{F}$. Let $(Y_i)_{1 \leq i \leq r}$ be an enumeration of Reg. Using 0, 1 and $1^{\mathcal{D} \times \mathcal{D}}$, we can transform γ into the form $\gamma' = \bigoplus_{i=1}^n (\beta_i \otimes s_i \otimes \bigotimes_{k=1}^r f_{ik}(Y_k, x))$ where $\beta_1, ..., \beta_n \in \mathsf{Bool}(\Sigma, \mathbb{D})$, $s_i \in S$ and $f_{ik} \in \mathcal{F}$. Note that the idea of the construction of [12] relies on the fact that $\beta_1, ..., \beta_n$ are mutually exclusive. However, in our situation this is not the case. We also may assume for simplicity that, for all $i \in \{1, ..., n\}$ and $k \in \{1, ..., r\}$, β_i has a subformula of the form $R(Y_k, x)$. This means that $\mathsf{Reg} \subseteq \mathsf{Free}(\beta_i)$ for all $i \in \{1, ..., n\}$. Let $\tilde{S} \subseteq S$ be the set of all s_i appearing in γ' and $\tilde{\mathcal{F}} \subseteq \mathcal{F}$ the set of all f_{ik} appearing in γ'. Consider the extended alphabet $\Delta = \Sigma \times \{1, ..., n\} \times \tilde{S} \times \tilde{\mathcal{F}}^r$. We construct a formula $\xi \in \mathsf{Bool}(\Delta, \mathbb{D})$ over the extended alphabet which demands that, for all positions of a data word, whenever the $\{1, ..., n\}$-component is i, then β_i (lifted to the extended alphabet) holds and the $\tilde{S}$-component is s_i and $\tilde{\mathcal{F}}^r$-component is $(f_{i1}, ..., f_{ir})$. By Theorem 5.3, there exists a deterministic visibly register automaton $\mathcal{A} = (Q, \mathsf{Reg}, I, T, F)$ over $\Delta \times \{0,1\}^{\mathsf{Free}(\xi) \backslash \mathsf{Reg}}$, Reg and $\mathbb{D}$ such that $\mathcal{L}(\mathcal{A}) = \mathcal{L}(\xi)$. Note that $\mathcal{A}$ can be considered as a register automaton over the alphabet $\Delta \times \{0,1\}^{\mathsf{Free}(\xi)}$ and $\mathbb{D}$. We construct a WRA $\mathcal{A}' = (Q, \mathsf{Reg}, I, T, F, \langle \mathsf{wt}_{\mathsf{trans}}, \mathsf{wt}_{\mathsf{data}} \rangle)$ over $\Delta \times \{0,1\}^{\mathsf{Free}(\xi)}$, $\mathbb{D}$ and $\mathbb{S}$ where $\mathsf{wt}_{\mathsf{trans}}$ and $\mathsf{wt}_{\mathsf{data}}$ are defined according to the auxiliary components $\tilde{S}$ and $\tilde{\mathcal{F}}^r$ of the extended alphabet Δ and obtain a WRA $\mathcal{A}'$ over the alphabet $\Delta \times \{0,1\}^{\mathsf{Free}(\xi)}$. Let $h : \Delta \times \{0,1\}^{\mathsf{Free}(\xi)} \to \Sigma \times \{0,1\}^{\mathsf{Free}(\xi)}$ be the projection. By Lemma 3.4 (b), there exists a WRA $\mathcal{B}$ with $[\![\mathcal{B}]\!] = h([\![\mathcal{A}']\!])$. Then one can show that $[\![\mathcal{B}]\!] = [\![\zeta]\!]$.

Then our Theorem 4.4 follows from Theorems 6.1 and 6.3.

7 Discussion

We introduced a model of weighted register automata and gave an expressively equivalent weighted logic. On the one hand, our results show the robustness of the automata-theoretic approach, help to understand better the behaviors of weighted register automata and can also find applications for the setting of timed words. On the other hand, our expressive equivalence result could be used as a basis for the quantitative verification of systems with data, e.g., for the study of quantitative extensions of temporal logics on data words [15]. An important open question concerns algorithmic properties of weighted register automata. We believe that the optimal reachability problem for weighted register automata is decidable for various examples considered in this paper. It could be interesting to extend our results to the setting of infinite data words and data trees and to investigate in the setting of data words the cases where the weight measure cannot be modelled using semirings (e.g., average or discounted costs, energy problems and weighted register automata with multiple cost parameters). Note that these nonclassical weight measures have been extensively studied in the setting of weighted timed automata. It could be also interesting to compare the expressive power of our register automata model with the data automata model of [9]. We believe that they are incomparable. An extension of class register automata and the logic captured by them [7], where data words have been considered as behavioral models of concurrent systems, to the weighted setting could be attractive, as well.

References

1. Alur, R., D'Antoni, L., Deshmukh, J., Raghothaman, M., Yuan, Y.: Regular functions and cost register automata. In: LICS 2013, pp. 13–22. IEEE Computer Society (2013)
2. Alur, R., Dill, D.L.: A theory of timed automata. Theor. Comput. Sci. **126**(2), 183–235 (1994)
3. Alur, R., Fix, L., Henzinger, T.: Event-clock automata: a determinizable class of timed automata. Theor. Comput. Sci. **211**(1–2), 253–273 (1999)

4. Alur, R., Madhusudan, P.: Visibly pushdown languages. In: STOC 2004, pp. 202–211. ACM (2004)
5. Alur, R., La Torre, S., Pappas, G.J.: Optimal paths in weighted timed automata. In: Di Benedetto, M.D., Sangiovanni-Vincentelli, A.L. (eds.) HSCC 2001. LNCS, vol. 2034, pp. 49–62. Springer, Heidelberg (2001)
6. Bojańczyk, M., David, C., Muscholl, A., Schwentick, T., Segoufin, L.: Two-variable logic on data words. ACM Trans. Comput. Logic 12(4), 27 (2011)
7. Bollig, B.: An automaton over data words that captures EMSO Logic. In: Katoen, J.-P., König, B. (eds.) CONCUR 2011. LNCS, vol. 6901, pp. 171–186. Springer, Heidelberg (2011)
8. Bollig, B., Gastin, P.: Weighted versus probabilistic logics. In: Diekert, V., Nowotka, D. (eds.) DLT 2009. LNCS, vol. 5583, pp. 18–38. Springer, Heidelberg (2009)
9. Bouyer, P.: A logical characterization of data languages. Inf. Process. Lett. 84(2), 75–85 (2002)
10. Bouyer, P., Petit, A., Thérien, D.: An algebraic characterization of data and timed languages. In: Larsen, K.G., Nielsen, M. (eds.) CONCUR 2001. LNCS, vol. 2154, pp. 248–261. Springer, Heidelberg (2001)
11. Büchi, J.R.: Weak second order arithmetic and finite automata. Zeitschrift für Mathematische Logik und Grundlagen der Informatik 6, 66–92 (1960)
12. Droste, M., Gastin, P.: Weighted automata and weighted logics. Theor. Comput. Sci. 380(1–2), 69–86 (2007)
13. Droste, M., Kuich, W., Vogler, H. (eds.): Handbook of Weighted Automata. EATCS Monographs on Theoretical Computer Science. Springer, Heidelberg (2009)
14. Droste, M., Perevoshchikov, V.: A Nivat theorem for weighted timed automata and weighted relative distance logic. In: Esparza, J., Fraigniaud, P., Husfeldt, T., Koutsoupias, E. (eds.) ICALP 2014, Part II. LNCS, vol. 8573, pp. 171–182. Springer, Heidelberg (2014)
15. Demri, S., Lazić, R.: LTL with the freeze quantifier and register automata. ACM Trans. Comput. Logic 10(3), 30 (2009)
16. Figueira, D.: Alternating register automata on finite data words and trees. Logical Methods Comput. Sci. 8(1: 22), 1–43 (2012)
17. Figueira, D., Hofman, P., Lasota, S.: Relating timed and register automata. In: EXPRESS 2010, EPTCS, vol. 41, pp. 61–75 (2010)
18. Kaminski, M., Francez, N.: Finite-memory automata. Theor. Comput. Sci. 134, 329–363 (1994)
19. Neven, F., Schwentick, T., Vianu, V.: Finite state machines for strings over infinite alphabets. ACM Trans. Comput. Logic 5(3), 403–435 (2004)
20. Quaas, K.: MSO logics for weighted timed automata. Formal Methods Syst. Des. 38(3), 193–222 (2011)
21. Wilke, T.: Specifying timed state sequences in powerful decidable logics and timed automata. In: Langmaack, H., Roever, W.-P., Vytopil, J. (eds.) FTRTFT 1994. LNCS, vol. 863, pp. 694–715. Springer, Heidelberg (1994). doi:10.1007/3-540-58468-4_191

Hybrid Automata as Coalgebras

Renato Neves[✉] and Luis S. Barbosa

HASLab (INESC TEC) & Universidade do Minho, Braga, Portugal
rjneves@inescporto.pt, lsb@di.uminho.pt

Abstract. Able to simultaneously encode discrete transitions and continuous behaviour, hybrid automata are the *de facto* framework for the formal specification and analysis of hybrid systems. The current paper revisits hybrid automata from a coalgebraic point of view. This allows to interpret them as state-based components, and provides a uniform theory to address variability in their definition, as well as the corresponding notions of behaviour, bisimulation, and observational semantics.

1 Introduction

1.1 Context

Consider a cruise control system. It comprises digital controllers, sensors, and actuators, that act in coordination to make the vehicle reach the intended speed. The system's behaviour, from an *external perspective*, is observed in the (continuous) evolution of a physical process (velocity). But at the same time we know that the controller, which has influence over this process, changes its *internal* state in a discrete manner.

Systems with this interaction pattern are often called *hybrid*. Their formal specification and analysis typically resorts to the theory of *hybrid automata* [Hen96], whose distinguishing feature is the ability of state variables to continuously evolve. This allows to express the evolution of physical processes, like movement, time, temperature, and pressure. In addition, there is syntactical machinery (guards, state invariants, and assignments) to facilitate the description of complex behaviour in a concise manner. For illustration purposes,

Example 1. Consider a (simplistic) system comprised of *a tank and a valve* connected to it. The valve allows water to flow in at a rate of $2\,\mathrm{cm/s}$ during intervals of c seconds; between these periods the valve is shut (also) for c seconds. We can describe this behaviour via the hybrid automaton below.

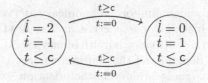

A. Sampaio and F. Wang (Eds.): ICTAC 2016, LNCS 9965, pp. 385–402, 2016.
DOI: 10.1007/978-3-319-46750-4_22

The variable l denotes the water level, which rises when the valve is open (differential equation $\dot{l} = 2$). Then, the differential equation $\dot{t} = 1$ defines the passage of time, which, along with invariant $t \leq \mathsf{c}$, forces the current state to be active for at most c seconds. On the other hand, the guards $t \geq \mathsf{c}$ and assignments $t := 0$ force the current state to be active at least c seconds before a switch. Finally, note that the guards $t \geq \mathsf{c}$ do not force transitions to happen, but only permit them. This means that if not for invariant $t \leq \mathsf{c}$, the valve could be open (or shut) indefinitely.

The semantics of hybrid automata is traditionally described in terms of labelled transition systems (LTS): each hybrid automaton yields an LTS whose edges encode both the discrete events and continuous evolutions (*cf.* [Hen96]). Edges in the latter are labelled by elements of $\mathbb{R}_{\geq 0}$ and reflect the difference of state variables with respect to the source and sink nodes. For example, denoting the left state (of the previous hybrid automaton) by m_1, the edge $(m_1, 1, 0.5) \xrightarrow{t} (m_1, 1 + 2t, 0.5 + t)$ exists in the underlying LTS iff $0.5 + t \leq \mathsf{c}$. This will be explained in more detail in Sect. 2.

For now, we emphasise that such a semantics 'collapses' both discrete assignments and continuous evolutions into the same relation, which makes difficult to distinguish the system's internal, thus hidden behaviour (typically its state changes), from what can be observed externally. Such a distinction, however, is at the very heart of the component-based paradigm, in which complex systems are verified through a suitable analysis of their (simpler) constituents (see *e.g.*, [Bar03, HJ11, Szy98]).

To understand hybrid automata as state-based components is an important step towards their coalgebraic characterisation in the spirit of [Bar03, HJ11]. Such an achievement would provide them several composition operators (with corresponding laws), refinement techniques, and synchronisation mechanisms.

Another relevant point is the existence of several variants of hybrid automata (*e.g.*, [Hen96, Spr00, LLK+99]), motivated by the need to capture different types of behaviour (*e.g.*, nondeterministic, probabilistic, faulty). To the best of our knowledge, a uniform, formal theory for different types of hybrid automata does not yet exist.

1.2 Contributions

This paper characterises hybrid automata as coalgebras of a specific type. This promotes the black-box perspective discussed above, where the (discrete) state transitions are internal, hidden from the environment, and the continuous evolutions are external, making up the observable behaviour. To be concrete,

- 'going coalgebraic' provides a uniform, canonical *observational semantics* that faithfully reflects the black-box perspective, and frames the behaviour into well known constructions (*e.g.*, streams, infinite binary trees), marking a separation between the discrete domain and the continuous one.
- Moreover, a generic (coalgebraic) characterisation of *bisimulation*, parametrised by a transition type (technically, a functor), emerges across different

sorts of hybrid automata in a uniform manner. Indeed, it is shown that different notions of bisimilarity (associated with variants of hybrid automata) are subsumed by the corresponding coalgebraic definition.

We will also see that the coalgebraic characterisation proposed in this paper facilitates the understanding of hybrid automata and helps to systematise the concept along a plethora of, often elaborated, definitions in the literature. In its most basic variant, a hybrid automaton becomes reduced to a machine that from a state (internally) jumps to another, and (externally) produces a continuous evolution. As expected, this implies that, even in the presence of both discrete and continuous behaviour, only the continuous part can be directly observed.

The coalgebraic characterisation paves the way to yet another contribution: a hierarchy of different types of hybrid automata organised with respect to their 'expressivity', a concept also to be here understood within the coalgebraic framework.

1.3 Roadmap

Section 2 provides a brief background on hybrid automata and coalgebras. Section 3 establishes the relation between classic hybrid automata (in a deterministic setting) and the corresponding coalgebras. In particular, it shows how to encode hybrid automata as coalgebras, explores the associated observational semantics, and reframes the classic notion of bisimulation (for hybrid automata) as a coalgebraic one.

Then, building on the coalgebraic perspective, Sect. 4 considers different types of functors in order to (re)discover several variants of hybrid automata. Two interesting cases are the ones that involve *probabilistic* [Spr00] and *replicating* behaviour, the latter being new to the best of our knowledge. Section 4 also establishes the hierarchy of hybrid automata mentioned above. Finally, Sect. 5 concludes and hints at future work directions.

We assume that the reader has some familiarity with elementary category theory and topology.

2   Background

2.1 Hybrid Automata

Introduced in the early nineties as an answer to the rapid emergence of hybrid systems, hybrid automata form an active research area that encompasses diverse topics. These span from decidability [Hen96], to extensions that cater for *input* mechanisms (*e.g.*, [AH97,LLK+99]), and *uncertainty* [Spr00]. Hybrid automata have also been considered as a modelling tool in life sciences [BCB+09,AMP+03]. Formally,

Definition 1 ([Hen96]). *A hybrid automaton is a tuple* $(M, E, \Sigma, X, \mathsf{init}, \mathsf{inv}, \mathsf{dyn}, \mathsf{asg}, \mathsf{grd})$ *where*

- M is a finite set of discrete states (often called control modes, or locations), E is a transition relation $E \subseteq M \times \Sigma \times M$, and Σ a set of labels. A triple $(m_1, l, m_2) \in E$ will often be written as $m_1 \overset{l}{\rightsquigarrow} m_2$.
- X is a finite set of real-valued variables $\{x_1, \ldots, x_n\}$.
- init and inv are functions that associate to each mode a predicate over the variables in X. Letter Z denotes the set $\{(m, v) \in M \times \mathbb{R}^n \mid v \models (\text{inv } m)\}$, where expression $v \models (\text{inv } m)$ means that predicate $(\text{inv } m)$ is satisfied by v.
- dyn is a function that associates to each state a predicate over the variables in $X \cup \dot{X}$, where $\dot{X} = \{\dot{x}_1, \ldots, \dot{x}_n\}$ represents the first derivatives of the variables in X. It is used to define the set of continuous evolutions that may occur at each state.
- asg is a function such that given an edge $(e \in E)$ returns a predicate over $X \cup X'$, where $X' = \{x'_1, \ldots, x'_n\}$ represents the variables in X immediately after a discrete jump. This provides an assignment to each edge. Finally, the function grd associates each edge with a guard, i.e., a predicate over X.

A classic example may help to illustrate this quite complex definition.

Example 2. Consider a bouncing ball dropped at some positive height p and with no initial velocity v. Due to the gravitational acceleration g, it falls into the ground but then bounces back up, losing part of its kinetic energy in the process. The following hybrid automaton sums up this behaviour.

$$
\begin{pmatrix} \dot{p} = v \\ \dot{v} = g \\ p \geq 0 \end{pmatrix} \circlearrowleft \begin{array}{l} p = 0 \wedge v > 0, \\ v' = v \times -0.5 \end{array}
$$

Note that only one mode exists; let us call it m. Also, there is exactly one discrete transition: $m \rightsquigarrow m \in E$, omitting its label for simplicity. Actually, in this example there is no need for labels. Then $X = \{p, v\}$, and $(\text{inv } m)$ is $p \geq 0$ – which entails $Z = \{m\} \times \mathbb{R}_{\geq 0} \times \mathbb{R}$, where the second $(\mathbb{R}_{\geq 0})$ and third components $(\mathbb{R})$ denote, respectively, position and velocity.

Finally, $\text{grd}(m \rightsquigarrow m)$ is $p = 0 \wedge v > 0$, $(\text{dyn } m)$ is $\{\dot{p} = v, \dot{v} = g\}$, and $\text{asg}(m \rightsquigarrow m)$ is $v' = v \times -0.5 \wedge p' = p$. Note that the right-hand side of the last predicate does not appear in the hybrid automaton above, a common practice to avoid a burdened notation.

In order to keep results simple and intuitive, we do not consider labels or initial states, as they can be accommodated later on in a straightforward manner.

Frequently it is assumed that, given any mode, function dyn returns a system of differential equations with exactly one solution (*e.g.*, [Jac00, ACH+95]). We adopt this approach as well. Such an assumption may seem too restrictive but, in fact, such is not the case for most hybrid systems described in the literature, as they rarely involve nonlinear differential equations. The important point is that this condition allows function dyn to induce a function,

$$
\text{flow} : (M \times \mathbb{R}^n) \times \mathbb{R}_{\geq 0} \to \mathbb{R}^n
$$

such that given a pair $(m, v) \in (M \times \mathbb{R}^n)$, flow $((m, v), -) : \mathbb{R}_{\geq 0} \to \mathbb{R}^n$ is a continuous function, which represents the solution to the system of differential equations; note that its domain ($\mathbb{R}_{\geq 0}$) represents time.

Assume also that an hybrid automaton cannot jump from a valid state $(m, v) \in (M \times \mathbb{R}^n)$ into an invalid one, where by valid we mean that $(m, v) \in Z$. In symbols, assume that for any pair $((m_1, v_1), (m_2, v_2)) \in (M \times \mathbb{R}^n)^2$ such that $m_1 \rightsquigarrow m_2$, $v_1 \models \mathsf{grd}(m_1 \rightsquigarrow m_2)$, and $(v_1, v_2) \models \mathsf{asg}(m_1 \rightsquigarrow m_2)$ we have $v_2 \models (\mathsf{inv}\ m_2)$.

As mentioned in Sect. 1, the semantics of hybrid automata is traditionally described in terms of LTSs.

Definition 2 ([Hen96]). *Consider a hybrid automaton. Its underlying LTS is a tuple (Z, L, T) such that $L = 1 + \mathbb{R}_{\geq 0}$ (1 is a singleton set), and $T \subseteq Z \times L \times Z$ is defined as $((m_1, v_1), l, (m_2, v_2)) \in T$ iff*

1. *if $l \in 1$ then $m_1 \rightsquigarrow m_2$, $v_1 \models \mathsf{grd}\left(m_1 \rightsquigarrow m_2\right)$, $(v_1, v_2) \models \mathsf{asg}\left(m_1 \rightsquigarrow m_2\right)$,*
2. *if $l \in \mathbb{R}_{\geq 0}$ then $m_1 = m_2$, flow $((m_1, v_1), l) = v_2$, and for all $t \in [0, l]$ flow $((m_1, v_1), t) \models (\mathsf{inv}\ m_1)$.*

We write a triple $(z_1, l, z_2) \in T$ as $z_1 \xrightarrow{l} z_2$.

Example 3. Recall the hybrid automaton from Example 2. The associated LTS (Z, L, T) is defined as follows: $Z = \{m\} \times \mathbb{R}_{\geq 0} \times \mathbb{R}$, $L = 1 + \mathbb{R}_{\geq 0}$, and $(m, p_1, v_1) \xrightarrow{l} (m, p_2, v_2)$ iff

1. if $l \in 1$ then $p_1 = 0 \wedge v_1 > 0$, and $v_2 = v_1 \times -0.5 \wedge p_1 = p_2$;
2. if $l \in \mathbb{R}_{\geq 0}$ then flow $((m, p_1, v_1), l) = (p_2, v_2)$, and for all $t \in [0, l]$, flow $((m, p_1, v_1), t) \geq 0$.

In this case the function flow, induced by dyn, describes the continuous evolution of position and velocity (between jumps).

Note that both discrete events and continuous evolutions are embedded in the relation T. Not only this makes difficult to adopt the black-box perspective mentioned above, but it also turns the verification of hybrid automata into a challenging task, as an infinite number of states and edges needs to be taken into consideration. The standard technique for overcoming the latter issue is to quotient by a bisimulation equivalence, *i.e.*, to collapse states that possess equivalent behaviour. The resulting states become then symbolic representations of (possibly infinite) regions, and verification techniques are applied to the reduced system instead.

Definition 3 ([Hen96]). *Consider the underlying labelled transition system (S, L, T) of a hybrid automaton, and an equivalence relation $\Phi \subseteq S \times S$ over the states. A Φ-bisimulation $R \subseteq S \times S$ is a relation such that $(s_1, q_1) \in R$ (or more concisely, $s_1\ R\ q_1$) entails the following cases:*

1. $s_1\ \Phi\ q_1$,

2. *for each label $l \in L$, if $s_1 \xrightarrow{l} s_2$ then there is a state q_2 such that $q_1 \xrightarrow{l} q_2$ and $s_2 R q_2$,*

3. *for each label $l \in L$, if $q_1 \xrightarrow{l} q_2$ then there is a state s_2 such that $s_1 \xrightarrow{l} s_2$ and $s_2 R q_2$.*

Two states $s_1, q_1 \in S$ are *Φ-bisimilar* (in symbols, $s_1 \equiv^{\Phi} q_1$) if they are related by a Φ-bisimulation.

We will start our (coalgebraic) rendering of hybrid automata in a deterministic setting, restricting Definition 1 with the following conditions:

1. Relation E is a function ($E : M \to M$).
2. Assignments are deterministic, *i.e.*, they take the form $x := \theta$, where θ is an expression with variables of X that denotes a real value, and $x \in X$. For example, in the case of the bouncing ball above, the assignment $v' = v \times -0.5$ is changed to $v := v \times -0.5$. Note that Example 1 (tank-and-valve) also adopted this approach.
3. As soon as an edge becomes enabled (*i.e.* the associated guard is satisfied) the current state must switch (a similar condition is adopted in [Nad97], where hybrid automata with this property are called time-deterministic). More concretely, each pair $(m, v) \in Z$ has exactly one duration ($\delta \in \mathbb{R}_{\geq 0}$) for its evolution flow$((m, v), -) : \mathbb{R}_{\geq 0} \to \mathbb{R}^n$, which, intuitively, corresponds to the time that the current mode takes to jump starting in (m, v). This happens, for example, in the hybrid automaton that describes the tank-and-valve (c seconds) and the bouncing ball system (the time the ball takes to reach the ground from a specific height and velocity).

   Unlike the two conditions above, this condition, which we refer to as *as-soon-as*, is assumed throughout the paper.

The three conditions together give no possibility for a hybrid automaton to choose between possible executions, and therefore induce a function $\mathsf{nxt} : Z \to Z$, which given a pair $(m, v) \in Z$, returns the pair that results from the corresponding evolution (given by function flow and associated duration δ) and subsequent discrete transition. Formally,

$$\mathsf{nxt}(m, v) = \big(E(m), \mathsf{asg}(m \rightsquigarrow E(m))\, u\big)$$

where $u = \mathsf{flow}((m, v), \delta)$. By a slight abuse of notation we denote the expression $\mathsf{asg}(m \rightsquigarrow E(m))$ as a function. Note also that the value u is the last point (in the evolution of (m, v)) before the jump.

2.2 Coalgebras

The theory of coalgebras [Rut00] establishes an abstract, categorial framework that promotes a uniform study of state-based transition systems[1]. The idea is

---

[1] We restrict ourselves to the concepts strictly necessary to the paper. The interested reader will find in document [Rut00] a comprehensive introduction to the theory of coalgebras.

that a functor $\mathcal{F} : \mathbf{C} \to \mathbf{C}$ over some category $\mathbf{C}$ (typically, $\mathbf{Set}$) gives 'shape' to a transition type, and arrows $S \to \mathcal{F}S$ in $\mathbf{C}$ ($\mathcal{F}$-coalgebras, or simply coalgebras) make up the family of corresponding transition systems.

Definition 4. *Consider a functor $\mathcal{F} : \mathbf{C} \to \mathbf{C}$. It gives rise to category $\mathbf{CoAlg}_{\mathcal{F}}$ whose objects are coalgebras $S \to \mathcal{F}S$, and morphisms between two coalgebras $\alpha : S \to \mathcal{F}S$, $\beta : Q \to \mathcal{F}Q$ are arrows $f : S \to Q$ in $\mathbf{C}$ such that the diagram below in the left commutes.*

$$
\begin{array}{ccc}
S & \xrightarrow{\ f\ } & Q \\
{\scriptstyle \alpha}\downarrow & & \downarrow{\scriptstyle \beta} \\
\mathcal{F}S & \xrightarrow[\mathcal{F}f]{} & \mathcal{F}Q
\end{array}
\qquad\qquad
\begin{array}{ccc}
S & \dashrightarrow^{[\![-]\!]} & \nu_{\mathcal{F}} \\
{\scriptstyle \alpha}\downarrow & & \downarrow{\scriptstyle \gamma} \\
\mathcal{F}S & \dashrightarrow[\mathcal{F}[\![-]\!]] & \mathcal{F}\nu_{\mathcal{F}}
\end{array}
$$

Under mild conditions, a category $\mathbf{CoAlg}_{\mathcal{F}}$ has a *final* object, *i.e.*, a coalgebra $\gamma : \nu_{\mathcal{F}} \to \mathcal{F}\nu_{\mathcal{F}}$ such that for any coalgebra $\alpha : S \to \mathcal{F}S$ there is a unique morphism $[\![-]\!] : S \to \nu_{\mathcal{F}}$ that makes the diagram above in the right to commute. A prime example is the final $(- \times A)$-coalgebra $\langle tl, hd \rangle : A^{\omega} \to A^{\omega} \times A$. Briefly, A^{ω} is the set of infinite lists (streams) of elements in A, and $\langle tl, hd \rangle$ is defined as,

$$
\langle tl, hd \rangle\, (a_0, a_1, \dots) = ((a_1, \dots), a_0).
$$

Since $\langle tl, hd \rangle$ is final, each coalgebra $\alpha : S \to S \times A$ has a unique morphism $[\![-]\!]_{\alpha} : S \to A^{\omega}$, called the behaviour or *coinductive* extension of α – whenever found suitable we will drop the subscript in $[\![-]\!]_{\alpha}$. Intuitively, $[\![-]\!]_{\alpha} : S \to A^{\omega}$ gives the observable behaviour of each state $(s \in S)$ of $\alpha : S \to S \times A$. Actually, final objects in categories of coalgebras provide the observational semantics mentioned in Sect. 1.

Bisimulation is another key concept in coalgebra theory.

Definition 5. *Consider two $\mathcal{F}$-coalgebras $\alpha : S \to \mathcal{F}S$, $\beta : Q \to \mathcal{F}Q$ in $\mathbf{Set}$, and a relation $R \subseteq S \times Q$. Then R is an $\mathcal{F}$-bisimulation (or simply bisimulation) if there is a third coalgebra $\gamma : R \to \mathcal{F}R$ that makes the following diagram to commute.*

$$
\begin{array}{ccccc}
S & \xleftarrow{\ \pi_1\ } & R & \xrightarrow{\ \pi_2\ } & Q \\
{\scriptstyle \alpha}\downarrow & & {\scriptstyle \gamma}\downarrow & & \downarrow{\scriptstyle \beta} \\
\mathcal{F}S & \xleftarrow[\mathcal{F}\pi_1]{} & \mathcal{F}R & \xrightarrow[\mathcal{F}\pi_2]{} & \mathcal{F}Q
\end{array}
$$

We say that states $s \in S$, and $q \in Q$ are *coalgebraically bisimilar* (in symbols, $s \sim q$) if they are related by some $\mathcal{F}$-bisimulation.

3   Deterministic Hybrid Automata as Coalgebras

3.1   The Model

In order to encode hybrid automata as coalgebras, recall the state-based, black-box perspective described in the introductory section: discrete transitions occur internally, hidden from the environment, whereas the observable behaviour (or output) corresponds to continuous evolutions. As explained before, for any given suitable pair $(m, v) \in (M \times \mathbb{R}^n)$, a hybrid automaton outputs a continuous evolution over $\mathbb{R}^n$, with a specific duration $\delta \in \mathbb{R}_{\geq 0}$. Formally, a continuous function $[0, \delta] \to \mathbb{R}^n$ where $[0, \delta]$ has the subspace topology induced by the Euclidean one, and $\mathbb{R}^n$ has the Euclidean topology – this requires a brief use of topological notions in the following construction.

Definition 6. *Generalising the output type from $\mathbb{R}^n$ to an arbitrary topological space (O, τ), the output of an hybrid automaton is defined as the sum of all continuous evolutions over (O, τ). In symbols,*

$$U \left(\coprod_{\delta \in \mathbb{R}_{\geq 0}} (O, \tau)^{[0,\delta]} \right)$$

*where $[0, \delta]$ is equipped with the subspace topology induced by the Euclidean one, and $U :$ **Top** $\to$ **Set** is the forgetful functor between the category of topological spaces and continuous functions (**Top**) and **Set**. We will denote the construction above by $\mathcal{H}(O, \tau)$, or simply $\mathcal{H}O$.*

For what follows, let us denote the curried version of a function $f : A \times B \to C$, by $\lambda f : A \to C^B$. Then, consider a hybrid automaton and recall that each pair $(m, v) \in Z$ defines $\mathsf{flow}((m, v), -) : \mathbb{R}_{\geq 0} \to \mathbb{R}^n$ whose domain can be restricted to duration $[0, \delta]$. This leads to a function $\lambda\,\mathsf{flow} : Z \to \mathcal{H}(\mathbb{R}^n)$, which by a slight abuse of notation, and for the sake of generality we type as

$$\mathsf{out} : Z \to \mathcal{H}O.$$

Finally, note that function $\mathsf{out} : Z \to \mathcal{H}O$, together with function $\mathsf{nxt} : Z \to Z$ (see Sect. 2), forms a $(- \times \mathcal{H}O)$-coalgebra

$$\langle \mathsf{nxt}, \mathsf{out} \rangle : Z \to Z \times \mathcal{H}O,$$

which (fully) characterises the behaviour of the hybrid automaton.

The intuition is that each state $(m, v) \in Z$ gives rise to an observable, continuous evolution ($e \in \mathcal{H}O$), and an internal, discrete transition to the next state ($z \in Z$). Let us illustrate this concept with a few examples.

Example 4. Recall the tank-and-valve system described in Sect. 1. The corresponding coalgebra $\langle \mathsf{nxt}, \mathsf{out} \rangle : Z \to Z \times \mathcal{H}O$ is defined as

$$\langle \mathsf{nxt}, \mathsf{out} \rangle (m_1, l, t) = ((m_2, l + 2\,\mathsf{c}, 0), f), \quad \langle \mathsf{nxt}, \mathsf{out} \rangle (m_2, l, t) = ((m_1, l, 0), g)$$

where the functions $f, g : [0, \mathsf{c}] \to \mathbb{R}^2$ are defined as

$$f\,r = (l + 2r,\ t + r), \quad g\,r = (l,\ t + r).$$

Example 5. Consider again the bouncing ball system and, for illustration purposes, take only its movement as the observable behaviour. The corresponding coalgebra $\langle \mathsf{nxt}, \mathsf{out} \rangle : Z \to Z \times \mathcal{H}O$ is given by

$$\langle \mathsf{nxt}, \mathsf{out} \rangle \, (m, p, v) = ((m, 0, v'), \mathsf{mov}(p, v, -))$$

where variable v' corresponds to the (abrupt) change of velocity due to the collision, function $\mathsf{mov}(p, v, -) : [0, \delta] \to \mathbb{R}$ describes the ball's movement between jumps, and δ denotes the time that the ball takes to reach the ground from state (p, v). In symbols,

$$v' = (v + g\delta) \times -0.5, \quad \mathsf{mov}(p, v, t) = p + vt + \tfrac{1}{2}gt^2, \quad \delta = \frac{\sqrt{2gp + v^2} + v}{g}$$

As mentioned in the previous section, each coalgebra $S \to S \times \mathcal{H}O$ yields a function $[\![-]\!] : S \to (\mathcal{H}O)^\omega$ which computes, for a given $s \in S$, a stream of (observable) continuous evolutions $[\![s]\!]$, which correspond to the (internal) states that are visited starting in s. For example,

Example 6. Consider again the bouncing ball system; the first three elements of $[\![(0, 5)]\!]$ are represented in the following plots.

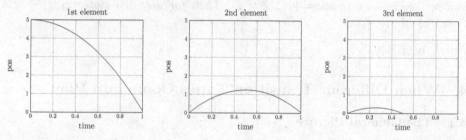

3.2 Bisimulation in the Deterministic Case

Recall from the previous section that bisimulation for hybrid automata (Definition 3) is parametrised by an equivalence relation over the state space. Let us see how to capture this coalgebraically.

Consider a coalgebra $\langle \mathsf{nxt}, \mathsf{out} \rangle : Z \to Z \times \mathcal{H}O$ (modelling a hybrid automaton) and an equivalence relation over its states $\Phi \subseteq Z \times Z$. We define a coalgebra $\langle \mathsf{nxt}, \mathsf{out} \rangle^\Phi : Z \to Z \times \mathcal{H}(Z/_\Phi)$ such that

$$\langle \mathsf{nxt}, \mathsf{out} \rangle^\Phi z = (\mathsf{nxt}\, z, q \cdot (\mathsf{ev}\, z))$$

where $\mathsf{ev} : Z \to \mathcal{H}Z$ is defined as $(\mathsf{ev}(m, v))\, t = (m, (\mathsf{out}(m, v))\, t)$, and $q : Z \to Z/_\Phi$ is the quotient map induced by Φ.

Technically, $\langle \mathsf{nxt}, \mathsf{out} \rangle^\Phi$ is a $\mathcal{F}_{Z/_\Phi}$-coalgebra where $\mathcal{F}_{Z/_\Phi} X = X \times \mathcal{H}(Z/_\Phi)$. Intuitively, coalgebra $\langle \mathsf{nxt}, \mathsf{out} \rangle^\Phi$ behaves like $\langle \mathsf{nxt}, \mathsf{out} \rangle$ but allows its internal states and continuous evolutions to be 'partially' observed; 'how much' one can observe, is dictated by equivalence relation Φ. Denoting $Z/_\Phi$ by Q,

Definition 7. *Consider a coalgebra* $\langle \mathsf{nxt}, \mathsf{out} \rangle^{\Phi} : Z \to Z \times \mathcal{H}Q$ *induced by a hybrid automaton and an equivalence relation* $\Phi \subseteq Z \times Z$. *A relation* $R \subseteq Z \times Z$ *is a* coalgebraic Φ-bisimulation *iff there is a* $\mathcal{F}_{Z/\Phi}$-coalgebra $\gamma : R \to R \times \mathcal{H}Q$ *that makes the following diagram to commute.*

$$
\begin{array}{ccccc}
Z & \xleftarrow{\quad \pi_1 \quad} & R & \xrightarrow{\quad \pi_2 \quad} & Z \\
{\scriptstyle \langle \mathsf{nxt}, \mathsf{out} \rangle^{\Phi}} \downarrow & & {\scriptstyle \gamma} \downarrow & & \downarrow {\scriptstyle \langle \mathsf{nxt}, \mathsf{out} \rangle^{\Phi}} \\
Z \times \mathcal{H}Q & \xleftarrow{\ \pi_1 \times id\ } & R \times \mathcal{H}Q & \xrightarrow{\ \pi_2 \times id\ } & Z \times \mathcal{H}Q
\end{array}
$$

We say that states $z_1, z_2 \in Z$ are *coalgebraically* Φ-*bisimilar* (in symbols, $z_1 \sim^{\Phi} z_2$) if they are related by a coalgebraic Φ-bisimulation.

Given two functions $f, g : A \to B$, and relation $R \subseteq B \times B$, denote the condition $\forall a \in A. (f\,a)\,R\,(g\,a)$ by $f\,R\,g$. Definition 7 tells that a relation R is a coalgebraic Φ-bisimulation iff $z_1\,R\,z_2$ implies

$$(\mathsf{ev}\ z_1)\,\Phi\,(\mathsf{ev}\ z_2), \text{ and } (\mathsf{nxt}\ z_1)\,R\,(\mathsf{nxt}\ z_2).$$

Theorem 1. *Let* $\langle \mathsf{nxt}, \mathsf{out} \rangle^{\Phi} : Z \to Z \times \mathcal{H}Q$ *be induced by a hybrid automaton and an equivalence relation* $\Phi \subseteq Z \times Z$. *Then for any two states* $z_1, z_2 \in Z$, $z_1 \equiv^{\Phi} z_2$ *iff* $z_1 \sim^{\Phi} z_2$.

Proof. In [NB16].

4   When Different Transition Types Come into Play

4.1   The General Picture

The previous section introduced a coalgebraic semantics for hybrid automata in a deterministic setting. The behaviour of digital controllers, however, is far more complex, often combining nondeterministic, or probabilistic features. This calls for variations in the definition of hybrid automata, and, consequently, for a more general coalgebraic semantics, able to capture such variants in a uniform manner. Therefore, we consider coalgebras,

$$\langle \mathsf{nxt}, \mathsf{out} \rangle : S \to (\mathcal{F}S \times \mathcal{H}O)^I$$

where $\mathcal{F}$ determines an internal transition type, and set I denotes an input type. Technically, such arrows can be decomposed into $\mathsf{nxt} : S \times I \to \mathcal{F}S$, $\mathsf{out} : S \times I \to \mathcal{H}O$ (again by a slight abuse of notation). This makes clear that variations in functor $\mathcal{F}$ correspond to variations on how the system (discretely) jumps to a next state. In regard to hybrid automata, we will see that these changes are essentially reflected in relation E and the assignment function asg (recall Definition 1).

Table 1 lists several variations of the functor $\mathcal{F}$ and input type I. Each variant corresponds to a specific definition of hybrid automata. Some of the latter are already well known (*e.g.* the nondeterministic case in row 4), but others are new and thus have not been studied before (*e.g.* the replicating case in row 3).

Table 1. Possible variants for $\mathcal{F}$

Coalgebra	Functor $\mathcal{F}$	Behaviour	Input
$S \to (S \times \mathcal{H}O)$	$Id\, X = X$	Deterministic	No
$S \to (S \times \mathcal{H}O)^I$	$Id\, X = X$	Deterministic	Yes
$S \to (\Delta S \times \mathcal{H}O)$	$\Delta X = X \times X$	Replicating	No
$S \to (\mathcal{P}S \times \mathcal{H}O)$	$\mathcal{P} X = \{A \subseteq X\}$	Nondeterministic	No
$S \to (\mathcal{D}S \times \mathcal{H}O)$	$\mathcal{D} X = \{\mu \in [0,1]^X \mid \mu[X] = 1\}^1$	Probabilistic	No
$S \to (\mathcal{P}\mathcal{D}S \times \mathcal{H}O)$	$\mathcal{P}\mathcal{D}$ —	Segala[2]	No

[1] $\mu[X] = \sum_{x \in X} \mu\, x$.
[2] Traditionally this expression refers to systems with both nondeterministic and probabilistic behaviour.

This illustrates the high level of genericity that coalgebras bring to the theory of hybrid automata: specific types of automata are captured in specific instantiations of functor $\mathcal{F}$, and global constructions and results are defined parametric on $\mathcal{F}$ once and for all. For example, such is the case of coalgebraic Φ-bisimulation, which we will discuss in Sect. 4.4.

The cases listed in Table 1 will be discussed in more detail in the following sections.

4.2   Reactive and Replicating Behaviour

The arrows $S \to (S \times \mathcal{H}O)$ were studied in the previous section. We saw that they provide a suitable coalgebraic semantics for deterministic hybrid automata. Hence, we pass directly to arrows typed as, $S \to (S \times \mathcal{H}O)^I$.

These correspond to a variant of hybrid automata, qualified as *open* (or *reactive*), that takes input/output into consideration (*cf.* [LLK+99]); thus extending the classical definition of hybrid automata (Definition 1) as follows:

Definition 8 ([LLK+99]). *Fix an input set I. Then, add I to the domain of functions* dyn, inv, grd, *and* asg, *keeping the remaining components equal.*

For example, while in the classic case each mode $m \in M$ gave rise to a predicate (inv m), now each pair $(m, i) \in M \times I$ induces a predicate (inv (m, i)).

Similarly, function flow : $(M \times \mathbb{R}^n) \times \mathbb{R}_{\geq 0} \to \mathbb{R}^n$, induced by dyn, has now the signature

$$\text{flow} : (M \times \mathbb{R}^n) \times I \times \mathbb{R}_{\geq 0} \to \mathbb{R}^n.$$

In order to encode open hybrid automata as coalgebras, the condition as-soon-as also needs to be slightly changed: while previously each pair $(m, v) \in (M \times \mathbb{R}^n)$ was associated with a duration δ (see Sect. 2), now we require the same for each triple $(m, v, i) \in (M \times \mathbb{R}^n \times I)$. As in Sects. 2 and 3, we also assume that an open hybrid automaton cannot jump from a valid state into an invalid one.

Then, let us define function $\mathsf{nxt} : Z \times I \to Z$ as

$$\mathsf{nxt}(m, v, i) = \big(E(m), \mathsf{asg}(m \rightsquigarrow E(m), i)\, u\big)$$

where $u = \mathsf{flow}(m, v, i, \delta)$. Finally, given functions $\mathsf{nxt} : Z \times I \to Z$, $\mathsf{out} : Z \times I \to \mathcal{H}O$, we form coalgebra

$$\langle \mathsf{nxt}, \mathsf{out} \rangle : Z \to (Z \times \mathcal{H}O)^I.$$

Let us illustrate the expressive power of $(- \times \mathcal{H}O)^I$-coalgebras via an example related to the bouncing ball system.

Example 7. Suppose we can set the instants of time at which the ball bounces – this may be interpreted, for example, as a foot that kicks the ball up. To define such a behaviour one can construct the following coalgebra:

$$\langle \mathsf{nxt}, \mathsf{out} \rangle\, (m, p, v, i) = \big((m, 0, v'), \mathsf{mov}(p, v, -)\big)$$

where $v' = (v + gi) \times -0.5$, and function $\mathsf{mov}(p, v, -) : [0, i] \to \mathbb{R}$ is defined as before.

The category of $(- \times \mathcal{H}O)^I$-coalgebras also has a final coalgebra. Formally, the following diagram commutes uniquely,

$$
\begin{array}{ccc}
S & \xdashrightarrow{\;[-]\;} & (\mathcal{H}O)^{I^+} \\
{\scriptstyle \alpha}\big\downarrow & & \big\downarrow{\scriptstyle \gamma} \\
(S \times \mathcal{H}O)^I & \xdashrightarrow[([-]\times id)^I]{} & ((\mathcal{H}O)^{I^+} \times \mathcal{H}O)^I
\end{array}
$$

where I^+ denotes the set of nonempty lists of elements in I, and

$$(\gamma\, f)\, i = (g,\, f\, [i]), \quad g\, is = f\, (i : is).$$

Intuitively, any coalgebra $\alpha : S \to (S \times \mathcal{H}O)^I$, induces a unique function $[-] : S \to (\mathcal{H}O)^{I^+}$ (*cf.* [Jac12]), such that $[s]$ associates to each nonempty list of inputs the *last* evolution observed in α, starting in $s \in S$.

Example 8. In Example 7 we considered a bouncing ball system that allows to choose the instants of time at which the ball bounces. Expressions $[(0, 5)]\, [0.8]$, $[(0, 5)]\, [0.8, 0.6]$, and $[(0, 5)]\, [0.8, 0.6, 0.6]$ denote the following sequence.

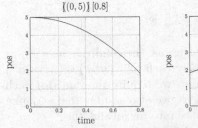
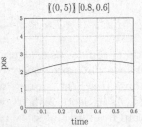
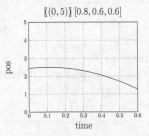

Functor Diagonal (Δ) gives rise to arrows of type $\langle \text{nxt}, \text{out} \rangle : S \to \Delta S \times \mathcal{H}O$. These correspond to deterministic hybrid automata, as studied in Sect. 3, but now able to jump to two different places at the same time. The intuition is that such systems replicate themselves at each discrete transition. For example, the bouncing ball would turn into two at each bounce. From a strict computer science point of view this may seem rather strange, but in other areas it is a common behaviour: *e.g.*, in biology, cells indeed replicate when a specific saturation point is reached. To the best of our knowledge, there is no variant of hybrid automata in the literature associated with this type of behaviour.

4.3   Nondeterministic and Probabilistic Behaviour

Let us now concentrate on the powerset functor ($\mathcal{P}$). Actually, for the sake of simplicity we will restrict to its *finitary* version, $\mathcal{P}_\omega$, which considers only the finite subsets of a given set X. As expected, it gives rise to arrows typed as,

$$Z \to (\mathcal{P}_\omega Z \times \mathcal{H}O)$$

which precisely correspond to a nondeterministic version of the hybrid automata explored in the previous section. More concretely,

- relation E, previously assumed to be a function, is now *finitely branching* (*i.e.*, each mode has a finite number of outgoing edges),
- the assignments are allowed to be *finitely non deterministic*, meaning that the value assigned to a variable is determined up to a finite number of possibilities.

Thus, function $\text{nxt} : Z \to \mathcal{P}_\omega Z$ is defined as

$$\text{nxt}(m, v) = \bigcup_{m' \in E(m)} \left(\{m'\} \times \text{asg}(m \rightsquigarrow m')(u) \right)$$

where $u = \text{flow}(m, v, \delta)$, and $\text{asg}(m \rightsquigarrow m')$ is regarded as a function that given a tuple of valuations $v \in \mathbb{R}^n$, returns the assignments that are possible to perform.

Consider now probabilistic branching by taking $\mathcal{F} = \mathcal{D}$, or $\mathcal{F} = \mathcal{P}_\omega \mathcal{D}$, in $S \to (\mathcal{F}S \times \mathcal{H}O)$. Interestingly, hybrid automata whose internal transition type corresponds to $\mathcal{P}_\omega \mathcal{D}$ were already introduced in document [Spr00]. The idea is that these systems are able to nondeterministically choose a distribution function over the states (which, intuitively, gives the probability of a given state being the next one). Actually, not only this allows to equip edges with probabilities, but also gives rise to probabilistic assignments: for example, one may say $x := x + 10$ *with probability* 0.9.

We refer the interested reader to this paper's extended version [NB16] for a more detailed overview of arrows $S \to \mathcal{D}S \times \mathcal{H}O$, $S \to \mathcal{P}_\omega \mathcal{D}S \times \mathcal{H}O$ and their correspondence to the probabilistic hybrid automata introduced in [Spr00].

4.4 Bisimulation and Observational Semantics

Let us now generalise the notion of coalgebraic Φ-bisimulation (Definition 7) to coalgebras typed as $\langle \text{nxt}, \text{out} \rangle : Z \to (\mathcal{F}Z \times \mathcal{H}O)^I$. As before, assume that $Z \subseteq M \times \mathbb{R}^n$. Then given an equivalence relation $\Phi \subseteq Z \times Z$, we define coalgebra $\langle \text{nxt}, \text{out} \rangle^\Phi : Z \to (\mathcal{F}Z \times \mathcal{H}Q)^I$ similarly to before. More concretely,

$$\langle \text{nxt}, \text{out} \rangle^\Phi(z, i) = (\text{nxt}(z, i), q \cdot (\text{ev}(z, i)))$$

where $\text{ev} : Z \times I \to \mathcal{H}Z$ is a function such that for any $z = (m, v) \in Z, i \in I$, $(\text{ev}(z, i))\, t = (m, (\text{out}(z, i))\, t)$, and $q : Z \to Z/\Phi$ is the quotient map induced by Φ. Then denoting Z/Φ by Q,

Definition 9. *Consider a coalgebra* $\langle \text{nxt}, \text{out} \rangle^\Phi : Z \to (\mathcal{F}Z \times \mathcal{H}Q)^I$ *induced by an equivalence relation* $\Phi \subseteq Z \times Z$. *A relation* $R \subseteq Z \times Z$ *is a coalgebraic* Φ-*bisimulation if there is a coalgebra* $R \to (\mathcal{F}R \times \mathcal{H}Q)^I$ *that makes the following diagram to commute.*

$$
\begin{array}{ccccc}
Z & \xleftarrow{\;\pi_1\;} & R & \xrightarrow{\;\pi_2\;} & Z \\
{\scriptstyle \langle \text{nxt,out} \rangle^\Phi} \downarrow & & \downarrow & & \downarrow {\scriptstyle \langle \text{nxt,out} \rangle^\Phi} \\
(\mathcal{F}Z \times \mathcal{H}Q)^I & \xleftarrow{(\mathcal{F}\pi_1 \times id)^I} & (\mathcal{F}R \times \mathcal{H}Q)^I & \xrightarrow{(\mathcal{F}\pi_2 \times id)^I} & (\mathcal{F}Z \times \mathcal{H}Q)^I
\end{array}
$$

We say that states $z_1, z_2 \in Z$ are *coalgebraically* Φ-*bisimilar* (in symbols, $z_1 \sim^\Phi z_2$) if they are related by a coalgebraic Φ-bisimulation.

Observe that a coalgebraic Φ-bisimulation R is, in fact, a coalgebraic bisimulation in the category of $(\mathcal{F} \times \mathcal{H}Q)^I$-coalgebras. Moreover, note that this definition coincides with Definition 7 when $\mathcal{F} = Id$ and $I = 1$. Actually, for $\mathcal{F} = \mathcal{P}_\omega$, $\mathcal{F} = \mathcal{P}_\omega \mathcal{D}$ (with $I = 1$) we have the following results relating classic and coalgebraic Φ-bisimilarity, $\equiv^\Phi$ and $\sim^\Phi$, respectively.

Theorem 2. *Consider a coalgebra* $\langle \text{nxt}, \text{out} \rangle^\Phi : Z \to (\mathcal{P}_\omega Z \times \mathcal{H}Q)$ *induced by a nondeterministic hybrid automaton and an equivalence relation* $\Phi \subseteq Z \times Z$. *Then for any two states* $z_1, z_2 \in Z$, $z_1 \equiv^\Phi z_2$ *iff* $z_1 \sim^\Phi z_2$.

Proof. In [NB16].

Theorem 3. *Consider a coalgebra* $\langle \text{nxt}, \text{out} \rangle^\Phi : Z \to (\mathcal{P}_\omega \mathcal{D}Z \times \mathcal{H}Q)$ *induced by a probabilistic hybrid automaton* [Spr00] *and an equivalence relation* $\Phi \subseteq Z \times Z$. *Then, for any two states* $z_1, z_2 \in Z$, $z_1 \equiv^\Phi z_2$ *iff* $z_1 \sim^\Phi z_2$.

Proof. In [NB16].

Another interesting aspect to mention concerns open hybrid automata and the apparent absence of a suitable notion of Φ-bisimulation for them (see the previous subsection and also [LLK+99]). However, instantiating Definition 9 with $\mathcal{F} = Id$, we obtain a suitable notion of Φ-bisimulation for such automata, which gives evidence to the generality of the coalgebraic framework.

In order to characterise the observational semantics associated with the arrows $S \to (\mathcal{F}S \times \mathcal{H}O)^I$, we need to guarantee the existence of a final $(\mathcal{F} \times \mathcal{H}O)^I$-coalgebra. In **Set**, the existence of an observational semantics (*i.e.*, a final coalgebra) for systems of type $S \to (\mathcal{F}S \times \mathcal{H}O)^I$ is ensured whenever functor $\mathcal{F}$ is *bounded* (*cf.* [Rut00]). This is not a strong condition. Actually, it holds for all polynomial functors, the finite powerset ($\mathcal{P}_\omega$), and all composites made up of these cases (the reader will find in [Rut00] a complete characterisation of this condition and corresponding proofs). Another case is the distribution functor with finite support ($\mathcal{D}_\omega$); more explicitly, the restriction of functor $\mathcal{D}$ that only considers distributions $\mu \in \mathcal{D}_\omega X$ with a finite number of elements $x \in X$ such that $\mu\, x > 0$ (see the proof, for example, in [Jac12], Theorem 4.6.9).

Therefore, all cases enumerated in Table 1 have a final coalgebra provided that functors $\mathcal{P}$ and $\mathcal{D}$ are restricted to their finitary versions.

4.5   A Hierarchy of Hybrid Automata

Natural transformations are a suitable mechanism to transform a coalgebra into another of a different transition type, because naturality entails preservation of bisimilarity [Sok05]. The case for reflection, however, is more complex: as described in [Sok05], in **Set** bisimilarity is reflected when the natural transformation is injective (*i.e.*, all its components are injective), and the underlying functor of the resulting system preserves weak pullbacks.

Fortunately, it is known that all polynomial functors, the powerset, and the distribution functor, preserve weak pullbacks (*cf.* [Sok05]). Moreover, preservation of weak pullbacks is closed by composition. Therefore, in many cases checking for reflectivity reduces to checking for injectivity. Actually, this is precisely the case for all variants of $S \to (\mathcal{F}S \times \mathcal{H}O)^I$ considered in this paper.

Observe that from a natural transformation $\tau : \mathcal{F} \to \mathcal{G}$ we can construct the natural transformation $(\tau \times id)^I : (\mathcal{F} \times \mathcal{H}O)^I \to (\mathcal{G} \times \mathcal{H}O)^I$.

Then, given a coalgebra $\alpha : S \to (\mathcal{F} \times \mathcal{H}O)^I$, via the natural transformation above, we define $((\tau \times id)^I)_S \cdot \alpha : S \to (\mathcal{G}S \times \mathcal{H}O)^I$.

Since all internal transition types (functors) considered in this paper preserve weak pullbacks, from the existence of injective natural transformations (between transition types), it is possible generate a hierarchy of systems in terms of their expressive power.

'To be more expressive' here means that looking at an $(\mathcal{F} \times \mathcal{H}O)^I$-coalgebra as a $(\mathcal{G} \times \mathcal{H}O)^I$-coalgebra – through the natural transformation $\tau : \mathcal{F} \to \mathcal{G}$ – never entails *loss of observable information*. In other words, if two states of a $(\mathcal{F} \times \mathcal{H}O)^I$-coalgebra are bisimilar when looking at the latter as a $(\mathcal{G} \times \mathcal{H}O)^I$-coalgebra, then the same is true before the application of τ (*i.e.* coalgebraic bisimilarity is reflected). The hierarchy is expressed in the following diagram of injective natural transformations,

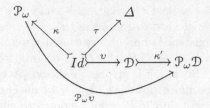

where for any set X, $\tau_X\, x = (x, x)$, $\upsilon_X\, x = \mu$ where $\mu\, x = 1$, $\kappa_X\, x = \{x\}$, and $\kappa'_X\, \mu = \{\mu\}$. Note that there is no injective natural transformation $\Delta \to \mathcal{P}_\omega$ as order is not preserved. Moreover observe that the obvious mapping $\mathcal{P}_\omega \to \mathcal{D}$ (which maps any finite set to the corresponding uniform distribution) does not respect naturality.

We conclude by mentioning the canonical injective natural transformation $(\mathcal{F} \times \mathcal{H}O) \to (\mathcal{F} \times \mathcal{H}O)^I$ (assuming that $I \neq \emptyset$), which, given an element, returns the constant function over it. This adds to the hierarchy the obvious relation between a family of systems and the corresponding extended version that harbours the input/output dimension.

5   Conclusions and Future Work

Even if hybrid automata are the standard formalism for hybrid systems, their definition often needs to cater for different computational behaviours found in practice. In order to make such a process systematic, this paper proposes a coalgebraic rendering of hybrid automata. This allows the study of several variants of the latter, as well as related notions, (*e.g.*, bisimulation, observational semantics) in a uniform manner, at the same time promoting a black-box perspective in which discrete actions are hidden from the environment while continuous evolutions make up the observable behaviour. Furthermore, this characterises hybrid automata as (coalgebraic) components, in the spirit of [Bar03,HJ11].

Interestingly, a somewhat dual perspective appears in the work of Jacobs [Jac00], where an object-oriented approach for hybrid systems is pursued. More concretely, hybrid systems are viewed there as coalgebras equipped with a monoid action (to represent time) that acts over the state space, forcing continuous evolutions to be hidden from the environment. Such a view allows to express physical processes that (continuously) evolve internally, and are possible to interact with at specific instants of time.

It is also relevant to mention the work of Haghverdi *et al.* [HTP05], whose aim is to provide an abstract notion of bisimulation for dynamical, control, and hybrid systems (the latter being understood as hybrid automata). To achieve this, they resort to the notion of an *open map*, which has a close relation to that of coalgebras. Variants of hybrid automata, however, are not taken into consideration.

As future work, we intend to further explore different variants of hybrid automata by varying the functor that gives shape to the internal transitions. For example, arrows of type $S \to (\mathcal{D}S \times \mathcal{H}O)^I$, giving rise to what we call '*reactive Markov* hybrid automata', deserve an independent study. Other interesting cases are replicating hybrid systems (which we briefly addressed here) and the arrows $S \to WS \times \mathcal{HO}$ ($WS = K^S$, for K a set of weights), which makes possible to prescribe costs to discrete transitions and assignments.

Going more generic, and in order to drop the condition as-soon-as (see Sect. 2), one may extend the internal transition type to the 'continuous part' by considering arrows of type $S \to (\mathcal{F}(S \times \mathcal{HO}))^I$ instead. In some cases, however, this may be problematic, as the transition type would need to have a continuous nature. For instance, probabilistic behaviour should be replaced with a stochastic counterpart instead.

On a different note, recall that the results established in this paper allow to define a general characterisation of bisimulation for (different types of) hybrid automata. Such results pave the way to do the same for other notions of bisimulation, one interesting example being *approximate bisimulation* for hybrid automata [GP11].

Finally, a coalgebraic characterisation of hybrid automata makes possible to see them as *hybrid components* (*cf.* [NBHM16]), in the spirit of [Bar03,HJ11]. Generally speaking, this sort of component reproduces the black-box perspective here adopted; and the associated calculus brings to hybrid automata several forms of composition operators (*e.g.*, parallel, pipelining, sum), refinement techniques, and wiring mechanisms, as well as the corresponding algebraic laws. We are currently studying the results brought by this development to the theory of hybrid automata.

Acknowledgements. This work is funded by ERDF - European Regional Development Fund, through the COMPETE Programme, and by National Funds through FCT within project PTDC/EEI-CTP/4836/2014. The first author is also sponsored by FCT grant SFRH/BD/52234/2013, and the second by FCT grant SFRH/BSAB/113890/2015

References

[ACH+95] Alur, R., Courcoubetis, C., Halbwachs, N., Henzinger, T.A., Ho, P.-H., Nicollin, X., Olivero, A., Sifakis, J., Yovine, S.: The algorithmic analysis of hybrid systems. Theor. Comput. Sci. **138**(1), 3–34 (1995)

[AH97] Alur, R., Henzinger, T.A.: Modularity for timed and hybrid systems. In: Mazurkiewicz, A., Winkowski, J. (eds.) CONCUR 1997. LNCS, vol. 1243, pp. 74–88. Springer, Heidelberg (1997). doi:10.1007/3-540-63141-0_6

[AMP+03] Antoniotti, M., Mishra, B., Piazza, C., Policriti, A., Simeoni, M.: Modeling cellular behavior with hybrid automata: bisimulation and collapsing. In: Priami, C. (ed.) CMSB 2003. LNCS, vol. 2602, pp. 57–74. Springer, Heidelberg (2003)

[Bar03] Barbosa, L.S.: Towards a calculus of state-based software components. J. Univ. Comput. Sci. **9**, 891–909 (2003)

[BCB+09] Bartocci, E., Corradini, F., Di Berardini, M.R., Entcheva, E., Smolka, S.A., Grosu, R.: Modeling, simulation of cardiac tissue using hybrid i, o automata. Theor. Comput. Sci. **410**(33–34), 3149–3165 (2009). Concurrent Systems Biology: To Nadia Busi (1968–2007)

[GP11] Girard, A., Pappas, G.J.: Approximate bisimulation: a bridge between computer science and control theory. Eur. J. Control **17**(5–6), 568–578 (2011)

[Hen96] Henzinger, T.A.: The theory of hybrid automata. In: Proceedings, 11th Annual IEEE Symposium on Logic in Computer Science, pp. 278–292. IEEE Computer Society (1996)

[HJ11] Hasuo, I., Jacobs, B.: Traces for coalgebraic components. Math. Struct. Comput. Sci. **21**(2), 267–320 (2011)

[HTP05] Haghverdi, E., Tabuada, P., Pappas, G.J.: Bisimulation relations for dynamical, control, and hybrid systems. Theor. Comput. Sci. **342**(2–3), 229–261 (2005)

[Jac00] Jacobs, B.: Object-oriented hybrid systems of coalgebras plus monoid actions. Theor. Comput. Sci. **239**(1), 41–95 (2000)

[Jac12] Jacobs, B.: Introduction to coalgebra. Towards mathematics of states and observations (2012)

[LLK+99] Liu, J., Liu, X., Koo, T.-KJ., Sinopoli, B., Sastry, S., Lee, E.A.: A hierarchical hybrid system model and its simulation. In: 38th IEEE Decision and Control, vol. 4, pp. 3508–3513. IEEE (1999)

[Nad97] Nadjm-Tehrani, S.: Time-deterministic hybrid transition systems. In: Antsaklis, P., Lemmon, M., Kohn, W., Nerode, A., Sastry, S. (eds.) HS 1997. LNCS, vol. 1567, pp. 238–250. Springer, Heidelberg (1999). doi:10. 1007/3-540-49163-5_13

[NB16] Neves, R., Barbosa, L.S.: Hybrid automata as coalgebras (extended version) (2016). http://alfa.di.uminho.pt/~nevrenato/pdfs/HAExtended. pdf

[NBHM16] Neves, R., Barbosa, L.S., Hofmann, D., Martins, M.A.: Continuity as a computational effect. CoRR, abs/1507.03219 (2016). To appear in J. Logical Algebraic Methods Programm

[Rut00] Rutten, J.: Universal coalgebra: a theory of systems. Theor. Comput. Sci. **249**(1), 3–80 (2000). Modern Algebra

[Sok05] Sokolova, A.: Coalgebraic analysis of probabilistic systems. Ph.D. thesis, Technische Universiteit Eindhoven (2005)

[Spr00] Sproston, J.: Decidable model checking of probabilistic hybrid automata. In: Joseph, M. (ed.) FTRTFT 2000. LNCS, vol. 1926, pp. 31–45. Springer, Heidelberg (2000). doi:10.1007/3-540-45352-0_5

[Szy98] Szyperski, C.: Component Software. Beyond Object-Oriented Programming. Addison-Wesley, New York (1998)

Temporal Logics

Temporal Logic Verification for Delay Differential Equations

Peter Nazier Mosaad$^{(\boxtimes)}$, Martin Fränzle, and Bai Xue

Department of Computer Science, Carl v. Ossietzky Universität, Oldenburg, Germany
{peter.nazier.mosaad,fraenzle,bai.xue}@informatik.uni-oldenburg.de

Abstract. Delay differential equations (DDEs) play an important role in the modeling of dynamic processes. Delays may arise in contemporary control schemes like networked distributed control and may cause deterioration of control performance, invalidating both stability and safety properties. This induces an interest in DDE especially in the area of modeling embedded control and formal methods for its verification. In this paper, we present an approach aiming at automatic safety verification of a simple class of DDEs against requirements expressed in a linear-time temporal logic. As requirements specification language, we exploit metric interval temporal logic (MITL) with a continuous-time semantics evaluating signals over metric spaces. We employ an interval-based Taylor over-approximation method to enclose the solution of the DDE. As the solution of the DDE gets represented as a timed state sequence in terms of the Taylor coefficients, we can effectively solve temporal-logic verification problems represented as time-bounded MITL formulae. We encode necessary conditions for their satisfaction as SMT formulae over polynomial arithmetic and use the iSAT3 SMT solver in its bounded model checking mode for discharging the resulting proof obligations, thus proving satisfaction of MITL specifications by the trajectories induced by a DDE. In contrast to our preliminary work in [34], we can verify arbitrary time-bounded MITL formulae, including nesting of modalities, rather than just invariance properties.

1 Introduction

Ordinary differential equations (ODEs) are traditionally used to model the continuous behavior within continuous- or hybrid-state feedback control systems. Significant research has consequently been pursued to achieve automatic verification for such dynamical systems, among it seamless integration of safe numeric ODE solving with satisfiability-modulo-theory solving [8,13]. In practice, delay is introduced into the feedback loop if components are spatially or logically distributed. Such delays may significantly alter the system dynamics and unmodeled delays in a control loop consequently have the potential to invalidate any stability and safety certificate obtained on the delay-free model. An appropriate

This research is funded by the Deutsche Forschungsgemeinschaft as part of the Research Training Group DFG GRK 1765 SCARE.

© Springer International Publishing AG 2016
A. Sampaio and F. Wang (Eds.): ICTAC 2016, LNCS 9965, pp. 405–421, 2016.
DOI: 10.1007/978-3-319-46750-4_23

generalization of ODE able to model the delay within the framework of differential equations is delay differential equations (DDEs), as suggested by [3].

DDEs play an important role in the modeling of natural or artificial processes with time delays in biology, physics, economics, engineering, etc. As a consequence, attention has gone to developing tools permitting their mechanical analysis. However, such tools still are mostly confined to numeric simulation, e.g. by Matlab's dde23 algorithm. Numerical simulation, despite being extremely useful in system analysis, fails to present reliable certificates of system properties due to numeric approximation. Techniques for safely enclosing set-based initial value problems of ODEs, be it safe interval enclosures [22,25,31], Taylor models [4,26], or flow-pipe approximations based on polyhedra [6], zonotopes [14], ellipsoids [19], or support functions [21], consequently need to be lifted to DDEs.

In [34], a safe enclosure method using interval-based Taylor forms is presented to enclose a set of functions by a parametric Taylor series with parameters in interval form. To avoid dimension explosion incurred by the ever-growing degree of the Taylor series along the time axis, the method depends on fixing the degree for the Taylor series and moving higher-degree terms into the parametric uncertainty permitted by the interval form of the Taylor coefficients. By using this data structure to iterate bounded degree Taylor over-approximations of the time-wise segments of the solution to a DDE, the approach identifies the operator that yields the parameters of the Taylor over-approximation for the next temporal segment from the current one. Employing constraint solving to analyze the properties of this operator, an automatic procedure is obtained to provide stability and safety verification for a simple class of DDEs of the form

$$\frac{\mathrm{d}}{\mathrm{d}t}\boldsymbol{x}(t) = f(\boldsymbol{x}(t-\delta)) \tag{1}$$

with linear or polynomial vector field $f : \mathbb{R}^N \to \mathbb{R}^N$, where the derivative at t is a function of the trajectory at $t - \delta$, i.e., the signal with the delay δ is applied. The method proposed in [34] dealt only with simple invariants as safety properties. Improving on the previous work in [34], the contribution of the current paper lies in verifying a class of safety requirement specified using linear-time temporal logic.

The method proposed in this article again addresses DDEs in the form of Eq. (1) and builds upon the safe enclosure method for DDEs presented in [34], yet addresses metric interval temporal logic (MITL) [1,10,28] with continuous semantics over signals. MITL is a linear temporal logic that is meaningful when the states evolve in metric spaces, an assumption met by continuous-state systems as in Eq. (1). It is considered as a real-time extension of linear temporal logic (LTL), where the modalities of LTL are constrained with timing bounds. In particular, given a continuous dynamical system (1) with its initial condition and a temporal logic specification expressed in time-constrained MITL, we employ the interval-based Taylor over-approximation method to enclose the solution of the given DDE. This facilitates effective reduction of the signal-based, continuous-time and continuous-state MITL verification problem to a related discrete-time MITL verification problem expressible in terms of timed state sequences. By using any bounded model checking (BMC) tool built on top of an arithmetic

SMT solver being able to address polynomial arithmetic, we obtain a procedure able to provide safety certificates for DDE relative to temporal logic specifications. In our case, we use the iSAT3[1] implementation of the iSAT algorithm [12] that provides techniques for bounded and unbounded verification problems like k-induction [30] and Craig interpolation [24].

For dealing with temporal properties expressed in MITL, the key step is to safely determine truth values of atomic propositions occurring in different polarities, i.e., to generate necessary or sufficient conditions for their validity over a time frame based on the Taylor approximation of the DDE (1). Based on this, the solver is able to verify more complex formulae of temporal logic also involving Boolean connectives and temporal modalities, like the *(bounded) until* operator. Our approach to constructing the necessary (or sufficient, resp.) conditions for the atomic predicates relies on a safe linear approximation method, namely a tangent line approximation over the Taylor model; hence, it encloses the atomic formula over a time frame. Our approach is characterized by the soundness guarantees obtained due to the over-approximation of the DDE and the over- or under-approximation —depending on polarity— of atomic predicates. The accuracy of approximation can be selected; an automatic refinement method dynamically adapting the accuracy in case of a negative verdict, however, remains to be developed.

We demonstrate how our approach works in practice and therefore present verification of temporal properties on example systems.

This paper is structured as follows. In Sect. 2, we first formulate the temporal verification problem on DDE in the form of Eq. (1) by defining syntax and continuous-time, continuous-state signal-based semantics of MITL formulae, our requirements specification language. Section 3 develops interval-based Taylor over-approximation as a safe time-wise discretization to the solution of the DDEs, providing a time-invariant operator generating a timed state sequence on Taylor coefficients. In Sect. 4, we adapt the interpretation of MITL to the timed state sequence such that it safely recovers the original semantics on the actual solution of the DDE in terms of conditions on the Taylor coefficients of the time-discrete model. In Sects. 4.2 and 4.3, we then show how to use the iSAT3 SMT solver in bounded model checking mode to discharge the resulting verification conditions, thus demonstrating our technique on pertinent examples. Section 5 finally presents some ideas for refinement of the method proposed and future directions of our work.

2   Problem Formulation

In this section, we formulate the verification problem of a simple class of DDEs in the form of Eq. (1) against a class of safety requirements specified using linear-time temporal logic. We define metric interval temporal logic (MITL) with continuous semantics as requirements specification language for such continuous-state systems.

---

[1] http://projects.informatik.uni-freiburg.de/projects/isat3/.

Let $\mathbb{R}$ be the set of the real numbers. Our time domain is the set of nonnegative real numbers $\mathbb{R}_{\geq 0}$. Also, the trajectory of the DDE of Eq. (1) on an initial condition $x([0,\delta]) \equiv c \in \mathbb{R}$ is a function $x(t)$ such that $x : \mathbb{R}_{\geq 0} \rightarrow \mathbb{R}^N$ satisfies the initial condition and $\forall t \geq \delta : \frac{d}{dt}\boldsymbol{x}(t) = f(\boldsymbol{x}(t-\delta))$, where the positive integer N denotes the dimension of the space. In order to specify the temporal properties of interest, we exploit MITL with continuous semantics that is meaningful when the states evolve in metric spaces as in Eq. (1). We say that $\mathcal{P}(\mathcal{C})$ denotes the powerset of a set $\mathcal{C}$ and assume that AP is a set of atomic propositions. Then, the predicate mapping $\mathcal{M} : AP \rightarrow \mathcal{P}(\mathbb{R}^N)$ is a set valued function that assigns to each atomic proposition $\rho \in AP$ a set of states $\mathcal{M}(\rho) \subseteq \mathbb{R}^N$. In this paper, we take the set of atomic propositions AP to be the simple bound constraints $x(t) \sim c$, where $\sim \in \{<, \leq, >, \geq\}$ and $c \in \mathbb{Q}$.

2.1   Metric Interval Temporal Logic

Metric interval temporal logic (MITL) [1] is a linear temporal logic that is meaningful when the states evolve in metric spaces, an assumption met by continuous-state systems as in Eq. (1). It is a real-time extension of linear temporal logic (LTL), where the modalities of LTL are constrained with timing bounds. Metric temporal logic (MTL) was first introduced by Koymans [17] to specify real-time properties. In order to address the undecidability problem of MTL, Alur et al. [1] relaxed the punctuality of the temporal operators s.t. they cannot constrain to singleton intervals. We employ MITL to formally characterize the desired behavior of DDEs. Along the following lines, we review and suitably adapt the syntax and the continuous semantics of MITL as presented in [1,28].

Definition 1 (Syntax of MITL). *An MITL formula φ is built from a set of atomic propositions AP using Boolean connectives and timed-constrained versions of the until operator. It is inductively defined according to the grammar*

$$\varphi :: = \top \mid \rho \mid \neg\varphi_1 \mid \varphi_1 \wedge \varphi_2 \mid \varphi_1 \mathcal{U}_{\mathcal{I}} \varphi_2$$

where $\rho \in AP$, $\top$ is the Boolean constant true and $\mathcal{I} \subseteq \mathbb{Q}_{\geq 0}$ is a nonsingular interval imposing timing bounds on the temporal operators, where $\mathbb{Q}_{\geq 0}$ is the set of non-negative rational numbers.

We can derive the constant *false* by $\bot \equiv \neg\top$. Also, we can define additional temporal operators such as *release* $\mathcal{R}_{\mathcal{I}}$, *eventually* $\Diamond_{\mathcal{I}}$, and *always* $\Box_{\mathcal{I}}$ as follows:

$$\varphi_1 \mathcal{R}_{\mathcal{I}} \varphi_2 \equiv \neg((\neg\varphi_1)\mathcal{U}_{\mathcal{I}}(\neg\varphi_2)),$$
$$\Diamond_{\mathcal{I}}\varphi \equiv \top\mathcal{U}_{\mathcal{I}}\varphi, \text{ and}$$
$$\Box_{\mathcal{I}}\varphi \equiv \bot\mathcal{R}_{\mathcal{I}}\varphi \equiv \neg\Diamond\neg\varphi.$$

Note that MITL has no *next* operator as the time domain is dense. When $\mathcal{I} = [0, \infty]$, we can remove the subscript $\mathcal{I}$ from the temporal operators, obtaining the traditional modalities of LTL. Finally, we would like to point out that the decidability problem of MITL in the continuous semantics for both model checking and satisfiability problems is out of the scope of this paper. For details about the decidability problem, refer to [1,27]. Also, having a decidable model property of DDE as being a model of MITL formula is an open issue.

Continuous-Time, Continuous-State Semantics of MITL. The continuous semantics of MITL formulae is used to express specifications on the desired temporal evolution to the solutions of DDEs in the form of Eq. (1). This semantics is based on real-valued signals $x : \mathbb{R}_{\geq 0} \to \mathbb{R}^N$ over time. We say that signal x satisfies atomic proposition $x_i \sim c$ at time $t \geq 0$, denoted $x, t \models x_i \sim c$, iff $x_i(t) \sim c$ holds, where x_i is one of the variables interpreted by the trajectory x. Based on this, semantics of arbitrary MITL formulae is defined inductively, with the semantics of Boolean connectives $\neg$ and $\wedge$ as well as the constant $\top$ being standard. The semantics of the time-constrained *until* operator is defined as follows: $x, t \models \varphi_1 \mathcal{U}_\mathcal{I} \varphi_2$ iff for some $t' \in \mathcal{I}$, $x, t + t' \models \varphi_2$ holds and furthermore $x, t \models \varphi_1$ for all $t \in (t, t + t')$.

By convention, we say that the DDE of Eq. (1) with an initial value $x([0, \delta]) \equiv c$ satisfies an MITL formula φ if its solution trajectory $x(t)$ satisfies φ in the sense of $x, 0 \models \varphi$. In what follows, we employ the interval-based Taylor over-approximation method [34] to enclose the solution of such a DDE, which in turn generates a timed state sequence of Taylor coefficients. Thereby we reduce a correctness problem over the continuous semantics into a corresponding problem of a time-invariant operator over discrete time. Later, we recover the continuous semantics on the actual solution of the DDE from the timed state sequence semantics on the Taylor coefficients.

3   Computing Enclosures for DDEs by Taylor Models

In this section, we review the bounded degree interval-based Taylor overapproximation method for a simple class of DDEs first presented in [34].[2] In order to compute an enclosure for the trajectory $x(t)$ defined by an initial value problem of the DDE (1), a template interval Taylor form of fixed degree k is defined as

$$f_n(t) = a_{n_0} + a_{n_1} t + \cdots + a_{n_k} t^k, \tag{2}$$

where f_n encloses the trajectory for time interval $[n\delta, (n+1)\delta]$, the constant δ is the feedback delay from Eq. (1), and $a_{n_0}, \ldots, a_{n_k}$ are interval-vector parameters. The trajectory induced by DDE (1) can be represented by a piece-wise function, with the duration of each piece being the feedback delay δ. To compute the enclosure for the whole solution of the DDE, we need to calculate the relation between the interval Taylor coefficients in successive time steps as pre-post-constraints on these interval parameters. For notational convenience, we denote the interval parameters $[a_{n_0}, \ldots, a_{n_k}]$ by a matrix $A(n)$ in $\mathbb{R}^{N \times (k+1)}$. The relation

---

[2] The corresponding prototype implementation of the interval Taylor over-approximation method for DDEs as well as some examples are available for download from https://github.com/liangdzou/isat-dde.

between $A(n)$ and $A(n + 1)$ can be computed, exploiting different orders of Lie derivatives $f_{n+1}^{(1)}, f_{n+1}^{(2)}, \ldots, f_{n+1}^{(k)}$, as follows:

$$f_{n+1}^{(1)}(t) = g(f_n(t)), f_{n+1}^{(2)}(t) = \frac{d f_{n+1}^{(1)}(t)}{dt}, \ldots, f_{n+1}^{(k)}(t) = \frac{d f_{n+1}^{(k-1)}(t)}{dt}, \qquad (3)$$

i.e., the first order is obtained directly from the given DDE (1) and the $(i+1)$-st order is computed from the i-th order by symbolic differentiation. Then, the Taylor expansion of $f_{n+1}(t)$ with fixed degree k is derived as follows:

$$f_{n+1}(t) = f_n(\delta) + \frac{f_{n+1}^{(1)}(0)}{1!}t + \cdots + \frac{f_{n+1}^{(k-1)}(0)}{(k-1)!}t^{k-1} + \frac{f_{n+1}^{(k)}(\xi_n)}{k!}t^k, \qquad (4)$$

where ξ_n is a vector ranging over $[0, \delta]^N$.

From Eq. (4), by comparing the coefficients of monomials with the same degree at the two sides and by replacing ξ_n by the interval vector $[0, \delta]^N$, we can obtain a time-invariant operator which represents the relation between $A(n)$ and $A(n + 1)$. The details of this construction can be found in [34] or retrieved from the example underneath. Hence, we safely enclose the trajectory induced by the DDE (1) by a discrete-time model providing a timed state sequence on a state space $\mathcal{S} \subseteq \mathbb{R}^N$.

3.1 Time-Wise Discretization of DDEs into Timed State Sequences

We demonstrate on a running example (taken from [34]) how to provide the discrete-time model that encloses the solution of a DDE like Eq. (1). The running example is the DDE

$$\dot{x}(t) = -x(t - 1) \qquad (5)$$

with the initial condition $x([0, 1]) \equiv 1$. Figure 1 shows the solution of ODE $\dot{x} = -x$ without delay (the dashed line) and with 1 second delay (the solid line). Obviously, the difference between the ODE and the DDE is substantial and necessitates analysing the behavior of the DDE.

The method, in [34], aims at over-approximating the solution of DDE (5) by iterating bounded degree interval-based Taylor over-approximations of the time-wise segments of the solution to the DDE. That way, we identify the operator that yields the parameters of the Taylor over-approximation for the next temporal segment from the current one. For instance, suppose we are trying to over-approximate the solution of DDE (5) by polynomials of degree 2. Then we can predefine a template Taylor form $f_n(t) = a_{n_0} + a_{n_1}t + a_{n_2}t^2$ on interval $[n, n + 1]$, where a_{n_0}, a_{n_1}, and a_{n_2} are interval parameters able to incorporate the approximation error eventually necessarily incurred by bounding the degree of the polynomial to (in this example) 2. Here, $f_n(t)$ corresponds to the solution x of DDE (5) at time $n + t$, i.e., $f_n(t)$ over-approximates $x(n + t)$ in the sense of $x(n + t) \in f_n(t)$.

In order to compute the Taylor model, the first and second derivative $f_{n+1}^{(1)}(t)$ and $f_{n+1}^{(2)}(t)$ of solution segment $n + 1$ based on the preceding segment (where

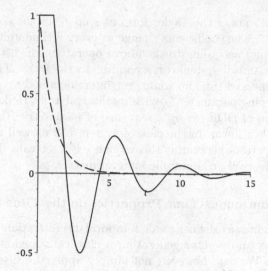

Fig. 1. Solutions to the ODE $\dot{x} = -x$ (dashed graph) and the related DDE $\dot{x}(t) = -x(t-1)$ (solid line), both on similar initial conditions $x(0) = 1$ and $x([0, 1]) \equiv 1$, respectively.

both segments are of duration 1 each) have to be calculated. The first derivative $f^{(1)}_{n+1}(t)$ is computed directly from Eq. (5) as

$$f^{(1)}_{n+1}(t) = -f_n(t) = -a_{n_0} - a_{n_1}t - a_{n_2}t^2 .$$

The second derivative $f^{(2)}_{n+1}(t)$ is computed based on $f^{(1)}_{n+1}(t)$ by

$$f^{(2)}_{n+1}(t) = \frac{d\left(f^{(1)}_{n+1}(t)\right)}{dt} = -a_{n_1} - 2a_{n_2}t .$$

By using a Lagrange remainder with fresh variable $\xi_n \in [0, 1]$, we obtain

$$f_{n+1}(t) = f_n(1) + \frac{f^{(1)}_{n+1}(0)}{1!}t + \frac{f^{(2)}_{n+1}(\xi_n)}{2!}t^2$$

$$= (a_{n_0} + a_{n_1} + a_{n_2}) - a_{n_0}t - \frac{a_{n_1} + 2a_{n_2}\xi_n}{2}t^2 .$$

Then, the operator expressing the relation between Taylor coefficients in the current and the next step can be derived by replacing both $f_n(t)$ and $f_{n+1}(t)$ with their parametric forms $a_{n_0} + a_{n_1}t + a_{n_2}t^2$ and $a_{n+1_0} + a_{n+1_1}t + a_{n+1_2}t^2$ in the above equation and pursuing coefficient matching. As a result, one obtains the operator

$$\begin{bmatrix} a_{n+1_0} \\ a_{n+1_1} \\ a_{n+1_2} \end{bmatrix} = \begin{bmatrix} 1 & 1 & 1 \\ -1 & 0 & 0 \\ 0 & -\frac{1}{2} & -\xi_n \end{bmatrix} \begin{bmatrix} a_{n_0} \\ a_{n_1} \\ a_{n_2} \end{bmatrix} \tag{6}$$

mapping the coefficients of the Taylor form at step f_n to the coefficients of the Taylor form of f_{n+1}. The coefficients change at every δ time units (every second in the given example) according to the above operator, which therefore defines a discrete-time dynamical system corresponding to the DDE. The discrete-time operator can be rendered time-invariant, yet interval-valued by substituting the uncertain time varying parameter ξ_n with its interval $[0, \delta]$. Hence, we can safely enclose the solution of DDE (5) by a sequence of parametric Taylor series with parameters in interval form. In the case of system (5), as well as for any other linear DDE, the operator generating this sequence is a set-valued linear operator definable by an effectively computable interval matrix.

3.2   Proving Continuous-Time Properties on the Time Discretization

Operator (6) straightaway defines a safe temporal discretization of the DDE system in Eq. (1), i.e., an operator generating a classical timed state sequence in the sense of [1,9]. We can, however, not simply apply the discrete-time interpretation of MITL to this timed state sequence, as it ranges over a different state space such that we have to translate forth and back between the state spaces and time models. The iterated execution of operator (6), starting from an initial vector $a_{0_0}, \ldots, a_{0_k}$ of Taylor coefficients encoding the initial solution segment $x([0, \delta])$, generates a timed state sequence over (interval) Taylor coefficients, with time stamps $t_i = i\delta$, rather than a signal over the state variables x_i. We do therefore need a translation step generating conditions over the timed sequence of Taylor coefficients from which we are able to recover the original continuous-time, continuous-state signal-based semantics on the actual solution x of the DDE, as defined in Sect. 2.1.

As has already been observed in [34], such a mapping is straightforward when invariance properties are to be dealt with, for which a sufficient —yet, in the light of over-approximation of the solution, obviously not necessary— condition can be obtained as follows. For an invariance requirement $\Box x \in Safe$, where $Safe$ is a safe set of states, the requirement in the n-th segment is translated to $\forall t \in [0, 1] : f_n(t) \in Safe$, where f_n is the Taylor form stemming from the n-th iteration of the operator (6). Hence, the safety property $S(x)$ for system (5) is translated to a safety property $\forall n \in \mathbb{N}, \forall t \in [0, \delta] : f_n(t) \in Safe$. As its violation is an existential statement both w.r.t. a step number n and an existentially quantified time point t, a solver for satisfiability modulo theory over the existential theory of polynomial arithmetic can be used to solve the safety verification problem. It requires polynomial constraint solving due to the Taylor forms, i.e., polynomial expressions involved in the statement $f_n(t) \in Safe$.

Different proof schemes can be implemented using such a solver: using k-induction [30] or interpolation-based unbounded proof schemes [24], absence of any sequence of valuations generated by operator (6) and satisfying $\exists n \in \mathbb{N}, \exists t \in [0, \delta] : f_n(t) \notin Safe$ can be shown, thereby rigorously showing safety of the DDE system under investigation. Bounded model checking of the same system could, on the other hand, generate counterexamples to safety, which however may be spurious due to the overapproximation involved in the Taylor enclosure.

4 Solving MITL Formulae with Continuous Semantics Over Time-Discrete Taylor-Based Approximations

In this section, we extend the above idea of generating sufficient conditions for MITL specifications on DDEs in terms of the sequences of enclosing (interval) Taylor coefficients. The aim is to cover a large fragment of MITL, rather than just invariance properties as in [34]. We therefore present a technique to recover the continuous semantics of MITL formulae on the actual solution of the DDEs in the form of Eq. (1) from timed state sequences representing interval Taylor approximations. As explained in the previous section, we have obtained a generator for a timed state sequence (the operator (6)) representing the solution of the DDE, yet ranging over a different state space: the Taylor coefficients. Hence, the continuous interpretation of the MITL formulae over DDE solutions has to be translated into a semantically appropriate discrete interpretation on a timed state sequence. This translation needs to restore, in the sense of providing sufficient conditions for satisfaction, the continuous semantics of the MITL formulae over the discrete model of the timed state sequence. In order to be able to deal with negations in (sub-)formulae, we will subsequently state pairs of sufficient and of necessary conditions.

4.1 Atomic Propositions

According to the MITL syntax of Sect. 2, atomic propositions are of the form $x_i \sim c$, where c is a constant and $\sim$ an inequational relational operator, i.e., one of $<, \leq, >, \geq$. Conditions for satisfaction of such a proposition over a time frame $[n\delta, (n+1)\delta]$ can obviously be expressed in terms of the values of x_i in the endpoints $t = n\delta$ and $t = (n+1)\delta$ plus knowledge about the curvature of x_i throughout $[n\delta, (n+1)\delta]$. All the aforementioned information items can readily be retrieved from the Taylor coefficients $a_{n_0}, \ldots, a_{n_k}$.

We therefore split between the cases where the Taylor polynomials $f_n(t)$ are being *concave up* (i.e., $f_n^{(2)}(t) \geq 0$) or *concave down* (i.e., $f_n^{(2)}(t) \leq 0$) throughout the interval. Note that due to the interval coefficients as well as the impact of higher-order derivatives, a third case actually is possible: some solutions may be left-bend, others right-bend, or even a single one may feature both curvature directions. We do not take up this case here explicitly, as constraints solving will later resolve it implicitly by in that case asserting the conjunction of conditions for both curvatures and —if that turns out to be unsatisfiable— subsequently splitting time and/or value intervals until a pure situation is achieved.

Furthermore, based on the polarities of the inequalities in the atomic propositions (i.e., whether it is an upper or a lower bound), we consider another two main cases in constructing the sufficient conditions on the over-approximated model.

Case 1. Assume the atomic proposition is of the form $x_i \lhd c$, with $c \in \mathbb{Q}$ and $\lhd \in \{<, \leq\}$.

Assume that f_n is the interval Taylor polynomial enclosing x_i over the time interval $[n\delta, (n+1)\delta]$ and that we want to construct a sufficient condition for validity of $x_i \lhd c$ over $[n\delta, (n+1)\delta]$.

When $f_n(t)$ is *concave up*, i.e., $\forall t \in [0, \delta] : f_n^{(2)}(t) \geq 0$, we formulate the condition $\mathcal{N}_{n,1}$ guaranteeing validity of $x_i \lhd c$ over $[n\delta, (n+1)\delta]$ as follows:

$$\mathcal{N}_{n,1} := (f_n(0) \lhd r) \wedge (f_n(\delta) \lhd r). \tag{7}$$

In case that $f_n(t)$ is *concave down*, i.e., $\forall t \in [0, \delta] : f_n^{(2)}(t) \leq 0$, we apply the linear approximation method to construct the condition $\mathcal{N}_{n,2}$ guaranteeing validity of $x_i \lhd c$ over $[n\delta, (n+1)\delta]$ as follows:

$$\mathcal{N}_{n,2} := \forall \varepsilon \in [0, \delta] : ((f_n(0) + f_n^{(1)}(0) \cdot \varepsilon \lhd r) \vee (f_n(\delta) - f_n^{(1)}(\delta) \cdot \varepsilon \lhd r)). \tag{8}$$

By considering all the cases, when $f_n(t)$ is *concave up* or *concave down* w.r.t. the interval Taylor coefficients, we construct the condition $\mathcal{N}_n$ for the over-approximation model of the DDE as follows:

$$\mathcal{N}_n := (f_n^{(2)}(t) \geq 0 \wedge \mathcal{N}_{n,1}) \vee (f_n^{(2)}(t) \leq 0 \wedge \mathcal{N}_{n,2}). \tag{9}$$

Note that the condition $\mathcal{N}_n$ is the sufficient condition for validity of $x_i \lhd r$ over time frame $[n\delta, (n+1)\delta]$. Its universally quantified form therefore converts into an existentially quantified one amenable to SMT solving as soon as we are using its negation, as usual in SMT-based analysis trying to falsify the given property, e.g. using the iSAT solver.

Proposition 1. *Let ψ be an MITL formula, $\dot{x}(t) = f(\boldsymbol{x}(t - \delta))$ be a DDE with its initial conditions, and $\mathcal{N}_n$ be the formula generated from ψ and the DDE. Then the satisfiability of $\mathcal{N}_n$ implies $\forall t^* \in [n\delta, (n+1)\delta] : x, t^* \models \psi$, where x is the solution to the DDE.*

Case 2. Assume the atomic proposition is of the form $x_i \rhd c$, with $c \in \mathbb{Q}$ and $\rhd \in \{>, \geq\}$.

In this case, when $f_n(t)$ is *concave up*, i.e., $\forall t \in [0, \delta] : f_n^{(2)}(t) \geq 0$, the condition $\mathcal{N}_{n,1}$ is formulated by applying the linear approximation method to guarantee validity of $x_i \rhd c$ over $[n\delta, (n+1)\delta]$ as follows:

$$\mathcal{N}_{n,1} := \forall \varepsilon \in [0, \delta] : ((f_n(0) + f_n^{(1)}(0) \cdot \varepsilon \rhd r) \wedge (f_n(\delta) - f_n^{(1)}(\delta) \cdot \varepsilon \rhd r)). \tag{10}$$

For $f_n(t)$ is *concave down*, i.e., $\forall t \in [0, \delta] : f_n^{(2)}(t) \leq 0$, the sufficient condition $\mathcal{N}_{n,2}$ guaranteeing validity of $x_i \rhd c$ over $[n\delta, (n+1)\delta]$ is formulated as follows:

$$\mathcal{N}_{n,2} := (f_n(0) \rhd r) \wedge (f_n(\delta) \rhd r). \tag{11}$$

As above in Eq. (9), the condition $\mathcal{N}_n$ for the over-approximation model is constructed as the disjunction of (10) and (11).

Note that for cases where the curvature is not fixed over a time interval, we may have the constraint solver split both the time and the Taylor coefficient intervals into sub-intervals, thus eventually fixing them to (locally) defined signs.

4.2 Boolean Connectives

Assume we have a compound formula of the form $\psi_1 = \phi_1 \wedge \phi_2$ or $\psi_2 = \neg\phi_1$ and are given translations of $\phi_{1,2}$ into corresponding sufficient conditions over the Taylor approximations. Then we can obviously handle ψ_1 by just conjoining the encodings of ϕ_1 and ϕ_2. As the same trick applies for disjunction rather than conjunction, we can handle ψ_2 by converting it into negation normal form, taking advantage of the immediate translation of complementary atomic propositions.

4.3 Until Operator

Assume we have a compound formula of the form $\psi = \phi_1 \mathcal{U}_\mathcal{I} \phi_2$, with the lower and upper bound of $\mathcal{I}$ for simplicity being integer multiples $l\delta$ and $u\delta$ of δ (including the unbounded case $u = \infty$). Then a sufficient condition for validity of ψ over time frame $[n\delta, (n+1)\delta]$ can be formulated as

$$\mathcal{N}_n := \bigvee_{i=l}^{u} \left(\mathcal{N}_i^{\phi_2} \wedge \bigwedge_{j=0}^{i-1} \mathcal{N}_j^{\phi_1} \right), \tag{12}$$

where $\mathcal{N}^{\phi_i}$ represents the respective encodings of subformulae.

Please note that in practice, we will not expand the above conjunctions and disjunctions, but will instead encode them by state bits and an appropriate transition relation within a transition system, leaving the unwinding to a bounded model-checking engine, like the one available in `iSAT3`.

4.4 Verification Examples

In this section, we use the `iSAT3` SMT solver in its bounded model checking (BMC) mode to verify/falsify the temporal verification problems. The `iSAT3` solver is a satisfiability checker for Boolean combinations of arithmetic constraints over real- and integer-valued variables as well as a bounded model-checker for transition systems over the same fragment of arithmetic [29]. It is a stable version implementation of the iSAT algorithm [12]. The solver can efficiently solve bounded and unbounded verification problems that involve polynomial (and, if needed, transcendental) arithmetic. Hence, it is a good option to solve our proposed problem due to the Taylor forms involved. Also, it allows us to verify/falsify a variety of MITL formulae built on atomic predicates defined over simple bounds, linear, and nonlinear constraints [18]. Due to the solving procedures of the solver, for handling existential constraints only, we provide necessary or sufficient conditions based on the over-approximation model of the DDE to recover the continuous semantics of the MITL formula $\varphi(x)$ on the actual solution x of the DDE from the timed state sequence semantics. In such a way, the solver is able to verify/falsify more complex formulae of temporal logic including, e.g., the *(bounded) until* operator.

We demonstrate that approach based on some examples of DDEs in the form of Eq. (1). In our examples, we consider the DDE (5) as presented in Sect. 3.1 with different MITL formulae to be verified.

Example 1. Consider the linear DDE $\dot{x}(t) = -x(t-1)$ with initial condition $x([0,1]) \equiv 1$ and a safety property of the form $\square_{[0,50]}(x \le 3)$ to be checked.

Based on the bounded degree interval Taylor models to over-approximate the solution by the polynomials of degree 2, we calculate the operator Eq. (6) as presented in Sect. 3.1 that expresses the relation between Taylor coefficients in the current and the next time segment. Consequently, the timed state sequence is obtained where the states range over the interval Taylor coefficients in turn representing polynomials of degree 2 associated with time intervals of duration 1 each, and a predicate mapping function. We encode the over-approximation model in the input language of the iSAT3 solver as shown in Listing (1.1). To achieve this, we need to define the variables of the dynamic system, Taylor coefficients of the Taylor over-approximation solution, the duration of each segment t (here $t \in [0,1]$), the uncertain time varying parameter ξ (here $\xi \in [0,1]$), and the *counter* for the timing bound of the MITL formula.

As explained above, the MITL formula is transformed into the Taylor domain interpreted over the timed state sequence in terms of the Taylor coefficients such that the property now is $\square_{[0,50]}(f_n(t) \le 3)$, where $f_n(t)$ is the Taylor polynomial of degree 2. The solver can solve the formula upholding its original continuous semantics as its violation (i.e., $f_n(t) > 3$) is an existential statement both w.r.t. the state s_i and an existentially quantified time point t. Note that in Listing (1.1), lines 16 and 24 provide a sufficient condition for the over-approximation to eventually reach $x > 3$ preserving the continuous semantics of the MITL formula. If the solver found a CEX, it means that the solver found a violation for the safety property w.r.t. the timing bounds $\mathcal{I}$ and the over-approximation model. Otherwise, in case there is no CEX, it means that the safety property has no violation in k_{depth} of the timing bound. In addition, the bounded depth k_{depth} can be set with --start-depth and --max-depth command line options in the iSAT3 solver instead of defining a *counter*.

In our example, the solver outputs that the system is *safe*, namely the target property is not reachable. In other words, the violation of the property (i.e., the target property) is never reachable in depth 50; hence, the given MITL formula holds in depth 50.

Example 2. Consider the same DDE equation as Example (1) with the same initial condition, but for solving the safety property $(x \ge 1)\, \mathcal{U}_{[0,50]}\, (x \ge 3)$.

For the MITL formula in the Taylor domain, we construct the condition (the *Boolean variable c* in our example) on $f_n(t) \ge 1$ for the over-approximation model of the DDE; hence, we safely determine the truth value of the atomic predicate $f_n(t) \ge 1$ in the course of checking the stop condition of the *until* operator. In other words, we recover the continuous semantics of the MITL formula on the actual solution of the DDE from the interpretation on the timed state sequence. To encode such problem in the input language of iSAT3 solver as shown in Listing (1.2), we define a fresh *Boolean variable, b* in our example, and initialize it with *false* value that remains as long as the second predicate (i.e., $f_n(t) \ge 3$) in the given formula does not occur. In order to get a sound answer that the MITL formula has no violation w.r.t. the over-approximation model, we encode the violation of the property as the target property, in a sense, we search for the violation of the MITL formula. In this vein, the generated condition, c in our example, is the necessary condition for the violation of the MITL formula.

```
1    DECL
2    -- The range of each variable has to be bounded
3       float [-1000, 1000] a0,a1,a2,x;
4       float [0,1] t,xi;
5    -- Define counter for bounded verification problem
6       int [0,1000] counter;
7    INIT
8    -- Initialize Taylor parameters
9       a0 = 1;
10      a1 = 0;
11      a2 = 0;
12      x  = 1;
13      counter = 0; --t_l of the timing bound
14   TRANS
15   -- Taylor polynomial form of degree 2
16      x'= a0 + a1*t + a2*(t^2);
17   -- The relation between current and next step
18      a0' = a0 + a1 + a2;
19      a1' = -a0;
20      a2' = -0.5*a1 - xi*a2;
21      counter' = counter + 1;
22   TARGET
23   -- The negation of the safety property and t_u
24      x > 3 and counter <= 50;
```

Listing 1.1. MITL formula with the *always* temporal operator

Note that we may use the semantics of the *weak until* operator, which is similar to the *until* operator but the stop condition is not required. This is commonly used in case of unbounded verification problems. In this example, the solver outputs that the system is *unsafe*, which means that the violation of the MITL formula is reachable w.r.t. the over-approximation model of the DDE. In this case, the solver provides a CEX that can be spurious due to the over-approximation model. We can do refinements on the over-approximation model by increasing the bounded degree of the Taylor form $f_n(t)$. Note that after increasing the bounded degree by 1, the solver still gives *unsafe* result.

Finally, along these lines, we should point out that all the above verification procedures for the temporal specifications of the simple class of DDE may fail due to two reasons. First, the excessive over-approximation for the solution of DDE, which would be induced by selecting an insufficient bound on the degree of the Taylor forms. In order to address this issue, we select a higher degree, however, it is unclear with a negative verdict that the failure is obtained from the excessive over-approximation or the property of the system under investigation. Aiming at disambiguating these two cases, methods using *counter-example guided abstraction refinement (CEGAR)* [7] remain to be developed for enhancing the over-approximation model. Second, the iSAT3 solver may not terminate especially in case of unbounded verification problems due to the excessive complexity in particular to the Taylor models with high degree. This can in principle be cured by trying to negate the bound constraints in the safety property which might help the solver to find fast a witness for solving the problem. Furthermore, efficient algorithms for *interpolation-based BMC* [24] are employed in the iSAT3 solver to efficiently solve the unbounded verification problems even for non-linear constraints [18].

```
1   DECL
2     float [-1000, 1000] a0, a1, a2, x;
3     float [0,1] t, xi;
4     int [0,50] counter;   --define the counter for the timing bound
5     boole b, c;  -- define c and b as Boolean variables
6   INIT
7   --initialize the the system wrt. the initial condition
8     a0 = 1;
9     a1 = 0;
10    a2 = 0;
11    x = 1;
12    c  <-> (x>=1);   --initialize c, the necessary condition of x>=1
13    b = 0;           --initialize b with false
14    counter = 0;     --counter for the timing bound of the formula
15  TRANS
16  -- description of the over-approximation DDE model
17    x'  = a0 + a1*t + a2*(t^2);
18    a0' = a0 + a1 +a2;
19    a1' = -a0;
20    a2' = -0.5*a1 - xi *a2;
21    counter' = counter +1;   --increment the counter by 1 for each state
22  -- Boolean variable b is false as long as x>=3 doesn't occur
23    b'  <-> (b or x>=3);
24  -- Boolean variable c of the necessary condition of x>=1
25    c'  <-> ((x>=1  && x'>=1      && a2<=0)) or
26           ((x>=1  && x+ a1>=1    && a2>=0) and (x'>=1 && x'-a1>=1  && a2>=0))
27  TARGET
28    counter <=50;
29    !c and !b;   --the target property is the violation of the MITL formula
```

Listing 1.2. MITL formula with the *(bounded) until* temporal operator

5   Conclusion and Future Work

In this paper, we have elaborated a method to verify/falsify temporal specifications of time-delay systems modeled by a simple class of delay differential equations (DDEs) with a single constant delay. Several dynamical systems can be modeled by DDEs with a single constant delay as in biology [15, 23], optics [16], economics [32, 33], ecology [11], to name just a few. As requirements specification language, we have exploited metric interval temporal logic (MITL) [1] with continuous semantics on the solutions of the DDEs. We have built our method on employing a fixed degree interval-based Taylor over-approximation technique [34] to provide a safe enclosure method for DDEs, thereby in turn obtaining timed state sequences. In this way, the continuous semantics of the MITL formulae is reduced to a time-discrete problem on timed state sequences in terms of Taylor coefficients. Then, we have encoded the interpretation of MITL on these timed state sequences in order to recover the continuous semantics on the actual solutions of the DDEs. Furthermore encoding the negation of the resulting sufficient conditions for MITL satisfaction in a bounded model checking (BMC) tool built on top of an arithmetic SMT solver addressing (a.o.) polynomial arithmetic, we hence have obtained an approach able to provide certificates of temporal properties for a class of DDEs. In our case, we have used the iSAT3 solver, which is the third implementation of the iSAT algorithm [12]. In very first experiments on a simple DDE, the iSAT3 solver proved able to solve the temporal properties expressed in MITL formulae, thereby safely determining satisfaction of the formulae in an over-approximation setting. We were able to verify formulae of temporal logic also involving Boolean connectives and temporal modalities, like the *(bounded) until* operator. Our approach to construct the sufficient (or necessary, after final complementation in the BMC encoding) conditions for the atomic predicates relies on the linear approximation method (tangent line approximation) over the Taylor over-approximation model; hence, we safely enclose the atomic formula over each time interval of the timed state sequence.

We have presented some examples to demonstrate our method. The soundness of the method is guaranteed due to the over-approximation employed in DDE enclosure by Taylor forms and the over- or under-approximation — depending on polarity— of the atomic predicates. Such over-approximation may, however, provide spurious counterexamples in case of a failing verification attempt, which ought to be disambiguated from true counterexamples.

To resolve that ambiguity in case of a negative verdict, as a future work, further techniques remain to be developed. We may build our idea on the general *counter-example guided abstraction refinement (CEGAR)* technique [7].

In control applications, one may furthermore want to combine delayed feedback, as imposed a.o. by networked control, with immediate state feedback modeled by ordinary differential equations (ODEs). Some algorithms are currently under development to handle such cases. The main idea is based on a layered combination of Taylor-model computation for ODE, e.g., [26], with the ideas imposed in [34] for DDE. In this way, we may extend our method exposed herein to verify the temporal properties of dynamical systems modeled by the combination of ODE and DDE. In subsequent steps, we plan to extend the method even further to more general kinds of DDE, like DDE with multiple different discrete delays, DDE with randomly distributed delay, or DDE with time-dependent or more generally state-dependent delay [20]. Finally, we would like to point out that in this paper, we essentially have presented a verification method based on model checking to design a time-delay continuous systems modeled by a simple class of DDEs. This method may also be used in interactive proofs and stepwise refinement of hybrid systems featuring delayed feedback, akin to the methods developed for traditional hybrid systems [2,5].

References

1. Alur, R., Feder, T., Henzinger, T.A.: The benefits of relaxing punctuality. J. ACM **43**(1), 116–146 (1996)
2. Babin, G., Aït-Ameur, Y., Nakajima, S., Pantel, M.: Refinement and proof based development of systems characterized by continuous functions. In: Li, X., Liu, Z., Yi, W. (eds.) SETTA 2015. LNCS, vol. 9409, pp. 55–70. Springer, Heidelberg (2015). doi:10.1007/978-3-319-25942-0_4
3. Bellman, R., Cooke, K.L.: Differential-difference equations. Technical report R-374-PR, The RAND Corporation, Santa Monica, California, January 1963

4. Berz, M., Makino, K.: Verified integration of ODEs and flows using differential algebraic methods on high-order Taylor models. Reliable Comput. **4**(4), 361–369 (1998)

5. Butler, M.J., Abrial, J.-R., Banach, R.: Modelling and refining hybrid systems in Event-B and Rodin. In: Petre, L., Sekerinski, E. (eds.) From Action Systems to Distributed Systems - The Refinement Approach, pp. 29–42. Chapman and Hall/CRC, Boca Raton (2016)

6. Chutinan, A., Krogh, B.H.: Computing polyhedral approximations to flow pipes for dynamic systems. In: Proceedings of the 37th International Conference on Decision and Control (CDC 1998) (1998)

7. Clarke, E., Grumberg, O., Jha, S., Lu, Y., Veith, H.: Counterexample-guided abstraction refinement. In: Emerson, E.A., Sistla, A.P. (eds.) CAV 2000. LNCS, vol. 1855, pp. 154–169. Springer, Heidelberg (2000). doi:10.1007/10722167_15

8. Eggers, A., Fränzle, M., Herde, C.: SAT modulo ODE: a direct SAT approach to hybrid systems. In: Cha, S.S., Choi, J.-Y., Kim, M., Lee, I., Viswanathan, M. (eds.) ATVA 2008. LNCS, vol. 5311, pp. 171–185. Springer, Heidelberg (2008). doi:10.1007/978-3-540-88387-6_14

9. Fainekos, G.E., Girard, A., Pappas, G.J.: Temporal logic verification using simulation. In: Asarin, E., Bouyer, P. (eds.) FORMATS 2006. LNCS, vol. 4202, pp. 171–186. Springer, Heidelberg (2006)

10. Fainekos, G.E., Pappas, G.J.: Robustness of temporal logic specifications for finite state sequences in metric spaces. Technical report MS-CIS-06-05, Dept. of CIS, Univ. of Pennsylvania (2006)

11. Fort, J., Méndez, V.: Time-delayed theory of the neolithic transition in Europe. Phys. Rev. Lett. **82**(4), 867 (1999)

12. Fränzle, M., Herde, C., Ratschan, S., Schubert, T., Teige, T.: Efficient solving of large non-linear arithmetic constraint systems with complex Boolean structure. J. Satisfiability, Boolean Model. Comput. - Special Issue on SAT/CP Integr. **1**, 209–236 (2007)

13. Gao, S., Kong, S., Clarke, E.M.: Satisfiability modulo ODEs. In: Formal Methods in Computer-Aided Design, FMCAD 2013, Portland, OR, USA, pp. 105–112. IEEE, 20–23 October 2013

14. Girard, A.: Reachability of uncertain linear systems using zonotopes. In: Morari, M., Thiele, L. (eds.) HSCC 2005. LNCS, vol. 3414, pp. 291–305. Springer, Heidelberg (2005)

15. Glass, L., Mackey, M.C.: From Clocks to Chaos: The Rhythms of Life. Princeton University Press, Princeton (1988)

16. Ikeda, K., Matsumoto, K.: High-dimensional chaotic behavior in systems with time-delayed feedback. Physica D **29**(1–2), 223–235 (1987)

17. Koymans, R.: Specifying real-time properties with metric temporal logic. Real-Time Syst. **2**(4), 255–299 (1990)

18. Kupferschmid, S., Becker, B.: Craig interpolation in the presence of non-linear constraints. In: Fahrenberg, U., Tripakis, S. (eds.) FORMATS 2011. LNCS, vol. 6919, pp. 240–255. Springer, Heidelberg (2011)

19. Kurzhanski, A.B., Varaiya, P.: Ellipsoidal techniques for hybrid dynamics: the reachability problem. In: Kurzhanski, A.B., Varaiya, P. (eds.) New Directions and Applications in Control Theory. LNCIS, vol. 321, pp. 193–205. Springer, Heidelberg (2005)

20. Lakshmanan, M., Senthilkumar, D.V.: Dynamics of Nonlinear Time-Delay Systems. Springer, Heidelberg (2011)

21. Le Guernic, C., Girard, A.: Reachability analysis of linear systems using support functions. Nonlinear Anal. Hybrid Syst. **4**(2), 250–262 (2010)
22. Lohner, R.: Einschließung der Lösung gewöhnlicher Anfangs- und Randwertaufgaben. Ph.D. thesis, Fakultät für Mathematik der Universität Karlsruhe, Karlsruhe (1988)
23. Mackey, M.C., Glass, L., et al.: Oscillation and chaos in physiological control systems. Science **197**(4300), 287–289 (1977)
24. McMillan, K.L.: Interpolation and SAT-based model checking. In: Hunt, W.A., Somenzi, F. (eds.) CAV 2003. LNCS, vol. 2725, pp. 1–13. Springer, Heidelberg (2003). doi:10.1007/978-3-540-45069-6_1
25. Moore, R.E.: Automatic local coordinate transformation to reduce the growth of error bounds in interval computation of solutions of ordinary differential equations. In: Ball, L.B. (ed.) Error in Digital Computation, vol. II, pp. 103–140. Wiley, New York (1965)
26. Neher, M., Jackson, K.R., Nedialkov, N.S.: On Taylor model based integration of ODEs. SIAM J. Numer. Anal. **45**(1), 236–262 (2007)
27. Ouaknine, J., Worrell, J.: On the decidability of metric temporal logic. In: Proceedings of the 20th IEEE Symposium on Logic in Computer Science (LICS 2005), Chicago, IL, USA, pp. 188–197. IEEE Computer Society, 26–29 June 2005
28. Ouaknine, J., Worrell, J.: Some recent results in metric temporal logic. In: Cassez, F., Jard, C. (eds.) FORMATS 2008. LNCS, vol. 5215, pp. 1–13. Springer, Heidelberg (2008)
29. Scheiber, K.: iSAT3 Manual, April 2014
30. Sheeran, M., Singh, S., Stålmarck, G.: Checking safety properties using induction and a SAT-solver. In: Hunt, W.A., Johnson, S.D. (eds.) FMCAD 2000. LNCS, vol. 1954, pp. 127–144. Springer, Heidelberg (2000). doi:10.1007/3-540-40922-X_8
31. Stauning, O.: Automatic validation of numerical solutions. Ph.D. thesis, Technical University of Denmark, Lyngby (1997)
32. Szydłowski, M., Krawiec, A.: The stability problem in the kaldor-kalecki business cycle model. Chaos, Solitons & Fractals **25**(2), 299–305 (2005)
33. Szydłowski, M., Krawiec, A., Toboła, J.: Nonlinear oscillations in business cycle model with time lags. Chaos, Solitons & Fractals **12**(3), 505–517 (2001)
34. Zou, L., Fränzle, M., Zhan, N., Nazier Mosaad, P.: Automatic verification of stability and safety for delay differential equations. In: Kroening, D., Păsăreanu, C.S. (eds.) CAV 2015, Part II. LNCS, vol. 9207, pp. 338–355. Springer, Heidelberg (2015)

Dynamic Logic with Binders and Its Application to the Development of Reactive Systems

Alexandre Madeira[1(✉)], Luis S. Barbosa[1], Rolf Hennicker[2],
and Manuel A. Martins[3]

[1] HASLab INESC TEC, University of Minho, Braga, Portugal
madeira@ua.pt
[2] Ludwig-Maximilians-Universität München, Munich, Germany
[3] CIDMA - Department of Mathematics, University of Aveiro, Aveiro, Portugal

Abstract. This paper introduces a logic to support the specification and development of reactive systems on various levels of abstraction, from property specifications, concerning e.g. safety and liveness requirements, to constructive specifications representing concrete processes. This is achieved by combining binders of hybrid logic with regular modalities of dynamic logics in the same formalism, which we call $\mathcal{D}^{\downarrow}$-logic. The semantics of our logic focuses on effective processes and is therefore given in terms of reachable transition systems with initial states. The second part of the paper resorts to this logic to frame stepwise development of reactive systems within the software development methodology proposed by Sannella and Tarlecki. In particular, we instantiate the generic concepts of constructor and abstractor implementations by using standard operators on reactive components, like relabelling and parallel composition, as constructors, and bisimulation for abstraction. We also study vertical composition of implementations which relies on the preservation of bisimularity by the constructions on labeleld transition systems.

1 Introduction

The quest for suitable notions of *implementation* and *refinement* has been for more than four decades on the research agenda for rigorous Software Engineering. It goes back to Hoare's paper on data refinement [16], which influenced the whole family of model-oriented methods, starting with VDM [18]. A recent reference [30] collects a number of interesting refinement case studies in the B method, probably the most industrially successful in the family.

Almost 30 years ago, D. Sannella and A. Tarlecki claimed, in what would become a most influential paper in (formal) Software Engineering [28], that *"the program development process is a sequence of implementation steps leading from a specification to a program"*. Being rather vague on what was to be understood either by specifications (*"just finite syntactic objects of some kind"* which *"describe a certain signature and a class of models over it"*) or programs (*"which for us are just very tight specifications"*), the paper focuses entirely on the development process, based on a notion of refinement. In model-oriented approaches it is consensual that a specification refines to another if every model of the latter is a model

© Springer International Publishing AG 2016
A. Sampaio and F. Wang (Eds.): ICTAC 2016, LNCS 9965, pp. 422–440, 2016.
DOI: 10.1007/978-3-319-46750-4_24

of the former. Sannella and Tarlecki's work complemented and generalised this idea with the notions of *"constructor"* and *"abstractor implementations"*. The idea of a constructor implementation is that for implementing a specification SP one may use one or several given specifications and apply a construction on top of them to satisfy the requirements of SP. Abstractor implementations have been introduced to deal with the fact that sometimes the properties of a requirements .specification are not literally satisfied by an implementation but only up to an abstraction which usually involves hiding of implementation details. Over time, many others contributed along similar paths, with Sannella and Tarlecki's specific view later consolidated in their landmark book [29]. All main ingredients were already there: (i) the emphasis on *loose* specifications; (ii) correctness by construction, guaranteed by vertical compositionality and (iii) genericity, as the development process is independent, or parametric, on whatever logical system better captures the requirements to be handled.

Our paper investigates this approach in the context of reactive software, *i.e.* systems which interact with their environment along the whole computation, and not only in its starting and termination points [1]. The relevance of such an effort is anticipated in Sannella and Tarlecki's book [29] itself: *"An example of an area for which a satisfactory, commonly accepted solution still seems to be outstanding (despite numerous proposals and active research) is the theory of concurrency"* (page 157). Different approaches in that direction have been proposed, of which we single out an extension to concurrency in K. Havelund's Ph.D. thesis [15]. The book, however, focused essentially on functional requirements expressed by algebraic specifications and implemented in a functional programming language.

On the other hand, the development of reactive systems, nowadays the norm rather than the exception, followed a different path. Typical approaches start from the construction of a concrete model (e.g. in the form of a transition system [31], a Petri net [26] or a process algebra expression [4,17]) upon which the relevant properties are later formulated in a suitable (modal) logic and typically verified by some form of model-checking. Resorting to old software engineering jargon, most of these approaches proceed by *inventing & verifying*, whereas this paper takes the alternative *correct by construction* perspective.

Our hypothesis is that also in the domain of reactive systems, loose specification has an important role to play, because they support the gradual addition of requirements and implementation decisions such that verification of the correctness of a complex system can be done piecewise in smaller steps. Thus also a documentation keeping trace of design decisions is available supporting maintenance and extensibility of systems. Therefore, our challenge was twofold. First to design a logic to support the development of reactive systems at different levels of abstraction. Second, to follow Sannella and Tarlecki's recipe according to which *"specific notions of implementation (...) corresponds to a restriction on the choice of constructors and abstractors which may be used"* [28]. The paper's contributions respond to such challenges:

– Borrowing modalities indexed by regular expressions of actions, from dynamic logic [14], and state variables and binders, from hybrid logic [6], a new logic,

$\mathcal{D}^{\downarrow}$, is proposed to express properties of computations of reactive systems. $\mathcal{D}^{\downarrow}$ is able to express abstract properties, such as liveness requirements or deadlock avoidance, but also to describe concrete, recursive process structures implementing them. Note that our focus is actually on computations, and therefore on transition structures over reachable states with an initial point, rather than on arbitrary relational structures with global satisfaction, as usual in modal logic. Symbol $\downarrow$ in $\mathcal{D}^{\downarrow}$ stands for the *binder* operator borrowed from hybrid logic: $\downarrow x.\phi$ evaluates ϕ and assigns to variable x the current state of evaluation.

– Then, a particular pallete of constructors and abstractors found relevant to the development of reactive systems, is introduced. Interestingly, it turns out that requirements of Sannella and Tarlecki's methodology for vertical composition of abstractor/constructor implementations is just the congruence property of bisimulation w.r.t. constructions on labelled transition systems, like parallel composition and relabelling.

The new $\mathcal{D}^{\downarrow}$ logic is introduced in Sect. 2. Then, the two following sections, 3 and 4, respectively, introduce the development method, with a brief revision of the relevant background, and its tuning to the design of reactive systems. Finally, Sect. 5 concludes and points out some issues for future work. To respect the page limit fixed for the Conference, all proofs were removed from the paper. They appear in the accompanying technical report [21].

2 $\mathcal{D}^{\downarrow}$ - A Dynamic Logic with Binders

2.1 $\mathcal{D}^{\downarrow}$-logic: Syntax and Semantics

$\mathcal{D}^{\downarrow}$ logic is designed to express properties of reactive systems, from abstract safety and liveness properties, down to concrete ones specifying the (recursive) structure of processes. It thus combines modalities with regular expressions, as originally introduced in Dynamic Logic [14], and binders in state variables. This logic retains from Hybrid Logic [6], only state variables and the binder operator first studied by V. Goranko in [11]. These motivations are reflected in its semantics. Differently from what is usual in modal logics, whose semantics are given by Kripke structures and the satisfaction evaluated globally in each model, $\mathcal{D}^{\downarrow}$ models are reachable transition systems with initial states where satisfaction is evaluated.

Definition 1 (Model). Models for a finite set of atomic actions A *are reachable A-LTSs, i.e. triples* (W, w_0, R) *where W is a set of states, $w_0 \in W$ is the initial state and $R = (R_a \subseteq W \times W)_{a \in A}$ is a family of transition relations such that, for each $w \in W$, there is a finite sequence of transitions* $R_{a^k}(w^{k-1}, w^k)$, $1 \le k \le n$, *with $w_k \in W$, $a^k \in A$, such that $w_0 = w^0$ and $w^n = w$.*

The *set of (structured) actions*, $\mathrm{Act}(A)$, induced by a set of atomic actions A is given by

$$\alpha ::= a \mid \alpha; \alpha \mid \alpha + \alpha \mid \alpha^*$$

where $a \in A$.

Let X be an infinite set of variables, disjoint with the symbols of the atomic actions A. A valuation for an A-model $\mathcal{M} = (W, w_0, R)$ is a function $g : X \to W$. Given such a g and $x \in X$, $g[x \mapsto w]$ denotes the valuation for $\mathcal{M}$ such that $g[x \mapsto w](x) = w$ and $g[x \mapsto w](y) = g(y)$ for any other $y \neq x \in X$.

Definition 2 (Formulas and sentences). *The set* $\mathrm{Fm}^{\mathcal{D}^{\downarrow}}(A)$ *of A-formulas is given by*

$$\varphi ::= \mathbf{tt} \mid \mathbf{ff} \mid x \mid \downarrow x.\, \varphi \mid @_x \varphi \mid \langle \alpha \rangle \varphi \mid [\alpha] \varphi \mid \neg \varphi \mid \varphi \wedge \varphi \mid \varphi \vee \varphi$$

where $x \in X$ and $\alpha \in \mathrm{Act}(A)$. $\mathrm{Sen}^{\mathcal{D}^{\downarrow}}(A) = \{\varphi \in \mathrm{Fm}^{\mathcal{D}^{\downarrow}}(A) \mid \mathrm{FVar}(\varphi) = \emptyset\}$ is the set of A-sentences, where $\mathrm{FVar}(\varphi)$ are the free variables of φ, defined as usual with $\downarrow$ being the unique operator binding variables.

$\mathcal{D}^{\downarrow}$ retains from Hybrid Logic the use of binders, but omits nominals: only state variables are used, even as parameters to the satisfaction operator ($@_x$). By doing so, the logic becomes restricted to express properties of reachable states from the initial state, i.e. processes.

To define the satisfaction relation we need to clarify how composed actions are interpreted in models. Let $\alpha \in \mathrm{Act}(A)$ and $\mathcal{M} \in \mathrm{Mod}^{\mathcal{D}^{\downarrow}}(A)$. The interpretation of an action α in $\mathcal{M}$ extends the interpretation of atomic actions by $R_{\alpha;\alpha'} = R_\alpha \circ R_{\alpha'}$, $R_{\alpha+\alpha'} = R_\alpha \cup R_{\alpha'}$ and $R_{\alpha^*} = (R_\alpha)^*$, with the operations $\circ$, $\cup$ and $\star$ standing for relational composition, union and Kleene closure.

Given an A-model $\mathcal{M} = (W, w_0, R)$, $w \in W$ and $g : X \to W$,

- $\mathcal{M}, g, w \models \mathbf{tt}$ is true; $\mathcal{M}, g, w \models \mathbf{ff}$ is false;
- $\mathcal{M}, g, w \models x$ iff $g(x) = w$;
- $\mathcal{M}, g, w \models\, \downarrow x.\, \varphi$ iff $\mathcal{M}, g[x \mapsto w], w \models \varphi$;
- $\mathcal{M}, g, w \models @_x \varphi$ iff $\mathcal{M}, g, g(x) \models \varphi$;
- $\mathcal{M}, g, w \models \langle \alpha \rangle \varphi$ iff there is a $w' \in W$ with $(w, w') \in R_\alpha$ and $\mathcal{M}, g, w' \models \varphi$;
- $\mathcal{M}, g, w \models [\alpha] \varphi$ iff for any $w' \in W$ with $(w, w') \in R_\alpha$ it holds $\mathcal{M}, g, w' \models \varphi$;
- $\mathcal{M}, g, w \models \neg \varphi$ iff it is false that $\mathcal{M}, g, w \models \varphi$;
- $\mathcal{M}, g, w \models \varphi \wedge \varphi'$ iff $\mathcal{M}, g, w \models \varphi$ and $\mathcal{M}, g, w \models \varphi'$;
- $\mathcal{M}, g, w \models \varphi \vee \varphi'$ iff $\mathcal{M}, g, w \models \varphi$ or $\mathcal{M}, g, w \models \varphi'$.

We write $\mathcal{M}, w \models \varphi$ if, for any valuation $g : X \to W$, $\mathcal{M}, g, w \models \varphi$. If φ is a sentence, then the valuation is irrelevant, i.e., $\mathcal{M}, g, w \models \varphi$ iff $\mathcal{M}, w \models \varphi$. For each sentence $\varphi \in \mathrm{Sen}^{\mathcal{D}^{\downarrow}}(A)$, we write $\mathcal{M} \models \varphi$ whenever $\mathcal{M}, w_0 \models \varphi$. Observe again the pertinence of avoiding nominals: if a formula is satisfied in the standard semantics of Hybrid Logic, then it is satisfiable in ours. Obviously, this would not happen in the presence of nominals.

The remaining of the section discusses the versatility of $\mathcal{D}^{\downarrow}$ claimed in the introductory section. Here and in the following sentences, in the context of a set of actions $A = \{a_1, \ldots, a_n\}$, we write A for the complex action $a_1 + \ldots + a_n$ and for any $a_i \in A$, we write $-a_i$ for the complex action $a_1 + \ldots + a_{i-1} + a_{i+1} + \ldots + a_n$.

By using regular modalities from Dynamic Logic [13,14], $\mathcal{D}^{\downarrow}$ is able to express liveness requirements such as *"after the occurrence of an action a, an action b can be eventually realised"* with $[A^*; a]\langle A^*; b\rangle\mathbf{tt}$ or *"after the occurrence of an action a, an occurrence of an action b is eventually possible if it has not occurred before"* with $[A^*; a; (-b)^*]\langle A^*; b\rangle\mathbf{tt}$. Safety properties are also captured by sentences of the form $[A^*]\varphi$. In particular, *deadlock freeness* is expressed by $[A^*]\langle A\rangle\mathbf{tt}$.

Example 1. As a running example we consider a product line with a stepwise development of a product for compressing files services, involving compressions of text and of image files. We start with an abstract requirements specification SP_0. It is built over the set $A = \{inTxt, inGif, outZip, outJpg\}$ of atomic actions $inTxt, inGif$ for inputting a txt-file or a gif-file, and actions $outZip, outJpg$ for outputting a zip-file or a jpg-file. Sentences (0.1)–(0.3) below express three requirements: (0.1) Whenever a txt-file has been received for compression, the next action must be an output of a zip-file, (0.2) whenever a gif-file has been received, the next action must be an output of a jpg-file, and (0.3) the system should never terminate.

(0.1) $[A^*; inTxt](\langle outZip\rangle\mathbf{tt} \wedge [-outZip]\mathbf{ff})$
(0.2) $[A^*; inGif](\langle outJpg\rangle\mathbf{tt} \wedge [-outJpg]\mathbf{ff})$
(0.3) $[A^*]\langle A\rangle\mathbf{tt}$

Obviously, SP_0 is a very loose specification of rudimentary requirements and there are infinitely many models which satisfy the sentences (0.1)–(0.3).   □

$\mathcal{D}^{\downarrow}$-logic, however, is also suited to directly express process structures and, thus, the implementation of abstract requirements. The binder operator is crucial for this. The ability to give names to visited states, together with the modal features to express transitions, makes possible a precise description of the whole dynamics of a process in a single sentence. Binders allow to express recursive patterns, namely loop transitions (from the current to some visited state). In fact we have no way to make this kind of specification in the absence of a feature to refer to specific states in a model, as in standard modal logic. For example, sentence

$$\downarrow x_0.(\langle a\rangle x_0 \wedge \langle b\rangle \downarrow x_1.(\langle a\rangle x_0 \wedge \langle b\rangle x_1)) \tag{1}$$

specifies a process with two states accepting actions a and b respectively. As discussed in the sequel, the stepwise development of a reactive system typically leads to a set of requirements defining concrete transition systems and expressed in the fragment of $\mathcal{D}^{\downarrow}$ which omits modalities indexed by the Kleene closure of actions, that can be directly translated into a set of FSP [22] definitions. Figure 1 depicts the translation of the formula above as computed by a proof-of-concept implementation of such a translator[1]. Note, however, that sentence (1) loosely specifies the purposed scenario (e.g. a single state system looping on a and b also satisfies this requirement). Resorting to full $\mathcal{D}^{\downarrow}$ concrete processes unique up to isomorphism, can be defined, i.e. we may introduce monomorphic specifications.

---

[1] See `translator.nrc.pt`.

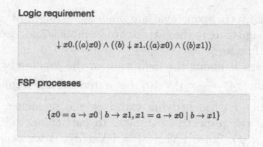

Fig. 1. D2FSP Translator: Translating $\mathcal{D}^{\downarrow}$ into FSP processes.

For this specific example, it is enough to consider, in the conjunction in the scope of x_1, the term $@_{x_1} \neg x_0$ (to distinguish between the states binded by x_0 and x_1), as well as to enforce determinism resorting to formula (det) in Example 2.

2.2   Turning $\mathcal{D}^{\downarrow}$-logic into an Institution

In order to fit the necessary requirements to adopt the Sannella Tarlecki development method, logic $\mathcal{D}^{\downarrow}$ has to be framed as a logical institution [10].

In this view, our first concern is about the *signatures category*. As suggested, signatures for $\mathcal{D}^{\downarrow}$ are finite sets A of *atomic actions*, and a signature morphism $A \xrightarrow{\sigma} A'$ is just a function $\sigma : A \to A'$. Clearly, this entails a category to be denoted by $\text{Sign}^{\mathcal{D}^{\downarrow}}$.

Our second concern is about the *models functor*. Given two models, $\mathcal{M} = (W, w_0, R)$ and $\mathcal{M}' = (W', w_0', R')$, for a signature A, a *model morphism* is a function $h : W \to W'$ such that $h(w_0) = w_0'$ and, for each $a \in A$, if $(w_1, w_2) \in R_a$ then $(h(w_1), h(w_2)) \in R_a'$. We can easily observe that the class of models for A, and the corresponding morphisms, defines a category $\text{Mod}^{\mathcal{D}^{\downarrow}}(A)$.

Definition 3 (Model reduct). *Let $A \xrightarrow{\sigma} A'$ be a signature morphism and $\mathcal{M}' = (W', w_0', R')$ an A'-model. The σ-reduct of $\mathcal{M}'$ is the A-model $\text{Mod}^{\mathcal{D}^{\downarrow}}(\sigma)(\mathcal{M}') = (W, w_0, R)$ such that*

- *$w_0 = w_0'$;*
- *W is the largest set with $w_0' \in W$ and, for each $v \in W$, either $v = w_0'$ or there is a $w \in W$ such that $(w, v) \in R_{\sigma(a)}'$, for some $a \in A$;*
- *for each $a \in A$, $R_a = R_{\sigma(a)}' \cap W^2$.*

Models morphisms are preserved by reducts, in the sense that, for each models morphism $h : \mathcal{M}_1' \to \mathcal{M}_2'$ there is a models morphism $h' : \text{Mod}^{\mathcal{D}^{\downarrow}}(\sigma)(\mathcal{M}_1') \to \text{Mod}^{\mathcal{D}^{\downarrow}}(\sigma)(\mathcal{M}_2')$, where h' is the restriction of h to the states of $\text{Mod}^{\mathcal{D}^{\downarrow}}(\sigma)(\mathcal{M}_1')$. Hence, for each signature morphism $A \xrightarrow{\sigma} A'$, a functor $\text{Mod}^{\mathcal{D}^{\downarrow}}(\sigma)$: $\text{Mod}^{\mathcal{D}^{\downarrow}}(A') \to \text{Mod}^{\mathcal{D}^{\downarrow}}(A)$ maps models and morphisms to the corresponding reducts. Finally, this lifts to a contravariant *models functor*, $\mathrm{Mod}^{\mathcal{D}^{\downarrow}} : (\mathrm{Sign}^{\mathcal{D}^{\downarrow}})^{op} \to \mathbb{C}at$, mapping each signature to the category of its models and, each signature morphism to its reduct functor.

The third concern is about the definition of the *functor of sentences*. Each signature morphism $A \xrightarrow{\sigma} A'$ can be extended to formulas' translation $\hat{\sigma} : \mathrm{Fm}^{\mathcal{D}^{\downarrow}}(A) \to \mathrm{Fm}^{\mathcal{D}^{\downarrow}}(A')$ identifying variables and replacing, symbol by symbol, each action by the respective σ-image. In particular, $\hat{\sigma}(\downarrow x.\varphi) = \downarrow x.\hat{\sigma}(\varphi)$ and $\hat{\sigma}(@_x\varphi) = @_x\hat{\sigma}(\varphi)$. Since $\mathrm{FVar}(\varphi) = \mathrm{FVar}(\hat{\sigma}(\varphi))$ we can assure that, for each signature morphism $A \xrightarrow{\sigma} A'$, we can define a translation of sentences $\mathrm{Sen}^{\mathcal{D}^{\downarrow}}(\sigma) : \mathrm{Sen}^{\mathcal{D}^{\downarrow}}(A) \to \mathrm{Sen}^{\mathcal{D}^{\downarrow}}(A')$, by $\mathrm{Sen}^{\mathcal{D}^{\downarrow}}(\sigma)(\varphi) = \hat{\sigma}(\varphi)$, $\varphi \in \mathrm{Sen}^{\mathcal{D}^{\downarrow}}(A)$. This entails the intended functor $\mathrm{Sen}^{\mathcal{D}^{\downarrow}} : \mathrm{Sign}^{\mathcal{D}^{\downarrow}} \to \mathbb{S}et$, mapping each signature to the set of its sentences, and each signature morphism to the corresponding translation of sentences.

Finally, our forth concern is on the agreement of the satisfaction relation w.r.t. *satisfaction condition*. This is established in the following result:

Theorem 1. *Let* $\sigma : A \to A'$ *be a signature morphism,* $\mathcal{M}' = (W', w_0', R') \in \mathrm{Mod}^{\mathcal{D}^{\downarrow}}(A')$, $\mathrm{Mod}^{\mathcal{D}^{\downarrow}}(\sigma)(\mathcal{M}') = (W, w_0, R)$ *and* $\varphi \in \mathrm{Fm}^{\mathcal{D}^{\downarrow}}(A)$. *Then, for any* $w \in W(\subseteq W')$ *and for any valuations* $g : X \to W$ *and* $g' : X \to W'$, *such that,* $g(x) = g'(x)$ *for all* $x \in \mathrm{FVar}(\varphi)$, *we have*

$$\mathrm{Mod}^{\mathcal{D}^{\downarrow}}(\sigma)(\mathcal{M}'), g, w \models \varphi \text{ iff } \mathcal{M}', g', w \models \hat{\sigma}(\varphi)$$

In order to get the satisfaction condition, we only have to note that for any $\varphi \in \mathrm{Sen}^{\mathcal{D}^{\downarrow}}(A)$, we have $\mathrm{FVar}(\varphi) = \emptyset$, and hence, by Theorem 1, for any $w \in W$, $\mathrm{Mod}^{\mathcal{D}^{\downarrow}}(\sigma)(\mathcal{M}'), w \models \varphi$ iff $\mathcal{M}', w \models \mathrm{Sen}^{\mathcal{D}^{\downarrow}}(\sigma)(\varphi)$. Moreover, by the definition of reduct, $w_0 = w_0' \in W$. Therefore, $\mathrm{Mod}^{\mathcal{D}^{\downarrow}}(\sigma)(\mathcal{M}') \models \varphi$ iff $\mathcal{M}' \models \mathrm{Sen}^{\mathcal{D}^{\downarrow}}(\sigma)(\varphi)$.

3   Formal Development *á la* Sannella and Tarlecki

Developing correct programs from specifications entails the need for a suitable logic setting in which meaning can be assigned both to specifications and their refinement. Sannella and Tarlecki have proposed a formal development methodology [28,29] which is presented in a generic way for arbitrary logical systems forming an institution. As already pointed out in the Introduction, Sannella and Tarlecki have studied various algebraic institutions to illustrate their methodology and they presume the lack of a satisfactory solution in the theory of concurrency. In this section we briefly summarize their crucial principles for formal program development over an arbitrary institution and we illustrate the case of simple implementations by examples of our $\mathcal{D}^{\downarrow}$-logic institution. The more involved concepts of constructor and abstractor implementations will be instantiated for the case of $\mathcal{D}^{\downarrow}$-logic later on in Sect. 4.

In the following we assume given an arbitrary institutions with category Sign of signatures and signature morphisms, with sentence functor Sen : Sign $\to \mathbb{S}et$, and with models functor Mod : $\text{Sign}^{op} \to \mathbb{C}at$ assigning to any signature $\Sigma \in$ |Sign| a category $\text{Mod}(\Sigma)$ whose objects in $|\text{Mod}(\Sigma)|$ are called Σ-models. As usual, the class of objects of a category C is denoted by $|C|$. If it is clear from the context, we will simply write C for $|C|$.

3.1 Simple Implementations

The simplest way to design a specification is by expressing the system requirements by means of a set of sentences over a suitable signature, i.e. as a pair $SP = (Sig(SP), Ax(SP))$ where $Sig(SP) \in$ |Sign| and $Ax(SP) \subseteq |Sen(Sig(SP))|$. The (loose) semantics of such a *flat* specification SP consists of the pair $(Sig(SP), Mod(SP))$ where

$$Mod(SP) = \{M \in |Mod(Sig(SP))| :\ M \models Ax(SP)\}.$$

In this context, a refinement step is understood as a restriction of an abstract class of models to a more concrete one. Following the terminology of Sannella and Tarlecki, we will call a specification which refines another one an *implementation*. Formally, a specification SP' is a *simple implementation* of a specification SP over the same signature, in symbols $SP \rightsquigarrow SP'$, whenever $Mod(SP) \supseteq Mod(SP')$. Transitivity of the inclusion relation ensures the *vertical composition* of simple implementation steps.

Example 2. We illustrate two refinement steps with simple implementations in the $\mathcal{D}^{\downarrow}$-logic institution. Consider the specification SP_0 of Example 1 which expresses some rudimentary requirements for the behavior of compressing files services. The action set A defined in Example 1 provides the signature of SP_0 and the axioms of SP_0 are the three sentences (0.1)–(0.3) shown in Example 1.

First refinement step $SP_0 \rightsquigarrow SP_1$. SP_0 is a very loose specification which would allow to start a computation with an arbitrary action. We will be a bit more precise now and require that at the beginning only an input (of a text or gif file) is allowed; see axiom (1.1) below. Moreover whenever an output action (of any kind) has happened then the system must go on with an input (of any kind); see axiom (1.4). This leads to the specification SP_1 with $Sig(SP_1) = Sig(SP_0) = A$ and with the following set of axioms $Ax(SP_1)$:

(1.1) $\langle inTxt + inGif \rangle \mathbf{tt} \wedge [outZip + outJpg]\mathbf{ff}$
(1.2) $[A^*; inTxt](\langle outZip \rangle \mathbf{tt} \wedge [-outZip]\mathbf{ff})$
(1.3) $[A^*; inGif](\langle outJpg \rangle \mathbf{tt} \wedge [-outJpg]\mathbf{ff})$
(1.4) $[A^*; (outZip + outJpg)](\langle inTxt + inGif \rangle \mathbf{tt} \wedge [outZip + outJpg]\mathbf{ff})$

It is easy to check that $SP_0 \rightsquigarrow SP_1$ holds: Axioms (0.1) and (0.2) of SP_0 occur as axioms (1.2) and (1.3) in SP_1. It is also easy to see that non-termination (axiom (0.3) of SP_0) is guaranteed by the axioms of SP_1.

The level of underspecification is, at this moment, still very high. Among the infinitely many models of SP_1, we can find, as an admissible model the LTS shown in Fig. 2 with initial state w_0 and with an alternating compression mode.

Second refinement step $SP_1 \rightsquigarrow SP_2$. This step rules out alternating behaviours as shown above. The first axiom (2.1) of the following specification SP_2

is equivalent to axiom (1.1) of SP_1. Alternating behaviours are ruled out by axioms (2.2) and (2.3) which require that after any text compression and after any image compression the initial state must be reached again. To express this we need state variables and binders which are available in $\mathcal{D}^{\downarrow}$-logic. In our example we introduce one state variable x_0 which names the initial state by using the binder at the beginning of axioms (2.2) and (2.3). Moreover, we only want to admit deterministic models such that in any (reachable) state there can be no two outgoing transitions with the same action. It turns out that $\mathcal{D}^{\downarrow}$-logic also allows to specify this determinism property with the set of axioms (det) shown below. This leads to the specification SP_2 with $Sig(SP_2) = Sig(SP_1) = A$ and with axioms $Ax(SP_2)$:

(2.1) $(\langle inTxt\rangle\mathbf{tt} \vee \langle inGif\rangle\mathbf{tt}) \wedge [outZip + outJpg]\mathbf{ff}$
(2.2) $\downarrow x_0.\,[inTxt](\langle outZip\rangle x_0 \wedge [-outZip]\mathbf{ff})$
(2.3) $\downarrow x_0.\,[inGif](\langle outJpg\rangle x_0 \wedge [-outJpg]\mathbf{ff})$
(det) For each $a \in A$, the axiom: $[A^*] \downarrow x.(\langle a\rangle\mathbf{tt} \Rightarrow (\langle a\rangle \downarrow y.\, @_x[a]y))$

Clearly SP_2 fulfills the requirements of SP_1, i.e. $SP_1 \rightsquigarrow SP_2$. SP_2 has three models which are shown in Fig. 3. (Remember that models can only have states reachable from the initial one.) The first model allows only text compression, the second one only image compression, and the third supports both. The signature of all models is A, though in the first two some actions have no transitions.

Let us still discuss some variations of SP_2 to underpin the expressive power of $\mathcal{D}^{\downarrow}$. If we want only the model where both text and image compression are possible, then we can simply replace in axiom (2.1) $\langle inTxt\rangle\mathbf{tt} \vee \langle inGif\rangle\mathbf{tt}$ by $\langle inTxt\rangle\mathbf{tt} \wedge \langle inGif\rangle\mathbf{tt}$. If we would like to require that text compression must be possible in any model but image compression is optional, i.e. we rule out the second model in Fig. 3, then we would simply omit $\vee\langle inGif\rangle\mathbf{tt}$ in axiom (2.1). This is an interesting case since this shows that $\mathcal{D}^{\downarrow}$-logic can express so-called "may"-transitions offered by modal transition systems [20] to specify options for implementations.   □

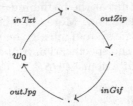

Fig. 2. A model of SP_1

3.2 Constructor Implementations

The concept of simple implementations is, in general, too strict to capture software development practice, along which, implementation decisions typically introduce new design features or reuse already implemented ones, usually entailing a change of signatures along the way. The notion of *constructor implementation* offers the necessary generalization. The idea is that for implementing a

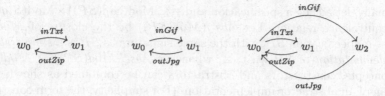

Fig. 3. Models of SP_2

specification SP one may use a given specification SP' and apply a construction to the models of SP' such that they become models of SP. More generally, an implementation of SP may be obtained by using not only one but several specifications $SP'_1, \ldots, SP'_n$ as a basis and applying an n-ary constructor such that for any tuple of models of $SP'_1, \ldots, SP'_n$ the construction leads to a model of SP. Such an implementation is called a *constructor implementation with decomposition* in [29] since the implementation of SP is designed by using several components. These ideas are formalized as follows, partially in a less general manner than the corresponding definitions in [29] which allow also partial and higher-order functions as constructors.

Given signatures $\Sigma_1, \ldots, \Sigma_n, \Sigma \in |\text{Sign}|$, a *constructor* is a total function $\kappa : \text{Mod}(\Sigma_1) \times \cdots \times \text{Mod}(\Sigma_n) \to \text{Mod}(\Sigma)$. Constructors compose as follows: Given a constructor $\kappa : \text{Mod}(\Sigma_1) \times \cdots \times \text{Mod}(\Sigma_n) \to \text{Mod}(\Sigma)$ and a set of constructors $\kappa_i : \text{Mod}(\Sigma_i^1) \times \cdots \times \text{Mod}(\Sigma_i^{k_i}) \to \text{Mod}(\Sigma_i)$, $1 \leq i \leq n$, the constructor $\kappa(\kappa_1, \ldots, \kappa_n) : \text{Mod}(\Sigma_1^1) \times \cdots \times \text{Mod}(\Sigma_1^{k_1}) \times \cdots \times \text{Mod}(\Sigma_n^1) \times \cdots \times \text{Mod}(\Sigma_n^{k_n}) \to \text{Mod}(\Sigma)$ is obtained by the usual composition of functions.

Definition 4 (Constructor implementation). *Given specifications* $SP, SP'_1, \ldots, SP'_n$, *and a constructor* $\kappa : \text{Mod}(Sig(SP'_1)) \times \cdots \times \text{Mod}(Sig(SP'_1)) \to Mod(Sig(SP))$, *we say that* $\langle SP'_1, \ldots, SP'_n \rangle$ *is a constructor implementation via* κ *of* SP, *in symbols* $SP \leadsto_\kappa \langle SP'_1, \ldots, SP'_n \rangle$, *if for all* $M_i \in Mod(SP'_i)$ *we have* $\kappa(M_1, \ldots, M_n) \in Mod(SP)$. *We say that the implementation involves a decomposition if* $n > 1$.

3.3 Abstractor Implementations

Another aspect in formal program development concerns the fact that sometimes the properties of a requirements specification are not literally satisfied by an implementation but only up to an admissible abstraction. Usually such an abstraction concerns implementation details which are hidden to the user of the system and which may, for instance for efficiency reasons, not be fully conform to the requirements specification. Then the implementation is still considered to be correct if it shows the desired observable behavior. In general this can be expressed by considering an equivalence relation $\equiv$ on the models of the abstract specification and to allow the implementation models to be only equivalent to models of the requirements specification.

Formally, let SP be a specification and $\equiv\, \subseteq \mathrm{Mod}(Sig(SP)) \times \mathrm{Mod}(Sig(SP))$ be an equivalence relation. Let $Abs_\equiv(Mod(SP))$ be the closure of $Mod(SP)$ under $\equiv$. A specification SP' with the same signature as SP is a *simple abstractor implementation* of SP w.r.t. $\equiv$, whenever $Abs_\equiv(Mod(SP)) \supseteq Mod(SP')$. Both concepts, constructors and abstractors can be combined as shown in the definition of an abstractor implementation. (For simplicity, the term constructor is omitted.)

Definition 5 (Abstractor implementation). *Let $SP, SP'_1, \ldots, SP'_n$ be specifications, $\kappa : \mathrm{Mod}(Sig(SP'_1)) \times \cdots \times \mathrm{Mod}(Sig(SP'_n)) \to \mathrm{Mod}(Sig(SP))$ a constructor, and $\equiv\, \subseteq \mathrm{Mod}(Sig(SP)) \times \mathrm{Mod}(Sig(SP))$ an equivalence relation. We say that $\langle SP'_1, \cdots, SP'_n \rangle$ is an* abstractor implementation *of SP via κ w.r.t. $\equiv$, in symbols $SP \rightsquigarrow_\kappa^\equiv \langle SP'_1, \cdots, SP'_n \rangle$, if for all $M_i \in Mod(SP'_i)$ we have $\kappa(M_1, \ldots, M_n) \in Abs_\equiv(Mod(SP))$.*

4 Reactive Systems Development with $\mathcal{D}^\downarrow$

4.1 Constructor Implementations in $\mathcal{D}^\downarrow$-logic

This section introduces a pallete of constructors to support the formal development of reactive systems with $\mathcal{D}^\downarrow$, instantiating the definitions in Sect. 3.2. The idea is to lift standard constructions on labelled transition systems (see, e.g. [31]) to constructors for implementations. We will illustrate most of the constructors introduced in the following with our running example.

Along the refinement process it is sometimes convenient to reduce the action set, for instance, by omitting some actions previously introduced as auxiliary actions or as options that are no longer needed. For this purpose we use the *alphabet extension constructor*. Remember that constructors always map concrete models to abstract ones. Therefore when omitting actions in a refinement step we need an alphabet extension on the concrete models to fit them to the abstract signature.

Definition 6 (Alphabet extension). *Let $A, A' \in |\mathrm{Sign}^{\mathcal{D}^\downarrow}|$ be signatures in $\mathcal{D}^\downarrow$, i.e. action sets, such that $A \subseteq A'$. The* alphabet extension constructor κ_{ext} : $\mathrm{Mod}^{\mathcal{D}^\downarrow}(A) \to \mathrm{Mod}^{\mathcal{D}^\downarrow}(A')$ *is defined as follows: For each $\mathcal{M} = (W, w_0, R) \in \mathrm{Mod}^{\mathcal{D}^\downarrow}(A)$, $\kappa_{ext}(\mathcal{M}) = (W, w_0, R')$ with $R'_a = R_a$ for all $a \in A$ and $R'_a = \emptyset$ for all $a \in A' \setminus A$.*

Example 3. The specification SP_2 of Example 2 has the three models shown in Fig. 3. Hence, it allows three directions to proceed further in the product line.

Third refinement step $SP_2 \rightsquigarrow_{\kappa_{ext}} SP_3$. We will consider here the simple case where we vote for a tool that supports only text compression. The following specification SP_3 is a direct axiomatisation of the first model in Fig. 3 considered over the smaller action set $A_3 = \{inTxt, outZip\}$. Hence, $Sig(SP_3) = A_3$ and the axioms in $Ax(SP_3)$ are:

(3.1) $\downarrow x_0. (\langle inTxt\rangle \downarrow x_1. (\langle outZip\rangle x_0 \wedge [inTxt]\mathbf{ff}) \wedge [outZip]\mathbf{ff})$
(det) For each $a \in A_3$, the axiom: $[A_3^*] \downarrow x.(\langle a\rangle \mathbf{tt} \Rightarrow (\langle a\rangle \downarrow y. @_x[a]y))$

Since the signature of SP_3 has less actions than the one of SP_2, we apply an alphabet extension constructor $\kappa_{ext} : \mathrm{Mod}^{\mathcal{D}^\downarrow}(A_3) \to \mathrm{Mod}^{\mathcal{D}^\downarrow}(A)$ which transforms the model of SP_3 into an LTS with the same states and transitions but with actions A and with an empty accessibility relation for the actions in $A \setminus A_3$. Then, trivially, $SP_2 \leadsto_{\kappa_{ext}} SP_3$ holds. Specification SP_3 is a simple example that shows how labeled transition systems can be directly specified in $\mathcal{D}^\downarrow$. This could suggest that we are already close to a concrete implementation. But this is not true, since SP_3 is in principle just an interface specification which specifies the system behavior "from the outside", i.e. its interactions with the user.   □

The standard way to build reactive systems is by aggregating in parallel smaller components. The following parallel composition constructor synchronising on shared actions caters for this.

Definition 7 (Parallel composition). *Given signatures A and A' the parallel composition constructor $\kappa_\otimes : \mathrm{Mod}^{\mathcal{D}^\downarrow}(A) \times \mathrm{Mod}^{\mathcal{D}^\downarrow}(A') \to \mathrm{Mod}^{\mathcal{D}^\downarrow}(A \cup A')$ is a function mapping models $\mathcal{M} = (W, w_0, R) \in \mathrm{Mod}^{\mathcal{D}^\downarrow}(A)$ and $\mathcal{M}' = (W', w_0', R') \in \mathrm{Mod}^{\mathcal{D}^\downarrow}(A')$, to the $A \cup A'$-model $\mathcal{M} \otimes \mathcal{M}' = (W^\otimes, (w_0, w_0'), R^\otimes)$ where $W^\otimes \subseteq W \times W'$ and $R^\otimes = (R_a^\otimes)_{a \in A \cup A'}$ are the least sets satisfying $(w_0, w_0') \in W^\otimes$, and, for each $(w, w') \in W^\otimes$,*

- *if $a \in A \cap A'$, $(w, v) \in R_a$, $(w', v') \in R_a'$, then $(v, v') \in W^\otimes$ and $((w, w'), (v, v')) \in R_a^\otimes$;*
- *if $a \in A \setminus A'$, $(w, v) \in R_a$, then $(v, w') \in W^\otimes$ and $((w, w'), (v, w')) \in R_a^\otimes$;*
- *if $a \in A' \setminus A$, $(w', v') \in R_a'$, then $(w, v') \in W^\otimes$ and $((w, w'), (w, v')) \in R_a^\otimes$.*

Since, up to isomorphism, parallel composition is associative, the extension of this constructor to the n-ary case is straightforward. Parallel composition is a crucial operator for constructor implementations with decomposition; see Definition 4. Remember again that constructors always go from concrete models to abstract ones, i.e. in the opposite direction as the development process. Therefore the parallel composition constructor justifies the implementation of reactive systems by decomposition.

Example 4. We are now going to construct an implementation for the interface specification SP_3. in Example 3. For this purpose, we propose a decomposition into two components, a controller component *Ctrl* and a component *GZip* which does the actual text compression. The controller has the actions $A_{Ctrl} = \{inTxt, txt, zip, outZip\}$. First, it receives (action $inTxt$) a txt-file from the user. Then it hands over the text, with action txt, to the *GZip* component and receives the resulting zip-file (action zip). Finally it provides the zip-file to the user (action $outZip$) and is ready to serve a next compression. Hence, the controller component has the signature $Sig(Ctrl) = A_{Ctrl}$ and the following axioms $Ax(Ctrl)$ specify a single model, shown in Fig. 4 (left), with the behavior as described above.

(4.1) $\downarrow x_0.\,(\langle inTxt\rangle \downarrow x_1.\,(\langle txt\rangle \downarrow x_2.\,(\langle zip\rangle \downarrow x_3.\,(\langle outZip\rangle x_0 \wedge [-outZip]\mathbf{ff})$
$$\wedge\,[-zip]\mathbf{ff})$$
$$\wedge\,[-txt]\mathbf{ff})$$
$$\wedge\,[-inTxt]\mathbf{ff})$$
(det) For each $a \in A^*_{Ctrl}$, the axiom: $[A^*_{Ctrl}] \downarrow x.(\langle a\rangle\mathbf{tt} \Rightarrow (\langle a\rangle \downarrow y.\,@_x[a]y))$

The *GZip* component has the actions $A_{Gzip} = \{txt, compTxt, zip\}$. First, it receives (action *txt*) the text to be compressed from the controller. Then it does the compression (action *compTxt*), delivers the zip-file (action *zip*) to the controller and is ready for a next round. The *GZip* component has the signature $Sig(Gzip) = A_{Gzip}$ and the axioms $Ax(Gzip)$ are similar to the ones of the controller and not shown here. They specify a single model, shown in Fig. 4 (right). To construct an implementation $\langle Ctrl, GZip\rangle$ by decomposition (see Definition 4), we use the synchronous parallel composition operator "$\otimes$" defined above. According to [29], Exercise 6.1.15, any constructor gives rise to a specification building operation. This means that we can define the specification $Ctrl \otimes GZip$ whose model class consists of all possible parallel compositions of the models of the single specifications. Since *Ctrl* and *GZip* have, up to isomorphism, only one model there is also only one model of $Ctrl \otimes GZip$ which is shown in Fig. 5. Therefore, we know by construction that $Ctrl \otimes GZip \leadsto_{\kappa_\otimes} \langle Ctrl, GZip\rangle$ is a constructor implementation with decomposition. It remains to fill the gap between SP_3 and $Ctrl \otimes GZip$ which will be done with the action refinement constructor to be introduced in Definition 9.   □

Two constructions which are frequently used and which are present in most process algebras are *relabelling* and *restriction*. They are particular cases of the reduct functor of the $\mathcal{D}^\downarrow$ institution.

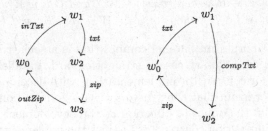

Fig. 4. Models of *Ctrl* and *GZip*

Definition 8 (Reduct, relabelling and restriction). *Let* $\sigma : A \to A'$ *be a signature morphism. The* reduct constructor $\kappa_\sigma : \mathrm{Mod}^{\mathcal{D}^\downarrow}(A') \to \mathrm{Mod}^{\mathcal{D}^\downarrow}(A)$ *maps any model* $\mathcal{M}' \in \mathrm{Mod}^{\mathcal{D}^\downarrow}(A')$ *to its reduct* $\kappa_\sigma(\mathcal{M}') = \mathrm{Mod}^{\mathcal{D}^\downarrow}(\sigma)(\mathcal{M}')$. *Whenever* σ *is a bijective function,* κ_σ *is a* relabelling *constructor. If* σ *is injective,* κ_σ *is a* restriction *constructor removing actions and transitions.*

A important refinement concept for reactive systems is *action refinement* where an abstract action is implemented by a combination of several concrete

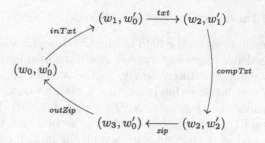

Fig. 5. Model of $Ctrl \otimes GZip$

ones (see [12]). It turns out that an action refinement constructor can be easily defined in $\mathcal{D}^{\downarrow}$-logic if we use the reduct functor for models over a signature consisting of structured actions built over atomic ones.

Definition 9 (Action refinement). *Let* $A, A' \in |\mathrm{Sign}^{\mathcal{D}^{\downarrow}}|$ *be signatures in* $\mathcal{D}^{\downarrow}$, *i.e. sets of actions. Let* D *be a finite subset of* $\mathrm{Act}(A')$ *considered as a signature in* $|\mathrm{Sign}^{\mathcal{D}^{\downarrow}}|$ *and let* $f : A \to D$ *be a signature morphism. The* action refinement *constructor* $|_f : \mathrm{Mod}^{\mathcal{D}^{\downarrow}}(D) \to \mathrm{Mod}^{\mathcal{D}^{\downarrow}}(A)$ *maps any model* $\mathcal{M}' \in \mathrm{Mod}^{\mathcal{D}^{\downarrow}}(D)$ *to its reduct* $\mathrm{Mod}^{\mathcal{D}^{\downarrow}}(f)(\mathcal{M}')$.

Example 5. Let us now establish a refinement relation between SP_3 (Example 3) and $Ctrl \otimes GZip$ (Example 4). The signature of SP_3 consists of the actions $A_3 = \{inTxt, outZip\}$, the signature of $Ctrl \otimes GZip$ is the set $A_4 = \{inTxt, txt, compTxt, zip, outZip\}$. To obtain an action refinement we define the signature morphism $f : A_3 \to \mathrm{Act}(A_4)$ by $f(inTxt) = inTxt; txt; compTxt$ and $f(outZip) = zip; outZip$. Then we use the action refinement constructor $|_f : \mathrm{Mod}^{\mathcal{D}^{\downarrow}}(A_4) \to \mathrm{Mod}^{\mathcal{D}^{\downarrow}}(A_3)$ induced by f. Clearly, the application of $|_f$ to the model of $Ctrl \otimes GZip$ leads to the model of SP_3 explained above. Hence, $SP_3 \rightsquigarrow_{|_f} Ctrl \otimes GZip$ and together with Example 4 we have also $Ctrl \otimes GZip \rightsquigarrow_{\kappa_\otimes} \langle Ctrl, GZip \rangle$ which completes our refinement chain

$$SP_0 \rightsquigarrow SP_1 \rightsquigarrow SP_2 \rightsquigarrow_{\kappa_{ext}} SP_3 \rightsquigarrow_{|_f} Ctrl \otimes GZip \rightsquigarrow_{\kappa_\otimes} \langle Ctrl, GZip \rangle.$$

Finally, let us discuss how we could implement the last specification of the chain in a concrete process algebra. Translation from $\mathcal{D}^{\downarrow}$ to FSP yields

```
Ctrl = (inTxt -> txt -> zip -> outZip -> Ctrl).
Gzip = (txt -> compTxt -> zip -> Gzip).
```

The FSP semantics of the two processes are just the two models of the *Ctrl* and *Gzip* specifications respectively. They can be put together to form a concurrent system (Ctrl || Gzip) by using the synchronous parallel composition of FSP processes. Since the semantics of parallel composition in FSP coincides with our constructor $\kappa_\otimes$, we have justified that the FSP system (Ctrl || Gzip) is a correct implementation of the interface specification SP_3.   □

4.2 Abstractor Implementations in $\mathcal{D}^{\downarrow}$-logic

Abstractor implementations in the field of algebraic specifications use typically observational equivalence relations between algebras based on the evaluation of terms with observable sorts. Interestingly, in the area of concurrent systems, abstractors have a very intuitive interpretation if we use bisimilarity notions. To motivate this, consider the specification $SP = (\{a\}, \{\downarrow x.\langle a \rangle x\})$. The axiom is satisfied by the first LTS in Fig. 6, but not by the second one. Clearly, however, both are bisimilar and so it should be irrelevant, for implementation purposes, to choose one or the other as an implementation of SP. We capture this with the principle of abstractor implementation using (strong) bisimilarity [24] as behavioural equivalence.

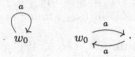

Fig. 6. Behavioural equivalent LTSs

Vertical composition of implementations refers to the situation where the implementation of a specification is further implemented in a next refinement step. For simple implementations it is trivial that two implementation steps compose. In the context of constructor and abstractor implementations the situation is more complex. A general condition to obtain vertical composition in this case was established in [28]. However, the original result was only given for unary implementation constructors. In order to adopt parallel composition as a constructor, we first generalise the institution independent result of [28] to the n-ary case involving decomposition:

Theorem 2 (Vertical composition). *Consider specifications* $SP, SP_1, \ldots,$ SP_n *over an arbitrary institution, a constructor* $\kappa : \mathrm{Mod}(Sig(SP_1)) \times \cdots \times$ $\mathrm{Mod}(Sig(SP_n)) \to \mathrm{Mod}(Sig(SP))$ *and an equivalence* $\equiv \subseteq \mathrm{Mod}(Sig(SP)) \times$ $\mathrm{Mod}(Sig(SP))$ *such that* $SP \rightsquigarrow_\kappa^{\equiv} \langle SP_1, \cdots, SP_n \rangle$. *For each* $i \in \{1, \ldots, n\}$, *let* $SP_i \rightsquigarrow_{\kappa_i}^{\equiv_i} \langle SP_i^1, \cdots, SP_i^{k_i} \rangle$ *with specifications* $SP_i^1, \ldots, SP_i^{k_i}$, *constructor* $\kappa_i :$ $\mathrm{Mod}(Sig(SP_i^1)) \times \cdots \times \mathrm{Mod}(Sig(SP_i^{k_i})) \to \mathrm{Mod}(Sig(SP_i))$, *and equivalence* $\equiv_i \subseteq$ $\mathrm{Mod}(Sig(SP_i)) \times \mathrm{Mod}(Sig(SP_i))$. *Suppose that* κ *preserves the abstractions* $\equiv_i$, *i.e. for each* $\mathcal{M}_i, \mathcal{N}_i \in \mathrm{Mod}(Sig(SP_i))$ *such that* $\mathcal{M}_i \equiv_i \mathcal{N}_i$, $\kappa(\mathcal{M}_1, \ldots, \mathcal{M}_n) \equiv$ $\kappa(\mathcal{N}_1, \ldots, \mathcal{N}_n)$. *Then,*

$$SP \rightsquigarrow_{\kappa(\kappa_1, \cdots, \kappa_n)}^{\equiv} \langle SP_1^1, \cdots, SP_1^{k_1}, \cdots, SP_n^1, \cdots, SP_n^{k_n} \rangle.$$

The remaining results establish the necessary compatibility properties between the constructors defined in $\mathcal{D}^{\downarrow}$ and behavioural equivalence $\equiv_A \subseteq |\mathrm{Mod}^{\mathcal{D}^{\downarrow}}(A)| \times$ $|\mathrm{Mod}^{\mathcal{D}^{\downarrow}}(A)|$, $A \in \mathrm{Sign}^{\mathcal{D}^{\downarrow}}$, defined as bisimilarity between LTSs.

Theorem 3. *The alphabet extension constructor κ_{ext} preserves behavioural equivalences, i.e. for any $\mathcal{M}_1 \equiv_A \mathcal{M}_2$, $\kappa_{ext}(\mathcal{M}_1) \equiv_{A'} \kappa_{ext}(\mathcal{M}_2)$.*

Theorem 4. *The parallel composition constructor $\kappa_\otimes$ preserves behavioural equivalences, i.e. for any $\mathcal{M}_1 \equiv_{A_1} \mathcal{M}'_1$ and $\mathcal{M}_2 \equiv_{A_2} \mathcal{M}'_2$, $\mathcal{M}_1 \otimes \mathcal{M}_2 \equiv_{A_1 \cup A_2} \mathcal{M}'_1 \otimes \mathcal{M}'_2$.*

Theorem 5. *Let $f : A \to \mathrm{Act}(A')$ be a signature morphism. The constructor $|_f$ preserves behavioural equivalences, i.e. for any $\mathcal{M}_1, \mathcal{M}_2 \in \mathrm{Mod}^{\mathcal{D}^\downarrow}(\mathrm{Act}(A'))$, if $\mathcal{M}_1 \equiv_{\mathrm{Act}(A')} \mathcal{M}_2$, then $|_f(\mathcal{M}_1) \equiv_A |_f(\mathcal{M}_2)$.*

5   Conclusions and Future Work

We have introduced the logic $\mathcal{D}^\downarrow$ suitable to specify abstract requirements for reactive systems as well as concrete designs expressing (recursive) process structures. Therefore $\mathcal{D}^\downarrow$ is appropriate to instantiate Sannella and Tarlecki's refinement framework to provide stepwise, correct-by-construction development of reactive systems. We have illustrated this with a simple example using specifications and implementation constructors over $\mathcal{D}^\downarrow$. We believe that a case was made for the suitability of both the logic and the method as a viable alternative to other, more standard approaches to the design of reactive software.

Related Work. Since the 80's, the formal development of reactive, concurrent systems has emerged as one of the most active research topics in Computer Science, with a plethora of approaches and formalisms. For a proper comparison with this work, the following paragraphs restrict to two classes of methods: the ones built on top of logics formalised as institutions, and the attempts to apply to the domain of reactive systems the methods and techniques inherited from the loose specification of abstract data types.

In the first class, references [7,9,25] introduce different institutions for temporal logics, as a natural setting for the specification of abstract properties of reactive processes. Process algebras themselves have also been framed as institutions. Reference [27] formalises CSP [17] in this way. What distinguishes our own approach, based on $\mathcal{D}^\downarrow$ is the possibility to combine and express in the same logic both abstract properties, as in temporal logics, and their realisation in concrete, recursive process terms, as typical in process algebras.

Our second motivation was to discuss how institution-independent methods, used in (data-oriented) software development, could be applied to the design of reactive systems. A related perspective is proposed in reference [23], which suggests the loose specification of processes on top of the CSP institution [27] mentioned above. The authors explore the reuse of institution independent structuring mechanisms introduced in the CASL framework [3] to develop reactive systems; in particular, process refinement is understood as inclusion of classes of models. Note that the CASL (in-the-large) specification structuring mechanisms can be also taken as specific constructors, as the ones given in this paper.

Future Work. A lot of work, however, remains to be done. First of all, logic $\mathcal{D}^{\downarrow}$ is worth to be studied in itself, namely proof calculi, and their soundness and completeness as well as decidability. In [2] it has been shown that nominal-free dynamic logic with binders is undecidable. Decidability of $\mathcal{D}^{\downarrow}$ is yet an open question: while [2] considers standard Kripke structures and global satisfaction, $\mathcal{D}^{\downarrow}$ considers reachable models and satisfaction w.r.t. initial states. On the other hand, in $\mathcal{D}^{\downarrow}$ modalities are indexed with regular expressions over sets of actions. It would also be worthwhile to discuss satisfaction up to some notion of observational equivalence, as done in [5] for algebraic specifications, thus leading to a behavioural version of $\mathcal{D}^{\downarrow}$.

The study of initial semantics (for some fragments) of $\mathcal{D}^{\downarrow}$ is also in our research agenda. For example, theories in the fragment of $\mathcal{D}^{\downarrow}$ that alternates binders with diamond modalities (thus binding all visited states) can be shown to have weak initial semantics, which becomes strong initial in a deterministic setting. The abstract study of initial semantics in hybrid(ised) logics reported in [8], together with the canonical model construction for propositional dynamic logic introduced in [19] can offer a nice starting point for this task. Moreover, for realistic systems, data must be included in our logic.

A second line of inquiry is more directly related to the development method. For example, defining an abstractor on top of some form of weak bisimilarity would allow for a proper treatment of *hiding*, an important operation in CSP [17] and some other process algebras through which a given set of actions is made non observable. Finally, our aim is to add a final step to the method proposed here in which any constructive specification can be translated to a process algebra expression, as currently done by our proof-of-concept translator D2CSP. A particularly elegant way to do it is to frame such a translation as an institution morphism into an institution representing a specific process algebra, for example the one proposed by M. Roggenbach [27] for CSP.

Acknowledgments. This work is financed by the ERDF European Regional Development Fund through the Operational Programme for Competitiveness and Internationalisation - COMPETE 2020 Programme and by National Funds through the Portuguese funding agency, FCT - Fundação para a Ciência Tecnologia within project POCI-01-0145-FEDER-016692 and UID/MAT/04106/2013 at CIDMA. A. Madeira and L. S. Barbosa are further supported by FCT individual grants SFRH/BPD/103004/2014 and SFRH/BSAB/113890/2015, respectively.

References

1. Aceto, L., Ingólfsdóttir, A., Larsen, K.G., Srba, J.: Reactive Systems: Modelling, Specification and Verification. Cambridge University Press, Cambridge (2007)
2. Areces, C., Blackburn, P., Marx, M.: A road-map on complexity for hybrid logics. In: Flum, J., Rodríguez-Artalejo, M. (eds.) CSL 1999. LNCS, vol. 1683, pp. 307–321. Springer, Heidelberg (1999)
3. Astesiano, E., Bidoit, M., Kirchner, H., Krieg-Brückner, B., Mosses, P.D., Sannella, D., Tarlecki, A.: CASL: the common algebraic specification language. Theor. Comput. Sci. **286**(2), 153–196 (2002)

4. Baeten, J.C.M., Basten, T., Reniers, M.A.: Process Algebra: Equational Theories of Communicating Processes. Cambridge University Press, Cambridge (2010)
5. Bidoit, M., Hennicker, R.: Constructor-based observational logic. J. Log. Algebr. Program. **67**(1–2), 3–51 (2006)
6. Braüner,' T.: Hybrid Logic and Its Proof-Theory. Applied Logic Series, vol. 37. Springer, Netherlands (2010)
7. Cengarle, M.V.: The temporal logic institution. Technical report 9805, LUM München, Institut für Informatik (1998)
8. Diaconescu, R.: Institutional semantics for many-valued logics. Fuzzy Sets Syst. **218**, 32–52 (2013)
9. Fiadeiro, J.L., Maibaum, T.S.E.: Temporal theories as modularisation units for concurrent system specification. Formal Asp. Comput. **4**(3), 239–272 (1992)
10. Goguen, J.A., Burstall, R.M.: Institutions: abstract model theory for specification and programming. J. ACM **39**(1), 95–146 (1992)
11. Goranko, V.: Temporal logic with reference pointers. In: Gabbay, D.M., Ohlbach, H.J. (eds.) ICTL 1994. LNCS, vol. 827, pp. 133–148. Springer, Heidelberg (1994). doi:10.1007/BFb0013985
12. Gorrieri, R., Rensink, A., Zamboni, M.A.: Action refinement. In: Handbook of Proacess Algebra, pp. 1047–1147. Elsevier (2000)
13. Groote, J.F., Mousavi, M.R.: Modeling and Analysis of Communicating Systems. MIT Press, Cambridge (2014)
14. Harel, D., Kozen, D., Tiuryn, J.: Dynamic Logic. MIT Press, Cambridge (2000)
15. Havelund, K.: The Fork Calculus -Towards a Logic for Concurrent ML. Ph.D. thesis, DIKU, University of Copenhagen, Denmark (1994)
16. Hoare, C.A.R.: Proof of correctness of data representations. Acta Inf. **1**, 271–281 (1972)
17. Hoare, C.A.R.: Communicating Sequential Processes. Series in Computer Science. Prentice-Hall International, Upper Saddle River (1985)
18. Jones, C.B.: Software Development - A Rigorous Approach. Series in Computer Science. Prentice Hall, Upper Saddle River (1980)
19. Knijnenburg, P., van Leeuwen, J.: On models for propositional dynamic logic. Theor. Comput. Sci. **91**(2), 181–203 (1991)
20. Larsen, K.G., Thomsen, B.: A modal process logic. In: Third Annual Symposium on Logic in Computer Science, pp. 203–210. IEEE Computer Society (1988)
21. Madeira, A., Barbosa, L., Hennicker, R., Martins, M.: Dynamic logic with binders and its applications to the developmet of reactive systems (extended with proofs). Technical report (2016). http://alfa.di.uminho.pt/~madeira/main_files/extreport. pdf
22. Magee, J., Kramer, J.: Concurrency - State Models and Java Programs, 2nd edn. Wiley, Hoboken (2006)
23. O'Reilly, L., Mossakowski, T., Roggenbach, M.: Compositional modelling and reasoning in an institution for processes and data. In: Mossakowski, T., Kreowski, H.-J. (eds.) WADT 2010. LNCS, vol. 7137, pp. 251–269. Springer, Heidelberg (2012)
24. Park, D.: Concurrency and automata on infinite sequences. In: Deussen, P. (ed.) GI-TCS 1981. LNCS, vol. 104, pp. 167–183. Springer, Heidelberg (1981). doi:10.1007/BFb0017309
25. Reggio, G., Astesiano, E., Choppy, C.: Casl-ltl: a casl extension for dynamic reactive systems version 1.0. - summary. Technical report disi-tr-03-36. Technical report, DFKI Lab Bremen (2013)

26. Reisig, W.: Petri Nets: An Introduction. EATCS Monographs on Theoretical Computer Science. Springer, Heidelberg (1985)
27. Roggenbach, M.: CSP-CASL - a new integration of process algebra and algebraic specification. Theor. Comput. Sci. **354**(1), 42–71 (2006)
28. Sannella, D., Tarlecki, A.: Toward formal development of programs from algebraic specifications: implementations revisited. Acta Inform. **25**(3), 233–281 (1988)
29. Sannella, D., Tarlecki, A.: Foundations of Algebraic Specification and Formal Software Development. Monographs on TCS, an EATCS Series. Springer, Heidelberg (2012)
30. Sekerinski, E., Sere, K.: Program Development by Refinement: Case Studies Using the B Method. Springer, Heidelberg (2012)
31. Winskel, G., Nielsen, M.: Models for concurrency. In: Abramsky, S., Gabbay, D.M., Maibaum, T.S.E. (eds.) Handbook of Logic in Computer Science, vol. 4, pp. 1–148. Oxford University Press, Oxford (1995)

Propositional Dynamic Logic for Petri Nets with Iteration

Mario R.F. Benevides[1], Bruno Lopes[2]([✉]), and Edward Hermann Haeusler[3]

[1] PESC/COPPE - Inst. de Matemática,
Universidade Federal do Rio de Janeiro, Rio de Janeiro, Brazil
mario@cos.ufrj.br
[2] Instituto de Computação, Universidade Federal Fluminense, Niterói, Brazil
bruno@ic.uff.br
[3] Departamento de Informática, Pontifícia Universidade
Católica do Rio de Janeiro, Rio de Janeiro, Brazil
hermann@inf.puc-rio.br

Abstract. This work extends our previous work [20] with the iteration operator. This new operator allows for representing more general networks and thus enhancing the former propositional logic for Petri Nets. We provide an axiomatization and a new semantics and prove soundness and completeness with respect with its semantics. In order to illustrate its usage, we also provide some examples.

Keywords: Propositional dynamic logic · Petri nets · Modal logic

1 Introduction

Propositional Dynamic Logic PDL plays an important role in formal specification and reasoning about programs and actions. PDL is a multi-modal logic with one modality for each program π $\langle \pi \rangle$. It has been used in formal specification to reasoning about properties of programs and their behaviour. Correctness, termination, fairness, liveness and equivalence of programs are among the properties usually desired. A Kripke semantics can be provided, with a frame $\mathcal{F} = \langle W, R_\pi \rangle$, where W is a set of possible program states and for each program π, R_π is a binary relation on W such that $(s, t) \in R_\pi$ if and only if there is a computation of π starting in s and terminating in t.

There are a lot of variations of PDL for different approaches [2]. Propositional Algorithmic Logic [24] that analizes properties of programs connectives, the interpretation of Deontic Logic as a variant of Dynamic Logic [23], applications in linguistics [17], Multi-Dimensional Dynamic Logic [27] that allows multi-agent [16] representation, Dynamic Arrow Logic [30] to deal with transitions in programs, Data Analysis Logic [7], Boolean Modal Logic [9], logics

M.R.F. Benevides, B. Lopes and E.H. Haeusler—This work was supported by the Brazilian research agencies CNPq, FAPERJ and CAPES.

A. Sampaio and F. Wang (Eds.): ICTAC 2016, LNCS 9965, pp. 441–456, 2016.
DOI: 10.1007/978-3-319-46750-4_25

for reasoning about knowledge [8], logics for knowledge representation [18] and Dynamic Description Logic [31].

Petri Net is a widely used formalism to specify and to analyze concurrent programs with a very nice graphical representation. It allows for representing true concurrency and parallelism in a neat way.

In [20], we present the logic Petri-PDL which uses Marked Petri Net programs as PDL programs. So if π is a Petri Net program with markup s, then the formula $\langle s, \pi \rangle \varphi$ means that after running this program with the initial markup s, φ will eventually be true (also possible a $\square$-like modality replacing the tags by brackets as an abbreviation for $\neg \langle s, \pi \rangle \neg \varphi$).

This work extends our previous work [20] with the iteration operator. This new operator allows for representing more general networks and thus enhancing the former propositional logic for Petri Nets. We provide a sound and complete axiomatization and also prove the finite model property, which together with the axiomatization yields decidability.

Our paper falls in the broad category of works that attempt to generalize PDL and build dynamic logics that deal with classes of non-regular programs. As examples of other works in this area, we can mention [11,12,19], that develop decidable dynamic logics for fragments of the class of context-free programs and [1,10,25,26] and [3], that develop dynamic logics for classes of programs with some sort of concurrency. Our logics have a close relation to two logics in this last group: Concurrent PDL [25] and Concurrent PDL with Channels [26]. Both of these logics are expressive enough to represent interesting properties of communicating concurrent systems. However, neither of them has a simple Kripke semantics. The first has a semantics based on *super-states* and *super-processes* and its satisfiability problem can be proved undecidable (in fact, it is Π_1^1-hard). Also, it does not have a complete axiomatization [26]. On the other hand, our logics have a simple Kripke semantics, simple and complete axiomatizations and the finite model property.

There are other approaches that use Dynamic Logic to reason about specifications of concurrent systems represented as Petri Nets [14,15,29]. They differ from our approach by the fact that they use Dynamic logic as a specification language for representing Petri Net, they do not encode Petri Nets as programs of a Dynamic Logic. They translate Nets into PDL language while we have a new Dynamic Logic tailored to reason about Petri Nets in a more natural way.

This paper is organized as follows. Section 2 presents all the background needed about (Marked) Petri Nets formalism and Propositional Dynamic Logic. Section 3, introduces our dynamic logic, with its language and semantics and also proposes an axiomatization and provides a prove of soundness and completeness. Section 5 illustrates the use of our logic with some examples. Finally, Sect. 6, presents some final remarks and future works.

2   Background

This section presents a brief overview of two topics on which the later development is based on. First, we make a brief review of the syntax and semantics of PDL [13].

Second, we present the Petri Nets formalism and its variant, Marked Petri Nets. Finally, the compositional approach introduced in [6] is briefly discussed.

2.1 Propositional Dynamic Logic

In this section, we present the syntax and semantics of the most used dynamic logic called PDL for regular programs.

Definition 1. *The PDL language consists of a set Φ of countably many proposition symbols, a set Π of countably many basic programs, the boolean connectives $\neg$ and $\wedge$, the program constructors ; (sequential composition), $\cup$ (non-deterministic choice) and * (iteration) and a modality $\langle \pi \rangle$ for every program π. The formulas are defined as follows:*

$$\varphi ::= p \mid \top \mid \neg\varphi \mid \varphi_1 \wedge \varphi_2 \mid \langle \pi \rangle \varphi, \ \text{ with } \ \pi ::= a \mid \pi_1; \pi_2 \mid \pi_1 \cup \pi_2 \mid \pi^\star,$$

where $p \in \Phi$ and $a \in \Pi$.

In all the logics that appear in this paper, we use the standard abbreviations $\bot \equiv \neg\top$, $\varphi \vee \phi \equiv \neg(\neg\varphi \wedge \neg\phi)$, $\varphi \to \phi \equiv \neg(\varphi \wedge \neg\phi)$ and $[\pi]\varphi \equiv \neg\langle \pi \rangle \neg\varphi$.

Each program π corresponds to a modality $\langle \pi \rangle$, where a formula $\langle \pi \rangle \alpha$ means that after the running of π, α eventually is true, considering that π halts. There is also the possibility of using $[\pi]\alpha$ (as an abbreviation for $\neg\langle \pi \rangle \neg\alpha$) indicating that the property denoted by α holds after every possible run of π.

The semantics of PDL is normally given using a transition diagram, which consists of a set of states and binary relations (one for each program) indicating the possible execution of each program at each state. In PDL literature a transition diagram is called a frame.

Definition 2. *A frame for PDL is a tuple $\mathcal{F} = \langle W, R_\pi \rangle$ where*

- *W is a non-empty set of states;*
- *R_a is a binary relation over W, for each basic program $a \in \Pi$;*
- *We can inductively define a binary relation R_π, for each non-basic program π, as follows*
  - *$R_{\pi_1; \pi_2} = R_{\pi_1} \circ R_{\pi_2}$,*
  - *$R_{\pi_1 \cup \pi_2} = R_{\pi_1} \cup R_{\pi_2}$,*
  - *$R_{\pi^\star} = R_\pi^\star$, where $R_\pi^\star$ denotes the reflexive transitive closure of R_π.*

Definition 3. *A model for PDL is a pair $\mathcal{M} = \langle \mathcal{F}, \mathbf{V} \rangle$, where $\mathcal{F}$ is a PDL frame and $\mathbf{V}$ is a valuation function $\mathbf{V} : \Phi \to 2^W$.*

The semantical notion of satisfaction for PDL is defined as follows:

Definition 4. *Let $\mathcal{M} = \langle \mathcal{F}, \mathbf{V} \rangle$ be a model. The notion of satisfaction of a formula φ in a model $\mathcal{M}$ at a state w, notation $\mathcal{M}, w \Vdash \varphi$, can be inductively defined as follows:*

- *$\mathcal{M}, w \Vdash p$ iff $w \in \mathbf{V}(p)$;*

- $\mathcal{M}, w \Vdash \top$ *always;*
- $\mathcal{M}, w \Vdash \neg\varphi$ *iff* $\mathcal{M}, w \nVdash \varphi;$
- $\mathcal{M}, w \Vdash \varphi_1 \wedge \varphi_2$ *iff* $\mathcal{M}, w \Vdash \varphi_1$ *and* $\mathcal{M}, w \Vdash \varphi_2;$
- $\mathcal{M}, w \Vdash \langle \pi \rangle \varphi$ *iff there is* $w' \in W$ *such that* $wR_\pi w'$ *and* $\mathcal{M}, w' \Vdash \varphi.$

For more details on PDL see [13].

2.2   Petri Nets

A Petri Net [28] is a tuple $\mathcal{P} = \langle S, T, W \rangle$, where S is a finite non-empty set of places, T is a finite set of transitions where S and T are disjoint and W is a function which defines directed edges between places and transitions and assigns a $w \in \mathbb{N}$ that represents a multiplicative weight for the transition, as $W: (S \times T) \cup (T \times S) \rightarrow \mathbb{N}$.

Marked Petri Nets. A markup function M is a function that assigns to each place a natural number, $M: S \rightarrow \mathbb{N}$. A Marked Petri Net is a tuple $\mathcal{P} = \langle S, T, W, M_0 \rangle$ where $\langle S, T, W \rangle$ is a Petri Net and M_0 as an initial markup. In the sequence, any reference to a Petri-Net means Marked Petri-Nets

The flow of a Petri Net is defined by a relation $F = \{(x, y) \mid W(x, y) > 0\}$; in this work we take the restriction that for all transitions $W(x, y) = 1$. Let $s \in S$ and $t \in T$. The preset of t, denoted by $^\bullet t$, is defined as $^\bullet t = \{s \in S: (s, t) \in F\}$; the postset of t, denoted by $t^\bullet$ is defined as $t^\bullet = \{s \in S: (t, s) \in F\}$. The preset of s, denoted by $^\bullet s$, is defined as $^\bullet s = \{t \in T: (t, s) \in F\}$; the postset of s, denoted by $s^\bullet$ is defined as $s^\bullet = \{t \in T: (s, t) \in F\}$.

Given a markup M of a Petri Net, we say that a transition t is enabled on M if and only if $\forall x \in {}^\bullet t, M(x) \geq 1$. A new markup generated by firing a transition which is enabled is defined as

$$M_{i+1}(x) = \begin{cases} M_i(x) - 1, \forall x \in {}^\bullet t \setminus t^\bullet \\ M_i(x) + 1, \forall x \in t^\bullet \cap {}^\bullet t \, . \\ M_i(x), \quad \text{otherwise} \end{cases} \tag{1}$$

A program behavior is described by the set $M = \{M_1, \ldots, M_n\}$ of a Petri Net markups.

A Petri Net may be interpreted in a graphical representation, using a circle to represent each $s \in S$, a rectangle to represent each $t \in T$, the relations defined by W as edges between places and transitions. The amount of tokens from M are represented as filled circles into the correspondent places. An example of a valid Petri Net is in Fig. 2.

Just as an example, the Petri Net on Fig. 1 represents the operation of an elevator for a building with five floors. A token in the place U indicates that the elevator is able to go up one floor; and, when T_1 fires, a token goes to D, so the elevator can go down a floor. If the elevator goes down a floor (i.e., T_2 fires) a token goes to the place U. Figure 1 illustrates the Petri Net with its initial markup.

Fig. 1. Petri Net for a simple elevator of five floors

Another example is in Fig. 2, which represents a SMS send and receive of two cellphones. When the user sends a SMS from his cellphone, it goes to his phone buffer (i.e., T_1 fires and the token goes to p_2). When the phone sends the message to the operator (i.e., T_2 fires) it goes to the operator buffer; so, the messages must be sent to the receiver, but the receiver is able to receive only one message at a time. If there is a message in the operator buffer and the receiver is not receiving other message (i.e., there is at least a token in p_3 and there is a token in p_4), the receiver can receive the message (i.e., T_3 fires). At this point the user can not receive other messages (i.e., there is no token in p_4, so T_3 is not enabled); but, after the complete receive of the message (i.e., T_4 fires), the receiver is able to receive messages again (i.e., there is a token in p_4 and when p_3 have at least a token, T_3 will be enabled again).

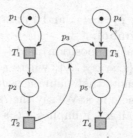

Fig. 2. Petri Net for a SMS send and receive

Basic Petri Nets. The Petri Net model used in this work is as defined by de Almeida and Haeusler [6]. It uses three basic Petri Nets to define all valid Petri Nets due to its compositions. These basic Petri Nets are as in Fig. 3.

To compose more complex Petri Nets from these three basic ones, it is used a gluing procedure described, and proved corrected in [6].

As an example, taking the Petri Net in Fig. 1 it can be modelled as a composition of two Petri Nets of type 1, where UT_1D is composed with DT_2U, generating the Petri Net $UT_1D \odot DT_2U$ where $\odot$ denotes the Petri Net composition symbol. The Petri Net from Fig. 2 can be modelled by composition of Petri Nets of the three basic types. The basic Petri Nets of this case are $p_2T_2p_3$, $p_5T_4p_4$ (type 1), $p_3p_4T_3p_5$ (type 2) and $p_1T_1p_1P_2$ (type 3), composing the Petri Net $p_2T_2p_3 \odot p_5T_4p_4 \odot p_3p_4T_3p_5 \odot p_1T_1p_1P_2$.

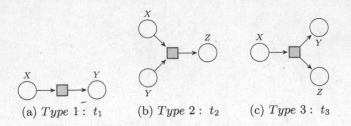

(a) *Type* 1 : t_1 (b) *Type* 2 : t_2 (c) *Type* 3 : t_3

Fig. 3. Basic Petri Nets.

3 Propositional Dynamic Logic for Petri Nets (Petri-PDL)

This section presents a Propositional Dynamic Logic that uses Petri Nets terms as programs (Petri-PDL) [20].

3.1 Language and Semantics

The language of Petri-PDL consists of

Propositional symbols: p, q...
Place names: e.g.: $a, b, c, d \ldots$
Transition types: $T_1 : xt_1y$, $T_2 : xyt_2z$ and $T_3 : xt_3yz$
Petri Net Composition symbol: $\odot$
PDL operators: ; (sequential composition) and ( )* (iteration)
Sequence of names: $S = \{\epsilon, s_1, s_2, \ldots\}$, where ϵ is the empty sequence. We use the notation $a \in s$ to denote that name a occurs in s. Let $\#(s, a)$ be the number of occurrences of name a in s. We say that sequence r is a sub-sequence of s, $r \preceq s$, if for any name a, if $a \in r$ implies $a \in s$ and $\#(r, a) \leq \#(s, a)$.

Definition 5. *Programs:*

Basic programs: $\pi :: =at_1b \mid abt_2c \mid at_3bc$ *where* t_i *is of type* $T_i, i = 1, 2, 3$ *and* a, b *and* c *are Place names.*
Petri Net Programs: $\eta :: =\pi \mid \pi \odot \eta \mid \eta^\star$

Definition 6. *A formula is defined as*

$$\varphi :: =p \mid \top \mid \neg\varphi \mid \varphi \wedge \varphi \mid \langle s, \eta \rangle \varphi.$$

We use the standard abbreviations $\bot \equiv \neg\top$, $\varphi \vee \phi \equiv \neg(\neg\varphi \wedge \neg\phi)$, $\varphi \to \phi \equiv \neg(\varphi \wedge \neg\phi)$ and $[s, \eta]\varphi \equiv \neg\langle s, \eta \rangle \neg\varphi$.

The definition below introduces the *firing* function. It defines how the marking of a basic Petri Nets changes after a firing.

Definition 7. *We define the firing function* $f : S \times \pi_b \to S$ *as follows*

$$- f(s, at_1 b) = \left\{ \begin{array}{c} s_1 b s_2, \; if \; s = s_1 a s_2 \\ \epsilon, \; if \; a \notin s \end{array} \right\}$$

$$- f(s, abt_2 c) = \left\{ \begin{array}{c} s_1 c s_2 s_3, \; if \; s = s_1 a s_2 b s_3 \\ \epsilon, \; if \; a \; or \; b \notin s \end{array} \right\}$$

$$- f(s, at_3 bc) = \left\{ \begin{array}{c} s_1 s_2 bc, \; if \; s = s_1 a s_2 \\ \epsilon, \; if \; a \notin s \end{array} \right\}$$

The definitions that follow of frame, model and satisfaction are similar to the one presented in Sect. 2.1 for PDL. Now, we have to adapt them to deal with the firing of basics Petri Nets.

Definition 8. *A frame for Petri-PDL is a 3-tuple* $\langle W, R_\pi, M \rangle$, *where*

- *W is a non-empty set of states;*
- *$M : W \to S$;*
- *R_π is a binary relation over W, for each basic program π, satisfying the following condition. Let $s = M(w)$*
  - *if $f(s, \pi) \neq \epsilon$, if $w R_\pi v$ then $f(s, \pi) \preceq M(v)$*
  - *if $f(s, \pi) = \epsilon$, $(w, v) \notin R_\pi$*
- *we inductively define a binary relation R_η, for each Petri Net program η as follows*
  - *$R_{\eta^*} = R_\eta^*$, where R_η^* denotes the reflexive transitive closure of R_η.*
  - *$\eta = \pi_1 \odot \pi_2 \odot \cdots \odot \pi_n$*
    $R_\eta = (R_{\pi_1} \circ R_{\eta^*}) \cup \cdots \cup (R_{\pi_n} \circ R_{\eta^*})$
Where $s = M(w)$, η_i are basic programs and $s_i = f(s, \eta_i)$, for all $1 \leq i \leq n$.

Definition 9. *A model for Petri-PDL is a pair* $\mathcal{M} = \langle \mathcal{F}, \mathbf{V} \rangle$, *where $\mathcal{F}$ is a Petri-PDL frame and $\mathbf{V}$ is a valuation function* $\mathbf{V} : \Phi \to 2^W$.

The semantical notion of satisfaction for Petri-PDL is defined as follows.

Definition 10. *Let* $\mathcal{M} = (\mathcal{F}, \mathbf{V})$ *be a model. The notion of satisfaction of a formula φ in a model $\mathcal{M}$ at a state w, notation $\mathcal{M}, w \Vdash \varphi$, can be inductively defined as follows:*

- *$\mathcal{M}, w \Vdash p$ iff $w \in \mathbf{V}(p)$;*
- *$\mathcal{M}, w \Vdash \top$ always;*
- *$\mathcal{M}, w \Vdash \neg\varphi$ iff $\mathcal{M}, w \nVdash \varphi$;*
- *$\mathcal{M}, w \Vdash \varphi_1 \wedge \varphi_2$ iff $\mathcal{M}, w \Vdash \varphi_1$ and $\mathcal{M}, w \Vdash \varphi_2$;*
- *$\mathcal{M}, w \Vdash \langle s, \eta \rangle \varphi$ if there exists $v \in W$, $w R_\eta v$, $s \preceq M(w)$ and $\mathcal{M}, v \Vdash \varphi$.*

If $\mathcal{M}, v \Vdash A$ for every state v, we say that A is *valid in the model* $\mathcal{M}$, notation $\mathcal{M} \Vdash A$. And if A is valid in all $\mathcal{M}$ we say that A is *valid*, notation $\Vdash A$.

3.2 Axiomatic System

We consider the following set of axioms and rules, where p and q are proposition symbols, φ and ψ are formulas, $\eta = \pi_1 \odot \pi_2 \odot \cdots \odot \pi_n$ is a Petri Net program and π_i are basic program.

(PL) Enough propositional logic tautologies
(K) $[s, \eta](p \to q) \to ([s, \eta]p \to [s, \eta]q)$
(Rec) $\langle s, \eta^* \rangle p \leftrightarrow p \vee \langle s, \eta \rangle \langle s, \eta^* \rangle p$
(FP) $p \wedge [s, \eta^*](p \to [s, \eta]p) \to [s, \eta^*]p$
(PC) $\langle s, \eta \rangle \varphi \leftrightarrow \langle s, \pi_1 \rangle \langle s_1, \eta^* \rangle \varphi \vee \langle s, \pi_2 \rangle \langle s_2, \eta^* \rangle \varphi \vee \cdots \vee \langle s, \pi_n \rangle \langle s_n, \eta^* \rangle \varphi$, where $s_i = f(s, \eta_i)$, for all $1 \leq i \leq n$.
(R$_\epsilon$) $[s, \pi] \bot$, if $f(s, \pi) = \epsilon$
(Sub) If $\Vdash \varphi$, then $\Vdash \varphi^\sigma$, where σ uniformly substitutes proposition symbols by arbitrary formulas.
(MP) If $\Vdash \varphi$ and $\Vdash \varphi \to \psi$, then $\Vdash \psi$.
(Gen) If $\Vdash \varphi$, then $\Vdash [s, \eta]\varphi$.

4 Soundness and Completeness

The axioms **(PL)** and **(K)** and the rules **(Sub)**, **(MP)** and **(Gen)** are standard in the modal logic literature.

Lemma 1. *Validity of Petri-PDL axioms*

Proof. 1. $\Vdash$ **Rec**

Proof. Suppose that there is a world w from a model $\mathcal{M} = \langle W, R_\pi, \mathbf{V}, M \rangle$ where Rec is false. For Rec to be false in w, there are two cases:

(a) Suppose $\mathcal{M}, w \Vdash \langle s, \pi^* \rangle p$ (1) and
$\mathcal{M}, w \nVdash p \vee \langle s, \pi \rangle \langle s, \pi^* \rangle p$ (2)
Applying Definition 10 in (1) we have that $\mathcal{M}, w \Vdash \langle s, \pi \rangle \langle s, \pi^* \rangle p$ (3).
Applying Definition 10 again we have that $\mathcal{M}, w \Vdash p \vee \langle s, \pi \rangle \langle s, \pi^* \rangle p$, which contradicts (2).
(b) Suppose $\mathcal{M}, w \nVdash \langle s, \pi^* \rangle p$ (1) and
$\mathcal{M}, w \Vdash p \vee \langle s, \pi \rangle \langle s, \pi^* \rangle p$ (2)
Applying Definition 10 in (1) we have $\mathcal{M}, w \Vdash p \vee \langle s, \pi \rangle p$ (3).
Using the axiom (Gen) and then (K) in (3) we have that $\mathcal{M}, w \Vdash [s, \pi]p \vee \langle s, \pi \rangle p$, then, using Definition 10, we have that $\mathcal{M}, w \Vdash \langle s, \pi \rangle p \vee \langle s, \pi \rangle p$, which by Definition 10 implies that $\mathcal{M}, w \Vdash \langle s, \pi \rangle p$. (4)
But by (1) and Definition 10 we can not have (4).
Then, there is a contradiction.

So, Rec is valid.   □

2. $\Vdash$ **FP**

Proof. Suppose that there is a world w from a model $\mathcal{M} = \langle W, R_\pi, \mathbf{V}, M \rangle$ where FP is false.

So, $\mathcal{M}_3^*, w \Vdash p \wedge [s, \pi^*](p \rightarrow [s, \pi]p)$ (1) and
$\mathcal{M}_3^*, w \nVdash [s, \pi^*]p$ (2).

By (1) and Definition 10 we have that $\mathcal{M}_3^*, w \Vdash p$ and $\mathcal{M}_3^*, w \Vdash [s, \pi^*](p \rightarrow [s, \pi]p)$ (4)

Applying (MP) in (4) we have that $\mathcal{M}_3^*, w \Vdash [s, \pi^*]([s, \pi]p)$, which contradicts (2).

So, FP is valid.

3. $\Vdash$ **PC**

Suppose that there is a world w from a model $\mathcal{M} = \langle W, R_\pi, \mathbf{V}, M \rangle$ where PC is false. For PC to be false in w, there are two cases:

(a) Suppose $\mathcal{M}, w \Vdash \langle s, \eta \rangle \varphi$ (1).
(1) iff there is a v such that $wR_\eta v$, $s \preceq M(w)$ and $\mathcal{M}, v \Vdash \varphi$ (2).
By Definition 8 $R_\eta = (R_{\pi_1} \circ R_{\eta^*}) \cup \cdots \cup (R_{\pi_n} \circ R_{\eta^*})$ which implies that for some $0 \leq i \leq n$, $w(R_{\pi_i} \circ R_{\eta^*})v$. Using Definition 10 twice we obtain.
$\mathcal{M}, w \Vdash \langle s, \pi_i \rangle \langle s_i, \eta^* \rangle \varphi$. This implies
$\mathcal{M}, w \Vdash \langle s, \pi_1 \rangle \langle s_1, \eta^* \rangle \varphi \vee \langle s, \pi_2 \rangle \langle s_2, \eta^* \rangle \varphi \vee \cdots \vee \langle s, \pi_n \rangle \langle s_n, \eta^* \rangle \varphi$.
(b) Suppose $\mathcal{M}, w \Vdash \langle s, \pi_1 \rangle \langle s_1, \eta^* \rangle \varphi \vee \langle s, \pi_2 \rangle \langle s_2, \eta^* \rangle \varphi \vee \cdots \vee \langle s, \pi_n \rangle \langle s_n, \eta^* \rangle \varphi$
(2), iff for some i $(1 \leq i \leq n)$, $\mathcal{M}, w \Vdash \langle s, \pi_i \rangle \langle s_i, \eta^* \rangle \varphi$ iff
there is a u such that $wR_{\eta_i} u$, $s \preceq M(w)$ and $\mathcal{M}, u \Vdash \langle s_i, \eta^* \rangle \varphi$,
iff there is a v such that $uR_\eta v$, $s_i \preceq M(u)$ and $\mathcal{M}, v \Vdash \varphi$ (3). But this implies that $w(R_{\pi_i} \circ R_{\eta^*})v$ and consequently $w((R_{\pi_1} \circ R_{\eta^*}) \cup \cdots \cup (R_{\pi_n} \circ R_{\eta^*}))v$(4)
By Definition 8, (3) and (4) we have $wR_\eta v$ and $s \preceq M(w)$ and $\mathcal{M}, v \Vdash \varphi$.
Thus, $\mathcal{M}, w \Vdash \langle s, \eta \rangle \varphi$

4. $\Vdash$ **R_ϵ,**

Suppose $f(s, \pi) = \epsilon$ and $\nVdash [s, \pi]\bot$, so there exists a model M and a state w such that $M, w \nVdash [s, \pi]\bot$ iff $M, w \Vdash \langle s, \pi \rangle \top$ (1)

From (1), there is a v such that $wR_\pi v$, $s \preceq M(w)$ and $\mathcal{M}, v \Vdash \top$
But $f(s, \pi) = \epsilon$ and thus by Definition 8 $(w, v) \notin R_\pi$, which is a contradiction.
Hence, the axiomatic system is consistent.

Definition 11. *(Fischer and Ladner Closure): Let Γ be a set of formulas. The* **closure** *of Γ, notation $C_{FL}(\Gamma)$, is the smallest set of formulas satisfying the following conditions:*

1. *$C_{FL}(\Gamma)$ is closed under subformulas,*
2. *if $\langle s, \eta^* \rangle \varphi \in C_{FL}(\Gamma)$, then $\langle s, \eta \rangle \varphi \in C_{FL}(\Gamma)$,*
3. *if $\langle s, \eta^* \rangle \varphi \in C_{FL}(\Gamma)$, then $\langle s, \eta \rangle \langle s, \eta^* \rangle \varphi \in C_{FL}(\Gamma)$,*
4. *if $\langle s, \eta \rangle \varphi \in C_{FL}(\Gamma)$, then $\langle s, \eta_i \rangle \langle s_i, \eta^* \rangle \varphi \in C_{FL}(\Gamma)$, where $\eta = \eta_1 \odot \eta_2 \odot \cdots \odot \eta_n$ and $s_i = f(s, \eta_i)$, for all $1 \leq i \leq n$.*
5. *if $\varphi \in C_{FL}(\Gamma)$ and φ is not of the form $\neg \psi$, then $\neg \varphi \in C_{FL}(\Gamma)$.*

We prove that if Γ is a finite set of formulas, then the closure $C_{FL}(\Gamma)$ of Γ is also finite. We assume Γ to be finite from now on.

Lemma 2. *If Γ is a finite set of formulas, then $C_{FL}(\Gamma)$ is also finite.*

Proof. This proof is standard in PDL literature [4].   □

Definition 12. *Let Γ be a set of formulas. A set of formulas $\mathcal{A}$ is said to be an* **atom of** Γ *if it is a maximal consistent subset of $C_{FL}(\Gamma)$. The set of all atoms of Γ is denoted by $At(\Gamma)$.*

Lemma 3. *Let Γ be a set of formulas. If $\varphi \in C_{FL}(\Gamma)$ and φ is consistent then there exists an atom $\mathcal{A} \in At(\Gamma)$ such that $\varphi \in \mathcal{A}$.*

Proof. We can construct the atom $\mathcal{A}$ as follows. First, we enumerate the elements of $C_{FL}(\Gamma)$ as $\phi_1, \cdots, \phi_n$. We start the construction making $\mathcal{A}_1 = \{\varphi\}$, then for $1 < i < n$, we know that $\vdash \bigwedge \mathcal{A}_i \leftrightarrow (\bigwedge \mathcal{A}_i \wedge \phi_{i+1}) \vee (\bigwedge \mathcal{A}_i \wedge \neg\phi_{i+1})$ is a tautology and therefore either $\mathcal{A}_i \wedge \phi_{i+1}$ or $\mathcal{A}_i \wedge \neg\phi_{i+1}$ is consistent. We take $\mathcal{A}_{i+1}$ as the union of $\mathcal{A}_i$ with the consistent member of the previous disjunction. At the end, we make $\mathcal{A} = \mathcal{A}_n$.   □

Definition 13. *Let Γ be a set of formulas and $\langle s, \eta \rangle \varphi \in At(\Gamma)$. The* **canonical relations over** Γ S_η^Γ *on $At(\Gamma)$ are defined as follows:*

$$\mathcal{A} S_\eta^\Gamma \mathcal{B} \text{ iff } \bigwedge \mathcal{A} \wedge \langle s, \eta \rangle \bigwedge \mathcal{B} \text{ is consistent.}$$

Definition 14. *Let $\{\langle s_1, \eta_1 \rangle \varphi_1, ..., \langle s_n, \eta_n \rangle \varphi_n\}$ be the set of all diamond formulas occurring in one atom $\mathcal{A}$. We define the* **canonical marking** *of $\mathcal{A}$ $M(\mathcal{A})$ as follows*

1. $M(\mathcal{A}) := s_1; s_2; ...; s_n;$
2. *for all basic programs π, if $\mathcal{A} S_\eta^\Gamma \mathcal{B}$ and $f(M(\mathcal{A}), \pi) \npreceq M(\mathcal{B})$, then add to $M(\mathcal{B})$ as few as possible names to make $f(M(\mathcal{A}), \pi) \preceq M(\mathcal{B})$.*

Definition 15. *Let Γ be a set of formulas. The* **canonical model over** Γ *is a tuple $\mathcal{M}^\Gamma = \langle At(\Gamma), S_\eta^\Gamma, M^\Gamma, \mathbf{V}^\Gamma \rangle$, where for all propositional symbols p and for all atoms $\mathcal{A} \in At(\Gamma)$ we have*

- $M^\Gamma : At(\Gamma) \mapsto S$, *called canonical marking;*
- $\mathbf{V}^\Gamma(p) = \{\mathcal{A} \in At(\Gamma) \mid p \in \mathcal{A}\}$ *is called canonical valuation;*
- S_η^Γ *are the canonical relations[1].*

Lemma 1. *For all basic programs π, let $s = M(\mathcal{A})$, S_π satisfies*

1. *if $f(s, \pi) \neq \epsilon$, if $\mathcal{A} S_\pi \mathcal{B}$ then $f(s, \pi) \preceq M(\mathcal{B})$*
2. *if $f(s, \pi) = \epsilon$, then $(\mathcal{A}, \mathcal{B}) \notin S_\pi$*

Proof. The proof of 1. is straightforward from the definition of canonical marking (Definition 14). The proof of 2. follows from axiom R_ϵ.

---

[1] For the sake of clarity we avoid using the Γ subscripts.

Lemma 2 (Existence Lemma for Canonical Models). *Let $\mathcal{A} \in At(\Gamma)$ and $\langle s, \eta \rangle \varphi \in C_{FL}$. Then,*

$\langle s, \eta \rangle \varphi \in \mathcal{A}$ *iff there exists $\mathcal{B} \in At(\Gamma)$ such that $\mathcal{AS}_\eta \mathcal{B}$, $s \preceq M(\mathcal{A})$ and $\varphi \in \mathcal{B}$.*

Proof. $\Rightarrow$: Suppose $\langle s, \eta \rangle \varphi \in \mathcal{A}$. By Definition 14 we know that $s \preceq M(\mathcal{A})$. By Definition 12, we have that $\bigwedge \mathcal{A} \wedge \langle s, \eta \rangle \varphi$ is consistent. Using the tautology $\vdash \varphi \leftrightarrow ((\varphi \wedge \phi) \vee (\varphi \wedge \neg \phi))$, we have that either $\bigwedge \mathcal{A} \wedge \langle s, \eta \rangle (\varphi \wedge \phi)$ is consistent or $\bigwedge \mathcal{A} \wedge \langle s, \eta \rangle (\varphi \wedge \neg \phi)$ is consistent. So, by the appropriate choice of ϕ, for all formulas $\phi \in C_{FL}$, we can construct an atom $\mathcal{B}$ such that $\varphi \in \mathcal{B}$ and $\bigwedge \mathcal{A} \wedge \langle s, \eta \rangle (\varphi \wedge \bigwedge \mathcal{B})$ is consistent and by Definition 13 $\mathcal{AS}_\eta \mathcal{B}$.

$\Leftarrow$: Suppose there is $\mathcal{B}$ such that $\varphi \in \mathcal{B}$ and $\mathcal{AS}_\eta \mathcal{B}$ and $s \preceq M(\mathcal{A})$. Then $\bigwedge \mathcal{A} \wedge \langle s, \eta \rangle \bigwedge \mathcal{B}$ is consistent and also $\bigwedge \mathcal{A} \wedge \langle s, \eta \rangle \varphi$ is consistent. But $\langle s, \eta \rangle \varphi \in C_{FL}$ and by maximality $\langle s, \eta \rangle \varphi \in \mathcal{A}$.

Lemma 3 (Truth Lemma for Canonical Models). *Let $\mathcal{M} = (W, S_\eta, M, \mathbf{V})$ be a finite canonical model constructed over a formula ϕ. For all atoms $\mathcal{A}$ and all $\varphi \in C_{FL}(\phi)$, $\mathcal{M}, \mathcal{A} \models \varphi$ iff $\varphi \in \mathcal{A}$.*

Proof. The proof is by induction on the construction of φ.

- Atomic formulas and Boolean operators: the proof is straightforward from the definition of $\mathbf{V}$.
- Modality $\langle x \rangle$, for $x \in \{\pi, \pi_1 \odot \cdots \odot \pi_n, \eta \star\}$.

  $\Rightarrow$: Suppose $\mathcal{M}, \mathcal{A} \models \langle s, x \rangle \varphi$, then there exists $\mathcal{A}'$ such that $\mathcal{AS}_x \mathcal{A}'$, $s \preceq M(\mathcal{A})$ and $\mathcal{M}, \mathcal{A}' \models \varphi$. By the induction hypothesis we know that $\varphi \in \mathcal{A}'$, and by Lemma 2 we have $\langle s, x \rangle \varphi \in \mathcal{A}$.

  $\Leftarrow$: Suppose $\mathcal{M}, \mathcal{A} \not\models \langle s, x \rangle \varphi$, by the definition of satisfaction we have $\mathcal{M}, \mathcal{A} \models \neg \langle s, x \rangle \varphi$. Then for all $\mathcal{A}'$, $\mathcal{AS}_x \mathcal{A}'$ and $s \preceq M(\mathcal{A})$ implies $\mathcal{M}, \mathcal{A}' \not\models \varphi$. By the induction hypothesis we know that $\varphi \notin \mathcal{A}'$, and by Lemma 2 we have $\langle s, x \rangle \varphi \notin \mathcal{A}$.

Lemma 4. *Let $\mathcal{A}, \mathcal{B} \in At(\Gamma)$. Then if $\mathcal{AS}_{\eta \star} \mathcal{B}$ then $\mathcal{AS}_\eta^\star \mathcal{B}$.*

Proof. Suppose $\mathcal{AS}_{\eta \star} \mathcal{B}$. Let $\mathbf{C} = \{\mathcal{C} \in At(\Gamma) \mid \mathcal{AS}_\eta^\star \mathcal{C}\}$. We want to show that $\mathcal{B} \in \mathbf{C}$. Let $\mathbf{C}_\vee^\wedge = (\bigwedge \mathcal{C}_1 \vee \cdots \vee \bigwedge \mathcal{C}_n)$ and $s = s_1 \ldots s_n$, where $s_i = M(\mathcal{C}_i)$.

It is not difficult to see that $\mathbf{C}_\vee^\wedge \wedge \langle s, \eta \rangle \neg \mathbf{C}_\vee^\wedge$ is inconsistent, otherwise for some $\mathcal{D}$ not reachable from $\mathcal{A}$, $\mathbf{C}_\vee^\wedge \wedge \langle s, \eta \rangle \bigwedge \mathcal{D}$ would be consistent, and for some $\mathcal{C}_i$, $\bigwedge \mathcal{C}_i \wedge \langle s_i, \eta \rangle \bigwedge \mathcal{D}$ was also consistent, which would mean that $\mathcal{D} \in \mathbf{C}$, which is not the case. From a similar reasoning we know that $\bigwedge \mathcal{A} \wedge \langle s, \eta \rangle \neg \mathbf{C}_\vee^\wedge$ is also inconsistent and hence $\vdash \bigwedge \mathcal{A} \to [s, \eta] \mathbf{C}_\vee^\wedge$ is a theorem.

As $\mathbf{C}_\vee^\wedge \wedge \langle s, \eta \rangle \neg \mathbf{C}_\vee^\wedge$ is inconsistent, so its negation is a theorem $\vdash \neg (\mathbf{C}_\vee^\wedge \wedge \langle s, \eta \rangle \neg \mathbf{C}_\vee^\wedge)$ and also $\vdash (\mathbf{C}_\vee^\wedge \to [s, \eta] \mathbf{C}_\vee^\wedge)$ (1), applying generalization $\vdash [s, \eta \star](\mathbf{C}_\vee^\wedge \to [s, \eta] \mathbf{C}_\vee^\wedge)$. Using Segerberg axiom (axiom 6), we have $\vdash ([s, \eta] \mathbf{C}_\vee^\wedge \to [s, \eta \star] \mathbf{C}_\vee^\wedge)$ and by (1) we obtain $\vdash (\mathbf{C}_\vee^\wedge \to [s, \eta \star] \mathbf{C}_\vee^\wedge)$. As $\vdash \bigwedge \mathcal{A} \to [s, \eta] \mathbf{C}_\vee^\wedge$ is a theorem, then $\vdash \bigwedge \mathcal{A} \to [s, \eta \star] \mathbf{C}_\vee^\wedge$. By supposition, $\bigwedge \mathcal{A} \wedge \langle s, \eta \star \rangle \bigwedge \mathcal{B}$ is consistent and so is $\bigwedge \mathcal{B} \wedge \mathbf{C}_\vee^\wedge$. Therefore, for at least one $\mathcal{C} \in \mathbf{C}$, we know that $\bigwedge \mathcal{B} \wedge \bigwedge \mathcal{C}$ is consistent. By maximality, we have that $\mathcal{B} = \mathcal{C}$. And by the definition of $\mathbf{C}_\vee^\wedge$, we have $\mathcal{AS}_\eta^\star \mathcal{B}$. $\square$

Definition 16. *Let Γ be a set of formulas. The* **proper canonical model over** Γ *is a tuple* $\mathcal{N}^\Gamma = \langle At(\Gamma), R_\eta^\Gamma, M^\Gamma, \mathbf{V}^\Gamma \rangle$, *where for all propositional symbols* p *and for all atoms* $\mathcal{A} \in At(\Gamma)$ *we have*

- $\mathbf{V}^\Gamma(p) = \{\mathcal{A} \in At(\Gamma) \mid p \in \mathcal{A}\}$ *is called canonical valuation;*
- M^Γ *is the canonical marking;*
- $R_\pi^\Gamma := S_\pi^\Gamma$, *for every basic program* π;
- *we inductively define a binary relation* R_η *is inductively define as follows,*[2]

- $R_{\eta^*} = R_\eta^*$;
- $\eta = \pi_1 \odot \pi_2 \odot \cdots \odot \pi_n$
  $R_\eta = (R_{\pi_1} \circ R_{\eta^*}) \cup \cdots \cup (R_{\pi_n} \circ R_{\eta^*})$

Lemma 4. *For all programs* η, $S_\eta \subseteq R_\eta$.

Proof. By induction on the length of programs η

- For basic programs π, $S_\eta = R_\eta$ (Definition 16)
- $\eta = \theta^\star$. We have that $R_{\theta^\star} = R_\theta^\star$. By induction hypothesis $S_\theta \subseteq R_\theta$. But we know that if $S_\theta \subseteq R_\theta$ then $S_\theta^\star \subseteq R_\theta^\star$. So $S_\theta^\star \subseteq R_\theta^\star$ (1).
  By Lemma 4, $S_{\theta^\star} \subseteq S_\theta^\star$, and thus $S_{\theta^\star} \subseteq S_\theta^\star \subseteq R_\theta^\star = R_{\theta^\star}$.
- $\eta = \pi_1 \odot \pi_2 \odot \cdots \odot \pi_n$. We have that $R_\eta = (R_{\pi_1} \circ R_{\eta^*}) \cup \cdots \cup (R_{\pi_n} \circ R_{\eta^*})$. By the previous item we know $S_{\theta^*} \subseteq R_{\theta^*}$, and by the induction hypothesis $S_{\pi_i} \subseteq R_{\pi_i}$ and thus $(S_{\pi_1} \circ S_{\eta^*}) \cup \cdots \cup (S_{\pi_n} \circ S_{\eta^*}) \subseteq R_\theta$ (0).
  Suppose $\mathcal{A} S_\eta \mathcal{B}$, iff $\bigwedge \mathcal{A} \wedge \langle s, \eta \rangle \bigwedge \mathcal{B}$ is consistent.
  Using axiom (PC)
  $\bigwedge \mathcal{A} \wedge \langle s, \pi_1 \rangle \langle s_1, \eta^* \rangle \bigwedge \mathcal{B} \vee \langle s, \pi_2 \rangle \langle s_2, \eta^* \rangle \bigwedge \mathcal{B} \vee \cdots \vee \langle s, \pi_n \rangle \langle s_n, \eta^* \rangle \bigwedge \mathcal{B}$
  is consistent. For at least one i, $\bigwedge \mathcal{A} \wedge \langle s, \pi_i \rangle \langle s_i, \eta^* \rangle \bigwedge \mathcal{B}$ is consistent.
  Using a forcing choices argument we can construct a $\mathcal{C}$ such that
  $\bigwedge \mathcal{A} \wedge \langle s, \pi_i \rangle \bigwedge \mathcal{C}$ is consistent (1) and
  $\bigwedge \mathcal{C} \wedge \langle s_i, \eta^* \rangle \bigwedge \mathcal{B}$ is consistent.
  Let $s' = M(\mathcal{C})$. As $s_i \preceq s'$, then
  $\bigwedge \mathcal{C} \wedge \langle s', \eta^* \rangle \bigwedge \mathcal{B}$ is consistent (2).
  From (1) and (2) we have $\mathcal{A} S_{\pi_i} \mathcal{C}$ and $\mathcal{C} S_{\eta^*} \mathcal{B}$, and
  $\mathcal{A}(S_{\pi_i} \circ S_{\eta^*}) \mathcal{B}$. Thus
  $\mathcal{A}(S_{\pi_1} \circ S_{\eta^*}) \cup \cdots \cup (S_{\pi_n} \circ S_{\eta^*}) \mathcal{B}$.
  By (0), $\mathcal{A} R_\eta \mathcal{B}$. Therefore, $S_\eta \subseteq R_\eta$.

Lemma 5 (Existence Lemma for Proper Canonical Models). *Let* $\mathcal{A} \in At(\Gamma)$ *and* $\langle s, \eta \rangle \varphi \in C_{FL}$. *Then,*
$\langle s, \eta \rangle \varphi \in \mathcal{A}$ *iff there exists* $\mathcal{B} \in At(\Gamma)$ *such that* $\mathcal{A} R_\eta \mathcal{B}$, $s \preceq M(\mathcal{A})$ *and* $\varphi \in \mathcal{B}$.

Proof. $\Rightarrow$: Suppose $\langle s, \eta \rangle \varphi \in \mathcal{A}$. By the Existence Lemma for Canonical Models, Lemma 2, we have then there exists $\mathcal{B} \in At(\Gamma)$ such that $\mathcal{A} S_\eta \mathcal{B}$ and $\varphi \in \mathcal{B}$. As by Lemma 4, $S_\eta \subseteq R_\eta$. Thus, there exists $\mathcal{B} \in At(\Gamma)$ such that $\mathcal{A} R_\eta \mathcal{B}$ and $\varphi \in \mathcal{B}$.

---

[2] For the sake of clarity we avoid using the Γ superscripts.

$\Leftarrow$: Programs x, for $x \in \{\pi, \pi_1 \odot \cdots \odot \pi_n, \eta\star\}$.

Suppose there exists $\mathcal{B} \in At(\Gamma)$ such that $\mathcal{A}R_x\mathcal{B}$ and $\varphi \in \mathcal{B}$. The proof follows by induction on the structure of x.

- $x = \pi$ (Basic programs): this is straightforward once $R_\pi = S_\pi$ and by the existence Lemma 2 for canonical models $\langle s, \pi \rangle \varphi \in \mathcal{A}$.
- $x = \eta\star$. By definition $R_{\eta\star} = (R_\eta)^*$
  Suppose that for some $\mathcal{B}$, $\mathcal{A}R_\eta^*\mathcal{B}$ and $\varphi \in \mathcal{B}$. Then, for some n,
  $\mathcal{A} = \mathcal{A}_1 R_\eta \cdots R_\eta \mathcal{A}_n = \mathcal{B}$. We can prove by sub-induction on $1 \le k \le n$.
  - $k = 1$: $\mathcal{A}R_\eta\mathcal{B}$ and $\mathcal{A} \in \mathcal{B}$. By induction hypothesis, $\langle s, \eta \rangle \varphi \in \mathcal{A}$. From axiom Rec, we know that $\vdash \langle s, \eta \rangle \varphi \to \langle s, \eta^* \rangle \varphi$ and by the definition of C_{FL} and maximality we have $\langle s, \eta^* \rangle \varphi \in \mathcal{A}$.
  - $k > 1$: By the sub-induction hypothesis $\langle s, \eta^* \rangle \varphi \in \mathcal{A}_2$ and $\langle s, \eta \rangle \langle s, \eta^* \rangle \varphi \in \mathcal{A}_1$. From axiom Rec, we know that $\vdash \langle s, \eta \rangle \langle s, \eta^* \rangle \varphi \to \langle s, \eta^* \rangle \varphi$ and by the definition of C_{FL} and maximality we have $\langle s, \eta^* \rangle \varphi \in \mathcal{A}$.
- $x = \pi_1 \odot \cdots \odot \pi_n$: $\mathcal{A}R_{\pi_1 \odot \cdots \odot \pi_n}\mathcal{B}$ and $\varphi \in \mathcal{B}$ iff $\mathcal{A}(R_{\pi_1} \circ R_{x^*}) \cup \cdots \cup (R_{\pi_n} \circ R_{x^*})\mathcal{B}$ and $\varphi \in \mathcal{B}$. For some $1 \le i \le n$ $\mathcal{A}(R_{\pi_i} \circ R_{x^*})\mathcal{B}$ and $\varphi \in \mathcal{B}$. There exists a $\mathcal{C}$ such that $\mathcal{A}R_{\pi_i}\mathcal{C}$ and $\mathcal{C}R_{x^*}\mathcal{B}$ and $\varphi \in \mathcal{B}$. By the previuos case $\langle s_i, x^* \rangle \varphi \in \mathcal{C}$ (where $s_i = f(s, \pi_i)$), and by the induction hypothesis $\langle s, \pi_i \rangle \langle s_i, x^* \rangle \varphi \in \mathcal{A}$. But this implies that $\langle s, \pi_1 \rangle \langle s_1, x^* \rangle \varphi \vee \ldots \vee \langle s, \pi_n \rangle \langle s_n, x^* \rangle \varphi \wedge \bigwedge \mathcal{A}$ is consistent. By axiom PC, $\langle s, \pi_1 \odot \cdots \odot \pi_n \rangle \varphi \wedge \bigwedge \mathcal{A}$ is consistent. By maximality, $\langle s, \pi_1 \odot \cdots \odot \pi_n \rangle \varphi \in \mathcal{A}$.

Lemma 6 (Truth Lemma for Proper Canonical Models). *Let* $\mathcal{N} = (W, R_\eta, \mathbf{V})$ *be a finite proper canonical model constructed over a formula* ϕ. *For all atoms* $\mathcal{A}$ *and all* $\varphi \in C_{FL}(\phi)$, $\mathcal{N}, \mathcal{A} \models \varphi$ *iff* $\varphi \in \mathcal{A}$.

Proof. The proof is by induction on the construction of φ.

- Atomic formulas and Boolean operators: the proof is straightforward from the definition of $\mathbf{V}$.
- Modality $\langle x \rangle$, for $x \in \{\pi, \pi_1 \odot \cdots \odot \pi_n, \eta\star\}$.

  $\Rightarrow$: Suppose $\mathcal{M}, \mathcal{A} \models \langle s, x \rangle \varphi$, then there exists $\mathcal{A}'$ such that $\mathcal{A}S_x\mathcal{A}'$ and $\mathcal{M}, \mathcal{A}' \models \varphi$. By the induction hypothesis we know that $\varphi \in \mathcal{A}'$, and by Lemma 2 we have $\langle s, x \rangle \varphi \in \mathcal{A}$.

  $\Leftarrow$: Suppose $\mathcal{N}, \mathcal{A} \not\models \langle s, x \rangle \varphi$, by the definition of satisfaction we have $\mathcal{N}, \mathcal{A} \models \neg \langle s, x \rangle \varphi$. Then for all $\mathcal{A}'$, $\mathcal{A}R_x\mathcal{A}'$ implies $\mathcal{N}, \mathcal{A}' \not\models \varphi$. By the induction hypothesis we know that $\varphi \notin \mathcal{A}'$, and by Lemma 5 we have $\langle s, x \rangle \varphi \notin \mathcal{A}$.

Theorem 1 (Completeness for Proper Canonical Models). *Propositional Dynamic Logic for PetriNets Programs is complete with respect to the class of Proper Canonical Models.*

Proof. For every consistent formula A we can build a finite proper canonical model $\mathcal{N}$. By Lemma 3, there exist an atom $\mathcal{A} \in At(A)$ such that $A \in \mathcal{A}$, and by the truth Lemma 6 $\mathcal{N}, \mathcal{A} \models A$. Therefore, our modal system is complete with respect to the class of finite proper canonical models.

5 Some Usage Examples

This section, presents some examples of the application of our logic.

Example 1. *We can prove that if we place a token at location b and leave the location a empty, then after the execution of the network we cannot obtain a configuration where there is a token at location c. This can be expressed by the formula* $\langle (b), abt_2c \rangle \top \rightarrow \neg \langle (c), abt_2c \rangle \top$.

We can use our proof system to prove this property.

Fig. 4. A Petri Net where a and b have one token each one

Example 2. *We illustrate the use of the iteration with the chocolate vending machine. It works as follows: we turn it on (l) and put one coin (m) and then it releases the chocolate (c).*

Its behavior can be specified by the Petri Net of Fig. 5. The upper left place (ℓ) is the power button of a vending machine; the bottom left is the coin inserted (m) and the bottom right is the chocolate output (c); if the vending machine is powered on, always when a coin is inserted you will have a chocolate released. This behavior must repeat forever and it is here that we need the iteration operator ()⋆. *We can express that once we have tuned the machine on and put one coin we can obtain a chocolate by the formula* $\langle (\ell, m), (\ell m t_2 x \odot x t_3 y c \odot y t_1 \ell)^\star \rangle \top \rightarrow \langle (\ell, c), (\ell m t_2 x \odot x t_3 y c \odot y t_1 \ell)^\star \rangle \top$.

In order to prove the above property we can use our proof system.

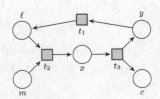

Fig. 5. A Petri Net for a chocolate vending machine

6   Conclusions and Further Work

The main contribution of this work is to extend the Propositional Dynamic Logic for Petri Nets presented in [20] with the iteration operator. This new operator allows for representing more general nets without loosing nice properties like decidability, soundness and completeness. Using iteration operator we can specify Petri Nets with loops and consequently, concurrent system with recursive behavior.

We propose an axiomatization and a semantics and prove its soundness and completeness and also prove the finite model property, which together with the axiomatization yields decidability.

As future work, we would like to extend our approach to other Petri Nets, like Timed and Stochastic Petri Nets (some initial work without the iteration operator is presented in [21,22]). Finally we would like to study and extend issues concerning Model Checking and Automatic Theorem Prover to Petri-PDL (some initial work without the iteration operator is presented in [5]).

References

1. Abrahamson, K.R.: Decidability and expressiveness of logics of processes. Ph.D. thesis, Department of Computer Science, University of Washington (1980)
2. Balbiani, P., Vakarelov, D.: PDL with intersection of programs: a complete axiomatization. J. Appl. Non-Classical Logics **13**(3–4), 231–276 (2003)
3. Benevides, M.R.F., Schechter, L.M.: A propositional dynamic logic for CCS programs. In: Hodges, W., de Queiroz, R. (eds.) Logic, Language, Information and Computation. LNCS (LNAI), vol. 5110, pp. 83–97. Springer, Heidelberg (2008)
4. Blackburn, P., de Rijke, M., Venema, Y.: Modal Logic. Theoretical Tracts in Computer Science. Cambridge University Press, Cambridge (2001)
5. Braga, C., Lopes, B.: Towards reasoning in dynamic logics with rewriting logic: the Petri-PDL case. In: Cornélio, M., Roscoe, B. (eds.) SBMF 2015. LNCS, vol. 9526, pp. 74–89. Springer, Heidelberg (2016). doi:10.1007/978-3-319-29473-5_5
6. de Almeida, E.S., Haeusler, E.H.: Proving properties in ordinary Petri Nets using LoRes logical language. Petri Net Newslett. **57**, 23–36 (1999)
7. del Cerro, L.F., Orlowska, E.: DAL - a logic for data analysis. Theoretical Comput. Sci. **36**, 251–264 (1985)
8. Fagin, R., Halpern, J.Y., Moses, Y., Vardi, M.: Reasoning About Knowledge. MIT Press, Cambridge (2004)
9. Gargov, G., Passy, S.: A note on boolean modal logic. In: Petkov, P.P. (ed.) Mathematical Logic, pp. 299–309. Springer, US, New York (1990)
10. Goldblatt, R.: Parallel action: concurrent dynamic logic with independent modalities. Stud. Logica. **51**, 551–558 (1992)
11. Harel, D., Kaminsky, M.: Strengthened results on nonregular PDL. Technical Report MCS99-13, Faculty of Mathematics and Computer Science, Weizmann Institute of Science (1999)
12. Harel, D., Raz, D.: Deciding properties of nonregular programs. SIAM J. Comput. **22**(4), 857–874 (1993)
13. Harel, D., Kozen, D., Tiuryn, J.: Dynamic Logic. Foundations of Computing Series. MIT Press, Cambridge (2000)

14. Hull, R.: Web services composition: a story of models, automata, and logics. In: Proceedings of the 2005 IEEE International Conference on Web Services (2005)
15. Hull, R., Jianwen, S.: Tools for composite web services: a short overview. ACM SIGMOD **34**(2), 86–95 (2005)
16. Khosravifar, S.: Modeling multi agent communication activities with Petri Nets. Int. J. Inf. Educ. Technol. **3**(3), 310–314 (2013)
17. Kracht, M.: Synctatic codes and grammar refinement. J. Logic Lang. Inform. **4**(1), 41–60 (1995)
18. Lenzerini, M.: Boosting the correspondence between description logics and propositional dynamic logics. In: Proceedings of the Twelfth National Conference on Artificial Intelligence, pp. 205–212. AAAI Press (1994)
19. Löding, C., Lutz, C., Serre, O.: Propositional dynamic logic with recursive programs. J. Logic Algebraic Programm. **73**(1–2), 51–69 (2007)
20. Lopes, B., Benevides, M., Haeusler, H.: Propositional dynamic logic for Petri nets. Logic J. IGPL **22**, 721–736 (2014)
21. Lopes, B., Benevides, M., Haeusler, E.H.: Extending propositional dynamic logic for Petri Nets. Electronic Notes Theoretical Comput. Sci. **305**(11), 67–83 (2014)
22. Lopes, B., Benevides, M., Haeusler, E.H.: Reasoning about multi-agent systems using stochastic Petri Nets. In: Bajo, J., Hernández, J.Z., Mathieu, P., Campbell, A., Fernández-Caballero, A., Moreno, M.N., Julián, V., Alonso-Betanzos, A., Jiménez-López, M.D., Botti, V. (eds.) Trends in Practical Applications of Agents, Multi-Agent Systems and Sustainability. AISC, vol. 372, pp. 75–86. Springer, Heidelberg (2015). doi:10.1007/978-3-319-19629-9_9
23. Meyer, J.-J.C.: A different approach to deontic logic: deontic logic viewed as a variant of dynamic logic. Notre Dame J. Formal Logic **29**(1), 109–136 (1987)
24. Mirkowska, G.: PAL - Propositional algorithmic logic. Fundam. Informaticæ **4**, 675–760 (1981)
25. Peleg, D.: Concurrent dynamic logic. J. Assoc. Comput. Mach. **34**(2), 450–479 (1897)
26. Peleg, D.: Communication in concurrent dynamic logic. J. Comput. Syst. Sci. **35**(1), 23–58 (1987)
27. Petkov, A.: Propositional Dynamic Logic in Two and More Dimensions. Mathematical Logic and its Applications. Plenum Press, New York (1987)
28. Petri, C.A.: Fundamentals of a theory of asynchronous information flow. Commun. ACM **5**(6), 319 (1962)
29. Tuominen, H.: Elementary net systems and dynamic logic. In: Rozenberg, G. (ed.) Advances in Petri Nets 1989. LNCS, pp. 453–466. Springer, Berlin Heidelberg (1990)
30. van Benthem, J.: Logic and Information Flow. Foundations of Computing. MIT Press, Cambridge (1994)
31. Wolter, F., Zakharyaschev, M.: Dynamic description logics. In: Proceedings of AiML1998, pp. 290–300. CSLI Publications (2000)

Tool and Short Papers

ML Pattern-Matching, Recursion, and Rewriting: From FoCaLiZe to Dedukti

Raphaël Cauderlier[1] and Catherine Dubois[2(✉)]

[1] Inria - Saclay and Cnam - Cedric, Paris, France
[2] ENSIIE - Cedric and Samovar, Évry, France
catherine.dubois@ensiie.fr

Abstract. The programming environment FoCaLiZe allows the user to specify, implement, and prove programs with the help of the theorem prover Zenon. In the actual version, those proofs are verified by Coq. In this paper we propose to extend the FoCaLiZe compiler by a backend to the Dedukti language in order to benefit from Zenon Modulo, an extension of Zenon for Deduction modulo. By doing so, FoCaLiZe can benefit from a technique for finding and verifying proofs more quickly. The paper focuses mainly on the process that overcomes the lack of local pattern-matching and recursive definitions in Dedukti.

1 Introduction

FoCaLiZe [15] is an environment for certified programming which allows the user to specify, implement, and prove. For implementation, FoCaLiZe provides an ML like functional language. FoCaLiZe proofs are delegated to the first-order theorem prover Zenon [3] which takes Coq problems as input and outputs proofs in Coq format for independent checking. Zenon has recently been improved to handle Deduction modulo [9], an efficient proof-search technique. However, the Deduction modulo version of Zenon, Zenon Modulo, outputs proofs for the Dedukti proof checker [17] instead of Coq [7].

In order to benefit from the advantages of Deduction modulo in FoCaLiZe, we extend the FoCaLiZe compiler by a backend to Dedukti called Focalide[1] (see Fig. 1). This work is also a first step in the direction of interoperability between FoCaLiZe and other proof languages translated to Dedukti [1,2,8].

This new compilation backend to Dedukti is based on the existing backend to Coq. While the compilation of types and logical formulae is a straightforward adaptation, the translation of FoCaLiZe terms to Dedukti is not trivial because Dedukti lacks local pattern-matching and recursive definitions.

In the following, Sect. 2 contains a short presentation of Dedukti, Zenon Modulo, and FoCaLiZe. Then Sect. 3 presents the main features of the compilation to Dedukti. In Sect. 4, the backend to Dedukti is evaluated on benchmarks. Section 5 discusses related work and Sect. 6 concludes the paper by pointing some future work.

---

[1] This work is available at http://deducteam.gforge.inria.fr/focalide.

© Springer International Publishing AG 2016
A. Sampaio and F. Wang (Eds.): ICTAC 2016, LNCS 9965, pp. 459–468, 2016.
DOI: 10.1007/978-3-319-46750-4_26

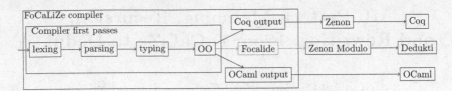

Fig. 1. FoCaLiZe compilation scheme

2   Presentation of the Tools

2.1   Dedukti

Dedukti [17] is a type checker for the $\lambda\Pi$-calculus modulo, an extension of a pure dependent type system, the $\lambda\Pi$-calculus, with rewriting. Through the Curry-Howard correspondence, Dedukti can be used as a proof-checker for a wide variety of logics [1,2,8]. It is commonly used to check proofs coming from the Deduction modulo provers Iprover Modulo [4] and Zenon Modulo [7].

A Dedukti file consists of an interleaving of declarations (such as `0 : nat`) and rewrite rules (such as `[n] plus 0 n --> n.`).

Declarations and rewrite rules are type checked modulo the previously defined rewrite rules. This mechanism can be used to perform proof by reflection, an example is given by the following proof of $2 + 2 = 4$ (theorem `two_plus_two_is_four` below):

```
nat : Type.        0 : nat.        S : nat -> nat.
def plus : nat -> nat -> nat.
[n] plus 0 n --> n
[m,n] plus (S m) n --> S (plus m n).
equal : nat -> nat -> Type.        refl : n : nat -> equal n n.
def two_plus_two_is_four :
  equal (plus (S (S 0)) (S (S 0))) (S (S (S (S 0)))).
[] two_plus_two_is_four --> refl (S (S (S (S 0)))).
```

For correctness, Dedukti requires this rewrite system to be confluent. It does not guarantee to terminate when the rewrite system is not terminating.

2.2   Zenon Modulo

Zenon [3] is a first-order theorem prover based on the tableaux method. It is able to produce proof terms which can be checked independently by Coq.

Zenon Modulo [9] is an extension of Zenon for Deduction modulo, an extension of first-order logic distinguishing computation from reasoning. Computation is defined by a rewrite system, it is part of the theory. Reasoning is defined by a usual deduction system (Sequent Calculus in the case of Zenon Modulo) for which syntactic comparison is replaced by the congruence induced by the rewrite system. Computation steps are left implicit in the resulting proof which has to be checked in Dedukti.

Zenon (resp., Zenon Modulo) accepts input problems in Coq (resp., Dedukti) format so that it can be seen as a term synthesizer: its input is a typing context and a type to inhabit, its output is an inhabitant of this type. This is the mode of operation used when interacting with FoCaLiZe because it limits ambiguities and changes in naming schemes induced by translation tools between languages.

2.3   FoCaLiZe and its Compilation Process

This subsection presents briefly FoCaLiZe and its compilation process (for details please see [15] and FoCaLiZe reference manual). More precisely we address here the focalizec compiler that produces OCaml and Coq code.

The FoCaLiZe (http://focalize.inria.fr) environment provides a set of tools to formally specify and implement functions and logical statements together with their proofs. A FoCaLiZe specification is a set of algebraic properties describing relations between input and output of the functions implemented in a FoCaLiZe program. For implementing, FoCaLiZe offers a pure functional programming language close to ML, featuring a polymorphic type system, recursive functions, data types and pattern-matching. Statements belong to first-order logic. Proofs are written in a declarative style and can be considered as a bunch of hints that the automatic prover Zenon uses to produce proofs that can be verified by Coq for more confidence [3].

FoCaLiZe developments are organized in program units called species. Species in FoCaLiZe define types together with functions and properties applying to them. At the beginning of a development, types are usually abstract. A species may inherit one or several species and specify a function or a property or implement them by respectively providing a definition or a proof. The FoCaLiZe language has an object oriented flavor allowing (multiple) inheritance, late binding and redefinition. These characteristics are very helpful to reuse specifications, implementations and proofs.

A FoCaLiZe source program is analyzed and translated into OCaml sources for execution and Coq sources for certification. The compilation process between both target languages is shared as much as possible. The architecture of the FoCaLiZe compiler is shown in Fig. 1. The FoCaLiZe compiler integrates a type checker, inheritance and late binding are resolved at compile-time (OO on Fig. 1), relying on a dependency calculus described in [15]. The process for compiling proofs towards Coq is achieved in two steps. First the statement is compiled with a hole for the proof script. The goal and the context are transmitted to Zenon. Then when the proof has been found, the hole is filled with the proof output by Zenon.

3   From FoCaLiZe to Focalide

As said previously, Focalide is adapted from the Coq backend. In particular it benefits from the early compilation steps. In this section, we briefly describe the input language we have to consider and the main principles of the translation and then focus on the compilation of pattern-matching and recursive functions. A more detailed and formal description can be found in [5].

3.1 Input Language

Focalide input language is simpler than FoCaLiZe, in particular because the initial compilation steps get rid of object oriented features (see Fig. 1). So for generating code to Dedukti, we can consider that a program is a list of type definitions, well-typed function definitions and proved theorems. A type definition defines a type *à la ML*, in particular it can be the definition of an algebraic datatype in which value constructors are listed together with their type. When applied, a function must receive all its parameters. So partial application must be named. FoCaLiZe supports the usual patterns found in functional languages such as OCaml or Haskell. In particular patterns are linear and tried in the order they are written. A logical formula is a regular first order formula where an atomic formula is a Boolean expression. Its free variables correspond to the functions and constants introduced in the program.

3.2 Translation

Basic types such as `int` are mapped to their counterpart in the target proof checker. However there is no standard library in Dedukti, so we defined the Dedukti counterpart for the different FoCaLiZe basic types. It means defining the type and its basic operations together with the proofs of some basic properties.

The compilation of types is straightforward. It is also quite immediate for most of the expressions, except for pattern-matching expressions and recursive functions because Dedukti, contrary to Coq, lacks these two mechanisms. Thus we have to use other Dedukti constructions to embed their semantics. The compilation of pattern-matching expressions and recursive functions is detailed in next sections. Other constructs of the language such as abstractions and applications are directly mapped to the same construct in Dedukti.

The statement of a theorem is compiled in the input format required by Zenon Modulo, which is here Dedukti itself [7].

3.3 Compilation of Pattern-Matching

Pattern-matching is a useful feature in FoCaLiZe which is also present in Dedukti. However pattern-matching in Dedukti is only available at toplevel (rewrite rules cannot be introduced locally) and both semantics are different. FoCaLiZe semantics of pattern-matching is the one of functional languages: only values are matched and the first branch that applies is used. In Dedukti however, reduction can be triggered on open terms and the order in which the rules are applied should not matter since the rewrite system is supposed to be confluent.

To solve these issues, we define new symbols called *destructors*, using toplevel rewrite rules and apply them locally.

If C is a constructor of arity n for some datatype, the *destructor* associated with C is $\lambda a, b, c.$ **match** a **with** $\mid C(x_1, \ldots, x_n) \Rightarrow b \; x_1 \; \ldots \; x_n \mid \_ \Rightarrow c.$ We say that a pattern-matching has *the shape of a destructor* if it is a fully applied destructor.

Each FoCaLiZe expression is translated into an expression where each pattern-matching has the shape of a destructor. This shape is easy to translate to Dedukti because we only need to define the destructor associated with each constructor. It is done in two steps: we first *serialize* pattern-matching so that each pattern-matching has exactly two branches and the second pattern is a wildcard, and we then *flatten* patterns so that the only remaining patterns are constructors applied to variables. Serialization and flattening terminate and are linear; moreover they preserve the semantics of pattern-matching.

3.4 Compilation of Recursive Functions

Recursion is a powerful but subtle feature in FoCaLiZe. When certifying recursive functions, we reach the limits of Zenon and Zenon Modulo because the rewrite rules corresponding to recursive definitions have to be used with parsimony otherwise Zenon Modulo could diverge.

In FoCaLiZe backend to Coq, termination of recursive functions is achieved thanks to the high-level `Function` mechanism [10]. This mechanism is not available in Dedukti. Contrary to Coq, Dedukti does not require recursive functions to be proved terminating *a priori*. We can postpone termination proofs.

As we did in a previous translation of a programming language in Dedukti [6], we express the semantics of FoCaLiZe by a non-terminating Dedukti signature.

In FoCaLiZe, recursive functions can be defined by pattern-matching on algebraic types but also with regular conditional expressions. For example, if lists are not defined but axiomatized, we might define list equality as follows:

```
let rec list_equal (l1, l2) = (is_nil (l1) && is_nil (l2)) ||
   (~ is_nil(l1) && ~ is_nil(l2) && head(l1) = head(l2) &&
      list_equal(tail(l1), tail(l2)))
```

In Dedukti, defining a recursive function f by a rewrite rule of the form `[x] f x --> g (f (h x)) x.` is not a viable option because it breaks termination and no proof of statement involving f can be checked in finite time.

What makes recursive definitions (sometimes) terminate in FoCaLiZe is the use of the **call-by-value** semantics. The idea is that we have to reduce any argument of f to a value before unfolding the recursive definition.

For efficiency reasons, we approximate the semantics by only checking that the argument starts with a constructor. This is done by defining a combinator CBV of type `A:Type -> B:Type -> (A -> B) -> A -> B` which acts as application when its last argument start with a constructor but does not reduce otherwise. Its definition is extended when new datatypes are introduced by giving a rewrite rule for each constructor. Here is the definition for the algebraic type **nat** whose constructors are 0 and S:

```
[B,f]   CBV nat B f O --> f O.
[B,f,n] CBV nat B f (S n) --> f (S n).
```

Local recursion is then defined by introducing the fixpoint combinator `Fix` of type `A:Type -> B:Type -> ((A -> B) -> (A -> B)) -> A -> B` defined by the rewrite rule `[A, B, F, x] Fix A B F x --> CBV A B (F (Fix A B F)) x`. This does not trivially diverge as before because the term `Fix` in the right-hand side is only partially applied so it does not match the pattern `Fix A B F x`.

If `f` is a FoCaLiZe recursive function, we get the following reduction behaviour: `f` alone does not reduce, `f v` (where `v` is a value) is fully reduced, and `f x` (where `x` is a variable or a non-value term) is unfolded once.

The size of the code produced by Focalide is linear wrt. the input, the operational semantics of FoCaLiZe is preserved and each reduction step in the input language corresponds to a bounded number of rewriting steps in Dedukti, so the execution time for the translated program is only increased by a linear factor.

4   Experimental Results

We have evaluated Focalide by running it on different available FoCaLiZe developments. When proofs required features which are not yet implemented in Focalide, we commented the problematic lines and ran both backends on the same input files; the coverage column of Fig. 2 indicates the percentage of remaining lines.

FoCaLiZe ships with three libraries: the standard library (stdlib) which defines a hierarchy of species for setoids, cartesian products, disjoint unions, orderings and lattices, the external library (extlib) which defines mathematical structures (algebraic structures and polynomials) and the user contributions (contribs) which are a set of concrete applications. Unfortunately, none of these libraries uses pattern-matching and recursion extensively. The other developments are more interesting in this respect; they consist of a test suite for termination proofs of recursive functions (term-proof), a pedagogical example of FoCaLiZe features with several examples of functions defined by pattern-matching (ejcp) and a specification of Java-like iterators together with a list implementation of iterators using both recursion and pattern-matching.

Results[2] in Figs. 2 and 3, show that on FoCaLiZe problems the user gets a good speed-up by using Zenon Modulo and Dedukti instead of Zenon and Coq. Proof-checking is way faster because Dedukti is a mere type-checker which features almost no inference whereas FoCaLiZe asks Coq to infer type arguments of polymorphic functions; this also explain why generated Dedukti files are bigger than the corresponding Coq files. Moreover, each time Coq checks a file coming from FoCaLiZe, it has to load a significant part of its standard library which often takes the majority of the checking time (about a second per file). In the end, finding a proof and checking it is usually faster when using Focalide.

These files have been developed prior to Focalide so they do not yet benefit from Deduction modulo as much as they could. The Coq backend going through

---

[2] The files can be obtained from http://deducteam.inria.fr/focalide.

Library	FoCaLiZe	Coverage	Coq	Dedukti
stdlib	163335	99.42%	1314934	4814011
extlib	158697	100%	162499	283939
contribs	126803	99.54%	966197	2557024
term-proof	24958	99.62%	227136	247559
ejcp	13979	95.16%	28095	239881
iterators	80312	88.33%	414282	972051

Fig. 2. Size (in bytes) comparison of Focalide with the Coq backend

Library	Zenon	ZMod	Coq	Dedukti	Zenon + Coq	ZMod + Dedukti
stdlib	11.73	32.87	17.41	1.46	29.14	34.33
extlib	9.48	26.50	19.45	1.64	28.93	28.14
contribs	5.38	9.96	26.92	1.17	32.30	11.13
term-proof	1.10	0.55	24.54	0.02	25.64	0.57
ejcp	0.44	0.86	11.13	0.06	11.57	0.92
iterators	2.58	3.85	6.59	0.27	9.17	4.12

Fig. 3. Time (in seconds) comparison of Focalide with the Coq backend

Zenon is not very efficient on proofs requiring computation because all reduction steps are registered as proof steps in Zenon leading to huge proofs which take a lot of time for Zenon to find and for Coq to check. For example, if we define a polymorphic datatype `type wrap ('a) = | Wrap ('a)`, we can define the isomorphism f : 'a -> wrap('a) by `let f (x) = Wrap(x)` and its inverse g : wrap('a) -> 'a by `let g(y) = match y with | Wrap (x) -> x`. The time taken for our tools to deal with the proof of $(g \circ f)^n(x) = x$ for n from 10 to 19 is given in Fig. 4; as we can see, the Coq backend becomes quickly unusable whereas Deduction modulo is so fast that it is even hard to measure it.

Value of n	Zenon	Coq	Zenon Modulo	Dedukti
10	31.48	4.63	0.04	0.00
11	63.05	11.04	0.04	0.00
12	99.55	7.55	0.05	0.00
13	197.80	10.97	0.04	0.00
14	348.87	1020.67	0.04	0.00
15	492.72	1087.13	0.04	0.00
16	724.46	> 2h	0.04	0.00
17	1111.10	1433.76	0.04	0.00
18	1589.10	> 2h	0.07	0.00
19	2310.48	> 2h	0.04	0.00

Fig. 4. Time comparison (in seconds) for computation-based proofs

5 Related Work

The closest related work is a translation from a fragment of Coq kernel to Dedukti [1]. Pattern-matching is limited in Coq kernel to flat patterns so it is possible to define a single `match` symbol for each inductive type, which simplifies greatly the compilation of pattern-matching to Dedukti. To handle recursion, filter functions playing the role of our `CBV` combinator are proposed. Because of dependent typing, they need to duplicate their arguments. Moreover, we define the `CBV` operator by ad-hoc polymorphism whereas filter functions are unrelated to each other.

Compilation techniques for pattern-matching to enriched λ-calculi have been proposed, see e.g. [12,13,16]. We differ mainly in the treatment of matching failure.

A lot of work has also been done to compile programs (especially functional recursive definitions [11,14]) to rewrite systems. The focus has often been on termination preserving translations to prove termination of recursive functions using termination checkers for term rewrite systems. However, these translations do not preserve the semantics of the programs so they can hardly be adapted for handling translations of correctness proofs.

6 Conclusion

We have extended the compiler of FoCaLiZe to a new output language: Dedukti. Contrary to previously existing FoCaLiZe outputs OCaml and Coq, Dedukti is not a functional programming language but an extension of a dependently-typed λ-calculus with rewriting so pattern-matching and recursion are not trivial to compile to Dedukti.

However, we have shown that ML pattern-matching can easily and efficiently be translated to Dedukti using destructors. We plan to further optimize the compilation of pattern-matching, in particular to limit the use of dynamic error handling. For recursion, however, efficiency comes at a cost in term of normalization because we can not fully enforce the use of the call-by-value strategy without loosing linearity. Our treatment of recursive definitions generalizes directly to mutual recursion but we have not implemented this generalization.

Our approach is general enough to be adapted to other functional languages because FoCaLiZe language for implementing functions is an ML language without specific features. FoCaLiZe originality comes from its object-oriented mechanisms which are invisible to Focalide because they are statically resolved in an earlier compilation step. Moreover, it can also easily be adapted to other rewriting formalisms, especially untyped and polymorphic rewrite engines because features specific to Dedukti (such as higher-order rewriting or dependent typing) are not used.

We have tested Focalide on existing FoCaLiZe libraries and have found it a decent alternative to the Coq backend whose adoption can enhance the usability of FoCaLiZe to a new class of proofs based on computation.

As Dedukti is used as the target language of a large variety of systems in the hope of exchanging proofs; we want to experiment the import and export of proofs between logical systems by using FoCaLiZe and Focalide as an interoperability platform.

Acknowledgements. This work has been partially supported by the BWare project (ANR-12-INSE-0010) funded by the INS programme of the French National Research Agency (ANR).

References

1. Assaf, A.: A framework for defining computational higher-order logics. Ph.D. thesis, École Polytechnique (2015)
2. Assaf, A., Burel, G.: Translating HOL to Dedukti. In: Kaliszyk, C., Paskevich, A. (eds.) Proceedings Fourth Workshop on Proof eXchange for Theorem Proving. EPTCS, vol. 186, Berlin, Germany, pp. 74–88 (2015)
3. Bonichon, R., Delahaye, D., Doligez, D.: Zenon: an extensible automated theorem prover producing checkable proofs. In: Dershowitz, N., Voronkov, A. (eds.) LPAR 2007. LNCS (LNAI), vol. 4790, pp. 151–165. Springer, Heidelberg (2007). doi:10.1007/978-3-540-75560-9_13
4. Burel, G.: A shallow embedding of resolution and superposition proofs into the $\lambda\Pi$-calculus modulo. In: Blanchette, J.C., Urban, J. (eds.) PxTP 2013. 3rd International Workshop on Proof Exchange for Theorem Proving. EasyChair Proceedings in Computing, vol. 14, Lake Placid, USA, pp. 43–57 (2013)
5. Cauderlier, R.: Object-oriented mechanisms for interoperability between proof systems. Ph.D. thesis, Conservatoire National des Arts et Métiers, Paris (draft)
6. Cauderlier, R., Dubois, C.: Objects and subtyping in the $\lambda\Pi$-calculus modulo. In: Post-proceedings of the 20th International Conference on Types for Proofs and Programs (TYPES 2014). Leibniz International Proceedings in Informatics (LIPIcs), Schloss Dagstuhl, Paris (2014)
7. Cauderlier, R., Halmagrand, P.: Checking Zenon Modulo proofs in Dedukti. In: Kaliszyk, C., Paskevich, A. (eds.) Proceedings 4th Workshop on Proof eXchange for Theorem Proving. EPTCS, vol. 186, Berlin, Germany, pp. 57–73 (2015)
8. Cousineau, D., Dowek, G.: Embedding pure type systems in the $\lambda\Pi$-calculus modulo. In: Rocca, S.R.D. (ed.) TLCA 2007. LNCS, vol. 4583, pp. 102–117. Springer, Heidelberg (2007). doi:10.1007/978-3-540-73228-0_9
9. Delahaye, D., Doligez, D., Gilbert, F., Halmagrand, P., Hermant, O.: Zenon Modulo: when Achilles outruns the tortoise using deduction modulo. In: McMillan, K., Middeldorp, A., Voronkov, A. (eds.) LPAR-19. LNCS, vol. 8312, pp. 274–290. Springer, Heidelberg (2013). doi:10.1007/978-3-642-45221-5_20
10. Dubois, C., Pessaux, F.: Termination proofs for recursive functions in FoCaLiZe. In: Serrano, M., Hage, J. (eds.) TFP 2015. LNCS, vol. 9547, pp. 136–156. Springer, Heidelberg (2016). doi:10.1007/978-3-319-39110-6_8
11. Giesl, J., Raffelsieper, M., Schneider-Kamp, P., Swiderski, S., Thiemann, R.: Automated termination proofs for Haskell by term rewriting. ACM Trans. Program. Lang. Syst. **33**(2), 7:1–7:39 (2011)
12. Kahl, W.: Basic pattern matching calculi: a fresh view on matching failure. In: Kameyama, Y., Stuckey, P.J. (eds.) FLOPS 2004. LNCS, vol. 2998, pp. 276–290. Springer, Heidelberg (2004). doi:10.1007/978-3-540-24754-8_20

13. Klop, J.W., van Oostrom, V., de Vrijer, R.: Lambda calculus with patterns. Theoret. Comput. Sci. **398**(1–3), 16–31 (2008). Calculi, Types and Applications: Essays in honour of M. Coppo, M. Dezani-Ciancaglini and S. Ronchi Della Rocca

14. Lucas, S., Peña, R.: Rewriting techniques for analysing termination and complexity bounds of Safe programs. In: LOPSTR 2008, pp. 43–57 (2008)

15. Pessaux, F.: FoCaLiZe: inside an F-IDE. In: Dubois, C., Giannakopoulou, D., Méry, D. (eds.) Proceedings 1st Workshop on Formal Integrated Development Environment, F-IDE 2014. EPTCS, vol. 149, Grenoble, France, pp. 64–78 (2014)

16. Peyton Jones, S.L.: The Implementation of Functional Programming Languages. Prentice-Hall International Series in Computer Science. Prentice-Hall, Inc., Upper Saddle River (1987)

17. Saillard, R.: Type checking in the $\lambda\Pi$-calculus modulo: theory and practice. Ph.D. thesis, MINES Paritech (2015)

Parametric Deadlock-Freeness Checking Timed Automata

Étienne André[1,2(✉)]

[1] Université Paris 13, Sorbonne Paris Cité, LIPN, CNRS,
UMR 7030, F-93430 Villetaneuse, France
ea.ndre13@lipn13.fr
[2] École Centrale de Nantes, IRCCyN, CNRS, UMR 6597, Nantes, France

Abstract. Distributed real-time systems are notoriously difficult to design, and must be verified, *e. g.*, using model checking. In particular, deadlocks must be avoided as they either yield a system subject to potential blocking, or denote an ill-formed model. Timed automata are a powerful formalism to model and verify distributed systems with timing constraints. In this work, we investigate synthesis of timing constants in timed automata for which the model is guaranteed to be deadlock-free.

1 Introduction

Distributed real-time systems are notoriously difficult to design due to the intricated use of concurrency and timing constraints, and must therefore be verified, *e. g.*, using model checking. Model checking is a set of techniques to formally verify that a system, described by a model, verifies some property, described using formalisms such as reachability properties or more complex properties expressed using, *e. g.*, temporal logics.

Checking the absence of deadlocks in the model of a real-time system is of utmost importance. First, deadlocks can lead the actual system to a blockade when a component is not ready to receive any action (or synchronization label). Second, a specificity of models of distributed systems involving time is that they can be subject to situations where time cannot elapse. This situation denotes an ill-formed model, as this situation of time blocking ("timelock") cannot happen in the actual system due to the uncontrollable nature of time.

Timed automata (TAs) [1] are a formalism dedicated to modeling and verifying real-time systems where distributed components communicate via synchronized actions. Despite a certain success in verifying models of actual distributed systems (using *e. g.*, UPPAAL [10] or PAT [12]), TAs reach some limits when verifying systems only partially specified (typically when the timing constants are not yet known) or when timing constants are known with a limited precision only (although the robust semantics can help tackling some problems, see *e. g.*, [11]). Parametric timed automata (PTAs) [2] leverage these drawbacks by

This work is partially supported by the ANR national research program PACS (ANR-14-CE28-0002).

A. Sampaio and F. Wang (Eds.): ICTAC 2016, LNCS 9965, pp. 469–478, 2016.
DOI: 10.1007/978-3-319-46750-4_27

allowing the use of timing parameters, hence allowing for modeling constants unknown or known with some imprecision.

We address here the problem of the deadlock-freeness, *i. e.*, the fact that a discrete transition must always be taken from any state, possibly after elapsing some time. TAs and PTAs are both subject to deadlocks: hence, a property proved correct on the model may not necessarily hold on the actual system if the model is subject to deadlocks. Deadlock checking can be performed on TAs (using *e. g.*, UPPAAL); however, if deadlocks are found, then there is often no other choice than manually refining the model in order to remove them.

We recently showed that the existence of a parameter valuation in a PTA for which a run leads to a deadlock is undecidable [4]; this result also holds for the subclass of PTAs where parameters only appear as lower or upper bounds [8]. This result rules out the possibility to perform exact deadlock-freeness synthesis.

In this work, we propose an approach to automatically synthesize parameter valuations in PTAs (in the form of a set of linear constraints) for which the system is deadlock-free. If our procedure terminates, the result is exact. Otherwise, when stopping after some bound (*e. g.*, runtime limit, exploration depth limit), it is an over-approximation of the actual result. In this latter case, we propose a second approach to also synthesize an under-approximation: hence, the designer is provided with a set of valuations that are deadlock-free, a set of valuations for which there exist deadlocks, and an intermediate set of unsure valuations. Our approach is of particular interest when intersected with a set of parameter valuations ensuring some property: one obtains therefore a set of parameter valuations for which that property is valid and the system is deadlock-free.

Outline. We briefly recall necessary definitions in Sect. 2. We introduce our approach in Sect. 3, extend it to the synthesis of an under-approximated constraint in Sect. 4, and validate it on benchmarks in Sect. 5. We discuss future works in Sect. 6.

2   Preliminaries

Throughout this paper, we assume a set $X = \{x_1, \ldots, x_H\}$ of *clocks*. A clock valuation is $w : X \to \mathbb{R}_+$. We write $\mathbf{0}$ for the valuation that assigns 0 to each clock. Given $d \in \mathbb{R}_+$, $w + d$ denotes the valuation such that $(w + d)(x) = w(x) + d$, for all $x \in X$. We assume a set $P = \{p_1, \ldots, p_M\}$ of *parameters*, *i. e.*, unknown constants. A parameter *valuation* v is $v : P \to \mathbb{Q}_+$. In the following, we assume $\bowtie \in \{<, \leq, \geq, >\}$. A *constraint* C (*i. e.*, a convex polyhedron) over $X \cup P$ is a conjunction of inequalities of the form $lt \bowtie 0$, where lt denotes a linear term of the form $\sum_{1 \leq i \leq H} \alpha_i x_i + \sum_{1 \leq j \leq M} \beta_j p_j + d$, with $x_i \in X$, $p_i \in P$, and $\alpha_i, \beta_j, d \in \mathbb{Z}$. Given a parameter valuation v, $v(C)$ denotes the constraint over X obtained by replacing each parameter p in C with $v(p)$. Likewise, given a clock valuation w, $w(v(C))$ denotes the expression obtained by replacing each clock x in $v(C)$ with $w(x)$. We say that v *satisfies* C, denoted by $v \models C$, if the set of clock valuations satisfying $v(C)$ is nonempty. We say that C is *satisfiable* if $\exists w, v$ s.t. $w(v(C))$ evaluates to true. We define the *time elapsing* of C, denoted by $C^\nearrow$, as the constraint over X and P obtained from C by delaying all clocks by an arbitrary amount of time. We define the *past* of C, denoted by $C^\swarrow$, as the constraint over X and P obtained from C by letting time pass backward by an arbitrary amount of time (see *e. g.*, [9]). Given $R \subseteq X$, we define the *reset* of C, denoted by $[C]_R$, as the constraint obtained from C by resetting the clocks in R, and keeping the other clocks unchanged. We denote by $C\downarrow_P$ the projection of C onto P, *i. e.*, obtained by eliminating the clock variables (*e. g.*, using Fourier-Motzkin).

A *guard* g is a constraint defined by inequalities $x \bowtie z$, where z is either a parameter or a constant in $\mathbb{Z}$. A *parametric zone* is a polyhedron in which all constraints are of the form $x \bowtie plt$ or $x_i - x_j \bowtie plt$, where $x_i \in X$, $x_j \in X$ and plt is a parametric linear term over P, *i. e.*, a linear term without clocks ($\alpha_i = 0$ for all i). Given a parameter constraint K, $\neg K$ denotes the (possibly non-convex) negation of K. We extend the notation $v \models K$ to possibly non-convex constraints in a natural manner. $\top$ (resp. $\bot$) denotes the constraint corresponding to the set of all (resp. no) parameter valuations.

Definition 1. *A PTA $\mathcal{A}$ is a tuple $\mathcal{A} = (\Sigma, L, l_0, X, P, I, E)$, where: (i) Σ is a finite set of actions, (ii) L is a finite set of locations, (iii) $l_0 \in L$ is the initial location, (iv) X is a set of clocks, (v) P is a set of parameters, (vi) I is the invariant, assigning to every $l \in L$ a guard $I(l)$, (vii) E is a set of edges $e = (l, g, a, R, l')$ where $l, l' \in L$ are the source and target locations, $a \in \Sigma$, $R \subseteq X$ is a set of clocks to be reset, and g is a guard.*

Given a parameter valuation v, we denote by $v(\mathcal{A})$ the non-parametric timed automaton where all occurrences of a parameter p_i have been replaced by $v(p_i)$.

Definition 2 *(Semantics of a TA). Given a PTA $\mathcal{A} = (\Sigma, L, l_0, X, P, I, E)$, and a parameter valuation v, the concrete semantics of $v(\mathcal{A})$ is given by the timed transition system $(S, s_0, \rightarrow)$, with*

- $S = \{(l, w) \in L \times \mathbb{R}_+^H \mid w(v(I(l)))$ *evaluates to true*$\}$, $s_0 = (l_0, \mathbf{0})$
- $\rightarrow$ *consists of the discrete and (continuous) delay transition relations:*
  - *discrete transitions: $(l, w) \xrightarrow{e} (l', w')$, if $(l, w), (l', w') \in S$, there exists $e = (l, g, a, R, l') \in E$, $\forall x \in X : w'(x) = 0$ if $x \in R$ and $w'(x) = w(x)$ otherwise, and $w(v(g))$ evaluates to true.*
  - *delay transitions: $(l, w) \xrightarrow{d} (l, w + d)$, with $d \in \mathbb{R}_+$, if $\forall d' \in [0, d], (l, w + d') \in S$.*

Moreover we write $(l, w) \xmapsto{e} (l', w')$ for a sequence of delay and discrete transitions where $((l, w), e, (l', w')) \in \mapsto$ if $\exists d, w'' : (l, w) \xrightarrow{d} (l, w'') \xrightarrow{e} (l', w')$. Given a TA $v(\mathcal{A})$ with concrete semantics $(S, s_0, \rightarrow)$, we refer to the states of S as the *concrete states* of $v(\mathcal{A})$. A *concrete run* (or simply a *run*) of $v(\mathcal{A})$ is an alternating sequence of concrete states of $v(\mathcal{A})$ and edges starting from the initial concrete state s_0 of the form $s_0 \xmapsto{e_0} s_1 \xmapsto{e_1} \cdots \xmapsto{e_{m-1}} s_m$, such that for

(a) PTA deadlocked for some valuations   (b) PTA deadlocked for all valuations

Fig. 1. Examples of PTAs with potential deadlocks

all $i = 0, \ldots, m - 1$, $e_i \in E$, and $(s_i, e_i, s_{i+1}) \in \mapsto$. Given a state $s = (l, w)$, we say that this state has no successor (or is deadlocked) if, in the concrete semantics of $v(\mathcal{A})$, there exists no discrete transition from s of from a successor of s obtained by taking exclusively continuous transition(s) from s. If no state of $v(\mathcal{A})$ is deadlocked, then $v(\mathcal{A})$ is deadlock-free.

Example 1. Consider the PTA in Fig. 1a (invariants are boxed): deadlocks can occur if the guard of the transition from l_1 to l_2 cannot be satisfied (when $p_2 > p_1 + 5$) or if the invariant of l_2 is not compatible with the guard (when $p_2 > 10$). In Fig. 1b, the system may risk a deadlock for any parameter valuation as, if the guard is "missed" (if a run chooses to spend more than p time units in l_1), then no transition can be taken from l_1.

3   Parametric Deadlock-Freeness Checking

Let us first recall the symbolic semantics of PTAs (from, *e.g.*, [9]). A symbolic state is a pair (l, C) where $l \in L$ is a location, and C its associated parametric zone. The initial symbolic state of $\mathcal{A}$ is $\mathbf{s}_0^{\mathcal{A}} = (l_0, (\bigwedge_{1 \leq i \leq H} x_i = 0)^{\nearrow} \wedge I(l_0))$.

The symbolic semantics relies on the Succ operation. Given a symbolic state $\mathbf{s} = (l, C)$ and an edge $e = (l, g, a, R, l')$, the successor of $\mathbf{s}$ via e is the symbolic state $\mathsf{Succ}(\mathbf{s}, e) = (l', C')$, with $C' = ([(C \wedge g)]_R)^{\nearrow} \cap I(l')$. We write $\mathsf{Succ}(\mathbf{s})$ for $\cup_{e \in E} \mathsf{Succ}(\mathbf{s}, e)$. We write Pred for Succ^{-1}. Given a set $\mathbf{S}$ of states, we write $\mathsf{Pred}(\mathbf{S})$ for $\bigcup_{\mathbf{s} \in \mathbf{S}} \mathsf{Pred}(\mathbf{s})$.

A symbolic run of a PTA is an alternating sequence of symbolic states and edges starting from the initial symbolic state, of the form $\mathbf{s}_0^{\mathcal{A}} \overset{e_0}{\Rightarrow} \mathbf{s}_1 \overset{e_1}{\Rightarrow} \cdots \overset{e_{m-1}}{\Rightarrow} \mathbf{s}_m$, such that for all $i = 0, \ldots, m - 1$, $e_i \in E$, and $\mathbf{s}_{i+1} = \mathsf{Succ}(\mathbf{s}_i, e_i)$. The symbolic states with the Succ relation form a *state space*, i. e., a (possibly infinite) directed graph, the nodes of which are the symbolic states, and there exists an edge from $\mathbf{s}_i$ to $\mathbf{s}_j$ labeled with e_i iff $\mathbf{s}_j = \mathsf{Succ}(\mathbf{s}_i, e_i)$. Given a concrete (respectively symbolic) run $(l_0, \mathbf{0}) \overset{e_0}{\mapsto} (l_1, w_1) \overset{e_1}{\mapsto} \cdots \overset{e_{m-1}}{\mapsto} (l_m, w_m)$ (respectively $(l_0, C_0) \overset{e_0}{\Rightarrow} (l_1, C_1) \overset{e_1}{\Rightarrow} \cdots \overset{e_{m-1}}{\Rightarrow} (l_m, C_m))$, its corresponding *discrete sequence* is $l_0 \overset{e_0}{\Rightarrow} l_1 \overset{e_1}{\Rightarrow} \cdots \overset{e_{m-1}}{\Rightarrow} l_m$. Two runs (concrete or symbolic) are said to be *equivalent* if their associated discrete sequences are equal.

Deadlock-Freeness Synthesis. We now introduce below our procedure PDFC, that makes use of an intermediate, recursive procedure DSynth. Both are written in a functional form in the spirit of, *e. g.*, the reachability and unavoidability synthesis algorithms in [9]. Given $\mathbf{s} = (l, C)$, we use $\mathbf{s}_C$ to denote C. The notation $g(\mathbf{s}, \mathbf{s}')$ denotes the guard of the edge from $\mathbf{s}$ to $\mathbf{s}'$.

$$\mathsf{DSynth}(\mathbf{s}, \mathsf{Passed}) = \begin{cases} \perp & \text{if } \mathbf{s} \in \mathsf{Passed} \\ \left(\bigcup_{\mathbf{s}' \in \mathsf{Succ}(\mathbf{s})} \mathsf{DSynth}(\mathbf{s}', \mathsf{Passed} \cup \{\mathbf{s}\})\right) & \\ \cup \left(\mathbf{s}_C \setminus \left(\bigcup_{\mathbf{s}' \in \mathsf{Succ}(\mathbf{s})} (\mathbf{s}_C \wedge g(\mathbf{s}, \mathbf{s}'))^{\swarrow} \wedge \mathbf{s}'_C {\downarrow}_P\right)\right){\downarrow}_P & \text{otherwise} \end{cases}$$

$\mathsf{PDFC}(\mathcal{A}) = \neg \mathsf{DSynth}(s_0^{\mathcal{A}}, \emptyset)$

First, we use a function DSynth($\mathbf{s}$, Passed) to recursively synthesize the parameter valuations for which a deadlock may occur. This function takes as argument the current state $\mathbf{s}$ together with the list Passed of passed states. If $\mathbf{s}$ belongs to Passed (*i. e.*, $\mathbf{s}$ was already met), then no parameter valuation is returned. Otherwise, the first part of the second case computes the union over all successors of $\mathbf{s}$ of DSynth recursively called over these successors; the second part computes all parameter valuations for which a deadlock may occur, *i. e.*, the constraint characterizing $\mathbf{s}$ minus all clock and parameter valuations that allow to exit $\mathbf{s}$ to some successor $\mathbf{s}'$, all this expression being eventually projected onto P.

Finally, PDFC ("parametric deadlock-freeness checking") returns the negation of the result of DSynth called with the initial state of $\mathcal{A}$ and an empty list of passed states.

We show below that PDFC is sound and complete. Note however that, in the general case, the algorithm may not terminate, as DSynth explores the set of symbolic states, of which there may be an infinite number.

Proposition 1. *Assume* PDFC($\mathcal{A}$) *terminates with result K. Let $v \models K$. Then $v(\mathcal{A})$ is deadlock-free.*

Proof (**sketch**). Consider a run of $v(\mathcal{A})$, and assume it reaches a state $s = (l, w)$ with no discrete successor. From [8], there exists an equivalent symbolic run in $\mathcal{A}$ reaching a state (l, C). As DSynth explores all symbolic states, (l, C) was explored in DSynth too; since s has no successor, then w belongs to the second part of the second line of DSynth. Hence, the projection onto P was added to the result of DSynth, and hence does not belong to the negation returned by PDFC. Hence $v \not\models K$, which contradicts the initial assumption.

Proposition 2. *Assume* PDFC($\mathcal{A}$) *terminates with result K. Consider a valuation v such that $v(\mathcal{A})$ is deadlock-free. Then $v \models K$.*

Proof (**sketch**). Following a reasoning dual to Proposition 1.

Finally, we show that PDFC outputs an over-approximation of the parameter set when stopped before termination.

Proposition 3. *Fix a maximum number of recursive calls in* DSynth. *Then* PDFC *terminates and its result is an over-approximation of the set of parameter valuations for which the system is deadlock-free.*

Algorithm 1. BwUS($K, \mathcal{G}$)

---

input : result K of DSynth, parametric state space $\mathcal{G}$
output: Constraint over the parameters guaranteeing deadlock-freeness

1 $K^+ \leftarrow K$
2 Marked $\leftarrow$ {s | s has unexplored successors in $\mathcal{G}$}
3 Disabled $\leftarrow \emptyset$
4 **while** Marked $\neq \emptyset$ **do**
5 $\quad$ **foreach** $(l, C) \in$ Marked **do** $K^+ \leftarrow K^+ \cup C{\downarrow}_P$;
6 $\quad$ preds $\leftarrow$ Pred(Marked) \ Disabled
7 $\quad$ Marked' $\leftarrow \emptyset$
8 $\quad$ **foreach** s $\in$ preds **do**
9 $\quad\quad$ $K^+ \leftarrow K^+ \cup \left(\mathbf{s}_C \setminus \left(\bigcup_{\mathbf{s}' \in \mathsf{Succ}(\mathbf{s})} (\mathbf{s}_C \wedge g(\mathbf{s}, \mathbf{s}'))^{\swarrow} \wedge \mathbf{s}'_C{\downarrow}_P\right)\right){\downarrow}_P$
10 $\quad\quad$ **if** $\mathbf{s}_C{\downarrow}_P \subseteq K^+$ **then** Marked' $\leftarrow$ Marked' $\cup \{\mathbf{s}\}$;
11 $\quad$ Disabled $\leftarrow$ Disabled $\cup$ Marked $\quad$; $\quad$ Marked $\leftarrow$ Marked' \ Disabled
12 **return** $\neg K^+$

Proof. Observe that, in DSynth, the deeper the algorithm goes in the state space (*i. e.,* the more recursive calls are performed), the more valuations it synthesizes. Hence bounding the number of recursive calls yields an under-approximation of its expected result. As PDFC returns the negation of DSynth, this yields an over-approximation.

4   Under-approximated Synthesis

A limitation of PDFC is that either the result is exact, or it is an over-approximation when stopped earlier than the actual fixpoint (from Proposition 3). In the latter case, the result is not entirely satisfactory: if deadlocks represent an undesired behavior, then an over-approximation may also contain unsafe parameter valuations. More valuable would be an under-approximation, as this result (although potentially incomplete) firmly guarantees the absence of deadlocks in the model.

Our idea is as follows: after exploring a part of the state space in PDFC, we obtain an over-approximation. In order to get an under-approximation, we can consider that any unexplored state is unsafe, *i. e.,* may lead to deadlocks. Therefore, we first need to negate the parametric constraint associated with any state that has unexplored successors. But this may not be sufficient: by removing those unsafe states, their predecessors can themselves become deadlocked, and so on. Hence, we will perform a backward-exploration of the state space by iteratively removing unsafe states, until a fixpoint is reached.

We give our procedure BwUS (backward under-approximated synthesis) in Algorithm 1. BwUS takes as input *(1)* the (under-approximated) result of DSynth, *i. e.,* a set of parameter valuations for which the system contains a deadlocked run, and *(2)* the part of the symbolic state space explored while running DSynth.

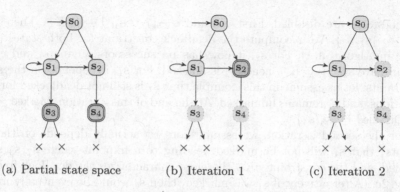

(a) Partial state space   (b) Iteration 1   (c) Iteration 2

Fig. 2. Application of Algorithm 1

The algorithm maintains several variables. Marked denotes the states that are potentially deadlocked, and the predecessors of which must be considered iteratively. Disabled denotes the states marked in the past, which avoids to consider several times the same state. K^+ is an over-approximated constraint for which there are deadlocks; since the negation of K^+ is returned (line 12), then the algorithm returns an under-approximation. Initially, K^+ is set to the (under-approximated) result of DSynth, and all states that have unexplored successors in the state space (due to the early termination) are marked.

While there are marked states (line 4), we remove the marked states by adding to K^+ the negation of the constraint associated with these states (line 5). Then, we compute the predecessors of these states, except those already disabled (line 6). By definition, all these predecessors have at least one marked (and hence potentially deadlocked) successor; as a consequence, we have to recompute the constraint leading to a deadlock from each of these predecessors. This is the purpose of line 9, where K^+ is enriched with the recomputed valuations for which a deadlock may occur from a given predecessor s, using the same computation as in DSynth. Then, if the constraint associated with the current predecessor s is included in K^+, this means that s is unreachable for valuations in $\neg K^+$ (recall that we will return the negation of K^+), and therefore s should be marked at the next iteration, denoted by the local variable Marked′ (line 10). Finally, all currently marked states become disabled, and the new marked states are all the marked predecessors of the currently marked states with the exception of the disabled states to ensure termination (line 11). The algorithm returns eventually the negation of the over-approximated K^+ (line 12), which yields an under-approximation.

Example 2. Let us apply Algorithm 1 to a (fictional) example of a partial state space, given in Fig. 2a. We only focus on the backward exploration, and rule out the constraint update (constraints are not represented in Fig. 2a anyway). s_3 and s_4 have unexplored successors (denoted by ×), and both states are hence unsafe as they might lead to deadlocks along these unexplored branches. Initially, Marked = $\{s_3, s_4\}$ (depicted in yellow with a double circle in Fig. 2a), and no states are disabled. First, we add $s_{3C}\downarrow_P \cup s_{4C}\downarrow_P$ to K^+. Then, preds is set to $\{s_1, s_2\}$. We recompute the deadlock constraint for both states (using line 9 in Algorithm 1). For s_2, it now has no successors anymore, and clearly we will have $s_{2C}\downarrow_P \subseteq K^+$, hence s_2 is marked. For s_1, it depends on the actual constraints; let us assume in this example that s_1 is still not deadlocked for some valuations, and s_1 remains unmarked. At the end of this iteration, Marked $= \{s_2\}$ and Disabled $= \{s_3, s_4\}$.

For the second iteration, we assume here (it actually depends on the constraints) that s_1 will not be marked, leading to a fixpoint where s_2, s_3, s_4 are disabled, and the constraint $\neg K^+$ therefore characterizes the deadlock-free runs in Fig. 2c. (Alternatively, if s_1 was marked, then s_0 would be eventually marked too, and the result would be $\bot$.)

First note that BwUS necessarily terminates as it iterates on marked states, and no state can be marked twice thanks to the set Disabled. In addition, the result is an under-approximation of the valuation set yielding deadlock-freeness: indeed, it only explores a part of state space, and considers states with unexplored successors as deadlocked by default, yielding a possibly too strong, hence under-approximated, constraint.

5   Experiments

We implemented PDFC in IMITATOR [3] (which relies on PPL [5] for polyhedra operations), and synthesized constraints for which a set of models of distributed systems are deadlock-free.[1] Our benchmarks come from teaching examples (coffee machines, nuclear plant, train controller), communication protocols (CSMA/CD, RCP), asynchronous circuits (and–or [7], flip-flop), a distributed networked automation system (SIMOP) and a Wireless Fire Alarm System (WFAS) [6].

If an experiment has not finished within 300 s, the result is still a valid over-approximation according to Proposition 3; in addition, IMITATOR then runs Algorithm 1 to also obtain an under-approximation.

We give in Table 1 from left to right the numbers of PTA components,[2] of clocks, of parameters, and of symbolic states explored, the computation time in seconds (TO denotes no termination within 300 s) for PDFC and BwUS (when necessary), the type of constraint (nncc denotes a non-necessarily convex constraint different from $\top$ or $\bot$) and an evaluation of the result soundness.

Analyzing the experiments, several situations occur: the most interesting result is when an nncc is derived and is exact; for example; the constraint synthesized by IMITATOR for Fig. 1a is $p_1 + 5 \geq p_2 \wedge p_2 \leq 10$, which is exactly the

---

[1] Experiments were conducted on Linux Mint 17 64 bits, running on a Dell Intel Core i7 CPU 2.67 GHz with 4 GiB. Binaries, models and results are available at www.imitator.fr/static/ICTAC16/.

[2] The synchronous product of several PTA components (using synchronized actions) yields a PTA. IMITATOR performs this composition on-the-fly.

Table 1. Synthesizing parameter valuations ensuring deadlock-freeness

Case study	$|\mathcal{A}|$	$|X|$	$|P|$	States	PDFC	BwUS	K	Soundness
Fig. 1a	1	1	2	3	0.012	-	nncc	exact
Fig. 1b	1	1	1	2	0.005	-	$\perp$	exact
and–or circuit	4	4	4	5,265	TO	171	$[\text{nncc}^-, \text{nncc}^+]$	under/over-app
coffee machine 1	1	2	3	9,042	TO	8.4	$[\text{nncc}^-, \text{nncc}^+]$	under/over-app
coffee machine 2	2	3	3	51	0.198	-	nncc	exact
CSMA/CD protocol	3	3	3	38	0.105	-	$\perp$	exact
flip-flop circuit	6	5	2	20	0.093	-	$\perp$	exact
nuclear plant	1	2	4	13	0.014	-	nncc	exact
RCP protocol	5	6	5	2,091	10.63	-	$\perp$	exact
SIMOP	5	8	2	22,894	TO	121	nncc	over-app
Train controller	1	2	3	11	0.025	-	nncc	exact
WFAS	3	4	2	14,614	TO	69.1	$[\text{nncc}^-, \text{nncc}^+]$	under/over-app

valuation set ensuring the absence of deadlocks. In several cases, the synthesized constraint is $\perp$, meaning that no parameter valuation is deadlock-free; this may not always denote an ill-formed model, as some case studies are "finite" (no infinite behavior), typically some of the hardware case studies (*e. g.,* flip-flop); this may also denote a modeling process purposely blocking the system (to limit the state space explosion) after some property (typically reachability) is proved correct or violated. When no exact result could be synthesized, our second procedure BwUS allows to get both an under-approximated and an over-approximated constraint (denoted by $[\text{nncc}^-, \text{nncc}^+]$). This is a valuable result, as it contains valuations guaranteed to be deadlock-free, others guaranteed to be deadlocked, and a third unsure set. An exception is SIMOP, where BwUS derives $\perp$, leaving the designer with only an over-approximation. This result remains valuable as the parameter valuations not belonging to the synthesized constraint necessarily lead to deadlocks, an information that will help the designer to refine its model, or to rule out these valuations.

Concerning the performances of BwUS, its overhead is significant, and depends on the number of dimensions (clocks and parameters) as well as the number states in the state space. However, it still remains smaller than the forward exploration (300 s) in all case studies, which therefore remains reasonable to some extent. It seems the most expensive operation is the computation of the deadlock constraint (line 9 in Algorithm 1); this has been implemented in a straightforward manner, but could benefit from optimizations (*e. g.,* only recompute the part corresponding to successor states that were disabled at the previous iteration of BwUS).

6   Perspectives

We proposed here a procedure to synthesize timing parameter valuations ensuring the absence of deadlocks in a real-time system; we implemented it in IMITATOR and we have run experiments on a set of benchmarks. When terminating, our procedure yields an exact result. Otherwise, thanks to a second procedure, we get both an under- and an over-approximation of the valuations for which the system is deadlock-free.

Our definition of deadlock-freeness addresses discrete transitions; however, in case of Zeno behaviors (an infinite number of discrete transition within a finite time), a deadlock-free system can still correspond to an ill-formed model. Hence, performing parametric Zeno-freeness checking is also on our agenda. Moreover, we are very interested in proposing distributed procedures for deadlock-freeness synthesis so as to take advantage of the power of clusters.

Finally, we believe our backward algorithm BwUS could be adapted to obtained under-approximated results for other problems such as the unavoidability synthesis in [9].

References

1. Alur, R., Dill, D.L.: A theory of timed automata. Theor. Comput. Sci. **126**(2), 183–235 (1994)
2. Alur, R., Henzinger, T.A., Vardi, M.Y.: Parametric real-time reasoning. In: STOC, pp. 592–601 (1993)
3. André, É., Fribourg, L., Kühne, U., Soulat, R.: IMITATOR 2.5: a tool for analyzing robustness in scheduling problems. In: Giannakopoulou, D., Méry, D. (eds.) FM 2012. LNCS, vol. 7436, pp. 33–36. Springer, Heidelberg (2012). doi:10.1007/978-3-642-32759-9_6
4. André, É., Lime, D.: Liveness in L/U-parametric timed automata (2016, submitted). https://hal.archives-ouvertes.fr/hal-01304232
5. Bagnara, R., Hill, P.M., Zaffanella, E.: The parma polyhedra library: toward a complete set of numerical abstractions for the analysis and verification of hardware and software systems. Sci. Comput. Program. **72**(1–2), 3–21 (2008)
6. Beneš, N., Bezděk, P., Larsen, K.G., Srba, J.: Language emptiness of continuoustime parametric timed automata. In: Halldórsson, M.M., Iwama, K., Kobayashi, N., Speckmann, B. (eds.) ICALP 2015. LNCS, vol. 9135, pp. 69–81. Springer, Heidelberg (2015). doi:10.1007/978-3-662-47666-6_6
7. Clarisó, R., Cortadella, J.: Verification of concurrent systems with parametric delays using octahedra. In: ACSD, pp. 122–131. IEEE Computer Society (2005)
8. Hune, T., Romijn, J., Stoelinga, M., Vaandrager, F.W.: Linear parametric model checking of timed automata. JLAP **52–53**, 183–220 (2002)
9. Jovanović, A., Lime, D., Roux, O.H.: Integer parameter synthesis for timed automata. IEEE Trans. Softw. Eng. **41**(5), 445–461 (2015)
10. Larsen, K.G., Pettersson, P., Yi, W.: UPPAAL in a nutshell. Int. J. Softw. Tools Technol. Transf. **1**(1–2), 134–152 (1997)
11. Markey, N.: Robustness in real-time systems. In: SIES, pp. 28–34. IEEE Computer Society Press (2011)
12. Sun, J., Liu, Y., Dong, J.S., Pang, J.: PAT: towards flexible verification under fairness. In: Bouajjani, A., Maler, O. (eds.) CAV 2009. LNCS, vol. 5643, pp. 709–714. Springer, Heidelberg (2009). doi:10.1007/978-3-642-02658-4_59

Author Index

Printed in the United States
by Book masters

Printed in the United States
By Bookmasters